Dynamical Systems

Dynamical Systems

A Differential Geometric Approach to Symmetry and Reduction

GIUSEPPE MARMO

Istituto Nazionale di Fisica, Nucleare, Italy

EUGENE J. SALETAN

Northeastern University, Boston, USA

ALBERTO SIMONI

Dipartimento di Fisica, Università di Napoli, Italy

BRUNO VITALE

Dipartimento di Fisica, Università di Napoli, Italy;
International Centre for Genetic Epistemology, University of Geneva, Switzerland

A Wiley–Interscience Publication

JOHN WILEY & SONS

Chichester · New York · Brisbane · Toronto · Singapore

***Library of Congress Cataloging in Publication Data*:**

Main entry under title:

Dynamical systems

'A Wiley–Interscience publication.'
Includes index.
1. Dynamical systems. 2. Vector fields.
3. Symmetry. I. Marmo, Giuseppe.
QA614.8.D 934 1985 515.3′5 84–25814

ISBN 0 471 90339 6

***British Library Cataloguing in Publication Data*:**

Dynamical systems
1. System analysis
I. Marmo, Giuseppe
003 QA402

ISBN 0 471 90339 6

Photosetting by Thomson Press (India) Ltd., New Delhi
Printed and bound in Great Britain

Contents

Preface

This book has four authors. As a result there are somewhere between six and fifteen opinions on each topic covered (the reader may wish to speculate as to how these numbers are arrived at), and this Preface is no exception. The proposals start at no preface at all, pass through a brief abstract of the entire book, and culminate in a deep philosophical tract. What we have here is a compromise of a sort.

One of the objects of this book is to present to the physicist reasonably conversant with classical mechanics some aspects of the more contemporary mathematical treatment of the subject. Our point of view is what is sometimes called global or intrinsic, consistently the point of view of differential geometry. We attempt at the beginning to show why it is convenient to adopt this point of view, why it is natural to model a dynamical system as a vector field Δ on a differentiable manifold M, and we lead the reader, in frequent chapters called Digressions, through the necessary mathematics.

Another object is to present some of the results of ongoing studies of dynamical systems, some properties that are now being investigated by researchers in the field. In particular, we are interested in how a dynamical system (that is, from our point of view, a vector field on some manifold of given dimension μ) is sometimes *reduced* to a system of lower dimension, often by taking into account its *symmetries* and *invariance*. We believe that the understanding of such ideas is useful, extendable, to other areas of theoretical physics, even to gauge theories and condensed matter physics. We hope, therefore, that this book will be of interest to researchers other than just those already working in the field.

In the real world what are encountered are not so much dynamical **systems** as dynamical **events**. Nevertheless, as it has been put by Wittgenstein (*Tractatus Logico-Philosophicus* 5.4541), 'Men have always had a presentiment that there must be a realm in which the answers to questions are symmetrically combined—a priori—to form a self-contained system. A realm subject to the law: Simplex sigillum veri.' The present book is in part a confirmation of this presentiment: we try to exhibit the details, often technical, of what we see as constituting Wittgenstein's 'self-contained' system, to bring it down to earth, to strip it of its mystical quality.

The concept of symmetry is an important one for the construction of this self-contained system. Symmetry is an idea, sometimes quite mystical, with a long history in the description of motion. For instance Plato (in *Timaeus* 34A) speaks of the movement of the Cosmos which is 'proper to its body, namely, that one of the seven motions which specially belongs to its reason and intelligence.' The Cosmos, he writes, is spun 'round uniformly in the same spot and within itself and ... revolving in a cycle.' Movement which partakes of the 'other six motions' he describes as 'disorderly and irrational.' But now man has learned to control motion and, perhaps as a result of such control, to describe it in more detail, and what has emerged is that order and symmetry are found in (or perhaps imposed on) movement which once seemed disorderly and irrational. The resulting modern notion of symmetry is more carefully defined (see for instance Chapter 15 on the Noether theorem) and is used as an important tool in the solution of dynamical systems. One of the implicit themes of this book is the modeling of dynamical systems in such a way that symmetry properties become apparent and exploitable.

In the book we generally adopt the view that symmetry is a tool which can be applied to each model system (M, Δ) after it has been found or constructed. This differs from another commonly held view according to which symmetry comes first, the abstract group of invariances is known, and the possible laws of nature, the possible dynamical systems, then follow (Houtappel, Van Dam and Wigner, 1965).

Part I of this book, on the foundations of mechanics, is meant to be self-contained. It describes our point of view on constructing the model system from the raw data (from what we have called dynamical events) and includes several digressions on the basic mathematics. The last chapter of Part I is on the Noether theorem; this is a bridge to Part II, for it deals with invariance and symmetry.

We should mention that we deal in this book almost exclusively with what are called time-independent dynamical systems, for instance systems whose Lagrangian or Hamiltonian functions do not depend explicitly on the time. At the end of Part I, however, we have included a brief Appendix on time-dependent Hamiltonian systems in order to indicate one way in which the formalism can be generalized.

Part II is concerned more specifically with invariance, symmetry and reduction. Section A, on reduction, introduces foliations and extends the concept of invariance from functions, vector fields and forms to foliations. It shows that there are geometrical objects other than functions which are constants of the motion whose invariance can be exploited, through reduction, in solving dynamical systems. In this first section such reduction is discussed without the additional structure of groups or algebras. In Section B this structure is added and the resulting enrichment is discussed. The book ends with a chapter in which many of the techniques are illustrated and applied to several dynamical systems.

We would like to thank the following colleagues for critically reading a preliminary version of Part I of this work and for many helpful suggestions: A. P. Balachandran, J. P. Cleave, G. F. Dell Antonio, G. Morandi, F. A. E. Pirani, T. Poston, G. Sawitzki, B. Tessier, F. Zaccaria.

Glossary of Mathematical Notation

$\varphi: M \to N$	mapping of a manifold M to another manifold N
$\varphi: m \mapsto n = \varphi(m)$	effect of φ on $m \in M$
$\varphi: M \to N: m \mapsto n$	mapping of M to N and its effect on m
$\mathbb{R}^\mu$	Euclidean space of dimension μ
Q	configuration manifold or space
$f \in \mathscr{F}(M)$	C^∞ functions on M; $f: M \to \mathbb{R}$
$X \in \mathfrak{X}(M)$	vector fields on M
$\alpha \in \mathfrak{X}^*(M)$	one-forms on M
$\beta \in \Omega^k(M)$	k-forms on M
TM	tangent bundle of M
$T^2 M \equiv TTM$	tangent bundle of TM
T^*M	cotangent bundle of M
$T_m M$	tangent space at $m \in M$
$Tf(m)$	tangent of f at $m \in M$
τ_M	projection from TM to M
$\varphi^* f$	pullback of a function by a mapping
$U \subset M$	open submanifold or neighborhood in M
$(U, \varphi), \varphi: U \to U' \subset \mathbb{R}^\mu$	local chart, $\mu = \dim(M)$
$\wedge$	exterior product
i_X	inner product, contraction
$\otimes$	tensor product
d	exterior derivative
L_X	Lie derivative with respect to (along) X
$[X, Y] \equiv L_X Y$	Lie bracket
$\mathrm{Diff}(M)$	group of diffeomorphisms of M
$\mathrm{Lin}(E, F)$	linear maps from E to F
F_t^X	flow of the vector field X
$X^T \in \mathfrak{X}(TQ)$	lift of $X \in \mathfrak{X}(Q)$ to TQ
$X^* \in \mathfrak{X}(T^*Q)$	lift of $X \in \mathfrak{X}(Q)$ to T^*Q
(M, ω)	symplectic manifold M with symplectic form ω

$v: TTQ \to TTQ$	vertical endomorphism
d_v	vertical derivative
$i_v: \Omega^k(TQ) \to \Omega^k(TQ)$	vertical insertion
Δ	dynamical vector field, dynamics
$\mathscr{L}$	Lagrangian function
$\theta_{\mathscr{L}} = d \vee \mathscr{L}$	Cartan or Lagrangian one-form
$\omega_{\mathscr{L}} = -d\theta_{\mathscr{L}}$	Lagrangian symplectic form
$F\mathscr{L}: TQ \to T^*Q$	fiber derivative
θ_0	canonical one-form on T^*Q
$\omega_0 = -d\theta_0$	canonical symplectic form on T^*Q
$\mathfrak{X}_{\mathscr{H}}$	Hamiltonian vector fields
$\mathfrak{X}_{\mathscr{L}\mathscr{H}}$	locally Hamiltonian vector fields
$\{f,g\}$	Poisson bracket
$\Phi: G \times M \to M$	action of a Lie group G on M
$\varphi: \mathfrak{G} \times M \to TM$	action of a Lie algebra $\mathfrak{G}$ on M
$\mu: M \to \mathfrak{G}^*$	momentum map from M to the dual of $\mathfrak{G}$ (μ is also often used for $\dim(M)$)
$\mathscr{D}$	distribution
$\vdash$	'such that', as in $\mathfrak{X}_{\mathscr{L}\mathscr{H}}(M) = \{X \in \mathfrak{X}(M) \vdash L_X\omega = 0\}$
$\vert$	restriction, as in $f(U) = f(M) \vert U \subset M$

PART I

Foundations of mechanics

This part is an extension of joint work with G. Caratù

1

Introduction to Part I

The first part is devoted to those aspects of the foundations of theoretical mechanics which are particularly relevant to this book, namely to the use of invariance properties in solving dynamical systems, in finding their motions. We therefore skip over many topics which are essential to any thorough exposition of the subject, such as rigid-body motion, and treat in detail many others, among which we particularly mention the mathematical tools. We are consistently interested in constructing as many different representations of a dynamical system as is possible, so as to maximize the number of invariance and symmetry properties that can be explored and used.

We have chosen or, more accurately, found it necessary to use modern differential geometry as our principal mathematical tool. It is true that many of the classical results of theoretical mechanics can be discussed without using the complicated differential geometrical approach (calculus on manifolds), but not with any efficiency when one is interested, as we are, in global rather than local properties. This stems at least in part from the fact that in regions of phase space which do not contain singular points of the system, all systems are canonically equivalent (in particular to the free particle). In a certain sense, therefore, systems differ essentially in their global properties, in their singular points and the relations between them. In this differential geometric approach, vector fields, rather than second-order differential equations, describe the dynamical system. Essentially what this provides is that there be only one orbit through each point of the space considered (the carrier space). It is equivalent locally to transforming a set of n second-order equations into a set of $2n$ first-order ones, but globally the vector field on the manifold aids significantly in visualizing and understanding the motion. It is for reasons like this that we consider the coordinate-free approach provided by differential geometry to be the most convenient and transparent one. On the other hand, we shall often refer to local properties and discuss most results in both contexts.

We have decided also to try to avoid the dry kind of approach characterized by openings such as: 'Let M be a manifold and X be a vector field on M.' Where this manifold comes from, how a vector field happens to exist on it and what the relation of this vector field is to the observations of an experimentalist are, in our view, among the important questions. How does the analysis of motion lead to such abstract concepts as field of force, potential, equations of motion?

We therefore start from observations of positions and trajectories and try to understand how the measurement of such physical properties leads to the construction of a manifold. According to our point of view, a particular configuration space cannot be separated from the dynamical system on it; we hope never to speak of such a space as given *a priori* and then of some possible dynamical system which can exist on it. To us, motion is inseparable from the very existence of a dynamical system, so that both its configuration space and its phase (or other carrier) space are only defined by the motion itself. A similar view was expressed centuries ago:

> It is a heretical doctrine to think that in essence water does not run, and the tree does not pass through vicissitude. The bloom of flowers and the fall of leaves are the conditions that exist. And yet unwise people think that in the world of essence there should be no bloom of flowers and no fall of leaves. (Dōgen (Sōtō Zen, 13th C.) from Shōbōgenzō (*The Nature of Dharma*))

To put it into modern terminology, we wish to avoid the customary division of mechanics into *kinematics* as opposed to *dynamics*, as stated for instance by Whittaker (1904): 'It is natural to begin this discussion by considering the various possible types of motion in themselves, leaving out of account for a time the causes to which the initiation of motion may be ascribed; this preliminary enquiry constitutes the science of *kinematics*.' Our aim will be to show how, when completely unfolded and suitably interpolated, the experimental information available on the time evolution of a dynamical system suggests a corresponding *model system* consisting of a differentiable manifold and a smooth vector field on it. Projections from this model system to manifolds of lower dimension will then provide several equivalent geometrical descriptions of the motion, within the geometry of the chosen configuration space; these we may call *kinematic* descriptions.

In order to have the tools available for discussing the construction of the configuration-space manifold, we make the first of our mathematical digressions in Chapter 2, this one on manifolds. Then Chapter 3 deals with the configuration space Q, its ambiguities and invariances. In Chapter 4 we discuss time, the privileged evolution parameter in our nonrelativistic work.

Chapters 5 to 9 develop the dynamical information which is initially described on Q, and the complex apparatus is constructed which carries the considerations from Q to a larger carrier space, which in these chapters is the tangent bundle TQ (configuration-velocity space). From the generally scanty information supplied by the pieces of trajectories measured on Q we construct, not without some ambiguity, the vector field on TQ. The main part of this work is done in Chapter 5. Chapter 6 is another mathematical digression, this one on vector fields and fiber bundles, and Chapter 7 presents some examples of dynamical systems together with a discussion of possible pathologies. Chapters 8 and 9 are again technical, and by Chapter 10 we are prepared to start a discussion of integrating the dynamics, that is, of finding the trajectories from the vector field.

At this point we may be accused of madness: we have started with trajectories, constructed the vector field, and now want to find the trajectories from the vector field. That this is not mad has to do with the ambiguity in constructing the vector field from the trajectories and the fact that the theoretical analysis starts from the abstraction of the field. This is discussed somewhat throughout all of Chapters 5–10. Here we add another centuries-old observation:

> Now the path of investigation must lie from what is more immediately cognizable and clear to us, to what is clearer and more intimately cognizable in its own nature; for it is not the same thing to be directly accessible to our cognition and to be intrinsically intelligible.... Now the things most obvious and immediately cognizable by us are concrete and particular, rather than abstract and general; whereas elements and principles are only accessible to us afterwards, as derived from the concrete data when we have analysed them. So we must advance from the concrete whole to the several constituents which it embraces.... (Aristotle, *Physics*, 184a)

The problem that this book concentrates on is that of integrating the dynamics, in particular by defining, recognizing and making explicit use of the invariance properties of the dynamical system. After posing this problem in its generality in Chapter 10, we discuss its solution on TQ, describe how one goes to phase space T^*Q in Chapter 11, and then discuss the solution on T^*Q in Chapter 12. Chapter 13 discusses the ambiguities in the usual formalisms on both TQ and T^*Q, and Chapter 14 introduces other commonly used carrier spaces. The last chapter (Chapter 15) of Part I plays the role of a bridge between the two parts of the present book. It is concerned with the Noether theorem and introduces, in the simplest possible framework, the interplay of invariance and conservation that will play a major role in Part II.

To this short summary of what Part I contains as an introduction to Part II (reduction, Lie algebra and group actions), we must add a few words describing what it does not contain. First, our description of mechanics (except near the very end of the book) is strictly nonrelativistic. Time plays a role which is essentially different from other parameters that enter the description, such as local coordinates in the carrier space. Second, we deal only with finite-dimensional systems, and generalizations to infinite-dimensional ones, and hence to classical field theory, are not always possible. Third, we deal with smooth (i.e. infinitely differentiable) systems, and the results do not invariably apply to finitely (k times) differentiable ones or to topological (i.e. continuous) ones. Finally, we do not address the important problem of measure and measurability in classical mechanics.

In spite of these limitations, Part I should be useful as an advanced introduction to the study of mechanics. It is reasonably self-consistent and has been kept free of the group-theoretical machinery that pervades Part II.

BIBLIOGRAPHY TO CHAPTER 1

The differential geometric approach to dynamics (in particular, the global approach to a dynamical system through the definition of a smooth vector field on a differentiable manifold) is presented in Abraham and Marsden (1978) (referred to throughout as AM). We have tried to keep our approach and our notation as consistent as possible with those of AM.

Along the same lines—at different levels of sophistication—are the following texts: MacLane (1968) (the first volume is concerned with the local coordinate approach, the second with the global, coordinate-free approach), Godbillon (1969), Souriau (1970) and Robbin (1974).

Modern textbooks on classical mechanics are Arnol'd (1978) and Thirring (1978).

There is still much to learn from classical books on dynamics. Among them, we have found particularly stimulating Whittaker (1904) and Wintner (1941).

The quotation from Dogen comes from H. Nakamura: *Ways of Thinking of Eastern People*, Honolulu 1964. Aristotle is quoted from Wicksteed's translation: *The Physics*, Loeb Classical Library, London, 1929.

2

A Digression on manifolds and diffeomorphisms

This is the first of a series of chapters called Digressions, which are meant to aid the reader with some mathematical concepts. This Digression is meant for the reader with little or no experience in differential geometry. Generally the Digressions will be limited to relevant definitions and to statements of some results, as there are many excellent books on the subjects treated at in any levels of sophistication (see Further reading and the bibliography at the end of this chapter).

The concept of a manifold arises naturally enough when one is modelling a mechanical system. Roughly speaking, an n-dimensional manifold is made up by patching together open subsets of an n-dimensional vector space. It has taken many people more than a century to make this vague idea precise and to construct the useful theory we know today. An important step in this development was due to Whitney (1936), who proved that any finite-dimensional manifold can be realized as a subset of a suitable $\mathbb{R}^n$. This theorem enables one, while dealing with the formal definitions, to keep in mind the example of a surface (e.g. a sphere) in the usual three-dimensional Euclidean space.

Consider a set S. A *local chart* (or simply a chart, or a *coordinate system*) for S is a one-to-one map φ from a subset U of S onto an open subset V of $\mathbb{R}^n$. One sometimes writes (U, φ) instead of φ alone to indicate explicitly the *domain* U of φ. Here $\mathbb{R}^n$ is understood to be equipped with the usual topological and real vector-space structures. Two charts φ, ψ for S are *compatible* iff (if and only if) they take their values in the same $\mathbb{R}^n$ and iff the *overlap maps* $\varphi \circ \psi^{-1}, \psi \circ \varphi^{-1}$ are C^∞. If the domain of $\varphi \circ \psi^{-1}$ is void, $\varphi \circ \psi^{-1}$ is conventionally assumed to be C^∞. An *atlas* on S is a set of compatible charts, (U_i, φ_i) with i in some index set I, such that $S = \bigcup_{i \in I} U_i$. Two atlases are compatible iff their union is an atlas. A *differentiable structure* on S is a maximal atlas, that is, one containing any other compatible atlas. Given an atlas A for S, a unique differentiable structure $\tilde{A}$ is determined, namely the union of all atlases compatible with A. The pair $(S, \tilde{A}) = M$ is a *differentiable manifold*. Its dimension is n iff the charts of $\tilde{A}$ take their values in $\mathbb{R}^n$.

The *manifold topology* is the unique topology on S for which all charts in $\tilde{A}$ are homeomorphisms.

It turns out, however, that this definition of a manifold is too broad for our purposes, for the manifold topology defined in this way may fail to be *Hausdorff* or to be *second countable.*

These two properties, described below, are very desirable, as the first one guarantees uniqueness for the solutions of differential equations, and the two together assure the existence of partitions of unity, which in turn, permit the construction of various important objects, such as Riemannian structures (hence metrics), hyperregular functions etc. Thus we shall include these topological requirements in our definition of a differentiable manifold.

A topological space (or a topology) is *Hausdorff* iff, given any two distinct points m_1, m_2 it is always possible to find two disjoint open sets $O_1 \ni m_1, O_2 \ni m_2$. It is *second countable* iff it has a countable basis, which is defined as follows. A topology τ for a set S is the set of open sets of S, and then a basis for a topology τ is a subset β of τ such that any element of τ can be expressed as a union of elements of β. A *partition of unity* is a set of functions on the manifold, each with compact support, such that only a finite number of these supports intersect at each point of the manifold and the sum of the values of the functions is everywhere unity. Our choice of Hausdorff manifolds (with the exception of non-Hausdorff quotient manifolds in some special cases, as in Part II, Section B) is motivated by the group theoretical machinery to be used in Part II: Lie groups are indeed both Hausdorff and second countable (Brickell and Clark, 1970, p. 214). The reader will appreciate the advantage of these properties while working, in particular, through Chapters 21 to 24.

Remark. The existence of partitions of unity is in fact equivalent to a weaker topological condition, namely paracompactness. A manifold is *paracompact* iff it is a union of Hausdorff, second countable manifolds. It can be shown that a manifold is second countable iff it has a countable atlas.

In the definition of compatible charts, the C^∞ condition could have been replaced by some other (stronger or weaker) condition. For instance, one can require C^ω (that is, real analytic) or C^k, $k \in N$ (N being the natural numbers), and then the word 'differentiable' would be replaced by the appropriate adjective. Now, it is clear that a C^k manifold is also $C^r\ \forall r \leqslant k$; the converse is almost true, for if a set can be given a C^1 differentiable structure, then it can be given a unique structure which is $C^r \forall r \in N$, and even C^∞ or C^ω. More precisely, any C^1 maximal atlas contains a unique nonvoid C^r atlas for $r \in N$ or $r = \infty$ or $r = \omega$. For dynamical systems we shall require some differentiability, and this theorem allows us to achieve simplicity by restricting the considerations to C^∞. To require C^ω would leave too little elbow room; see Chillingworth (1976), p. 86.

Continuity is the weakest condition one can impose on the overlap maps if the set is to be endowed with both a topology and a dimension. With this condition alone, one obtains a C^0 or topological manifold. One might hope that by starting with a C^0 manifold one could construct a unique differentiable manifold, as one can, according to the above assertion, when starting with a C^1 manifold. This turns out, however, to be impossible. In fact an example has been given by Kervaire (1960) of a locally Euclidean topological space that admits no differentiable structure at all. Moreover, when it is possible to construct a differentiable from a C^0 manifold, it is generally not unique.

There are two aspects to this lack of uniqueness. First, two different differentiable manifolds can be what is called *diffeomorphic*. We start with an example.

Take $S = \mathbb{R}$ and consider the two one-chart atlases $(\mathbb{R}, \varphi_1 : x \mapsto x)$ and $(\mathbb{R}, \varphi_2 : x \mapsto x^3)$. These are not compatible since $\varphi_1 \circ \varphi_2^{-1}$ is not differentiable at the origin. They define the same topological manifold, but different differentiable manifolds $M_1 = (\mathbb{R}, \tilde{A}_1)$, $M_2 = (\mathbb{R}, \tilde{A}_2)$, where $\tilde{A}_1$ and $\tilde{A}_2$ are the C^∞ differentiable structures containing φ_1 and φ_2 respectively. One has the feeling, however, that these manifolds are not very different, that they are in fact closely related. This feeling is made precise by the concept of a diffeomorphism, that is an isomorphism of differentiable structures.

Let $M_1 = (S_1, \tilde{A}_1)$, $M_2 = (S_2, \tilde{A}_2)$ be differentiable manifolds. *A* map $\chi : S_1 \to S_2$ is a *diffeomorphism* of M_1 to M_2 iff (a) it is a bijection, i.e. one-to-one and onto (that is, it is an 'isomorphism' for the sets), (b) it is a homeomorphism (that is, an isomorphism for the topological structures), (c) for all $\varphi_1 \in \tilde{A}_1$, $\varphi_2 \in \tilde{A}_2$ it follows that $\varphi_2 \circ \chi \in \tilde{A}_1$ and $\varphi_1 \circ \chi^{-1} \in \tilde{A}_2$. For such diffeomorphisms we shall usually write $\chi : M_1 \to M_2$, identifying the manifolds with the underlying sets. Differentiable or *smooth* will always mean C^∞ unless otherwise specified, but it should be kept in mind that most of our statements hold true also for C^k structures. It may be remarked that conditions (b) and (c) can be replaced by (b') for all $x \in S_1$, there exist charts $(U_1, \varphi_1) \in \tilde{A}_1$, $(U_2, \varphi_2) \in \tilde{A}_2$ such that $x \in U_1$, $\chi(x) \in U_2$ and the maps $\varphi_2 \circ \chi \circ \varphi_1^{-1}$ and $\varphi_1 \circ \chi \circ \varphi_2^{-1}$ are differentiable.

Two manifolds M_1 and M_2 are said to be *diffeomorphic* iff there exists a diffeomorphism $\chi : M_1 \to M_2$. In the example above the map $f : \mathbb{R} \to \mathbb{R} : x \mapsto x^{1/3}$ is a diffeomorphism of M_1 to M_2, as $\varphi_2 \circ f \circ \varphi_1^{-1} = \mathbb{1}_{\mathbb{R}}$. Note also that $\mathbb{1}_{\mathbb{R}}$ is not a diffeomorphism of M_1 to M_2. This example on $\mathbb{R}$ is probably the simplest meaningful one of two differentiable manifolds which are different but diffeomorphic and which have the same underlying open sets and therefore the same underlying topological space or C^0 manifold. Simple examples of diffeomorphic manifolds with different underlying sets in $\mathbb{R}^3$ are: two spheres, either nonconcentric or with different radii; a sphere and an ellipsoid; a sphere with a handle and a torus; a circle and a circle with a number of knots.

The second lack of uniqueness in constructing a differentiable from a C^0 manifold lies in the fact that there exist nondiffeomorphic differentiable manifolds with the same underlying topological space. Otherwise stated, there are manifolds which are homeomorphic but not diffeomorphic (Milnor, 1956, 1959).

Diffeomorphism between manifolds is, of course, an equivalence relation and it is customary to identify manifolds in the same class, calling the class itself the manifold. Then a particular element in the class is seen as a particular 'realization' of an abstractly given manifold. With this agreement, one speaks of the circle S^1, of the n-dimensional torus T^n, of the projective n-plane $\mathbb{R}P^n$ etc., with smooth structures understood, at least when no ambiguity can arise.

Remark. It is in this context that the problem of classifying the possible manifolds can appropriately be stated. For this, however, the interested reader is referred to the bibliography.

By definition, the composition of two diffeomorphisms is always a diffeomorphism (by the convention stated above, even if the composition has empty domain). In particular, for any manifold M, the set $\underline{\text{Diff}(M)}$ of diffeomorphisms from M to itself, obviously nonvoid, is a group, and it turns out that this group is extremely large. A reflection of this is the important property that, if M is connected, Diff(M) acts *transitively* on M; that is, for all $x, y \in M$, there is a $\varphi \in \text{Diff}(M)$ such that $\varphi(x) = y$. Moreover, this property is actually shared by many proper subgroups of Diff(M). We shall be often concerned with diffeomorphisms in this book. They arise naturally in studying dynamical systems, in order both to discriminate what is and what is not choice-dependent in the mathematical models which are constructed and to solve the differential equations which will eventually be obtained. Generally there will be additional structures on the manifolds, like projections, fields, foliations etc., so from time to time we shall select and study particular classes of diffeomorphisms.

At this point we present some examples to give some idea of the kind of work generally needed to handle the objects we have been discussing.

Example 2.1: The circle in the plane. As a first example consider the unit circle in the plane, centered at the origin, that is the set $S^1 = \{(x, y) \in \mathbb{R}^2 \vdash x^2 + y^2 = 1\}$. The maps (stereographic projections) φ_+ and φ_- defined by

$$\varphi_+ : S^1 - \{(0, -1)\} \to \mathbb{R} : (x, y) \mapsto \left(\frac{1-y}{1+y}\right)^{1/2} \text{sign}(x),$$

$$\varphi_- : S^1 - \{(0, 1)\} \to \mathbb{R} : (x, y) \mapsto \left(\frac{1+y}{1-y}\right)^{1/2} \text{sign}(x),$$

provide a C^∞ atlas A_1 for S^1. Their inverses are

$$\varphi_+^{-1} : \mathbb{R} \to S^1 - \{(0, -1)\} : t \mapsto \left(\frac{2t}{1+t^2}, \frac{1-t^2}{1+t^2}\right),$$

$$\varphi_-^{-1} : \mathbb{R} \to S^1 - \{(0, 1)\} : t \mapsto \left(\frac{2t}{1+t^2}, \frac{t^2-1}{1+t^2}\right),$$

so that (see Fig. 2.1)

$$\varphi_+ \circ \varphi_-^{-1} = \varphi_- \circ \varphi_+^{-1} : \mathbb{R} - \{0\} \to \mathbb{R} - \{0\} : t \mapsto \frac{1}{t}.$$

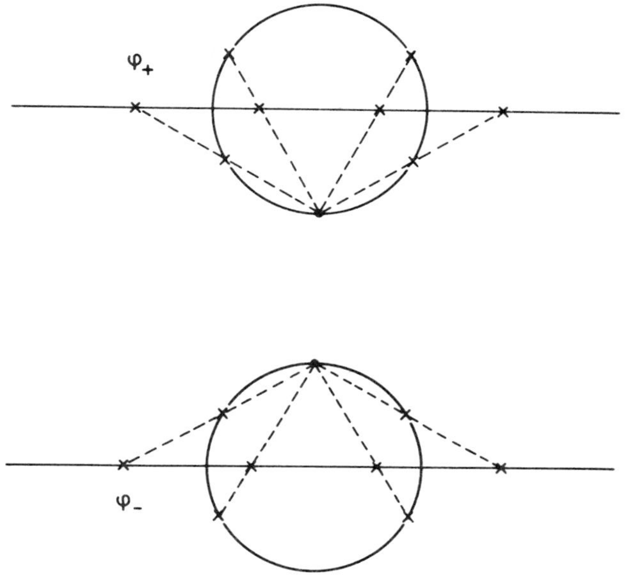

Fig. 2.1. The stereographic projections φ_+ and φ_- from S^1 to $\mathbb{R}$

It is an easy exercise to check that the following are also C^∞ atlases for S^1, and that all of them are compatible with A_1.

$$A_2 = \{(x)_+, (x)_-, (y)_+, (y)_-\}, \text{ where for instance,}$$
$$(y)_-:\{(x, y) \in S^1 \vdash -x > 0\} \to]-1, 1[:(x, y) \mapsto y;$$

$$A_3 = \left\{\left(\frac{y}{x}\right)_+, \left(\frac{y}{x}\right)_-, \left(\frac{x}{y}\right)_+, \left(\frac{x}{y}\right)_-\right\}, \text{ where, for instance,}$$
$$\left(\frac{y}{x}\right)_-:\{(x, y) \in S^1 \vdash -x > 0\} \to \mathbb{R}:(x, y) \mapsto \frac{y}{x};$$

$$A_4 = \{\varphi_1, \varphi_2\}, \text{ with}$$
$$\varphi_1^{-1}:]0, 2\pi[\to S^1 - \{(1, 0)\}:t \mapsto (\cos t, \sin t),$$
$$\varphi_2^{-1}:]-\pi, \pi[\to S^1 - \{(-1, 0)\}:t \mapsto (\cos t, \sin t).$$

It should be noted that atlases A_1 and A_2 are easily generalized to higher dimensions, and that all of the above atlases are easily generalized to circles of arbitrary radius and centered other than at the origin.

Example 2.2: Four circular arcs in the plane. As a second example, consider the

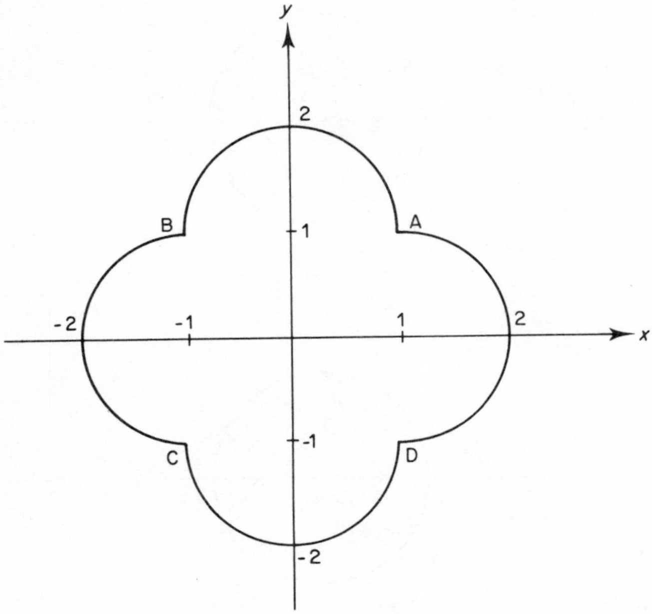

Fig. 2.2. Four circular arcs

subset S of the plane represented in Figure 2.2. The equations of the arcs are:

$$\begin{aligned} AB&: y = 1 + \sqrt{1 - x^2}, \\ BC&: x = -1 - \sqrt{1 - y^2}, \\ CD&: y = -1 - \sqrt{1 - x^2}, \\ DA&: x = 1 + \sqrt{1 - y^2}. \end{aligned}$$

It is clear that stereographic projections from the 'poles' do not provide charts for S, because they are not one-to-one. One may attempt an atlas similar to A_2. Define a set of maps $B_2 = \{(x)_+, (x)_-, (y)_+, (y)_-\}$ by $(x)_+ : \{(x, y) \in S \vdash y > 0\} \to]-2.2[: (x, y) \mapsto x$, and similarly for the others. It is clear that these maps are charts and that their domains together cover S. As for compatibility, for instance of $(y)_+$ and $(x)_+$,

$$(y)_+ \circ (x)_+^{-1} :]0, 2[\to]0, 2[: t \to \begin{cases} 1 + \sqrt{1 - t^2}, & 0 < t \leqslant 1 \\ \sqrt{1 - (t-1)^2}, & 1 \leqslant t < 2. \end{cases}$$

This function is continuous, but not differentiable at the point $t = 1$, as there the left derivative does not exist. Hence, with B_2, S is only a topological manifold. However, a set of maps B_3 defined like A_3 in the first example turns out to be

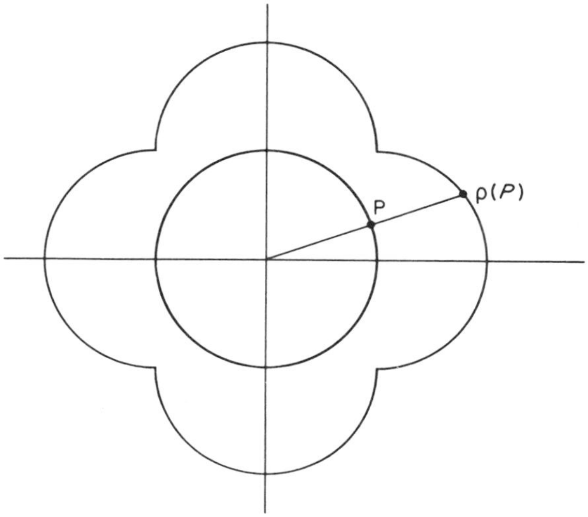

Fig. 2.3. Projection of S^1 onto S

a smooth atlas for S. For instance,

$$\left(\frac{y}{x}\right)_+ \circ \left(\frac{x}{y}\right)_+^{-1} = \left(\frac{x}{y}\right)_+ \circ \left(\frac{y}{x}\right)_+^{-1} :]0, +\infty[\to]0 + \infty[: t \mapsto \frac{1}{t}$$

Now, S should be diffeomorphic to S^1, for it is known that there exists only one class of diffeomorphic compact 1-manifolds, namely the class of S^1. Indeed, as is easily verified directly, the projection ρ through the origin, say, of S^1onto S, is a diffeomorphism (see Fig. 2.3).

Another characterization of diffeomorphisms which sometimes provides a useful test is the following. First, a *function* is defined as any real-valued map and then if M is a manifold, $\mathscr{F}(M)$ denotes the set of C^∞ functions whose domain is M. Then $\varphi: M_1 \to M_2$ is a diffeomorphism iff for any $f \in \mathscr{F}(M_2)$ and $g \in \mathscr{F}(M_1)$ it composes to yield functions, i.e. iff $f \circ \varphi \in \mathscr{F}(M_1)$ and $g \circ \varphi^{-1} \in \mathscr{F}(M_2)$.

BIBLIOGRAPHY TO CHAPTER 2

As an introduction to manifolds and differential calculus on them, besides AM, see Chillingworth (1976). For more details, see Bishop and Crittenden (1964), Brickell and Clark (1970); Guillemin and Pollack (1974) and Boothby (1975).

3

Construction of Q: from observables to the configuration space manifold

We would like at this point to begin a formal definition of the kind of phenomena which are the concern of this book. To do so thoroughly would, however, involve nontrivial epistemological problems, concerning which none of us are expert. Moreover the differences between us would lead us to unreconcilable positions which would make it almost impossible to proceed. It is thus perhaps best to avoid a very deep analysis of these questions, which we recognize as being important in principle, but which would be of little influence on the rest of the book.

Interactions through the senses or measuring instruments with the environment tend to become organized by the mind into separate events of varying complexity, events which take place at different times. Each interaction is itself an event (e.g. a sound or a flash of light), but such elementary events become associated, at first in simple ways into relatively simple compound events (e.g. such a sound and such a flash may be associated to become an explosion), and then in more complex ways into more complex events.

The organization may change: too many or too few events may be associated, these things changing as theories evolve. 'Event' is an open concept. Some examples of events from different spheres of experience are: collision, photon emission, transition, purchase, company merger, price change, war, collapse of government, revolution.

The kind of event we are interested in has *position* as one of its aspects. This may seem to exclude, for instance, transition or price change, but one could also define position more generally, as that aspect of the event in which we are most interested (provided, as we shall see, that it have certain properties). The other aspects then become auxiliary tools, which are used in constructing the theory.

Sometimes events can be grouped around systems or objects on the basis of some aspects that remain unchanged. For example, the two events which may be called a table in one position at one time and a table in another position at another time have within them certain invariant shapes, colors, etc., which are identified as the table, an object involved in both of these events. To identify objects or physical systems in this way by invariances is by no means a trivial operation, for the situation is not always clear and identifications change as theories evolve.

As we said at the start of this brief discussion, we are avoiding too formal a definition of the subject of our studies, that is, of localizable physical systems. Identifying such a system is a complex procedure intrinsically connected with the kind of theory one is hoping to construct and with the available phenomenological knowledge.

We turn now to the properties of whatever aspect it is that we wish to call position; this aspect should have some of the structure that is possessed by ordinary position, and thus we shall analyze in some detail how the position measurements of a physical system at a fixed time are equipped with structure.

3.1 CONSTRUCTION OF THE CONFIGURATION SPACE Q

Consider a physical system at a fixed time, and the position measurements on this system by all possible observers. It must be assumed that each observer knows at least roughly what is meant by position, so that all can at least agree that they are measuring the same aspect. Such common understanding of position must come from some sort of common experience with nearness and with motion, which is the result of complex mental processes that lead in all of us to the concept of a three-dimensional space in which objects occupy positions. Whatever this experience and this process may be, here we shall assume only that position can be distinguished from other aspects, or observables, and that all observers agree about which aspect it is. Then each observer may use whatever measuring instruments he pleases and may process the data in any way he sees fit to come up eventually with a set of numbers which he relates to position. By interpolating and extrapolating from a finite set of position measurements performed in different situations, each observer constructs his set of possible positions of the system, a subset of $\mathbb{R}^n$ with a suitable n, but without any particular topology. Note that in general an observer's measuring device is not even able to measure all possible positions, for instance because it is too small or too big.

Assume now that observer i has constructed his position subset $U_i \subset \mathbb{R}^{n_i}$, and that this has been done by all observers. Then it is natural (and in fact necessary) to suppose that there exist transition mappings that link U_i and U_j wherever observers i and j agree that they are measuring the same position. That is, if observer i measures a position which is available, according to the common understanding of position, to the instruments of observer j, there should be a way to tell from the measurement according to i what would be the measurement according to j.

Let f_{ij} be the transition mapping that goes from a subset of U_j to a subset of U_i. If f_{ij} is not one-to-one, observer i (or j) identifies positions that j (or i) measures as distinct. It shall be assumed, however, that all the f_{ij} are one-to-one, and hence there will be no attempt to link observers that use different 'levels' of description.

This does not mean, on the other hand, that the observers that happen to be linked are using the 'highest level' of description; there may be other possible observers who distinguish positions which these observers call identical.

There are two equivalent ways of looking at these transition mappings: they are defined from a knowledge of what is meant by 'same position', or they define it. In other words, if one assumes that observer i obtains his position subset U_i as the range of a one-to-one mapping f_i from a given (already known) set Q, then the transition mappings f_{ij} are completely determined by the formula

$$f_{ji} = f_j \circ f_i^{-1}, \tag{1}$$

and satisfy

(a) $f_{ii} = \mathbb{1}_{U_i}$,
(b) f_{ij} is one-to-one, $f_{ji} = f_{ij}^{-1}$,
(c) $f_{ij} \circ f_{jk}$ is a restriction of f_{ik}.

Remark. The *composition* $A \circ B$ is defined by $(A \circ B)(x) = A(B(x))$. Its domain $\mathrm{dom}(A \circ B)$ is $\mathrm{dom}(B) \cap B^{-1}(\mathrm{dom}(A))$ and, or course, may be void, which gives rise to no ambiguity.

Conversely, if only the U_i and a family f_{ij} of mappings satisfying conditions (a), (b), (c) are known, then the following construction will easily give a uniquely determined set Q and one-to-one mappings f_i satisfying Eq. (1). Let $\tilde{Q}$ be the set of pairs (x,i) with $x \in U_i$. Then (a), (b), (c) imply that the relation $\sim$ defined by $(x,i) \sim (y,j)$ iff $y = f_{ji}(x)$ is an equivalence. Hence the requirements will be met by $Q = \tilde{Q}/\sim$ (the set of equivalence classes) and the mappings $f_i:(x,i) \in Q \mapsto x \in U_i$.

Positions defined in this way lack any structure whatsoever. In order to make use of the formalism of differential and integral calculus, we shall now make three preliminary topological and differential assumptions about the f_{ij}, analyzing, when possible, their physical relevance.

(1) In $\mathbb{R}^{n_i}$ introduce the usual topology and the requirement that U_i be an open set. This hypothesis, permitting one eventually to take derivatives in any direction at every point, has the following two aspects. First, it restricts the kind of physical system under consideration. For example, systems with boundaries are excluded unless this assumption is modified. Second, observers are required to organize their measurements in a certain way. For example, if the position set of an observer is an open disk, he may not organize it as an open disk without center plus an isolated point.

(2) The f_{ij} are homeomorphisms. Then $n_i = n_j$ whenever a transition between U_i and U_j is actually defined by experiment. If this is the case $\forall i,j$ the common value n of the integers n_i will be called the number of degrees of freedom of the system. Otherwise such a number will be assumed defined uniquely. This is not a seriously restrictive assumption, but it is necessary if one is to avoid considerable technical difficulties.

(3) The f_{ij} are C^∞ diffeomorphisms. This is because all observers should have, roughly speaking, the same mathematical description of phenomena. That is, if

one observer finds a singularity or lack of differentiability at some position, so should every other.

When these requirements are fulfilled, there results a mathematical object, in fact a differentiable manifold, already well suited for developing differential calculus and doing a lot of other things. We shall now impose two other somewhat technical conditions on the position set Q, *the configuration space* or *configuration manifold*, in order that it have additional needed structure.

3.2 TOPOLOGICAL AND DIFFERENTIAL NATURE OF Q

It is said conventionally, and we shall often say so as well, that Q is the space *in which* the motion takes place. This kind of statement would seem to imply that Q has an existence independent of the motion. Yet when we stated in the Introduction that kinematics and dynamics are inseparable, we essentially denied this independent existence. In fact we view the construction of Q as just one of the steps in analyzing a dynamical system, and we should now explain why we do, how the construction of Q is linked to the motion itself. This means that we must show how the dynamical system plays a role in determining choices between different manifolds, differentiable structures on them, and atlases. We shall make use of the technical tools described in Chapter 2 and try to show how they help to make the required choices. The kind of principles we propose are tentative, for we do not believe that there exist compelling logical reasons to accept one structure instead of another, especially since the choices are generally between alternatives which are equally consistent with experiment. It is technical requirements that are paramount, like continuity and differentiability. These are tied to the mathematical tools available and are related to common-sense notions of space and motion. They are conventional, but not arbitrary.

Recall that in §3.1 it was assumed that the observers have some sort of common experience with motion that allows them to construct the concept of ordinary space. In the sense of such an assumption the position set Q is unambiguous. But what comes next, imbuing this set with topological and differential properties, is more ambiguous, and the manifold finally arrived at, its dimension and differentiable structure, depend largely on the assumptions we make concerning the transition maps f_{ij}.

In observing the motion of a given dynamical system, the observers make many different measurements. The present discussion concerns those measurements whose values by common consent they call the positions of the system. These will determine the space in which that particular motion takes place, but it does not follow that this space is necessarily the seat of all possible motion, like Newton's absolute space. It is, nevertheless, a configuration space, and it is desirable therefore that it have a number of simple, fundamental structures which common experience associates with such a space. Among these are that it be metric (or metrizable), in some sense continuous, smooth, locally homogeneous and isotropic and that it have a well-defined dimension. If, as may sometimes occur, technical requirements on the trajectories lead to a configuration manifold Q

which does not possess such common-sense structures, the technical requirements shall be allowed to be determining.

Recall that according to the definition of a differentiable manifold the only topology on Q is that induced by the differentiable structure. This leads to two restrictions. First, one is not free to impose *a priori* requirements on the topological properties of Q. Second, the only topologies that can be generated are those which are compatible with some differentiable structure. Significant properties of Q can therefore always be analyzed in terms of its differentiable structure, even when those properties are in reality topological (e.g. compactness, connectedness, dimension) and would survive even only under homeomorphisms on Q, as opposed to diffeomorphisms.

Every classification of manifolds starts with one of the most important properties, namely dimension. Do position measurements determine dim Q unambiguously? The answer to this depends in part on how one defines dimension, as will be seen in the examples which follow, examples which show that dim Q is not in fact unambiguous until further assumptions are made.

We start with an example which violates the assumption made in Chapter 2 which included second countability in the definition of a differentiable manifold. Nevertheless, this example exhibits the kind of trouble one can get into. Consider a dynamical system whose position set Q, as yet without a topology, is the plane. Now, the plane can be given the usual C^∞ structure of $\mathbb{R} \times \mathbb{R}$, or it can be thought of as the union of parallel lines and given the less usual structure of a one-dimensional manifold. So long as the observers do not worry about the continuity of certain trajectories (those that are not parallel to the lines into which the plane is decomposed) they can view their configuration space as either one- or two-dimensional.

Remark. As we mentioned above, this depends on how one defines dimensions. Here we are not using the usual physicist's definition, according to which the dimension of a space is the number of things one must name in order to determine a point in it. Even in the parallel-line foliation the plane is two-dimensional in this sense, for one must name the line and the distance along it. The topological definition of dimension (or, as mentioned above, the one determined by differentiability) is in terms of maps of open neighborhoods into open neighborhoods of $\mathbb{R}^n$ for some n. In the parallel-line topology the open neighborhoods are (maps into) intervals of $\mathbb{R}$, and points which are far apart in this topology may be close together in the $\mathbb{R} \times \mathbb{R}$ topology.

It is the motion, the dynamics, in combination with the topological concept of continuity, which forces the choice between these two topologies when one requires that trajectories be continuous even if they are not parallel to the foliation lines. Thus it is seen how common-sense and technical requirements can combine to determine the dimension of the configuration space manifold Q. This does not mean, on the other hand, that there will always be a manifold on which the motion is continuous, but that when it is possible to choose the manifold in this way, it shall be so chosen.

The pathology introduced by the loss of second countability is not necessary; we used it because of the simplicity of the example. It can be eliminated, for instance, by projecting an open square of $\mathbb{R}^2$ in one of many ways by a bijection onto an open interval of $\mathbb{R}$. No set theoretical information is lost by this bijection, and again the choice is between a one-dimensional manifold and a two-dimensional one. It would turn out again that considerations of continuity would determine which of the two parameterizations to choose.

Now suppose that the dimension of Q has been established, and the problem is to find a suitable manifold of the given dimension. The topological properties of Q are probably simply related to the position measurements, but its differentiable properties would seem to require more detailed analysis. There are essentially three different stages to this analysis. After establishing the dimension, (a) one must choose among homeomorphic manifolds because they are not all diffeomorphic; (b) one must choose among diffeomorphic manifolds, because they do not all have equivalent differentiable structures; and (c) one may choose on a given differentiable manifold those atlases (or that atlas) which most conveniently describe the dynamical system.

(a) For manifolds which are Hausdorff (as, by assumption, are all those we shall deal with) it was thought until recently that exotic structures start at dimension seven. Specifically, the lowest-dimensional case of the pathology mentioned in Chapter 2 (manifolds which are homeomorphic but not diffeomorphic) known is at the seven-dimensional sphere. It has been shown recently by Donaldson (1983a) that $\mathbb{R}^4$ admits differentiable structures inequivalent to the standard one. On the other hand, the standard smooth structure on $\mathbb{R}^n$ is in fact unique up to diffeomorphisms for all n. If a configuration-space manifold Q of dimension seven or greater is required it may be necessary to find some way to choose among nondiffeomorphic alternatives. Trajectories which are smooth on one manifold will not necessarily be smooth on another which is not diffeomorphic, and if a manifold exists on which the trajectories are smooth, that one will be chosen as configuration space.

(b) The same kind of requirement of continuity can determine the choice between inequivalent differentiable structures. It will be seen in Example 6.3 how this works when the choice is between the $x \to x$ and $x \to x^3$ atlases for $\mathbb{R}$ which were discussed in Chapter 2.

(c) The choice of an atlas is best illustrated by examples.

Consider first $Q = \mathbb{R}^2$ with the usual atlas A_1 and the compatible atlas A_2 defined by

$$\varphi(q) = (2 + \cos 4\theta)r, \quad r \in \mathbb{R}^2.$$

A given point q with coordinates (x, y) in A_1 has coordinates

$$(\xi, \eta) = ((2 + \cos 4\theta)x, \quad (2 + \cos 4\theta)y)$$

in A_2, where θ is the usual polar coordinate thought of as a function of x only. If the dynamical system contains an orbit which looks circular in A_1, it will of course look very different in A_2, as illustrated in Fig. 3.1. Suppose all orbits are

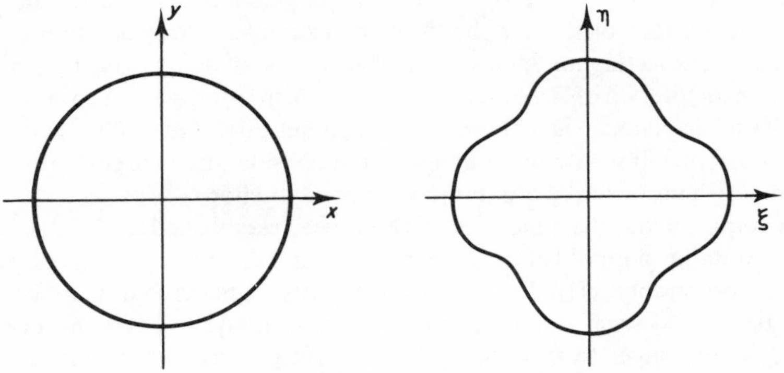

Fig. 3.1. A given orbit as seen in two charts

circular in A_1. Then one could say immediately that the dynamical system is rotationally invariant. This intrinsic property of the system is not eliminated when one goes from A_1 to A_2, but in the second atlas it becomes hidden and the symmetry becomes much harder to describe.

Again, consider $Q = \mathbb{R}^2$. Even when $\mathbb{R}^2$ is imbued with its usual differentiable structure it is not equipped with a vector-space structure as well, for different compatible but not linearly related atlases would define different composition laws. Think, for example, of $A_1 : q \mapsto (x, y)$ and $A_3 : q \mapsto (\sinh x, \sinh y)$. To provide $\mathbb{R}^2$ with a particular definite vector-space structure, which might be especially convenient for some calculations, a particular family of atlases would have to be chosen consistently.

It is clear that the choice of one atlas over another depends on the problems with which one is concerned. One atlas may make explicit invariance properties of the dynamics, another could help in recognizing at a glance some qualitative properties of the orbits, a third could simplify the mathematical description, etc.

This kind of consideration will prove even more important when the transitions are made from the configuration-space manifold Q to the tangent bundle TQ and to the contangent bundle T^*Q. In those manifolds, important will be what are called *natural* charts and *symplectic* charts, whose convenience and importance will become evident in Chapters 9 and 12.

The choices made in the construction of Q leave traces in all that follows and although they often seem unavoidable and sometimes trivial, it is well to be aware of them.

3.3 INVARIANTS OF A SYSTEM UNDER Q TRANSFORMATIONS

As we have pointed out in §3.2, there exist ambiguities in the definition of Q. Of course there are limits to such ambiguities, limits which place restrictions on the ways a given motion can be described. This subsection treats some aspects of these limits. It should not be thought, incidentally, that no ambiguity is left once it has been eliminated in the discussion of Q. There are other ambiguities which appear at the next level, in the carrier manifold (having to do, for example, with transformations, with the choice of Lagrangian or Hamiltonian function), but these will be discussed later.

What, then, are the limits on the ambiguities? What is it that does *not* change when one goes from one representation to another? How can one recognize that a fixed dynamical system is being represented, irrespective of the choice of a model representing it? Those aspects which do not change may be called the *invariants* of the dynamical system, and then it is these invariants which are being sought.

One can get around the problem of ambiguities by defining a dynamical system as the collection of all of the *equivalent* representations (whatever one may mean by 'equivalent'), for then the definition would contain all of the ambiguities. The problem with this is that it may hide some interesting invariants which are qualitative. Let us proceed in this way: we take two representations to be equivalent if they are related by a C^∞ diffeomorphism, and we proceed to a discussion of qualitative invariants.

Now, it is clear that Q alone does not determine a dynamical system, for observations allow one to construct not only Q, but the trajectories on it, which parametrize the observed motions in time. Thus there may exist invariants related to Q, and others related to the trajectories. These two kinds of invariants will be discussed separately.

The invariants related to Q are relatively few. One is the dimension, which has already been discussed. When the dimension and the differentiable structure have been established, remaining invariants are compactness and connectedness. In addition, a set of measure zero in one representation (under some suitable definition of measure) will remain of measure zero in any other related to it by a diffeomorphism (see Abraham and Robbin, 1967, p. 37).

Some of the invariants associated with trajectories arise because the image on Q of each trajectory, or rather its closure, is usually a submanifold of Q and therefore has the same kind of topological and differential invariants as does Q. In particular, if an orbit is closed (i.e. compact in the topology of Q) in one representation, it is closed in all, even though the orbit itself may have very different shapes in different representations. An example of this is illustrated in Fig. 3.1. Every closed trajectory has a *period*, the time it takes for it to close, and this period is also an invariant.

There are invariants related to the way in which an orbit is imbedded in Q. For example a circle and a circle with a knot are diffeomorphic, so that one might expect a knot to become unknotted in the transition from one representation to another. This is not, however, true. Having a knot is a special case of the problem of *placement* (in this case for a simple curve in $\mathbb{R}^3$), which can be stated as follows. Let A_1 and A_2 be subspaces of a topological space Q, and assume that there exists a homeomorphism $h: A_2 \to A_1$. Does there then also exist a homeomorphism $f: Q \to Q$ such that $f A_2 = A_1$? If f exists, A_1 and A_2 are said to be *placements* of a given A in Q, or to be of the same *type*. Otherwise they are of different types. In

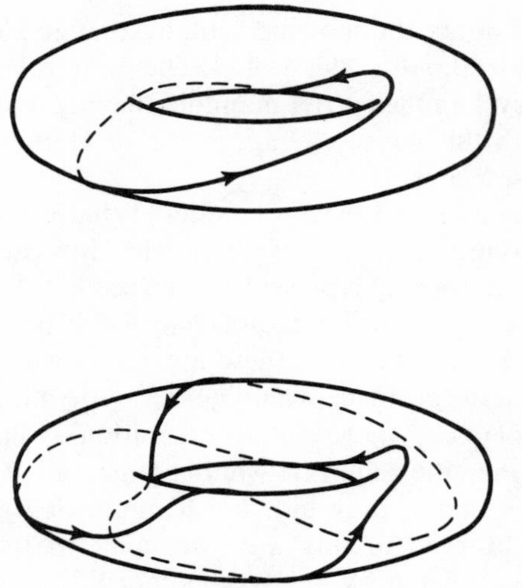

Fig. 3.2. Two orbits of different type on T^2

particular, for $Q = \mathbb{R}^3$, the circle and the circle with the knot are of different type, and there is no homeomorphism of Q into Q which will transform one into the other. A similar situation arises on the torus T^2, which is the configuration-space manifold for the double pendulum (see Arnol'd, 1978). The two closed orbits in Fig. 3.2, both of which are in the dynamics, are of different types.

Other trajectory-related invariants are intersections of orbits. If two orbits intersect in n points in one representation, they will do so also in any other. The number of *equilibrium points* in Q (points at which the system can remain at rest, or the seats of one-point orbits) is also an invariant. It may be noted that there exist *differential invariants*, left invariant by diffeomorphisms, which are not *topological invariants*, not necessarily left invariant by general homeomorphisms (see, for instance, the *characteristic multipliers* defined in AM, p. 520).

Invariants will be discussed again in the more usual context of carrier manifolds for the dynamics, in which the orbits are separated so that only one passes through each point. It is useful to remember, however, that there exist invariants also at the level of configuration space.

BIBLIOGRAPHY TO CHAPTER 3

A useful introduction to some of the problems related to the construction of Q can be found in Sklar (1974), in particular in the chapter: 'The epistemology of geometry'. Sklar uses the recursive definition of dimension given by Brouwer, Menger and Urysohn, but points out that there are several other definitions, all equivalent for metrizable spaces (and therefore for manifolds) but often inequivalent for more general spaces. On the problem of the definition of dimensionality see also Cartan (1957); Willard (1968) and Hartshorne (1977).

The fact that manifolds can be homeomorphic but not diffeomorphic was first proved (in seven dimensions) by Milnor (1956). Brickell and Clark (1970) give an example of two differentiable structures on S^1 leading to the different topological structures (p. 37), an example of two differentiable structures on $\mathbb{R}$ (distinct but diffeomorphic) leading to the same topology (p. 37), an example of two differentiable structures on a non-Hausdorff topological space which are homeomorphic but not diffeomorphic (p. 64; the non-Hausdorff space is needed to keep the example simple, avoiding the recourse to the seven dimensions needed by Milnor's result). For more recent discussions of this question, see Donaldson (1983a, 1983b) and Kreck (1984). Note that Kreck's paper, though published in 1984, was received in 1981.

What is meant by equivalence in the description of Q can be seen from different contexts: a topological one (see for instance Chillingworth (1976, p. 228)) or a differential one (same page in Chillingworth (1976)). It can also be discussed usefully in the special case of linear systems (see, for instance, Hirsch and Smale (1974, p. 39)). The problem of knots and placements is discussed, for instance, in Fort (1962), in particular in the contribution by R. H. Fox: 'A quick trip through knot theory'. See also Rolfsen (1976).

4

Time and transformations on time

The dynamical systems we wish to construct are supposed to describe the one-parameter evolution of several *dynamical variables*, and the parameter in which they evolve is what is called the *time*. It is therefore reasonable at this point to discuss some of the characteristics of time as we conceive it. The brief discussion which follows is not meant to be complete, but only to mention some of the problems and to exhibit our point of view and, we may hope, its consistency.

Scientific data, like all human experience, lie within a historical framework, which depends for each observer on his past experience, social, linguistic, certainly scientific. Recall, for instance, that when describing the construction of the configuration-space manifold Q, we postulated that different observers could at least agree on what they would call position and assumed that this agreement comes from some common experience with *motion*. Without some idea of what it means to distinguish between (and measure) positions at a *fixed time*, the construction of Q in Chapter 3 could not have been performed.

Nevertheless, the structure of time itself remains undetermined. For example, just as Q may be locally Euclidean while possessing a more complicated global structure, so time may have a more complicated structure in the large. The concept of space itself, based on the motion of local objects (within several billion kilometers of the Earth), used to be that it could be identified with $\mathbb{R}^3$. Relatively recent astronomical observations have, as is very well known, shown that this identification is not at all necessary. Similarly, there was a time when the inability to perform any but relatively short-range measurements led to the belief that the Earth was as flat as $\mathbb{R}^2$. In the same way it is conceivable that although every interval in time is (or at any rate will be in this book) identified with an interval of $\mathbb{R}$, time may have some other global structure, say that of S^1. That is, it is conceivable that time is cyclic with a period of several billion years, a period which would be essentially imperceptible, if not irrelevant, in terms of human experience. In any case, we do not pursue the idea that time may be cyclic; in this book it will always be identified with $\mathbb{R}$.

Remark. Irrelevance is postulated here, because of the class of dynamical systems treated in the present book. Remember, however, that we emphatically do not assume that the global structure of Q is irrelevant. On the other hand, cyclic time plays a role in the discussion of intrinsically periodic (e.g. circadian) systems and this can lead to a special, explicitly nonlinear representation of structure in time, which prompted A. T. Winfree (1980) to speak of 'time crystals'.

Time is associated with change, with motion. It is 'not identical with motion, but is that in terms of which motion is counted, and even if a thing is at rest, it may be countable by the same count as motion.' It is 'the calculable measure or dimension of motion with respect to before-and-afterness; time, then, is not movement, but that by which movement can be numerically estimated' (Aristotle, *Physics*, 221b, 219b). For dynamics the numerical aspect is essential, the transformation of qualitative ideas about change into a quantitative description of one-parameter evolution. It is not always easy to go from the qualitative notions to the numerical description, and we do not mean to imply that the latter is somehow contained in the former. In fact some results of genetical epistemology (e.g. a child's construction of the concept of velocity independent of the metrical structure of time) show that qualitative notions must often be altered, if not actually eliminated, in constructing the quantitative theory. In any case, the quantitative aspects of time are required at the formal level of describing dynamics.

Although we are trying to discuss time here in a way that is not too dissimilar from the way we discussed configuration space Q, there is a fundamental difference. A dynamical system is defined by its invariants under all sorts of diffeomorphisms allowed for Q, rather than by its particular description in terms of one of many diffeomorphically equivalent Qs, but the same kind of freedom will not be used in dealing with time. Not all possible diffeomorphisms on $\mathbb{R}$ will be allowed which can replace time with other diffeomorphically equivalent evolution parameters. The very limited kind of (linear) transformation which has ordinarily been allowed on the time t, and the kind we shall allow, replaces t by a new time

$$t' = at + b. \tag{1}$$

This permits freedom in the definition of unit and origin only. The reasons for this restriction are not entirely clear. If forces, lengths, and masses are taken as fundamentally measurable, then Newton's second law $\mathbf{F} = m\mathrm{d}^2\mathbf{s}/\mathrm{d}t^2$ defines time essentially up to the inhomogeneous linear transformation (1). But forces are at the best measurable only in inertial systems, and thus the definition of time depends on the existence of inertial systems, a notion necessarily of local character whose definition even locally is restricted by the imperfection of measuring devices. A similar way to define time involves laying out equal distances on the path of a free particle in an inertial frame; again one comes across the same local notion, and again time is defined up to an inhomogeneous linear transformation.

An insight into the restriction of Eq. (1) on transformations in t is obtained by considering momentum conservation in n-particle interactions, according to which for each such interaction there exists a constant (time independent) vector $\mathbf{P}$ such that

$$m_i\mathbf{v}_i(t) = \mathbf{P}, \quad i \in \{1, \ldots, n\} \tag{2}$$

for all time t, where m_i and $\mathbf{v}_i$ are the mass and velocity of particle i. If t is now replaced by a more or less arbitrary new 'time' T such that $t = f(T)$, where $f \in C^1$ is monotonic increasing (to preserve chronological ordering), Eq. (2) becomes

$$m_i\mathbf{V}_i(T) = f'(T)\mathbf{P}, \tag{3}$$

where $\mathbf{V}_i$ is the velocity of particle i in terms of the new time T. The conservation of momentum in terms of T would state that there exists a universal function $f'(T)$ of the time such that in n-particle interactions Eq. (3) is satisfied. A redefinition of time according to $t = \int f'(T)\,\mathrm{d}T = f(T) + K$, for arbitrary constant K, then makes the conservation of momentum a time-independent law. But again, momentum conservation holds only in inertial systems.

Thus one is led to plausible, perhaps even convincing explanations for the restriction on permissible time transformations, but not without some doubts. The choice of the time variable seems to depend on more general considerations than strict axioms and logical requirements for formalizing dynamics.

In spite of these restrictions, however, it is often convenient for specific requirements to redefine a new T which is not linearly related to t. An explicit case is given by Wintner (1941, p. 127). Let a conservative dynamical system in n degrees of freedom be described by a Hamiltonian function $H(x)$, where $x = (q_1, \ldots, q_n; p_1, \ldots, p_n)$ represents the point in phase space, and let $G(x)$ be any continuous nonvanishing function. Then a new time variable T may be defined for each integral curve $g(t)$ of the system by the integral

$$T(t) = \int_{t_0}^{t} \frac{\mathrm{d}\tau}{G(g(\tau))}. \tag{4}$$

Let h be some value of the energy, and consider the new Hamiltonian function

$$\bar{H}(x;t) = [H(x) - h]G(x).$$

Then a simple calculation will show that those integral curves which belong to energy $H = h$ with time parameter t belong to new energy $\bar{H} = 0$ with time parameter T. The practical use of this nonlinear time transformation is the following. Wintner chooses G locally so that T is essentially one of the generalized coordinates, say q_n, and so that the corresponding p_n can be written $p_n = -K(\tilde{x}, T; h)$ (this is always possible locally), where $\tilde{x} = (q_1, \ldots, q_{n-1}; p_1, \ldots, p_{n-1})$. Then he shows that the integral curves corresponding to $\bar{H} = 0$ are the same as the integral curves $\tilde{x} = \tilde{g}(T)$ obtained in the $(2n-2)$-dimensional phase space for the system described by the nonconservative Hamiltonian $K(\tilde{x}, T; h)$. These can be used in obtaining the integral curves $x = g(t)$ which belong to energy h.

Although such nonlinear time transformations would probably be useful in discussing the most general invariance properties of dynamical systems and although one may like to think that the concepts of dynamical system and evolution parameter are easily extended to fields other than classical mechanics (to biology, for instance, or economics), so that 'time' could mean many other things in abstract systems, we shall allow only the transformations of Eq. (1). We make this choice not because we think it is logically necessary and in spite of the unnatural restrictions it may lead to in other abstract systems, partly for simplicity and partly for the kind of considerations we have already given. Nevertheless, one should remain open to cases and examples where this restriction may be violated for good reasons. A good reason will be application to Lie group and algebra actions (Chapters 21 to 25), where the interest will be in going from incomplete to complete vector fields on a manifold (see Chapter 6).

A chapter on time cannot close without mention of relativity. This book is rather strictly nonrelativistic (although we may on occasion refer to relativistic particles), so that considerations about the relativistic structure of space–time are not really relevant. Our understanding of time is a classical one, and we shall disregard (perhaps even sweep under the rug) problems involving the intimate connection of space with time, as well as the fact that time is different for different observers and different particles and the fact that in general these different times cannot be matched in order to define an absolute 'unfolding' time for the universe in terms of local notions.

BIBLIOGRAPHY TO CHAPTER 4

The bibliography on time is immense. We shall give here only some references on problems discussed in this chapter. The importance of cyclic time in mythological and philosophical thinking is discussed, for instance, in Mandal (1968) and Needham (1962). The most complete and recent analysis of its relevance in biological modelling can be found in Winfree (1980).

The epistemological results are described in Piaget (1973) and in several volumes of the Etudes d'Epistémologie Génétique, Presses Universitaires de France, Paris.

Relevant to our discussion is also Grünbaum (1973), in particular Chapter 2 on the significance of alternative time measures in Newtonian mechanics and general relativity.

The fact that in general relativity different local times can not possibly be matched to define an absolute unfolding time is pointed out by Gödel, in his: *Static Interpretation of Space* in Čapek (1976).

5

A digression on calculus on manifolds, bundles, vector fields and forms

Throughout this book we make use of the formal apparatus of calculus on manifolds and, in particular, on tangent bundles. The present chapter will, therefore, be a discussion of some of the essential elements of this calculus, starting with bundles (both tangent bundles and more general ones), going on to vector fields and forms, and finishing with remarks and reminders concerning calculus on manifolds (e.g. Lie derivatives, pullbacks) and the relation between vector fields and one-parameter groups of transformations.

5.1 BUNDLES

1. The first step will be to arrive at a definition of tangent bundles through a definition of *local* tangent bundles. Let E be a vector space and $U \subset E$ be open, and let $I \subset \mathbb{R}$ also be open, $O \in I$. From a curve $c: I \to U$ (of class at least C^1) one can define the mapping

$$Tc: I \times \mathbb{R} \to U \times E: (i, r) \mapsto (c(i), Dc(i) \cdot r), \tag{1}$$

where $Dc(i) \in \mathrm{Lin}(\mathbb{R}, E)$ is the *derivative* of c at i (AM, p. 21). In this particular case, this derivative can be thought of as the line (linear curve) which is tangent to c at i. In fact, the notation Lin $(\mathbb{R}, E)$ denotes the set of linear maps from $\mathbb{R}$ to E, and thus $Dc(i) \cdot r$ is this map applied to $r \in \mathbb{R}$. The map Tc defined by Eq. (1) is called the *tangent* of c.

Consider two curves c_1 and c_2 such that $c_1(i_1) = c_2(i_2) = u$, that is, two curves which pass through a common point u of U. One says that c_1 and c_2 are tangent at u iff $Tc_1(i_1, r) = Tc_2(i_2, r)$ for each $r \in \mathbb{R}$, and the equivalence class of all curves tangent to them at u is then denoted by $[c_1]_u = [c_2]_u$. Actually, since $Dc(i)$ is linear, tangency of c_1 and c_2 at u is assured even if $Dc_1(i) \cdot r = Dc_2(i_2) \cdot r$ for only one value of $r \neq 0$, so one can set $r = 1$ in the definition of tangency. At every point $u \in U$ there is thus a set of equivalency classes of curves. In this way, one can establish a bijection from such tangency classes to $U \times E$, mapping $[c]_{c(i)}$ into $(c(i), Dc(i)\cdot 1) \in U \times E$.

Remarks. 1. A weaker definition of tangency, perhaps more common in elementary mathematics, defines two curves as tangent if there exist $r_1, r_2 \in \mathbb{R}$ such that $Tc_1(i_1, r_1) = Tc_2(i_2, r_2)$. Then each *ray* in $\{u\} \times E$ is identified with an equivalence class of curves at u. We shall not use this weaker definition.

2. In each tangency class at u there are, of course, curves such that $c(0) = u$. These will usually be taken as representatives of the tangency classes.

Cartesian products like $U \times E$ are examples of *local vector bundles.* In the present example, because the points in E are identified with tangency classes, what has been constructed is a local *tangent* bundle, denoted TU. Then U is called the *base space* and $T_u U \equiv \{u\} \times E$ is called the *fiber* over u. The map $\tau_U : U \times E \to U : (u, e) \mapsto u$ is called the *projection* of $U \times E$ onto the base space. It should be clear that $T_u U = \tau_U^{-1}(u)$ and that $TU = \cup_u T_u U$.

The mapping Tc can now be thought of as a mapping between the two local tangent bundles $TI = I \times \mathbb{R}$ and $TU = U \times E$ with the diagram

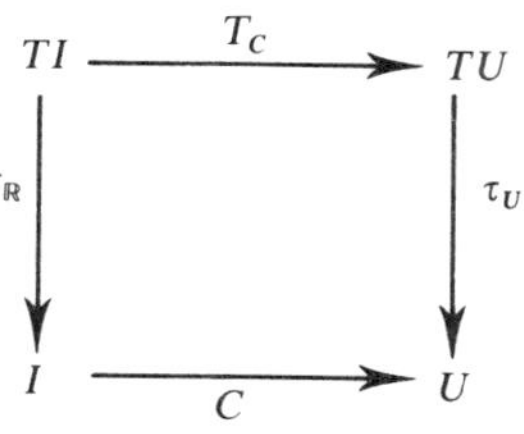

Remark. Unless otherwise stated, all diagrams commute. In this diagram this means that $\tau_U \circ Tc = c \circ \tau_{\mathbb{R}}$.

This map $Tc: TI \to TU$ is an example of a *local tangent bundle mapping.* Given a C^k mapping $g: U \to V$, the tangent Tg of g is a C^{k-1} mapping defined by

$$Tg: TU \equiv U \times E \to TV \equiv V \times F : (u, e) \mapsto (g(u), Dg(u)\cdot e)$$

(both $U \times E$ and $V \times F$ are endowed with natural manifold structures), or equivalently by

$$Tg: [c]_u \mapsto [g \circ c]_{g(u)}. \tag{2}$$

As will be seen, the definition of Eq. (2) is better suited for globalization. Note that Tg maps the fiber over u linearly into the fiber over $g(u)$. If g is a diffeomorphism, the restriction $T_u g$ of Tg to the fiber over u is a linear isomorphism for each $u \in U$; then Tg is called a *tangent-bundle isomorphism.* This is the first example in this chapter of a vector bundle isomorphism (see § 5.2).

In what follows we will use the fact that the construction, as it has been carried out to this point, could have been carried out not only on a vector space E, but also on any subset of a manifold which lies entirely within the domain of a chart. This is because a chart is modelled on a vector space.

2. The next step is to construct the definition of *global* tangent bundles, which proceeds in a rather smooth way from the treatment above of local tangent bundles. Let M be a differentiable manifold, and consider a curve $c_m: I \subset \mathbb{R} \to M$ of class C^1 such that $c_m(0) = m \in M$ (a curve at m). The equivalence class $[c]_m$ of *tangent curves* to c_m is then defined as above for a neighborhood $U \subset E$, but now for a local chart (U, φ) at m. The local representatives of this equivalence class are all tangent at $\varphi(m)$. The separation obtained in this way into equivalence classes (that is, the definition of tangent curves) is chart independent.

Then the *fiber* over m is defined as $T_m M \equiv \{[c]_m\}$, and the *tangent bundle* of M as $TM = \cup_m T_m M$. The projection is also defined as for local tangent bundles by $\tau_M: TM \to M: [c]_m \mapsto m$.

For each atlas (U, φ) on M there is a *natural atlas* $(TU, T\varphi)$ on TM, and then from a maximal atlas on M one can obtain a natural maximal atlas on TM, which gives TM a manifold structure in a natural way. If M is Hausdorff and second countable, then so is TM in the topology induced on it in this way, and with this differentiable structure τ_M is smooth.

If M and N are differentiable manifolds and $\chi: M \to N$ is a C^k map, the *tangent-bundle map* $T\chi: TM \to TN$, of class C^{k-1}, is defined by its action on each fiber, i.e. by

$$T\chi: [c]_m \mapsto [\chi \circ c]_{\chi(m)}. \tag{3}$$

Then $T\chi$ is a mapping between the two tangent bundles with the diagram

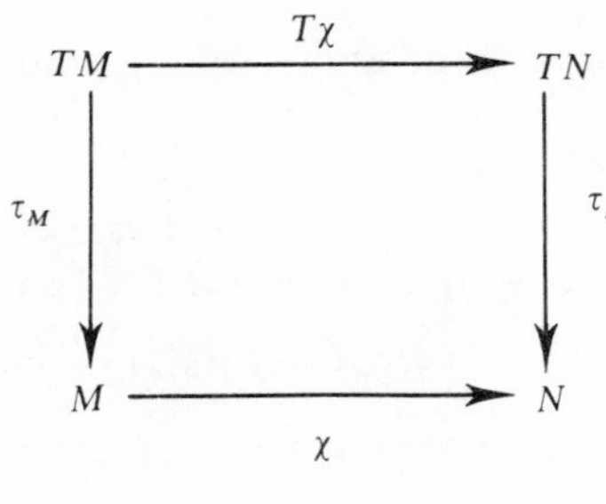

Note that $T\chi$ maps the fiber over m linearly into the fiber over $\chi(m)$. If χ is a diffeomorphism, $T\chi$ is an isomorphism on each fiber and is called a *tangent-bundle isomorphism*. Two tangent bundles are called *isomorphic* if there exists a tangent-bundle isomorphism between them.

If $\chi: M \to N$ and $\chi: N \to Q$ are two C^∞ maps (M, N, Q are all differentiable manifolds), then it is easily shown that the tangent-bundle map of the composition $\psi \circ \chi$ satisfies

$$T(\psi \circ \chi) = T\psi \circ T\chi. \tag{4}$$

Remark. A special case of a tangent-bundle map occurs when the N of the diagram is $\mathbb{R}$. Then the χ of Eq. (3) is a function $f:M \to \mathbb{R}$.

Some global tangent bundles can look very much like local ones. A tangent bundle is called *trivial* when there exists a diffeomorphism $\sigma: TM \to M \times E$, where E is a vector space, such that σ maps T_mM onto $\{m\} \times E$ by a linear isomorphism for each $m \in M$ in accordance with the diagram

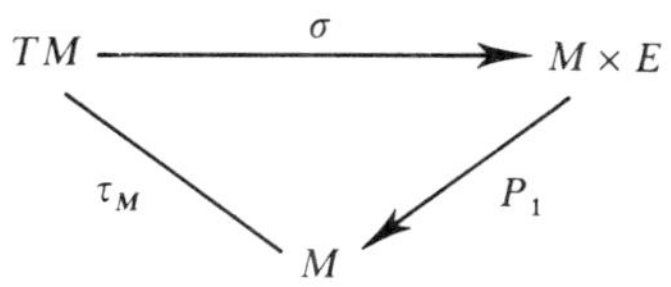

where P_1 denotes the *first projection* (or the projection onto the first factor), i.e. $P_1:(m, e) \mapsto m$. It should be clear from the definition that every local tangent bundle is trivial, but global ones need not be. For example, TS^1 is diffeomorphic to $S^1 \times \mathbb{R}$ and is therefore trivial, but TS^2 is not diffeomorphic to $S^2 \times \mathbb{R}^2$ and is therefore not trivial (see Example 6.4).

3. (a) *Tangent bundles* are an example of the more general idea of vector bundles, and these in turn are an example of general *bundles.* A general bundle can be described by generalizing the idea of the projection $\tau_M: TM \to M$ in the following way. The tangent bundle TM will be replaced by an arbitrary differentiable manifold N and τ_M by a differentiable mapping π. Note that τ_M is surjective and observe that if $v \in T_mM$ (i.e. if v is in the fiber over m), then $T_v\tau_M: T_v(TM) \to T_mM$ is surjective for each $m \in M$ [replace M by TM and N by M in the diagram which follows Eq. (3)]. In the same way the triple (N, π, M) is defined as a *bundle* [in this notation TM would be written (TM, τ_M, M)] if both π and $T_n\pi: T_nN \to T_{\pi(n)}M$ are surjective (the latter for each $n \in N$). This is a bundle of the most general type. If, now, all of the $\pi^{-1}(m)$ are diffeomorphic, that is if for every pair $m, m' \in M$ the *fibers* $\pi^{-1}(m)$ and $\pi^{-1}(m')$ are diffeomorphic, and if moreover about each $m \in M$ there exists a neighborhood $U \subset M$ such that $\pi^{-1}(U)$ and $U \times \pi^{-1}(m)$ are diffeomorphic, then the bundle is a *fiber bundle.*

Different types of fibers define fiber bundles of different types. For example, if each fiber F is a vector space, the fiber bundle is a *vector bundle.* If each fiber F is a tensor space, it is a *tensor bundle.* If each fiber is the quotient of a group G with respect to a subgroup H, it is a G bundle, where G stands for 'group'. Many of these bundles will arise in the course of this book, but sometimes only in asides and under particular conditions, so without going into detail we direct the reader here to the bibliography.

Let (N, π, M) and (N', π', M') be two bundles. Then $\chi: N \to N'$ is a bundle map if there exists a $\chi_0: M \to M'$ which enters the commuting diagram

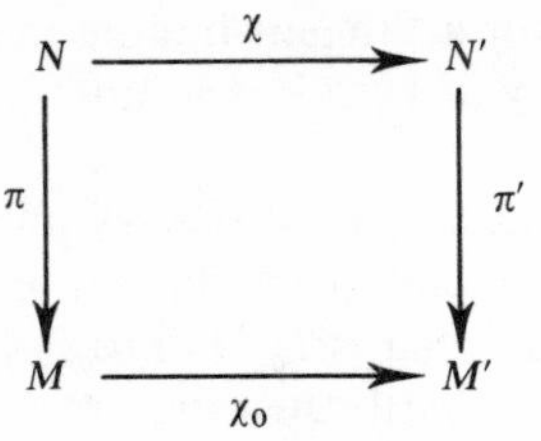

The map χ, restricted to fibers of the bundle, may or may not preserve structural properties of the fibers; for example, (N, π, M) may be a vector bundle, but (N', π', M') need not be. Thus a fiber bundle (or fibered manifold) may be diffeomorphic to another even if the diffeomorphism does not preserve the particular structures of the fibers. Note, however, that if the fibers are vector spaces and χ is linear on them, the vector space structure will be preserved.

It is sometimes helpful to understand a vector bundle, as it may be to understand any property of a manifold, from a local point of view. Let $\{\varphi_k, U_k\}$ be an atlas for a manifold M. Each local chart looks like a neighborhood of $\mathbb{R}^\mu$, where $\mu = \dim M$, and one may ask under what conditions on the transition functions $\varphi_i \circ \varphi_j^{-1}: \mathbb{R}^\mu \to \mathbb{R}^\mu$ some property which holds true in the chart will carry over to the entire manifold. An example of such a property is the following. As the transition function can be considered a matrix whose elements are C^∞ functions, it may occur that for each pair i,j there is a basis in $\mathbb{R}^\mu$ such that this matrix is completely reducible to the form

$$\psi_{ij}\begin{vmatrix} x \\ y \end{vmatrix} \equiv \varphi_i \circ \varphi_j^{-1}\begin{vmatrix} x \\ y \end{vmatrix} = \begin{vmatrix} A & 0 \\ 0 & B \end{vmatrix}\begin{vmatrix} x \\ y \end{vmatrix}, \tag{5}$$

where A is a κ-by-κ matrix and B is a $(\mu-\kappa)$-by-$(\mu-\kappa)$ matrix, both with elements that are C^∞ functions. If A and B depend only on x, not on y, where x represents the point in $\mathbb{R}^\kappa$ and y the point in $\mathbb{R}^{(\mu-\kappa)}$, then M is a vector bundle over $\mathbb{R}^\kappa$ with fibers of dimension $\mu-\kappa$. If, moreover, $\mu = 2\kappa$ and B is the Jacobian matrix of A, then M is a tangent bundle.

Remarks. 1. A less specific case is where the transition function is of the form

$$\psi_{ij}\begin{vmatrix} x \\ y \end{vmatrix} = \begin{vmatrix} A & 0 \\ D & B \end{vmatrix}\begin{vmatrix} x \\ y \end{vmatrix},$$

with A and D functions only of x, and B a function of both x and y. This is a sort of semidirect splitting of M and is characteristic of foliated manifolds (see Chapter 17).

2. In general, any operation on $\mathbb{R}^\mu$ that is compatible with the given atlas, i.e. whose result is independent of whether it is performed before or after the

transition function is applied, can be considered an operation on the manifold, not just on the local chart. This is true also for proofs of assertions, so that, with this proviso, global proofs may be performed in local charts.

(b) *Cotangent bundles* are of special importance for dynamical systems. They can be constructed in a way which is quite similar to the way the tangent bundle was constructed in §2. That construction began with the curves $c:I \subset \mathbb{R} \to U \subset E$, and those curves were then associated with their tangents $Tc:I \times \mathbb{R} \to U \times E$. The construction of the cotangent bundle with now begin with the C^∞ functions $f:U \subset E \to \mathbb{R}$ and the compositions $f \circ c:I \subset \mathbb{R} \to \mathbb{R}$. These are associated with their tangents $T(f \circ c):TI \to T\mathbb{R}$ whose action is defined by

$$T(f \circ c)(i, r) \equiv (f(c(i)), Df(c(i)) \cdot Dc(i) \cdot r). \tag{6}$$

For the tangent bundle, equivalence classes of curves were defined; for the cotangent bundle equivalence classes of functions are defined. Two functions f_1 and f_2 are equivalent at $u \in U$ iff for all curves c such that $c(0) = u$ it follows that

$$T(f_1 \circ c)(0, r) = T(f_2 \circ c)(0, r).$$

Let $[f]_u$ be the equivalence class at u which contains the function f. It is easy to check that $Df(c(0)) \cdot Dc(0)$ is a linear map from $\mathbb{R}$ to $\mathbb{R}$ and depends only on the equivalence classes to which f and c belong, not on f or c individually, and therefore it is possible to define the product $[f]_u \cdot [c]_u \equiv Df(u) \cdot Dc(0)$. The number so obtained is $\mathbb{R}$-linear with respect to the linear structures of both equivalence classes, and it is then seen that $[f]_u$ is in the dual space of $T_u U$, or of E.

This procedure therefore associates with each point $u \in U$ the vector space E^* of (linear) mappings in $\mathrm{Lin}(T_u U, \mathbb{R})$, called the *cotangent space* at u. The Cartesian product $U \times E^*$ is the *local cotangent bundle*, written $T^* U$.

The *global cotangent bundle* can now be constructed by analogy with the generalization from the local to the global tangent bundles. We shall not go into the details of the construction, which is very similar to the one for the tangent bundle, but from it one arrives in this way at the cotangent bundle $T^* M$ of the manifold M, together with the projection $\tau_M^*: T^* M \to M$. [As has been mentioned, the cotangent bundle is more correctly written $(T^* M, \tau_M, M)$.]

Recall how the tangent bundle map $T\chi: TM \to TN$ (the *tangent* of χ) was defined at Eq. (3) in terms of the map $\chi: M \to N$. The *cotangent* $T^* \chi$ of χ is somewhat harder to define, for it is defined in terms of $T\chi$ and the *duality* between TM and $T^* M$. It then turns out, perhaps surprisingly, that $T^* \chi$ is a map not from $T^* M$ to $T^* N$, but in the other direction from $T^* N$ to $T^* M$. For each $w_m \in T_m M$ and $\rho_m \in T_m^* M$ let $\sigma_{\chi(m)} \in T_{\chi(m)}^* N$ be given by

$$\rho_m \cdot w_m = \sigma_{\chi(m)} \cdot (T_{\chi(m)} \chi(w_m)). \tag{7}$$

Then $T^* \chi: T^* N \to T^* M$ is defined by its action on fibers through

$$T_{\chi(m)}^* \chi: T_{\chi(m)}^* N \to T_m^* M : \sigma_{\chi(m)} \mapsto \rho_m. \tag{8}$$

Then the cotangent map enters into the commuting diagram

$$\begin{array}{ccc} T^*M & \xleftarrow{\;T^*\chi\;} & T^*N \\ \chi_M^* \big\downarrow & & \big\downarrow \tau_N^* \\ M & \xrightarrow[\;\chi\;]{} & N \end{array}$$

and it has the dual of the composition property Eq. (4) of the tangent map, namely

$$T^*(\chi \circ \psi) = T^*\psi \circ T^*\chi. \tag{9}$$

Remark. Tangent bundles and cotangent bundles can be used to construct *tensor bundles.* Although, on occasion, we shall make use of tensor bundles, we shall not discuss their construction in any detail (see AM, p. 42). Roughly speaking, in the tensor bundle T_s^rM the fiber over $m \in M$ consists of the vector space obtained from the Cartesian product of r copies of T_mM and s copies of T_m^*M. Tensor bundle mappings are discussed in AM, p. 55. T_s^rM is said to be *contravariant* of order r and *covariant* of order s.

As was pointed out around Eq. (5), it is sometimes helpful to understand properties of manifolds from a local point of view. In terms of the transition function matrix of Eq. (5), a manifold is a cotangent bundle if $\mu = 2\kappa$, the matrices A and B depend only on x, and B is the inverse transpose of the Jacobian matrix of A.

5.2 VECTOR FIELDS

1. The next chapter will be devoted largely to constructing and illustrating the vector fields one associates with dynamical systems constructed from pieces of the trajectories (from trajectory germs) by first defining a carrier space for the dynamics. For this reason we now turn to vector fields, which will be discussed here in terms of the tangent bundle.

Consider a differentiable manifold M and its tangent bundle TM. As both are C^∞ manifolds, there exist C^∞ maps $\chi: M \to TM$. The maps which have the property that

$$\tau_M \circ \chi = \mathbb{1}_M \tag{10}$$

are called *vector fields*. To each point $m \in M$ the map χ assigns a point in TM which lies in the fiber above M, thus yielding a sort of slice through TM. In general, if (N, π, M) is a bundle, C^∞ maps $\chi: M \to N$ with the similar property $\pi \circ \chi = \mathbb{1}_M$ are called *sections* of π, and thus vector fields are C^∞ sections of τ_M. The set of these C^∞ sections of τ_M is written $\mathfrak{X}(M)$, so that a vector field on M is an element of $\mathfrak{X}(M)$.

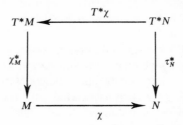

Let $X \in \mathfrak{X}(M)$ be a vector field. Its smoothness can be expressed in terms of the smoothness of its local representatives in an atlas over M. That is, X is C^∞ if in every chart (U, φ) of such an atlas, $X_\varphi \equiv T\varphi \cdot X \cdot \varphi^{-1}$ is C^∞. Then the following diagram commutes:

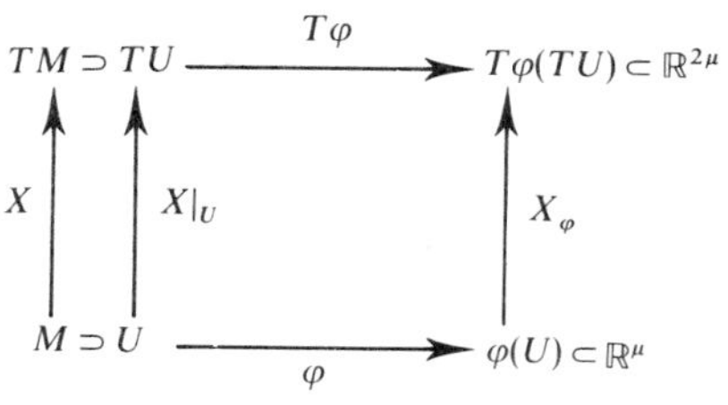

Remarks. 1. In a similar way, one can define $\mathfrak{X}^*(M)$, the set of all C^∞ sections of τ_M^*. An element of $\mathfrak{X}^*(M)$ is called a *one-form*. One-forms and more general *p-forms* will be discussed in §5.3.

2. From tensor bundles one can construct C^∞ sections in the same way. These are then called *tensor fields*.

A curve $g: I \subset \mathbb{R} \to M$ is called an *integral curve* of a vector field $X \in \mathfrak{X}(M)$ iff it is a solution of the differential equation

$$X(g(t)) = Tg(t, 1). \tag{11}$$

A *maximal* integral curve (or *trajectory*) of X is an integral curve $g_{\max}: I_{\max} \to M$ if for every I for which an integral curve is defined, $I \subset I_{\max}$ and $g_{\max|I} = g$. In the sequel we shall usually omit the adjective 'maximal'. Much of the problem of dynamics is to find the integral curves once a vector field X is given.

In a large number of applications, M will be TQ (configuration-velocity space, also called velocity phase space), itself a tangent bundle. Then vector fields $X \in \mathfrak{X}(TQ)$ will be C^∞ sections of τ_{TQ}, slices through the tangent bundle (of the tangent bundle). Among these vector fields are those said to be of *second order*, those whose integral curves $g: I \to TQ$ are *lifts* of *base integral curves* $h: I \to Q$, i.e. those which satisfy

$$g(t) = Th(t, 1) \text{ with } h(t) = \tau_Q g(t). \tag{12}$$

It is not possible to define second-order vector fields in $\mathfrak{X}(M)$ unless M is a tangent bundle. Indeed, suppose (M, π, Q) is any fiber bundle over the base space Q, and let the curve $g: I \to M$ project down to $h: I \to Q$. Then $g(t)$ is a point in M and $Th(t, 1)$ is a point in TQ and they cannot in general satisfy Eq. (12).

Second-order vector fields can be understood also in terms of their relation to projection. If $X \in \mathfrak{X}(TQ)$ is of second order, then from Eq. (12) and the definition of Eq. (11) one obtains

$$g(t) = Th(t, 1) = T(\tau_Q g)(t, 1) = T\tau_Q \cdot Tg(t, 1) = T\tau_Q \cdot X(g(t)).$$

For a second-order vector field, therefore,

$$T\tau_Q \cdot X = \mathbb{1}_{TQ}. \tag{13}$$

2. There are several ways to represent a vector field in local coordinates. Let $X \in \mathfrak{X}(M)$ and let (U, φ) be a local chart. Then since $\varphi: U \to \mathbb{R}^\mu$ is a diffeomorphism, $T\varphi$ is a bijection and $T(\varphi^{-1}) = (T\varphi)^{-1} \equiv T\varphi^{-1}$. Let $\mathbf{e}_k, k \in \{1, \ldots, \mu\}$ be the standard basis on $\mathbb{R}^\mu$, and use $T\varphi^{-1}$ to transfer this basis in a chart-dependent way to $T_u M$, $u \in U$:

$$e_k(u) = T_{\varphi(u)}\varphi^{-1}(\varphi(u), \mathbf{e}_k) \in T_u M, \tag{14}$$

where $T_{\varphi(u)}\varphi^{-1} = T\varphi^{-1}|_{T\varphi(u)}$. This establishes a basis $e_k(u)$ of vectors at each point $u \in U$, and each vector of this basis then defines a vector field $e_k \in \mathfrak{X}(U)$. The set of these vector fields then can serve as a basis for $\mathfrak{X}(U)$ in the following sense. Given a vector field $X \in \mathfrak{X}(M)$, one can write

$$X(u) = X^k(u) e_k(u), \quad X^k \in \mathscr{F}(U) \tag{15}$$

for each $u \in U$. Such a vector field will usually be represented in the form $(u, X^k(u))$, or even (u, X), in which form X represents the collection of the $X^k(u)$. Finally, on $\mathbb{R}^\mu$ the vector field can be represented by

$$X_\varphi(r) = (r, X^k(\varphi^{-1}(r)\mathbf{e}_k), \quad r \in \varphi(U).$$

In this notation it follows that

$$\tau_M X(u) = u. \tag{16}$$

Suppose now that M is the tangent bundle TQ of a manifold Q of dimension ν. Then one may write

$$X(q, \dot{q}) = (q, \dot{q}; X_q(q, \dot{q}), X_{\dot{q}}(q, \dot{q})), \tag{17}$$

where $q, \dot{q}$ labels the point in TQ, X_q represents the collection X_q^k of the first ν components of the field, and $X_{\dot{q}}$ the collection $X_{\dot{q}}^k$ of the second ν. Another form of writing a vector field, which is more in keeping with Eq. (15) and is particularly convenient for calculation, is

$$X(q, \dot{q}) = X_q^k(q, \dot{q}) \frac{\partial}{\partial q^k} + X_{\dot{q}}^k(q, \dot{q}) \frac{\partial}{\partial \dot{q}^k}. \tag{18}$$

Here the partial derivatives represent the e_k of Eq. (15).

On TQ, Eq. (16) becomes

$$\tau_{TQ} X(q, \dot{q}) = (q, \dot{q}).$$

But there is also another projection on $T(TQ)$, namely $T\tau_Q$, as already mentioned at Eq. (13). It is easily seen that

$$T\tau_Q X(q, \dot{q}) = (q, X_q(q, \dot{q})),$$

so that a vector field is of second order according to Eq. (13) iff it is of the form

$$X(q, \dot{q}) = (q, \dot{q}; \dot{q}, X_{\dot{q}}(q, \dot{q})). \tag{19}$$

The appearance of $\dot{q}$ in both the second and third entries on the right-hand side of Eq. (19) is suggestive, and it leads to another way to represent vector fields in $\mathfrak{X}(TQ)$, and indeed almost to represent T^2Q itself. For this purpose, let q represent a point of Q (rather than the local coordinates of a point in some local chart). In the same spirit, a point in TQ can be represented by $(q, \dot{q}) \in TQ$, where $\dot{q}$ now represents a vector in the tangent space T_qQ, i.e. an equivalence class of tangent curves in Q, all of which pass through the point $q \in Q$. Still in the same spirit, a point in T^2Q can be represented by $(q, \dot{q}; v_{q\dot{q}}) \in T^2Q$, where $v_{q\dot{q}}$ now represents a vector in the tangent space $T_{q\dot{q}}(TQ)$, i.e. an equivalence class of tangent curves in TQ all of which pass through the point $(q, \dot{q}) \in TQ$. The appearance of $\dot{q}$ in the second and third entries of Eq. (19) comes from the possibility of further decomposing the vector $v_{q\dot{q}}$, or rather the vector space, in the following way. Let $\pi_0: TQ \to TQ: (q, \dot{q}) \mapsto (q, 0)$ be the projection onto the *zero section*, whose definition is the following. The zero section of a tangent bundle TM is the zero vector field, which assigns to each point $m \in M$ the zero vector. As the zero section is used here, we are thinking of it as a submanifold of TM, and as such it has a more or less obvious identification with M itself. Then the equivalence class $v_{q\dot{q}}$ of tangent curves is projected by π_0 onto an equivalence class of tangent curves which lie in the zero section and pass through the point $(q, 0) \in TQ$. Let q' label the equivalence classes of vectors $v_{q\dot{q}}$ (equivalence classes of tangent curves) with respect to this projection. If the point $(q, \dot{q}) \in TQ$ is held fixed, the set $\{q'\}$ has an obvious linear structure and can be thought of as a vector space. For example, the curve $c: I \to TQ: t \mapsto (q, \dot{q}(t))$, that is, any curve which lies entirely in the fiber above a fixed point $q \in Q$, belongs to the equivalence class $q' = 0$. The identification of TQ with the zero section of T^2Q then makes it possible to identify the $\dot{q}$ vectors with the q' vectors. In this way, $v_{q\dot{q}}$ can be designated $(q', \dot{q}')$, where $\dot{q}'$ singles out the vector within the equivalence class q', and points in T^2Q can then be written in the form $(q, \dot{q}; q', \dot{q}') \in T^2Q$. The identification of the two vector spaces $\dot{q}$ and q' allows one to contemplate mappings defined, for example, by $(q, \dot{q}; q', \dot{q}') \mapsto (q, q'; \dot{q}, \dot{q}')$.

The last entry $\dot{q}'$ in these arrays cannot be given any coordinate independent vector meaning, except in the very special case in which $q' = 0$. In that case a representative curve in TQ (necessarily in the fiber above q, as mentioned above) may be written in the form $c: t \mapsto (q, \dot{q} + tw)$, where both $\dot{q}$ and w are (fixed) vectors in T_qQ, the fiber above q. Then w specifies the equivalence class of tangent vectors to which the curve belongs and thus plays the role of $\dot{q}'$. The particular set of curves belonging to this equivalence class, i.e. this particular point in T^2Q, can thus be written in the form $(q, \dot{q}; 0, \dot{q}')$, and in this way $\dot{q}'$ can be identified as a vector in $\{\dot{q}\}$ (or, equivalently, in $\{q'\}$) so long as q', the third entry is zero. This allows one to contemplate mappings defined, for instance, by $(q, \dot{q}; q', \dot{q}') \mapsto (q, \dot{q}; 0, q')$.

3. We close this subsection with some remarks about properties of vector fields that will be useful in the sequel.

Consider a local tangent bundle $TU \equiv U \times \mathbb{R}^\mu$ and a natural chart on it of the form $(\varphi, D\varphi)$. Then a section, or vector field, $U \to U \times \mathbb{R}^\mu$ has two parts, the identity $\mathbb{1}_U : U \to U$ on the base neighborhood and a map $X: U \to \mathbb{R}^\mu$. Let x^k be local coordinates on U; the $\partial/\partial x^k$ may be chosen as a basis for vector fields on $\mathbb{R}^\mu$, and the second part X can be written in the form (see Eq. (18))

$$X : u \equiv \{x^1, \ldots, x^\mu\} \mapsto X^k(u)\frac{\partial}{\partial x^k}, \tag{20}$$

where $X^k \in \mathscr{F}(U)$. This local way of writing a vector field can be extended to all of M only if $TM \equiv M \times \mathbb{R}^\mu$, for only then can the $\partial/\partial x^k$ be replaced by a global basis for vector fields on M and Eq. (20) extended to all $X \in \mathfrak{X}(M)$ in terms of such a basis. Thus if TM is a trivial bundle (i.e. if $TM \equiv M \times \mathbb{R}^\mu$), there exists at least one vector field which is not zero anywhere. It follows that if $\mathfrak{X}(M)$ contains no vector field without zeros, TM is not a trivial vector bundle.

Now, TM is defined canonically by M, and therefore whether TM is or is not a trivial vector bundle, i.e. whether or not a vector field with no zeros can be found in $\mathfrak{X}(M)$, is an intrinsic property of M. This is an example of the connection between properties of vector fields and properties of M. There are many such connections. They are usually studied and characterized by the machinery of algebraic topology, and we shall not go into them in this book, but when any of them are needed they will be explained separately or references will be given to the literature. Nevertheless, we mention one useful connection of this type here: if M is compact, any trajectory (that is, any maximal integral curve) $c: I \subset \mathbb{R} \to M$ of any vector field X is in fact defined on all of $\mathbb{R}$. In other words, trajectories are maps $c: \mathbb{R} \to M$. A vector field with this property is called *complete*: vector fields on compact manifolds are complete.

A vector field X can be multiplied by a function $f \in \mathscr{F}(M)$ to yield a new vector field fX defined by

$$(fX)(m) = f(m)X(m) \; \forall \, m \in M.$$

With this definition of multiplication by a function, it becomes possible to discuss two aspects of the structure of $\mathfrak{X}(M)$.

The first of these aspects is *linear independence*, which will be important throughout. A set of κ vector fields $X_j \in \mathfrak{X}(M)$ is said to be *independent* if the κ vectors $X_j(m) \in T_m(M)$ are linearly independent at all $m \in M$. In the special case of $\kappa = 1$, a single vector field is said to be independent if it is nowhere zero (see Brickell and Clark, 1970, p. 116). This may be called *pointwise independence*; equivalent to it is *linear independence*. A set of κ vector fields $X_j \in \mathfrak{X}(M)$ is said to be linearly independent if there exists no set of κ non-zero functions $f^j \in \mathscr{F}(M)$ such that $f^j X_j$ vanishes at some $m \in M$, i.e. if $(f^j X_j)(m) = 0$ implies that $f^j(m) = 0 \; \forall \, j$. It is clear that for κ vector fields to be independent, κ cannot be greater than μ. Unlike the situation for finite-dimensional vector spaces, however, the maximum number of independent vector fields that a manifold M can support may be less than μ (this will be of importance in Section B of Part II), and may even be zero. In fact, this number is zero for S^2, for on S^2 there can exist no vector field which is everywhere non-zero.

Remark. A manifold M is called *parallelizable* if there exists on it a set of μ independent vector fields. Thus if TM is a trivial vector bundle, M is parallelizable. On the other hand, the sphere S^2 is not parallelizable.

The second aspect of $\mathfrak{X}(M)$ is that $\mathfrak{X}(M)$ has the structure of an $\mathscr{F}$-module, for vector fields can be multiplied by functions and added (see, for instance, AM, p. 58). Finally, $\mathfrak{X}(M)$ has in addition another structure, resembling that of a Lie algebra. From two vector fields $X, Y \in \mathfrak{X}(M)$ with local representations $X^k\partial/\partial x^k$ and $Y^k\partial/\partial x^k$ a third can be constructed, called their *commutator* $Z = [X, Y]$, whose local representation is $Z^k\partial/\partial x^k$, where the Z^k are given by

$$Z^k = X^j\frac{\partial Y^k}{\partial x^j} - Y^j\frac{\partial X^k}{\partial x^j}. \tag{21}$$

Clearly the commutator is antisymmetric and linear under multiplication by numbers, but it is not $\mathscr{F}$-linear. In fact it follows simply from Eq. (21) that

$$[X, fY] = f[X, Y] + (X^k\partial f/\partial x^k)Y. \tag{22}$$

This means that even if two vector fields have the same *orbits*, as do Y and fY (that is, if the images of their trajectories are the same even though the $\mathbb{R}$-parametrization may be different), they have different commutation properties. Later in this chapter we shall return to the commutator in connection with the Lie derivative and intrinsic calculus on manifolds. There the commutator will be defined in a more intrinsic way.

5.3 FORMS

1. In Remark 1 above Eq. (11), one-forms were introduced, defined as elements of the set $\mathfrak{X}^*(M)$ of C^∞ sections of τ_M^*. Here, one-forms will be introduced by first constructing certain special ones from functions by means of a new operation d called the *exterior derivative* and then extending this construction in a local way to more general one-forms (for a more general intrinsic construction, see AM, p. 109).

Let $f \in \mathscr{F}(M)$ and consider the mapping

$$Tf: TM \to \mathbb{R} \times \mathbb{R}: (m, v) \mapsto (f(m), Df(m)\cdot v),$$

where v is a vector in the fiber above m. For each $m \in M$ let $T_m f$ be the restriction of Tf to the fiber $\tau_m^{-1}(m)$, and consider the mapping

$$df: M \to T^*M: m \mapsto P_2 \circ T_m f, \tag{23}$$

where P_2 is the *projection onto the second component* defined by $(P_2 \circ T_m f)(v) = Df(m)\cdot v$. The mapping df defined in Eq. (23), clearly a section of τ_M^* and therefore a one-form, is called the *differential* of f.

It was established following Eq. (6) that at each point $m \in M$ the vector space T_m^*M is dual to $T_m M$. Given a one-form such as df, then, and a vector field $X \in \mathfrak{X}(M)$, they can be *contracted* to yield a function $g \in \mathscr{F}(M)$ which is written

$g = i_X df$ and is defined by

$$g(m) = df(m) \cdot X(m). \tag{24}$$

A local representation of df can be obtained from Eq. (23): in a local chart each coordinate x^k is a function mapping each $m \in M$ on $\mathbb{R}$, and then df may be represented by

$$df = \frac{\partial f}{\partial x^k} dx^k. \tag{25}$$

If in a local chart X is written $X = X^k \partial/\partial x^k$, then locally

$$i_X df \equiv df(X) = X^k \frac{\partial f}{\partial x^k}, \tag{26}$$

which is a very convenient representation for calculation.

The one-forms defined in terms of functions by Eq. (23) are only a special case of one-forms. For example, given two functions $g, h \in \mathscr{F}(M)$, the expression $\alpha = g dh$ also clearly represents a section of τ_M^* and hence a one form, yet in general there is no $f \in \mathscr{F}(M)$ such that $\alpha = df$. (Those one-forms which can be written df for some $f \in \mathscr{F}(M)$ are called *exact*.) If the functions $x^k, k = 1, \ldots, \mu$, form a coordinate system in a local chart, the one-forms dx^k provide a basis for $T_m^* M$ at each $m \in U \subset M$. Thus the most general one-form $\alpha \in \mathfrak{X}^*(M)$ can be written locally as

$$\alpha|_U = f_k dx^k, \quad f_k \in \mathscr{F}(U).$$

Locally, then,

$$i_X \alpha = f_k X^k.$$

From one-forms one can define *two-forms* by using the *exterior* or *wedge product*, defined in the following way. Let α and β be one-forms. Then their wedge product is the two-form $\alpha \wedge \beta$ which is the section through $T_2^0(M)$ (see the Remark following Eq. (9)) defined by

$$i_X \alpha \wedge \beta = (i_X \alpha)\beta - (i_X \beta)\alpha. \tag{27}$$

More general two-forms are made by multiplying such exterior products of one-forms by functions and summing them. The definition of the exterior derivative d, which led from functions (zero-forms) to one-forms, can be extended to lead from one-forms to two-forms. Let $\alpha = f dh$ be a one-form. Then $d\alpha$ is the two-form

$$d\alpha = df \wedge dh.$$

More generally, let β be a one-form which is represented locally by

$$\beta = f_k dx^k.$$

Then $d\beta$ is the two-form which is represented locally by

$$d\beta = \frac{\partial f_k}{\partial x^i} dx^i \wedge dx^k. \tag{28}$$

Remark. If the tensor product had been used instead of the wedge product to define two-forms from one-forms and to extend the exterior derivative, that is if $d\beta$ had been defined by

$$d\beta = \frac{\partial f_k}{\partial x^j} dx^k \otimes dx^j$$

instead of by Eq. (28), this expression would not have maintained its form under a change of local coordinates. If $\beta = f'_k dx'^k$ in the coordinates x'^k, then the definition of Eq. (28) yields

$$d\beta = \frac{\partial f'_k}{\partial x'^i} dx'^i \wedge dx'^k,$$

while the tensor-product definition would have yielded

$$d\beta = \frac{\partial f'_k}{\partial x'^j} dx'^k \otimes dx'^j + f'_k \frac{\partial^2 x'^k}{\partial x^j \partial x^i} \frac{\partial x^j}{\partial x'^r} \frac{\partial x^i}{\partial x'^s} dx'^r \otimes dx'^s.$$

These arguments concerning the exterior product and the exterior derivative can be extended to define $(n+1)$-forms from n-forms also for n greater than one. We do not go into detail but refer the reader to AM, p. 109. Note in any case that the antisymmetry implied by the minus sign in Eq. (27) and in the extension of the wedge product to forms of higher order will force $d^2 = d \cdot d$ to vanish, so that one cannot obtain a two-form simply by applying d twice to a function, or indeed obtain an $(n+2)$-form by applying d twice to an n-form.

The operation of contraction introduced at Eq. (24) and extended at (27) can be further extended to forms of higher order. For example, let α be an n-form and β a p-form. Then

$$i_X(\alpha \wedge \beta) = (i_X \alpha) \wedge \beta + (-1)^n \alpha \wedge (i_X \beta).$$

Note that $i_X(\alpha \wedge \beta)$ is itself a form, of order $n + p - 1$, so that contraction, or *inner multiplication*, is an order-lowering operation on forms. The exterior derivative operates similarly on exterior products:

$$d(\alpha \wedge \beta) = (d\alpha) \wedge \beta + (-1)^n \alpha \wedge (d\beta).$$

For more details, see the last-named reference.

Just as a one-form is called exact if it can be obtained from a function by the application of d to a function, so an n-form is called exact if it can be written in the form $d\alpha$, where α is an $(n-1)$-form. An n-form ω is called *closed* if $d\omega = 0$. It is clear that all exact forms are closed, for $d^2 = 0$; however, not all closed forms are exact. For example, consider the one-form on the cylinder which is represented locally by $d\theta$. This is only a local representation because no (continuous) function θ exists on the cylinder, and this shows immediately that $d\theta$ is not exact. Nevertheless, its local representation shows just as immediately that in every neighborhood its exterior derivative is zero, so that it is closed. This illustrates the fact that exactness is not a property that can be studied locally, although there are other properties of n-forms, such as whether or not they are closed, which can be.

Independence can be defined for one-forms in the same way as it was defined for vector fields in §5.2. A set of κ one-forms α_j is said to be *independent* if the κ covectors $\alpha_j(m) \in T^*_m(M)$ are linearly independent at all $m \in M$. In the special case of $\kappa = 1$, a single one-form is said to be independent if it is nowhere zero. This may be called *pointwise independence*; equivalent to it is *linear independence*. A set of one-forms α_j is said to be linearly independent if there exists no set of κ non-zero functions $f^j \in \mathscr{F}(M)$ such that $f^j\alpha_j$ vanishes at some $m \in M$, i.e. if $(f^j\alpha_j)(m) = 0$ implies that $f^j(m) = 0$. It is clear that for κ one-forms to be independent, κ cannot be greater than μ.

It should be pointed out that because the dx^k form a basis for one-forms in each local chart, the general n-form can be represented locally by

$$\omega = f_{k_1 \ldots k_n} dx^{k_1} \wedge \ldots \wedge dx^{k_n}.$$

The set of all n-forms on M is called $\Omega^n(M)$.

An n-form ω can be used to define a submodule of $\mathfrak{X}(M)$ called the *kernel* of ω. The kernel is defined by

$$\ker \omega = \{X \in \mathfrak{X}(M) \vdash \omega(X, Y^1, \ldots, Y^{n-1}) = 0 \; \forall \; Y^j \in \mathfrak{X}(M)\}.$$

The restriction of a kernel to any $m \in M$ is a subspace. For example, the kernel of an independent (i.e. nowhere zero) one-form, restricted to some $m \in M$, is a $(\mu - 1)$-dimensional subspace of $T_m M$.

In general, the *kernel* of a geometric object or of an operation is the subset which annihilates or is annihilated by that object or operation. (This definition differs in some respects from other common definitions.) If φ is a linear mapping of one linear space E into another F, for example, the kernel of φ is the subspace of E which is mapped into the null vector of F. In this book we shall often refer to kernels of different kinds (for instance kernels of a two-form).

2. It is often necessary to transfer vector fields and forms from a manifold M to another N through a given diffeomorphism $\varphi: M \to N$. Let $T^r_s(M)$ be the set of tensor fields on M contravariant of order r and covariant of order s (see the last Remark in §5.1 and the first in §5.2), and consider $A_M \in T^r_s(M)$. Then the *push-forward* of A_M by φ is the tensor field $A_N \in T^r_s(N)$ given by

$$A_N(n) \equiv (\varphi_* A_M)(n) = (T\varphi)^r_s \circ A_M \circ \varphi^{-1}(n) \; \forall \; n \in N. \tag{29a}$$

The *pullback* of a tensor field $A_N \in T^r_s(N)$ by φ is defined similarly. It is the tensor field $A_M \in T^{rs}(M)$ which is given by

$$A_M(m) \equiv (\varphi^* A_N)(m) = [(\varphi^{-1})_* A_N](m) \; \forall \; m \in M. \tag{29b}$$

The push-forward and pullback can be defined also if φ is not a diffeomorphism, but even if it is only C^1. For more details on the pullback and push-forward, see AM, p. 108*ff.*

5.4 THE LIE DERIVATIVE

As was the case for forms, the following discussion of the Lie derivative and of calculus on manifolds is meant mostly as an outline touching on some of the points that will be useful in this book and attempting to give some intuitive geometric idea of the results.

The *Lie derivative* of a field (a function, a vector field, an n-form) on M will be defined here first if the field is a function and then by extending the definition to other fields. Let $X \in \mathfrak{X}(M)$ and $f \in \mathscr{F}(M)$. The Lie derivative $L_X f$ of f *along* or *with respect to* the vector field X is the function defined by (see Eq. (26))

$$L_X f = i_X df. \tag{30}$$

Let $c(t)$ be an integral curve of X. Then from the definition of df it follows that

$$(L_X f)(c(t)) = (X \cdot df)(c(t)) = df_{c(t)} \cdot Tc(t, 1)$$
$$= \frac{d}{dt} f \circ c(t) = \frac{d}{dt} f(c(t)),$$

which is sometimes written simply df/dt. That is, the Lie derivative of f with respect to X tells how f varies along the integral curve of X. If X is the dynamical vector field, for instance, and $L_X f = 0$, then f is a constant of the motion.

Remark. A sort of inverse statement can be made. Namely, just as X defines the Lie derivative $L_X f$ acting on any $f \in \mathscr{F}(M)$, so a *derivation on* $\mathscr{F}(M)$ defines a vector field through its Lie derivative. A derivation is a linear map $\delta : \mathscr{F}(M) \to \mathscr{F}(M)$ such that $\delta(fg) = (\delta f)g + f(\delta g)$. Then $\delta = L_X$ defines a vector field X. This can be seen in a local chart by applying δ first to the coordinate functions x^k and then to more general functions by writing them in a Taylor's series (the first two terms with remainder). It turns out that $L_X x^k = X^k$. For more details, see AM, p. 83.

The Lie derivative can now be extended to other fields, that is to vector fields and forms, by linearity and the Leibnitz rule and by using commutation with the exterior derivative. For instance, for exact one-forms

$$L_X(df) = d(L_X f);$$

the Lie derivative of an exact one-form is itself an exact one-form. As another example, consider $\alpha = f\,dg$; then

$$L_X \alpha = (L_X f) dg + f L_X(dg).$$

This can also be written in the form

$$L_X(f\,dg) = (i_X df) dg + f d(i_X dg).$$

Consider also the operation

$$\begin{aligned}(i_X d + di_X) f\,dg &= i_X(df \wedge dg) + d(f\,i_X dg) \\ &= (i_X df)dg - (i_X dg)df + (df)i_X dg + f\,d(i_X dg) \\ &= L_X(f\,dg).\end{aligned}$$

This illustrates the useful formula, valid quite generally (sometimes called *Cartan's identity*),

$$L_X = i_X d + di_X. \tag{31}$$

The Lie derivative is extended to act on vector fields by applying the Leibnitz rule to contractions as though they were products, i.e. by writing

$$L_X(i_Y \alpha) = i_{L_X Y}\alpha + i_Y(L_X \alpha), \tag{32}$$

where $X, Y \in \mathfrak{X}(M)$ and α is an n-form on M. For instance, if $\alpha = df, f \in \mathscr{F}(M)$, then

$$L_X i_Y df = i_{L_X Y} df + i_Y d(L_X f),$$

or

$$i_{L_X Y} df = [L_X L_Y - L_Y L_X] f \equiv [i_X d i_Y d - i_Y d i_X d] f.$$

Calculation in local coordinates and comparison with Eqs. (21) and (26) then leads to the result

$$L_X Y = [X, Y]. \tag{33}$$

It should be emphasized that this equation, like Eq. (31), was calculated by applying the Lie derivative to functions and simple one-forms, but that nevertheless it is valid in general.

At this point it may be useful to attempt to describe the Lie derivative in a more intuitive geometric way. Let A be any field on M. Then $L_X A$ is a field of the same kind as A and can be visualized as follows. At $m \in M$ the field A is 'attached' to X, and then X (or rather its *flow* F_t^X) is allowed to transport $A(m)$ to a neighboring point $m' \in M$. (The *flow* F_t^X is the mapping $M \to M$ defined by the set of all trajectories of X for each value of the evolution parameter t. See §5.5 and AM, p. 70.) This concept of 'attaching' A to X is kept deliberately vague in this intuitive description, but if A is a vector field one can think of attaching the tail of the arrow $A(M)$ to the point m and the tip more or less to the point $m + A(m)$, which could perhaps make some sense in a local chart. Then $A(m')$ is compared with the transported $A(m)$. The transported $A(m)$ will differ from $A(m')$ in two ways: first because $A(m)$ is different from $A(m')$, and second because $A(m)$ has been changed by being transported. The difference between the transported $A(m)$ and $A(m')$, in the limit as m' approaches m (or as the value of t in F_t^X approaches zero), is $(L_X A)(m)$. In particular, if $A(m')$ is the same as the transported $A(m)$ for all $m \in M$, then $L_X A = 0$, and A is said to be *an invariant for* X (or *invariant with respect to* X).

A picture can illustrate this for the case in which A is a vector field, so that $L_X Y = [X, Y]$. In Fig. 5.1 the line Y_1 represents the vector $Y(m)$, and Y_2 represents $Y(m')$. If the transported Y_1 is the Y_3 shown in the figure, then X and

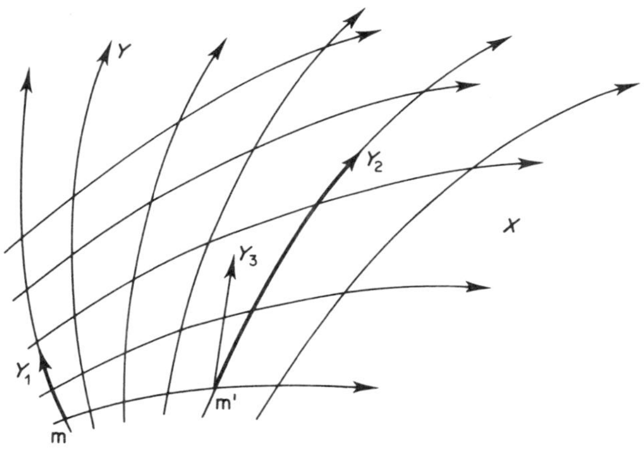

Fig. 5.1. The Lie derivative of a vector field. $Y_3 \neq Y_2$, so $[X, Y] \neq 0$

Y do not commute. Another way to understand the commutator pictorially is shown in Fig. 5.2. Starting from $m \in M$, go along the integral curve $g(t)$ of X for which $g(0) = m$ to the point $m' = g(\tau)$. Then go along the integral curve $h(t)$ of Y for which $h(0) = m'$ to the point $m'' = h(\tau')$. Then go along another integral curve $g^-(t)$ of X for which $g^-(0) = m''$ to the point $m''' = g^-(-\tau)$. Finally, go along a second integral curve $h^-(t)$ of Y for which $h^-(0) = m'''$ to the final point $m_f = h^-(-\tau')$. If $Y$and X are linearly independent in a neighborhood of m and if $[X, Y] = 0$ there, then $m = m_f$ as in Fig. 5.2(b). More generally, the situation is that of Fig. 5.2(a). A comparison of Figs. 5.1 and 5.2 shows how this second view is related to the general description. It is seen that whether or not $(L_X Y)(m)$ vanishes depends not only on the integral curve of X which passes through m, as in the case of $(L_X f)(m)$, $f \in \mathscr{F}(M)$, but also on the relation between the integral curves of the two fields in the neighborhood of m. This has to do with the deliberately vague concept of 'attaching'.

For more general geometric objects, like forms and tensor fields, the picture is less clear (see AM, p. 90). One can try to clarify it in the spirit of the intuitive description in the following way. Let F_t^X be the flow of X, and note that $F_t^X(m) = g_m(t)$, where $g_m(t)$ is the integral curve of X for which $g_m(0) = m$. Then if $A \in T_s^r$, define the curve $G_m(t)$ in the fiber $(T_s^r)_m M$ above m by

$$G_m(t) = (T_s^r F_{-t}^X) A(g_m(t)).$$

In other words, $G_m(t) \in (T_s^r)_m M$ is obtained by calculating A at $g_m(t)$ and then transporting it back to $g_m(0) = m$ by using the *lifted flow*. Then $L_X A(m)$ is the derivative of this curve:

$$(L_X A)(m) = T G_m(0, 1) = G'_m(0).$$

If $G_m(t)$ does not depend on t, then $(L_X A)(m) = 0$. If $L_X A = 0$ for all m, then A is said to be *an invariant for* X (or *invariant with respect to* X).

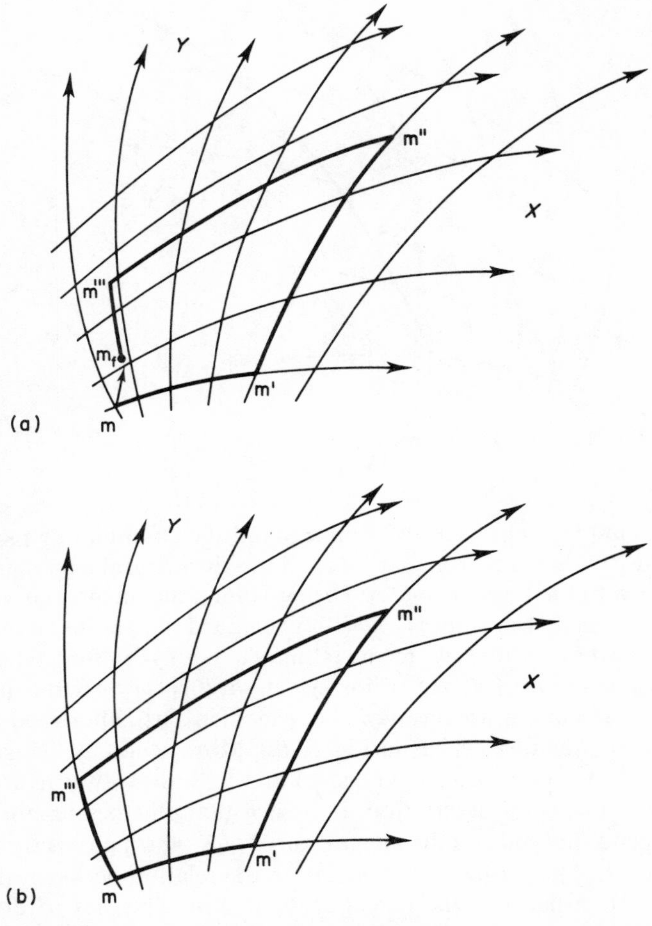

Fig. 5.2. The commutator of two vector fields in terms of integral curves. (a) $[X, Y] \neq 0$. (b) $[X, Y] = 0$

To finish off this section, we give four additional formulas that will often prove useful. The inherent antisymmetry of forms which comes from the antisymmetry of the wedge product implies that

$$i_X i_X = 0.$$

From the definition of contraction as a kind of (inner) multiplication, it follows that for any form α

$$i_{fX}\alpha = f(i_X\alpha).$$

On the other hand, straightforward application of Eq. (31) yields

$$L_{fX}\alpha = f(L_X\alpha) + (df) \wedge (i_X\alpha).$$

With the aid of Eq. (33) the Leibnitz rule (32) can be written

$$i_{[X,Y]} = L_X i_Y - i_Y L_X. \tag{34}$$

Finally, Eq. (22) can now be rewritten in the form

$$[X, fY] = f[X, Y] + (L_X f)Y.$$

5.5 VECTOR FIELDS AS GENERATORS OF TRANSFORMATIONS

1. The flow of a vector field has already been mentioned in the previous chapter. It can be defined somewhat more rigorously as follows. Let $X \in \mathfrak{X}(M)$ and consider an arbitrary point $m \in M$. If $c_m : I \subset \mathbb{R} \to M$ is the integral curve of X such that $c_m(0) = m$, then for $t \in I$ the *flow* $F_t^X : M \to M$ is defined by its action on each point $m \in M$ by

$$F_t^X : m \mapsto c_m(t)$$

If X is complete, its integral curves are defined for all of $t \in R$, and F_t^X turns out to be a diffeomorphism (AM, p. 70) such that (we suppress the superscript X) $F_t^{-1}(m) = F_{-t}(m)$ and $F_t \circ F_{t'}(m) = F_{t+t'}(m)$. With these two properties the set $\{F_t^X \equiv \varphi^X(t)\} \subset \mathrm{Diff}(M)$ (here $\mathrm{Diff}(M)$ is the set of all diffeomorphisms on M) can be interpreted as a one-parameter Lie group of transformations on M. More correctly, φ^X is an *action* of the one-dimensional Lie group $\mathbb{R}$ on M associated with the vector field X. The *orbits* of the action are the orbits (i.e. the images of the trajectories) of the vector field. In this way, every complete vector field corresponds to a unique one-parameter group of diffeomorphisms on M.

The converse is true: to every one-parameter group of diffeomorphisms on M (i.e. to every action of the Lie group $\mathbb{R}$ on M) corresponds a complete vector field on M, called the *infinitesimal generator* of the transformation group. If φ_t is the one-parameter group of diffeomorphisms, this infinitesimal generator $X \in \mathfrak{X}(M)$ is defined by giving the Lie derivative L_{X^φ} of each $f \in \mathscr{F}(M)$, namely by

$$(L_{X^\varphi} f)(m) = \lim_{t \to 0} \frac{f(\varphi_t(m)) - f(\varphi_0(m))}{t}, \quad m \in M. \tag{35}$$

This is obviously a derivation of $\mathscr{F}(M)$ and thus defines X^φ. Another way of defining X^φ is by equating the orbits of φ_t with the integral curves of the vector field: $\varphi_t(m) \equiv c_m(t)$ and then $X^\varphi(c_m(t)) = Tc_m(t, 1)$ gives the vector $X^\varphi(m)$ at each $m \in M$.

From these two versions of the definition, especially the second one, it is clear that X^φ is invariant under the action of φ_t, for it just slides M along the integral curves of X^φ. Specifically,

$$(\varphi_t)_* X^\varphi = X^\varphi.$$

The group property of φ_t implies that X^φ is complete. Finally, if X^φ and X^ψ are the generators of two one-parameter group actions φ_t and ψ_t on M, then $[X^\varphi, X^\psi] = 0$ iff $\varphi_t \circ \psi_s = \psi_s \circ \varphi_t$ for all $t, s \in \mathbb{R}$. In a sense, Fig. 5.2 illustrates this fact.

2. The association of a vector field X with its one-parameter group of transformations makes it possible to perform an operation called *lifting* a vector field $X \in \mathfrak{X}(Q)$ from Q to TQ or to T^*Q. Here we will do it only for TQ, postponing the analogous operation onto T^*Q to Chapter 11.

Given a vector field $X \in \mathfrak{X}(Q)$, one constructs its one-parameter group of diffeomorphisms φ_t^X. The family $T\varphi_t^X$ of tangents to φ_t^X is a one-parameter group of diffeomorphisms on TQ. Let the infinitesimal generator of $T\varphi_t^X$ be called $X^T \in \mathfrak{X}(TQ)$, the *lift* of $X \in \mathfrak{X}(Q)$. Schematically, this sequence of operation may be represented by

$$X \xrightarrow{\text{flow}} \varphi_t^X \xrightarrow{T} T\varphi_t^X \xrightarrow{\text{inf. gen.}} X^T.$$

In local coordinates, if X is written in the form $(q, X(q))$, then

$$X^T(q, \dot{q}) = (q, \dot{q}; X(q), X'(q, \dot{q})), \tag{36}$$

where $X'(q, \dot{q})$ is given by

$$X'^k(q, \dot{q}) = \frac{\partial X^k}{\partial q^j} \dot{q}^j. \tag{37}$$

It is easily checked that the following diagram commutes:

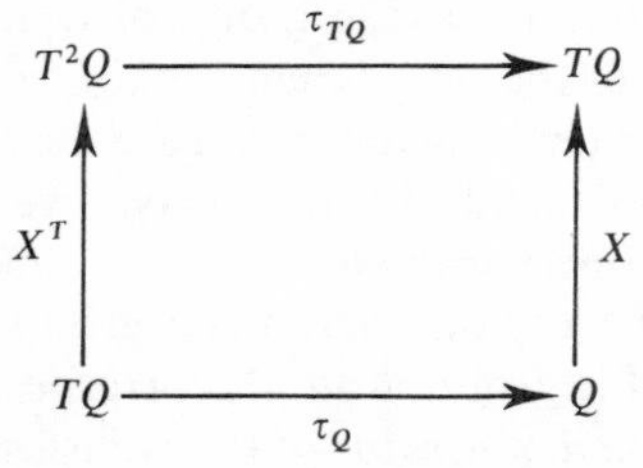

BIBLIOGRAPHY TO CHAPTER 5

On fiber bundles, see Steenrod (1951), Yano and Ishihara (1973), Eells (1974), and Husemoller (1975).

Vector fields are discussed in Brickell and Clark (1970, Ch. 7); second-order vector fields are introduced in Dieudonné (1971, Ch. 18).

There are several introductions to differential forms, for instance, Schreiber (1977), which provides a 'heuristic introduction'. More formal and richer in applications to physics is Flanders (1969).

Calculus on manifolds is the specific subject of Brickell and Clark (1970), that can be consulted on almost all topics discussed in this chapter. See also Chillingworth (1976) for the transition from calculus on vector spaces to calculus on manifolds.

For one-parameter groups of diffeomorphisms and their related vector fields see, as an introduction, Dodson and Poston (1977, Ch. 5) and Boothby (1975).

6

From trajectories to the vector field

We now turn to the problem of constructing the vector field from experimental data. It will be seen that the field cannot in general be constructed from the data alone, but that other considerations are often required for a unique vector field to result. The construction, a form of model-building, is the work of the theorist; it is assumed that the experimentalist has already collected the data. This distinction between the experimental and theoretical work may perhaps not be very well defined, but it is conventional and convenient for our purposes.

1. In particular, it is assumed that for a mechanical system the data consist of a manifold Q, which is the configuration space, and *a set S of curves on Q* which are possible trajectories or pieces of them. Generally S does not include all possible trajectories. Recall that a curve in a topological space T is defined as a continuous map from an open interval I or $\mathbb{R}$ into T. The $\mathbb{R}$ here is the time axis, and then continuity reflects the basic physical condition that an object does not disappear at one point and reappear simultaneously at another. To require that the interval in time be open is technical, but in any case it leads to no loss of generality. On a manifold, curves are taken to be at least C^1 (curves are more than piecewise differentiable). In fact this limits the formalism to exclude rigid-body (e.g. hardsphere) collisions, but this would seem reasonable, as all fundamental interactions seem to be smooth. There will be reason, in fact, to require even more: at least C^2 and eventually C^∞. In any case, whatever conditions of smoothness will be found necessary for the curves will be assumed satisfied.

Remark. The experimental data will not always be given in the form stated here (see the example of the Kepler problem later in this chapter), but it will be assumed that they can always be manipulated into this form.

The problem is now to translate the information contained in the pair (Q, S), that is, all the available information concerning the dynamical system, into a set of differential equations. In keeping with the vector-field formalism, the differential equations should be of first order on a suitable manifold, which will be called the *carrier space* of the system, so that one and only one solution passes through each point. Clearly Q will usually not be suitable, for generally more than one trajectory passes through each point of Q. If the system under consideration is time-independent and Newtonian, that is, if the position and velocity determine the trajectory uniquely, TQ will be a suitable carrier space. For a radiating charged particle, whose equations are of third order in the position, T^2Q is needed. Later we shall discuss T^nQ, but in any case, whether the construction is possible and which carrier space is needed depends on (Q, S). Thus we start from (Q, S).

Let $h: I \to Q$ and $h': I' \to Q$ be curves in S, and let $I'' \neq \varnothing$ be an open interval of $\mathbb{R}$ contained in both I and I'. Then it will be assumed that S has the following property:

$$(h|I'' = h'|I'') \Rightarrow [h|I \cap I') = h'|I \cap I')] \Rightarrow I = I'. \tag{1}$$

This means that curves in S neither bifurcate nor overlap. Thus S cannot contain, for instance, two C^∞ functions which are equal on a finite interval and unequal elsewhere, examples of which are well known, and Eq. (1) therefore represents a nontrivial requirement (see Fig. 6.1).

There are two aspects to this requirement, one involving bifurcations and the other involving overlaps and restrictions. We shall assume quite rigidly that S contains no bifurcating curves, for if it did, the system would either be nondeterministic or would have additional (internal) variables. Nondeterministic

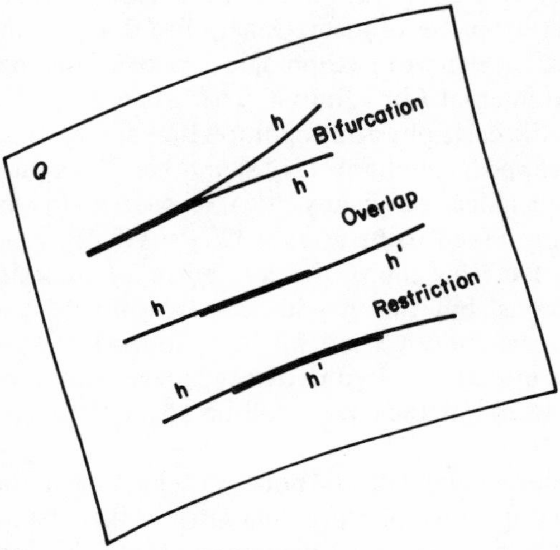

Fig. 6.1. The time is not indicated on this figure, but it is to be understood that where h and h' overlap they are simultaneous. That is, if they overlap at some point q, then $q = h(\tau) = h'(\tau)$. Curves such as those in this figure do not occur in S

systems will be excluded from consideration, and additional variables, if they have appropriate structure, can be added to Q, thereby enlarging the configuration manifold. When this cannot be done, additional variables may be used like the electric charge of a particle, to break the system into several separated ones, but even this is not always possible. For example, such an additional variable may take on discrete values and change in a way that depends on the trajectory on Q, like the spin. Such systems will also be excluded from consideration.

If only this bifurcation aspect of the restriction represented by (1) is assumed, on the other hand, so that S may contain nonbifurcating overlaps, it is a simple matter to construct a new S which contains not even these. Consider, for instance, curves $h: I \to Q$ and $h': I' \to Q$ such that $I \cap I' \subset \mathbb{R}$ and $h|I \cap I' = h'|I \cap I'$, but $I \neq I'$. Then h and h' can be replaced by the curve

$$h'': I \cap I' \to Q$$

defined by

$$\begin{aligned} h''(t) &= h(t), t \in I, \\ h''(t) &= h(t) = h'(t), t \in I \cap I', \\ h''(t) &= h'(t), t \in I'. \end{aligned}$$

In this way eventually a new S is formed, one which satisfies the requirements. Note, incidentally, that S will in general contain orbits which close on themselves and overlap, but at different times. In the Kepler problem, for instance, each ellipse is traced out over and over again.

2. If $q \in Q$ and $\tau \in \mathbb{R}$ we shall say that h is a curve *through q at τ* if $h(\tau) = q$. Let F be the subset of $Q \times \mathbb{R}$ defined by

$$(q, \tau) \in F \Leftrightarrow \exists : h \in S \text{ such that } h(\tau) = q. \tag{2}$$

The structure of F depends on (Q, S), and this structure will determine the degree of ambiguity in the vector field which will eventually be constructed. If the experimental information is very poor, F might even be a discrete set of points (and the vector field may be ambiguous indeed). In the historical Kepler problem, for instance, F included only those points in $q \in Q = \mathbb{R}^3$ through which the elliptical orbits of the planets actually passed, and for each such q only those times $\tau \in \mathbb{R}$ when a planet passed through q.

Now consider the set $G_{(q,\tau)}$ of all those curves $h \in S$ which pass through q at τ, for $(q, \tau) \in F$. It should be emphasized that the sparseness of F can be so severe that $G_{(q,\tau)}$ may contain very few curves, often only one. On the other hand, each curve h will belong to many sets $G_{(q,\tau)}$ with different q, τ. We now define the *set G of trajectory germs* as the disjoint unions of the $G_{(q,\tau)}$; that is, $G = \dot{\bigcup}_{(q,\tau) \in F} G_{(q,\tau)}$. Then G is a sort of fiber space over F and is related to what is called in mathematics the set of germs.

Each element or *germ* $g \in G$ may be represented by a triplet (q, τ, w), where $(q, \tau) \in F$ designate the $G_{(a.\tau)}$ to which g belongs, and w is an index which distinguishes between the different elements in the fiber over (q, τ). In order to be rigorous, in fact, one should write $w_{(q,\tau)}$ for this index, since it may not be possible to use the same set of ws for each (q, τ), and a given w over one (q, τ) may have no relation to the w over another. As may be expected, however, since G is not given arbitrarily, but is constructed from the set S of curves, there is in fact a natural relation between the indices on different fibers, a relation that will make it possible to define something like a flow on G.

Each germ $g = (q, \tau, w)$ determines a certain curve $h_g \in S$, namely the curve through q at τ which is indexed by w. A new map W will now be defined by

$$W: G \to S: g \mapsto h_g. \tag{3}$$

and W is then surjective. Let $I_g \subset \mathbb{R}$ be the interval on which h_g is defined. Then because I_g is open and $\tau \in I_g$, there exists a $t \in \mathbb{R}$ such that $\tau' = \tau + t \in I_g$ and thus $W(g) = h_g$ is a curve through $q' = h_g(\tau')$ at τ' (where $(q', \tau') \in F$). Thus one may write

$$h_g = W(g) = W(g') = h_{g'} \tag{4}$$

although $g \neq g'$, where

$$\begin{aligned} g &= (h_g(\tau), \tau, w), \\ g' &= (h_g(\tau + t), \tau + t, w) = (h_{g'}(\tau'), \tau', w). \end{aligned} \tag{5}$$

Strictly speaking w should be replaced by w' in the expression for g', for the index may not be the same in both fibers and rigor requires in any case that w be labeled with the (q, τ) to which the fiber belongs. But we shall continue to write simply w. Now with the aid of each $h: I \to Q$ in S a new function $\bar{h}: I \to G$ can be defined by

$$\bar{h}: t \mapsto g \equiv (h(t), t, w). \tag{6}$$

Then it follows for all $g \in G$ that

$$W(\bar{h}_g(t)) = h_g \forall t \in I_g.$$

The point of this construction is to make G a *carrier space* for the dynamics; on it, or actually on (a part of) $\mathbb{R} \times G$, one defines a kind of flow mapping $H: \mathbb{R} \times G \to G$ (see Eq. (5)) by

$$H: (t, g) \mapsto \bar{h}_g(\tau + t) = g' \tag{7}$$

if $g = (h_g(\tau), \tau, w)$. Then (wherever either one of the two sides of the equation exists) one obtains

$$H(t + s, g) = H(s, H(t, g)), \tag{8}$$

for

$$\begin{aligned} H(s, H(t, g)) &= H(s, g') = \bar{h}_{g'}(\tau' + s) = \bar{h}_g(\tau' + s) \\ &= (h_g(\tau + t + s), \tau + t + s, w) \end{aligned}$$

where use has been made of (5) and (4) and therefore of the fact that $h_g = h_{g'}$.

3. We now return to the problem of indexing the germs. Once $g = (q, \tau, w)$ is given, a curve h is determined together with its derivatives up to all orders allowed. It is natural to ask whether such derivatives will suffice conversely to determine g, for

in that case they would provide a canonical way to index the germs. Generally speaking, of course, they will not suffice, but under certain conditions they may, for instance if the curves are analytic (although that would involve indexing them with an infinite number of derivatives). We shall in fact be interested in the situation which arises when a finite number of derivatives will suffice, which is the case for most physical systems. Assume, therefore, that there exists a positive integer m such that the curves in S are at least C^m and that

$$h_1, h_2 \in S, h_2^{(i)}(\tau) = h_2^{(i)}(\tau) \ \forall \ i \in \{0, \ldots, m\} \Rightarrow h_1 = h_2.$$

Let n be the lowest integer m satisfying this relation. Then G can be immediately identified with a subset of $T^nQ \times \mathbb{R}$. That is, $g = (q, \tau, w)$ becomes $g = (T^nh(\tau), \tau)$, where $h = W(g)$. Then $W, \bar{h}$, and H become

$$W: T^nQ \times \mathbb{R} \to S: (x, \tau) \mapsto h \text{ if } T^nh(\tau) = x; \tag{3'}$$

$$\bar{h} = T^nh \times \mathbb{1}: I \to T^nQ \times \mathbb{R}: t \mapsto (T^nh(t), t); \tag{6'}$$

$$H: \mathbb{R} \times [(T^nQ \times \mathbb{R})] \to T^nQ \times \mathbb{R}: (t; x, \tau) \to (T^nh(t+\tau), t+\tau), \quad \text{if } W(x, \tau) = h. \tag{7'}$$

Recall that $W(x, \tau)$ like $H(t; x, \tau)$ does not always exist, for these are not defined for all possible values of their arguments.

Note that

$$T^nh(\tau + t) \equiv T[W(x, \tau)](\tau + t)$$

has a double dependence on the 'initial time' τ. It turns out in many cases of interest that in fact this point in T^nQ is independent of τ, so that

$$T^nh_1(\tau_1) = T^nh_2(\tau_2) \Rightarrow T^nh_1(\tau_1 + t) = T^nh_2(\tau_2 + t)$$

for all t such that both expressions on the right-hand side exist. Such a system is said to be *autonomous* or *time-independent*. In that case the τ can be dropped from the definition of H, which leads to the new mapping

$$\tilde{H}: \mathbb{R} \times T^nQ \to T^nQ: (t, x) \mapsto T^n[W(x, \tau)](\tau + t) \tag{9}$$

for arbitrary τ such that the last expression exists.

Now note that for each fixed $g = (x, \tau) \in G$, the mapping H of Eq. (7′) yields a curve in $T^nQ \times \mathbb{R}$, the only one passing through g. If this curve is derivable, one may define

$$X(g) = T[H|_{\mathbb{R} \times \{g\}}](0, 1) = \left(g; \frac{\partial}{\partial t} T^n[W(g)](\tau + t)|_{t=0} \cdot 1 \right). \tag{10a}$$

If the system is time independent this may be written in the form

$$X(x) = \left(x, \frac{\partial}{\partial t} \tilde{H}(t, x)|_{t=0} \right). \tag{10b}$$

4. We have almost arrived at our goal, namely a vector field on $T^nQ \times \mathbb{R}$ (or, in

the time-independent case, on T^nQ). But X is not a vector field because it is defined only for some points of $T^nQ \times \mathbb{R}$, and the next problem is to fill it out, to define X, or equivalently H, for other points of the manifold.

Now, in general there is no unique way to extend X to the entire manifold, to obtain a unique maximal extension. This is a situation which seems almost logically necessary, for this is the point at which the theorist 'makes science'. The final theoretical construction should be such that the experimental observations are derivable from it by logical deduction, and if in addition the theory were deducible from the experimental data, the two would be logically equivalent. One would have had to be somewhat mad to go through all this construction to obtain a vector field if from the vector field one could only get back the already well established experimental data, which lead inexorably to the vector field. Thus one may expect that at some point the theorist must leave the experimental observations and construct a model which includes not just them, but significantly more. There is no recipe for this procedure. It is inherently ambiguous. The theorist is asked to find a theory that has certain desirable structural properties and which includes the observations, and then this theory can be tested in situations not already contained among the observations, that is, on *new* observations.

In the present situations the structural properties which have already been accepted as (or at any rate, declared to be) desirable, are essentially the requirement that the dynamics be in the form of a vector field on a suitable manifold, or in the form of first-order differential equations. Sometimes the experimental observations include enough curves on Q for the extension to all of Q to be more or less obvious. For instance, the data may be given at so many $q \in Q$ and at so many velocities (in a given range, perhaps), that one may assume that the data are known, are in some sense 'the same' as the experimentally observed data, on a dense set in TQ. Then one may end up not with what we have called 'a kind of a flow mapping' H, but a *flow box* and one can then try to extend this in some analytic way. This is not to say that there is no other way of extending X or H to the entire manifold, but this way is related to many familiar examples.

5. To indicate some of the difficulties and to exhibit some of the kinds of possibilities and ambiguities, we take as an example a simplified version of the historical Kepler problem. This simplification will show more clearly, perhaps, than the historical problem what some of the difficulties are and how some of the choices are made.

Example 6.1. The Kepler problem

Let Q be $\mathbb{R}^2$, the two-dimensional plane. The chart to be used for the description of the experimental data will be polar coordinates r, θ with the origin at the 'sun'. Assume the following given data for the motion of a finite (integer) number $N > 1$ of particles (the planets).

K-I. The orbits are N coaxial nonintersecting ellipses with their common focus at *the origin.* Specifically, the orbits are given by the N equations

$$r=(a_k+b_k\cos\theta)^{-1},\quad a_k>b_k>0,\qquad k\in\{1,\ldots,N\}. \tag{11}$$

For each ellipse the major and minor semiaxes R_k and ρ_k are given by

$$R_k=\frac{a_k}{a_k^2-b_k^2},\quad \rho_k=(a_k^2-b_k^2)^{-1/2}=\sqrt{R_k/a_k}.$$

There is a time (which we call $\tau=0$) *when all N planets are at their points of closest approach, given by* $\theta=0$, $r=(a_k+b_k)^{-1}$.

K-II. On each orbit, area is swept out at a constant rate J_k (or angular momentum is conserved):

$$\tfrac{1}{2}r^2\dot{\theta}=J_k. \tag{12}$$

K-III. The period T_k *and the major semiaxis* R_k *of each orbit are related by*

$$R_k^3=KT_k^2, \tag{13}$$

where K is a 'universal' experimentally measured constant, i.e. the same for all orbits.

Law K-III relates the J_k for different k: since the area of the kth ellipse is swept out uniformly, this area is just J_kT_k, and thus

$$J_kT_k=\pi R_k\rho_k=\pi R_k^{3/2}a_k^{-1/2}=\pi T_k\sqrt{K/a_k},$$

or

$$J_k=\tfrac{1}{2}r^2\dot{\theta}=\pi\sqrt{K/a_k}. \tag{14}$$

In principle S can now be constructed. For the kth ellipse a_k and b_k are observed, J_k is calculated according to (14), and the result is inserted into (12). Then (12) is integrated with the aid of (11), and an expression for $\theta_k(\tau)$ is obtained:

$$\int_0^{\theta_k}\frac{d\theta}{(a_k+b_k\cos\theta)^2}=2\pi\sqrt{K/a_k}\,\tau. \tag{15}$$

Finally, $\theta_k(\tau)$ is inserted into (11), and an expression for $r_k(\tau)$ is obtained. Thus S contains N curves $h_k(\tau)=(r_k(\tau),\theta_k(\tau))$.

The set $F\subset\mathbb{R}^2\times\mathbb{R}$ contains only the points on the N ellipses and, for each point, the times τ when the planet is at the point. If $(q,\tau)\in F$, then $(q,\tau+nT_k)\in F$ for all integers n if q lies on the kth ellipse. The set G is in this case essentially the same as F, for there is, in the experimental data, only one curve through each point. Thus each $g=(q,\tau,w)\in G$ is of the form $(q,\tau,1)$, for the index set to which w belongs contains only one element. The experimental data in this case are separated on G in $Q\times\mathbb{R}=\mathbb{R}^2\times\mathbb{R}$, and there is thus no need to go to $T^nQ\times\mathbb{R}$ with $n>0$ in order to separate the orbits.

That there is a flow on $\mathbb{R}\times G$ is more or less clear, as well as that it is time-independent. In any case, that is not as interesting as the vector fields that can be obtained. To construct the vector fields, first take derivatives along the N curves with respect to τ, that is, derivatives of the $r_k(\tau)$ and $\theta_k(\tau)$. These derivatives may be written in many forms as functions on Q, for they are defined only along the orbits and not for arbitrary points. For example, one may write

$$\dot{\theta}=2\pi\sqrt{K/a_k}r^{-2}, \text{ or } \dot{\theta}=2\pi\sqrt{K/a_k}r^{-1}(a_k+b_k\cos\theta),$$

or

$$\dot{\theta}=2\pi\sqrt{K/a_k}(a_k+b_k\cos\theta)^2, \tag{16}$$

or many other forms, all of which will be the same along the kth ellipse, but not in general at other points of Q. Moreover, any one of these forms will contribute (for they are equations for one component) to a vector field on Q, each field differing from every other, except along the kth ellipse. That is, the vectors along the kth ellipse will coincide with the tangent vectors of each of the fields, so that the kth ellipse, but none other, is an integral curve of all the vector fields. Now consider a vector field constructed in this way about the kth ellipse and another constructed in the same way about another. It is generally possible to distort these vector fields in the regions between the orbits, even in C^∞ ways, in order to create one field whose tangent vectors coincide on both ellipses, and then to extend this procedure to all N orbits. In any case, the differential equations which must then be solved, of the form, for instance,

$$\dot{\theta}=2\pi\sqrt{K/a_k}r^{-2}, \quad \dot{r}=2\pi\sqrt{K/a_k}b_k\sin\theta, \tag{17}$$

involve the a_k and b_k. This requires solving N pairs of differential equations, each yielding solutions of the form

$$r=(A_k+b_k\cos\theta)^{-1}, \int_{\theta_{0k}}^{\theta_k}\frac{d\theta}{(A_k+b_k\cos\theta)^2}=2\pi\sqrt{K/a_k}\tau, \tag{18}$$

and each generalizing its orbit (now A_k is arbitrary, and so is $\theta_k(0)=\theta_{0k}$) to include even hyperbolas and a parabola. That $A_k=a_k$ in the experimental data (essentially the validity of Law K-III) can then be interpreted as an accident of observation: since K and a_k appear in 'the theory' only in the form K/a_k, there is no special universal meaning to K. There are, however, N separate dynamical systems, corresponding to the N observed values of a_k (or K/a_k, if you will) and b_k. What one arrives at, then, is N separate dynamics, N separate theories, which can be rather artificially sewn together.

In an attempt to obtain one unified dynamics by eliminating the a_k and b_k these considerations could be lifted to TQ. That is, in TQ the curves Th_k are, of course, still separated, and the tangent vectors there can be calculated and vector fields constructed. But in TQ the a_k and b_k can be eliminated because Eqs. (17) or any equivalent equations, such as those of (16), can be used to calculate the a_k and b_k in terms of the coordinates of the point in TQ at those points through which the curves pass. The resulting expressions, however, are again not unique as functions on TQ, but only on the curves Th_k. The resulting dynamics is hence also not unique.

We give two examples of the resulting vector fields (or differential equations).

i. Independent of K. From Eq. (12) it follows immediately that

$$\ddot{\theta} = -2\dot{r}\dot{\theta}/r. \tag{19}$$

From the second of Eqs. (17) (using it also to express b_k),

$$\ddot{r} = 2\pi\sqrt{K/a_k}\, b_k \dot{\theta}\cos\theta = \dot{r}\dot{\theta}\cot\theta. \tag{20}$$

The solutions are

$$r = (A + B\cos\theta)^{-1}, \quad \int_{\theta_0}^{\theta} \frac{d\theta'}{(A + B\cos\theta')^2} = C\tau,$$

where A, B, C, θ_0 are arbitrary. What is then obtained is all coaxial conic sections with arbitrary velocities. This generalization, like the previous one, implies that Law K-III is an accident of observation. The two theories differ, however, in that this one yields a unified dynamics. Moreover, in the dynamics of Eq. (18) there was one and only one curve passing through each point of Q (for fixed k, that is); the velocity was thus determined by the position. In the present dynamics the position does not determine the velocity; almost all points of TQ are available for initial conditions, the one restriction being that all trajectories cross the x axis (the line $\theta = 0$ and $\theta = \pi$) at right angles. We put off further discussion of this example in order to compare it with the next one.

ii. Dependent on K (Newton's dynamics). This time r is written in accordance with the first equality of (20), and (11) rather than (17) is used to express b_k. Then by using Eq. (14) one finds that $\ddot{r}$ is given by

$$\ddot{r} = 2\pi\sqrt{K/a_k}\,\dot{\theta}\left(\frac{1}{r} - a_k\right) = r\dot{\theta}^2 - 4\pi^2 K/r^2, \tag{21}$$

and $\ddot{\theta}$ again by Eq. (19). This is the familiar gravitational force. The solutions are well known: all conic sections, each covered with a definite velocity determined by the shape of the orbit (essentially by a_k and b_k), but no longer coaxial. The accident of observation according to this theory is coaxiality. Notice that in polar coordinates it is not obvious that the force is a function on Q, as it is, of course, in Cartesian coordinates.

What we have demonstrated is that there is usually more than one way to generalize the experimental data, and that other considerations are needed if one is to choose among the possibilities. Such other considerations would make the choice of the K-independent theory rather odd. For one thing its differential equations (19)–(20) are singular for $\theta = 0$ and $\theta = \pi$. If written in nonstandard form in Cartesian coordinates, the equations become

$$\ddot{x}yr^2 = x\dot{x}J \quad \text{and} \quad \ddot{y}r^2 = \dot{x}J, \tag{22}$$

where $r^2 = x^2 + y^2$ and $J = x\dot{y} - y\dot{x}$. The solutions, as has been mentioned, are all conic sections whose axes lie on the x axis. But the singularity for $y = 0$ does not allow all possible velocities, as has also been mentioned, and causes there to be two solutions with given admissible velocity passing through each point on the x axis (both the conic section and the straight line parallel to the y axis at constant velocity). Thus the singular nature of the equations gives rise to physically undesirable consequences. Moreover, the equations are not symmetric under interchange of x and y, and the force obtained is velocity dependent.

It is thus seen that such considerations as isotropy, the availability of all of TQ, symmetry, an insight into which experimental constants are fundamental, a preference for second-order equations, etc., play a role in the final extension. For example, for Newton the existence of other experimental data, such as the Earth–Moon system, and probably K-III were crucial. But there is no deductive procedure for constructing the dynamical theory which encompasses the experimental data. The theory is probably never unique.

It might be of interest to generalize such considerations, or at least to consider some other examples. One could, for instance consider N nonintersecting harmonic-oscillator ellipses. Or one could generalize the artificial Kepler example to noncoaxial orbits or to intersecting orbits or even to the set of all Kepler ellipses (i.e. to all the negative-energy orbits). The example which follows is in this spirit.

Example 6.2. The harmonic oscillator

Consider the following given data in one dimension (here $Q = \mathbb{R}$, and x is the coordinate of $q \in Q$):

$$x = A \sin(B\tau + C), \quad A \in \mathbb{R}; \quad B > 0; \quad 0 \leqslant C \leqslant 2\pi. \tag{23}$$

This unrealistically rich 'experimental' data can be interpreted immediately as all harmonic oscillators centered at $x = 0$, parametrized by their frequencies B, but it will be seen that this interpretation is not logically necessary. In this case all of Q is available for all τ, so that $F = \mathbb{R}^2$.

By taking the derivative, one obtains the first-order equation

$$\dot{x} = AB \cos(B\tau + C).$$

If A or B is eliminated from this equation and (23), one obtains a time-dependent relation between x and $\dot{x}$, or a time-dependent first-order dynamical system. Let us concentrate on time-independent systems, which can be obtained by eliminating C, in which case the C appearing in (23) will be interpreted as a constant of integration. The resulting first-order dynamical systems (vector fields on Q) are then given by

$$\dot{x} = B\sqrt{A^2 - x^2}.$$

The solutions to this differential equation are

$$x = A \quad \text{and} \quad x = A \sin(B\tau + \delta),$$

which extends (or generalizes) the experimental data to include $B = 0$. The result is thus a set of many physical systems (dynamics) parametrized by A and B, in each of which δ is a constant of integration.

To obtain second-order dynamical systems take the derivative again: $\ddot{x} = -B^2 A \sin(B\tau + C)$. There are two possible ways of proceedings:

$$\ddot{x} = -B^2 x \text{ or } \ddot{x} = -B\sqrt{A^2B^2 - \dot{x}^2}. \tag{24}$$

The first of these (in which A has also been eliminated) yields the harmonic oscillators already mentioned,

$$x = \alpha \sin(B\tau + \delta),$$

with constants of integration α and δ. The second, still parametrized with both A and B, yields

$$x = AB\tau + k \quad \text{and} \quad x = AB\sin(B\tau + \delta) + \lambda.$$

In second order A, B, and C cannot all be eliminated, so that one again obtains parameter-dependent sets of dynamics.

One can proceeds to third order: $\dddot{x} = -B^3 A\cos(B\tau + C)$. There are then several possible equations. First, consider

$$\dddot{x} = -B^2\dot{x}, \tag{25}$$

which is just the time derivative of the first of (24). Its solution merely adds a constant to that solution:

$$x = \alpha\sin(B\tau + \delta) + \mu.$$

This sort of trivial differentiation, substituting the $(n+1)$st derivative everywhere for the nth, is always possible. Again one obtains a parameter-dependent set of dynamics. Second, an equation similar to the second of (24) is also available:

$$\dddot{x} = -B\sqrt{A^2B^4 - \ddot{x}^2},$$

whose solutions are

$$x = \frac{1}{2}AB^2\tau^2 + k\tau + \mu$$

and

$$x = AB^2\sin(B\tau + \delta) + \lambda\tau + \nu.$$

Third, if one solves for B^2 from (24) and insert it into (25), what results is

$$\dddot{x} = \dot{x}\ddot{x}/x,$$

whose solutions are

$$x = \gamma\tau + \varepsilon \quad \text{and} \quad x = \alpha e^{\mu\tau} + \beta e^{-\mu\tau}.$$

This is the only true single dynamical system that has been obtained, independent of parameters. It is described, as was the system of Eqs. (19)–(20), by singular differential equations (the experimental data is reproduced when α and β are complex conjugates, μ imaginary).

Remark. By stopping at this point we do not mean to imply that these are the only differential equations available. For example, $\dddot{x} = -B^3\sqrt{A^2 - x^2}$ is another possibility.

It is only in third order that all those experimental constants can be eliminated, and thus only in third order that a single dynamical system is obtained. If one insists on second-order systems, the set of harmonic oscillators with all frequencies does indeed seem to be the one depending on fewest experimental constants.

Example 6.3. The free particle in one dimension

For the simple example of a free particle in one dimension we shall now go pedantically through the general construction of the dynamical vector field. Then we shall restate the 'experimental' data in order to show that the particular way in which it is formulated can influence the outcome.

Let $Q = \mathbb{R}$ as in Example 6.2, and let S be given by all curves of the form

$$h: \mathbb{R} \to \mathbb{R}: \tau \mapsto x = a\tau + b, \forall a, b \in \mathbb{R}. \tag{26}$$

Since there is a curve (in fact many curves) through each pair (x, τ), F is all of $\mathbb{R}^2$. Then $G_{(x,\tau)}$ consist of all those $h(\tau)$ for which a and b satisfy the equality of (26) for fixed (x, τ), so that $G_{(x,\tau)}$ is a line in the (a, b) plane (see Fig. 6.2). Thus one may write $G_{(x,\tau)} = \mathbb{R}$, and then the set G of trajectory germs, defined below Eq. (2), is

$$G = \dot{\bigcup_{(x,\tau)\in\mathbb{R}\times\mathbb{R}}} \mathbb{R} \equiv \mathbb{R}^3.$$

To index the germs $g \in G$ one must choose a chart in $G_{(x,\tau)} = \mathbb{R}$. Since the line

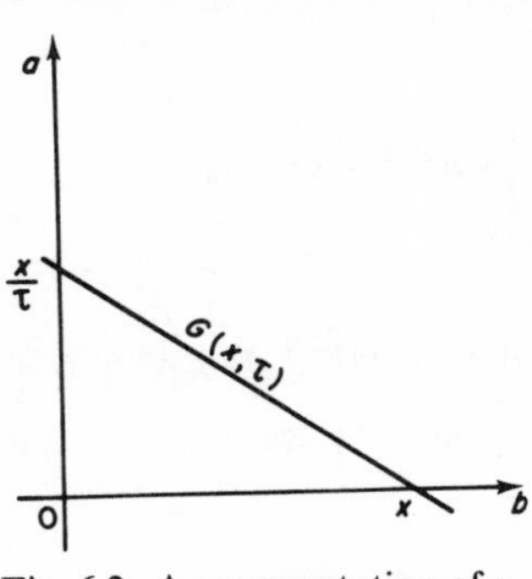

Fig. 6.2. A representation of a typical $G_{(x,\tau)}$. All $G_{(x,0)}$ are represented by vertical lines, and $G_{(0,\tau)}$, $\tau \neq 0$, by lines through the origin of slope $-1/\tau$.

representing $G_{(x,\tau)}$ in Fig. 6.2 can never be horizontal, the vertical coordinate a of each point of this line can be used as the index of the germ it represents. This amounts to indexing the germs by the velocity: $w = a = dh/d\tau$. Thus G can be identified with $TQ \times \mathbb{R}$.

Each element $g = (x, \tau, a)$ in G corresponds in this way to the curve $h_g : t \to at + x - a\tau$ that passes through x at time τ, which can be used to define the function $\bar{h}$: $\mathbb{R} \to G$ of Eq. (6), namely $\bar{h}_g : t \mapsto (at + x - a\tau, t, a)$. The flow mapping H, now strictly a flow (for it is defined on all of $\mathbb{R} \times G$) may be written

$$H : (t, (x, \tau, a)) \mapsto (at + x, \tau + t, a).$$

Since this system is time independent, the 'initial time' can be dropped, and one can now write down

$$\tilde{H} : \mathbb{R} \times \mathbb{R}^2 \to \mathbb{R}^2 \equiv T\mathbb{R} \equiv TQ$$

(recall that a is the velocity). One obtains

$$\tilde{H} : (t, (x, a)) \mapsto (at + x, a).$$

Thus we have finally arrived at the vector field (what is called x in Eq. (12) is here called (x, a))

$$X(x, a) = (x, a; a, 0).$$

The differential equation for the integral curves $g : \mathbb{R} \to TQ$ of this vector field (see Chapter 5) is then

$$X(g(t)) \equiv (g_1(t), g_2(t); g_2(t), 0) = Tg(t, 1) = (g_1(t), g_2(t); g_1'(t), g_2'(t)), \quad (27)$$

whose solutions $g_1(t) = at + b$, $g_2(t) = a$ project down on Q to the initial data $h(t) = at + b$.

We now restate the 'experimental' data for an observer using $y = x^3$ rather than x as a chart on $\mathbb{R} = Q$. Then S is given by all curves of the form

$$h : \mathbb{R} \to \mathbb{R} : \tau \mapsto y = (a\tau + b)^3. \quad (28)$$

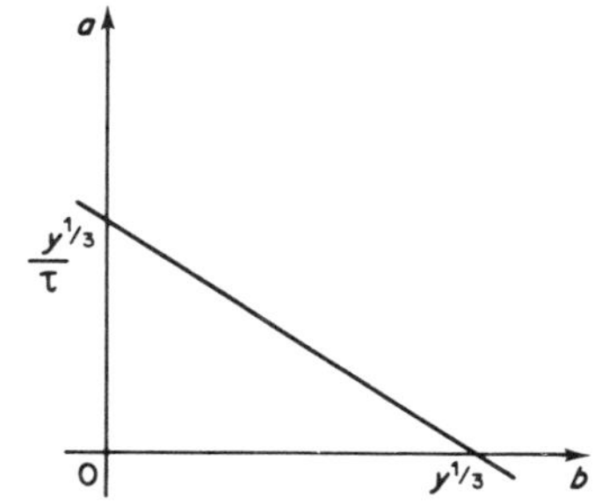

Fig. 6.3. A representation of a typical $G_{(y,\tau)}$. All $G_{(y,0)}$ are represented by vertical lines, and all $G_{(0,\tau)}$, $\tau \neq 0$, by lines through the origin of slope $-(1/\tau)^{1/3}$.

Again it is found that F is all of $\mathbb{R} \times \mathbb{R}$ and that $G(y, \tau)$ can be represented by a straight line (see Fig. 6.3), and again the germs can be indexed by their a coordinates. But now a is not the velocity, which is $dh/d\tau = 3ay^{2/3}$, so that G cannot be identified with $TQ \times \mathbb{R}$. For the flow one now obtains

$$\tilde{H}:(t,(y,a)) \mapsto ((y^{1/3} + at)^3, a),$$

and for the vector field

$$X(y,a) = (y, a; 3ay^{2/3}, 0).$$

The differential equation for the integral curve g becomes

$$(g_1(t), g_2(t); 3g_2(t)g_1(t)^{2/3}, 0) = (g_1(t), g_2(t); g_1'(t), g_2'(t)), \tag{29}$$

whose solutions are $g_1(t) = (at + b)^3, g_2(t) = a$. These project down on Q to the initial data $h(t) = (at + b)^3$.

The difference between the procedures following from (26) and (28) is that the first case makes use of TQ and the second of a different fiber space over Q. The vector field in the first case is of second order but is not so in the second case; in other words in Eq. (27) $g_1' = g_2$, while in Eq. (29) $g_1' \neq g_2$.

Could a second-order vector be obtained with the second chart if the germs were indexed differently? For that, one would need to use

$$v = dh/d\tau = 3y^{2/3}a \tag{30}$$

as an index, but then $dv/d\tau = 6y^{1/3}a^2 = 2/3v^2/y$. The differential equation obtained in this way, namely

$$(g_1, g_2; g_2, 2g_2^2/3g_1) = (g_1, g_2; g_1', g_2')$$

is singular at $g_1 = 0$. This is a reflection of the definition (30) of v, which forces all curves with $y = 0$ to pass through $v = 0$. Thus v (the y-velocity) is not a good index for labelling the germs at the origin.

It is easily seen that using T^3Q rather than TQ would not solve this problem, for then the fiber over each $q \in Q = \mathbb{R}$ would be three-dimensional, whereas it is clear from the structure of G that it must be one-dimensional.

Example 6.4. The tangent bundle of the sphere

We now give a simple example in which the geometry is not trivial, that is, in which G is $\mathbb{R} \times TQ$ but not $\mathbb{R}^n$ for any n.

Consider the free particle constrained to move on the sphere. For this case $Q = S^2$, and the 'experimental' data consist of all geodesics, covered at constant velocity. When the germs are indexed by the velocities, G becomes $\mathbb{R} \times TS^2$, and TS^2 is not diffeomorphic to $S^2 \times \mathbb{R}^2$. Yet when one looks at the experimental data, one sees that the germs are parametrized by two numbers at each point. Where in the analysis, one may ask, does it become evident that the carrier manifold is TS^2 rather than $S^2 \times \mathbb{R}^2$?

To answer this question, we start by remarking that in parametrizing the germs

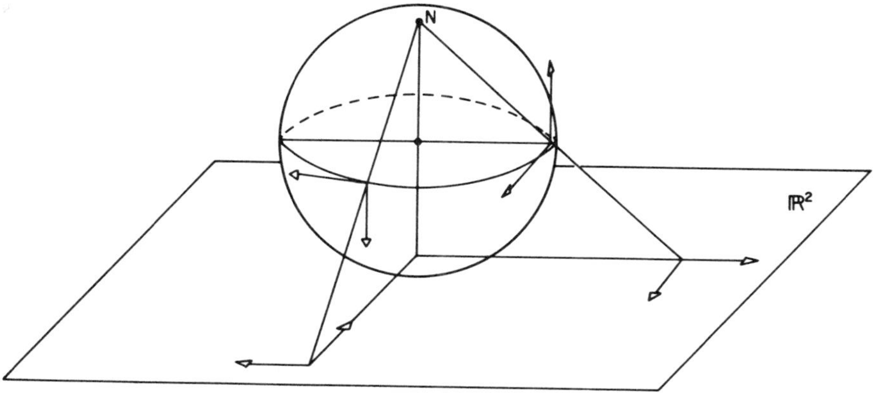

Fig. 6.4. A way to choose a basis of indepenent vectors at each point of S^2 by pulling back from $\mathbb{R}^2$

one must choose a pair of independent tangent vectors at each point of S^2. This is always possible to do, and once done it gives a map of the set of germs into $S^2 \times \mathbb{R}^2$. For example, one way to choose the vector pairs on S^2 is by using a polar projection and pulling back two basis vectors parallel to a fixed pair of axes on $\mathbb{R}^2$ (see Fig. 6.4). This works at every point of S^2 except the North pole N. At N, therefore, one must choose an arbitrary pair of vectors. But it is then clear that the flow will turn out to be discontinuous on $S^2 \times \mathbb{R}^2$ because the choice of axes is discontinuous at $N \in S^2$.

To obtain continuity one must choose the pair of axes in a continuous way on open subsets $U_i \subset S^2$, for each of which the germs can then be described in a continuous way on each $U_i \times \mathbb{R}^2$.

The parametrizations of a given germ in two open subsets U_i and $U_j, i \neq j$, will generally have to be different, however, even where U_i and U_j intersect. They will be connected by transformations depending on $T\varphi_{ij}$, where φ_{ij} connects the local coordinates in U_i and U_j. This leads to a set of germs parametrized in a globally smooth way on TS^2, but not on $S^2 \times \mathbb{R}^2$. Note that this does not imply that there is no one-to-one map between TS^2 and $S^2 \times \mathbb{R}^2$, but that there is no one-to-one bicontinuous map.

It is seen, then, that the requirement of continuity can force one to choose a certain carrier manifold over another.

7

Lifting to a carrier space: canonical lifting

Some of the difficulties and ambiguities of constructing the vector field from the experimental data were exhibited in Chapter 6. Let us now assume that the data contain enough information to allow for a unique construction of the vector field, as in most of the examples of Chapter 6, and let us proceed to study this construction in detail. We shall do this mostly with examples which exhibit some common types of pathologies, and which will suggest ways of dealing with them when they occur. Thus one may hope to learn what kind of assumptions are usefully made concerning the carrier manifold and vector field, and to see what kind of difficulties can arise when such restrictions are relaxed.

Just how much information is in fact enough to allow for a unique construction of the vector field depends first on the carrier space on which the orbits are separated. Usually Q itself is not suitable because in general more than one orbit passes through each $q \in Q$, or in terms of Chapter 6, the set $\{w\}$ at each point of Q contains more than one element. Probably the simplest example of a suitable carrier space is TQ, but even for TQ the question of the minimum information necessary has not, to our knowledge, been answered. Kasner has discussed this problem in terms of orbits on Q for the case of velocity-independent forces in two dimensions (Kasner, 1913). Among his interesting results are the following. First, not all possible triply infinite systems of curves can represent the orbits of such a dynamical system in the plane. 'There is, for instance, no field of force which produces as its trajectories all the circles in the plane.' Note that these are orbits, not trajectories (time is not taken into consideration). Note also that a velocity-dependent counterexample presents itself immediately: a charged particle in a uniform magnetic field. Second, a field of force is in general determined 'if we know $4\infty^1$ out of the totality of ∞^3 orbits', so long as one curve in each of the four systems of ∞^1 orbits passes through each point of the plane. These results can be extended to Q of dimension higher than two.

We shall not make use of Kasner's results, however, for we shall assume that the trajectories dealt with are all related to a dynamics, even if, as will be seen, the force must often be taken as velocity dependent. Moreover, in this chapter it will be assumed that essentially all of the trajectories are known and that therefore the final vector field contains no more data than those from which it is constructed. The data are put in the form of a vector field, as we have said elsewhere, because of the formal properties of the vector field, because it lends itself easily to global analysis, and because it provides a basis for logical extensions.

Let us start with a simple example, that of the one-dimensional simple harmonic oscillator. In this case $Q = \mathbb{R}$, and the (base) trajectories are assumed known and given by

$$\{h\} = \{A\cos(\omega t + \alpha), \quad 0 \leqslant A < \infty, \quad 0 \leqslant \alpha < 2\pi\}. \tag{1}$$

The family of trajectories passing through a point q_0 at time $t = 0$ is given by choosing A appropriately:

$$\{h_{q_0}\} = \left\{\frac{q_0}{\cos\alpha}\cos(\omega t + \alpha)\right\}. \tag{2}$$

In this subset of $\{h\}$ a trajectory can be further uniquely defined by specifying the initial velocity $\dot{q}_0 = h'(0)$, where $h' = dh/dt$:

$$\{h_{q_0\dot{q}_0}\} = \left\{\left(q_0^2 + \frac{\dot{q}_0^2}{\omega^2}\right)^{1/2}\cos(\omega t + \cos^{-1}[q_0(q_0^2 + \dot{q}_0^2/\omega^2)^{-1/2}])\right\}. \tag{3}$$

To each pair $q_0, \dot{q}_0$ there now corresponds a unique trajectory with a given A and α, and thus the one-dimensional harmonic oscillator is separated on $\mathbb{R} \times \mathbb{R} = \mathbb{R}^2$, the carrier space for this particular S.

The next step is to write the vector field for S on this carrier space, which may be done by going speedily and loosely through the construction of Chapter 6. We shall do this by first finding a set of first-order differential equations for the integral curves $g(t)$ on $TQ = \mathbb{R}^2$ whose projections $h(t)$ down to Q give the base trajectories of Eqs. (1)–(3). Since this is to yield a second-order field (see Chapter 5), half of the set of equations (in this case, one equation) is merely the definition of h'. The equations are

$$h'(t) = dh(t)/dt, \quad -\omega^2 h(t) = dh'(t)/dt. \tag{4}$$

Let $(q, \dot{q})$ be the general point in the carrier space $\mathbb{R} \times \mathbb{R}$, where q is the position coordinate in the first $\mathbb{R}$ and $\dot{q}$ the velocity coordinate in the second. With $g(t) = (h(t), h'(t))$, Eq. (4) may be put in the form

$$\Delta(g(t)) = Tg(t, 1) = (d/dt)g(t), \tag{5}$$

where the vector field Δ is given by

$$\Delta(q, \dot{q}) = (\dot{q}, -\omega^2 q), \tag{6}$$

or equivalently by

$$\Delta(q, \dot{q}) = \dot{q}\frac{\partial}{\partial q} - \omega^2 q\frac{\partial}{\partial \dot{q}}. \tag{7}$$

Remark. Ordinarily we use the notation X, Y or Z for vector fields, but for the special case of the vector field which describes the dynamics, and only for this one, we shall consistently use the symbol Δ. Note also that in Eq. (6) we suppress the first $q, \dot{q}$ on the right-hand side of (5.17) when we write Δ. We shall do this kind of thing often.

This is the explicit expression for the vector field of the harmonic oscillator in one dimension. The way one now sees the problem, it is to solve (5) for the integral curves $g(t)$ and then to project them down to the base integral curves $h(t)$ on Q. In this case, as in most cases of physical interest, the carrier space (on which the trajectories are separated) is TQ: each integral curve $g(t)$ is completely given by the initial condition $g(0) = (q_0, \dot{q}_0)$.

We now want to generalize from the specific example we have been discussing. Going from Q to a suitable carrier space, as was done above, will be called *lifting*. The specific way the lifting to TQ was performed will be called *canonical* lifting, and the vector field Δ defined in this way is called the *canonical vector field* on TQ. Given the dynamical system consisting of Q and $\{h\}$, the canonical lifting is obtained by defining the map

$$h(t) \mapsto g(t) = Th(t, 1) \in T_{h(t)}Q \tag{8}$$

and then defining the canonical vector field on TQ by Eq. (5). Then Δ is a second-order vector field as defined in Eq. (5), and

$$h(t) = \tau_Q g(t). \tag{9}$$

In terms of diagrams this can be put in the following form:

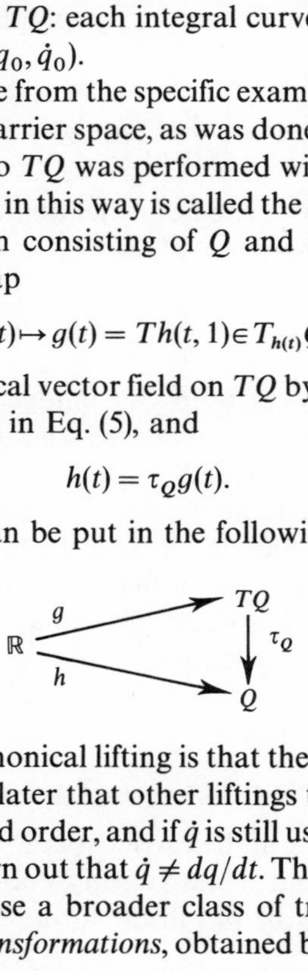

What characterizes the canonical lifting is that the vector field Δ obtained is of second order. It will be seen later that other liftings to TQ are possible in which the vector field is not of second order, and if $\dot{q}$ is still used for the coordinates in the fiber over $q \in Q$, it can then turn out that $\dot{q} \neq dq/dt$. This is reflected, as will become evident, in the freedom to use a broader class of transformations in TQ than those, usually called *point transformations*, obtained by lifting transformations on Q (Chapter 9).

The *canonical lifting* ψ will be defined in general through its local properties as a map from the family of trajectories $\{h\}$ on Q to the tangent bundle TQ. Let $\{h_q\}$ be the set of trajectories passing through $q \in Q$ at time $t = 0$. Then $\psi_q = \psi | h_q$ is defined by

$$\psi_q: \{h_q\} \to T_q Q: h_q \mapsto Th_q(0, 1) = (q, h_q'(0)) = (q, \dot{q}). \tag{10}$$

The system is assumed to be time independent, so that ψ can be defined from its local properties at any time t.

We now turn to several examples.

Example 7.1. Friction and gravity in one dimension

Let $Q = \mathbb{R}$ and the trajectories be given by

$$h_{q_0\dot{q}_0}(t) = a(t)\dot{q}_0 + b(t) + q_0, \tag{11}$$

where

$$a(t) = (1 - \mathrm{e}^{-Kt})/K,$$
$$b(t) = -G(1 - Kt - \mathrm{e}^{-Kt})/K^2.$$

The limits $G = 0$ (friction only) and $K \to 0$ (gravity only) are both well defined:

$$h^{\mathrm{fr}}_{q_0\dot{q}_0}(t) = (1 - \mathrm{e}^{-Kt})\dot{q}_0/K + q_0, \tag{12}$$

$$h^{\mathrm{gr}}_{q_0\dot{q}_0}(t) = Gt^2/2 + \dot{q}_0 t + q_0. \tag{13}$$

As in Eq. (4) one now writes $h'(t) = dh(t)/dt$ and calculates $dh'(t)/dt$. By eliminating q_0 and $\dot{q}_0$ from the subsequent equations, one arrives, in analogy with (4), at (we suppress the time dependence)

$$h' = dh/dt,$$
$$-Kh' + G = dh'/dt. \tag{14}$$

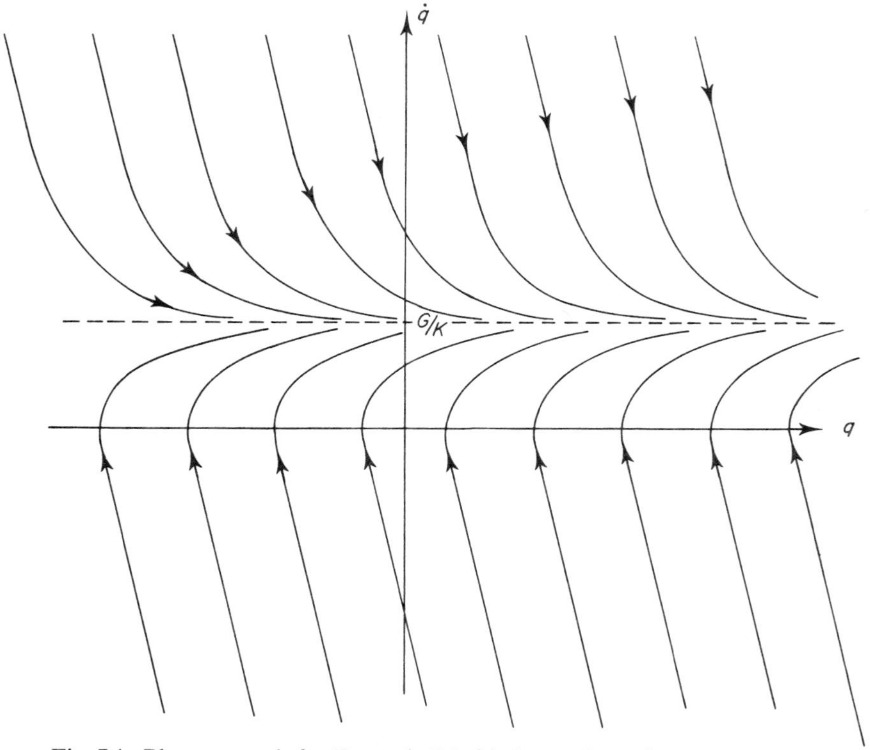

Fig. 7.1. Phase portrait for Example 7.1, friction and gravity. $K > 0$, $G > 0$

The canonical vector field is then

$$\Delta(q, \dot{q}) = (\dot{q}, -K\dot{q} + G). \tag{15}$$

Fig. 7.1 is a *phase portrait* of the dynamical system on $TQ = \mathbb{R}^2$ for $K > 0, G > 0$. The phase portrait is a representation of some of the trajectories $g(t) = (h(t), h'(t))$, but the time parameter itself is not indicated on these curves. Arrows nevertheless indicate the direction of increasing time. A line tangent to any one of these curves at any point is parallel to the tangent vector at that point; in fact one way to draw these curves without integrating the equations is to draw the tangent vector at every point $(q, \dot{q}) \in \mathbb{R}^2$ [in the present case that would mean vectors with components $(\dot{q}, -K\dot{q} + G)$] and then to construct the envelopes of these vectors. Any point $(q_0, \dot{q}_0)$ can be chosen as the initial point for the motion in TQ, and the system then develops along the curve passing through this point in the direction of the arrows. It is clear in this case that the trajectories are separated on TQ. The map ψ_q of Eq. (10) is thus injective (that is, one-to-one) for each q.

Consider now the limit $G = 0$, that is, the friction-only case of Eq. (12). The canonical vector field is then given by (15) with G set equal to zero, and the phase portrait becomes that of Fig. 7.2. All trajectories either approach $\dot{q} = 0$ as $t \to \infty$ or are zero-dimensional, points on the $\dot{q} = 0$ axis. (In Fig. 7.2, as in others, we shall use points to indicate such zero-dimensional trajectories. It is as true of these points as it is of the curves in the figures that they indicate only representative trajectories, not all of them. They have not been distributed the way they are in the figure for any deep reason.) Because in the canonical lifting $\dot{q} = dq/dt$ is the velocity, this means physically that the particle either slows down to zero velocity (in an infinite time) or has zero velocity, remaining at its initial position. Since the slowing-down trajectories never actually reach the $\dot{q} = 0$ axis, there is just one

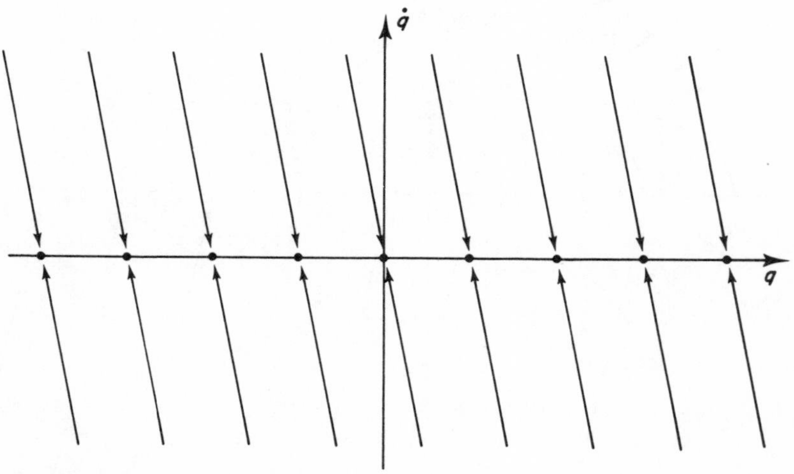

Fig. 7.2. Phase portrait for Example 7.1, friction without gravity. Same K as in Fig. 7.1, $G = 0$

trajectory through each point of TQ (the trajectories are separated) and ψ_q is again injective for each q.

Example 7.2. The four-pronged star (linear repulsive force)

Another example which is separated on TQ and contains zero-dimensional trajectories is the following. Let $Q = \mathbb{R}$ and the trajectories be given by

$$h_{q_0\dot{q}_0}(t) = a(t)\dot{q}_0 + b(t)q_0, \tag{16}$$

where

$$a(t) = \frac{1}{K}\sinh Kt, \quad b(t) = \cosh Kt$$

with $K > 0$. The second of Eqs. (4) (by now it should be clear that in Newtonian terms this is just the acceleration) now becomes

$$dh'/dt = K^2 h. \tag{17}$$

Remark. Comparison of this equation with (4) shows that it is closely related to the harmonic oscillator. Only the sign is different. It is left to the reader to draw the phase portrait for the oscillator, but in any case, there are certain similarities that are easily observed. For example, in both cases the vector field is zero at the origin $(q, \dot{q}) = (0, 0)$ of TQ, which is reflected by the fact that the origin is a zero-dimensional trajectory.

Fig. 7.3 is the phase portrait of this system. Two trajectories approach the

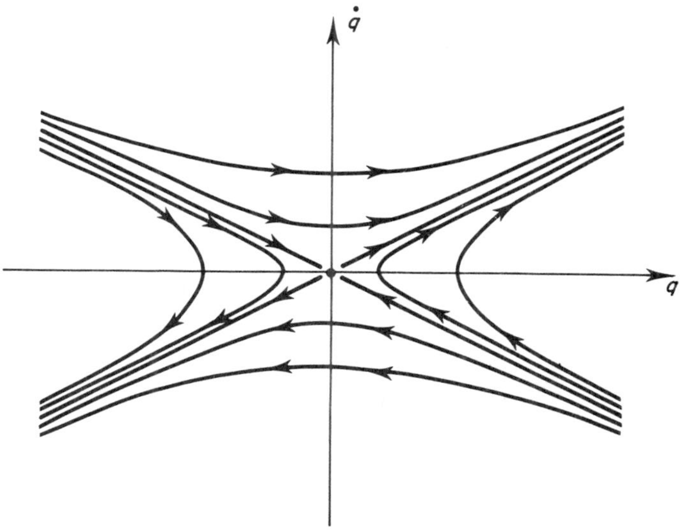

Fig. 7.3. Phase portrait for Example 7.2, linear repulsive force

origin as $t \to +\infty$, and two as $t \to -\infty$. As in the $G = 0$ case of Example 7.1, however, none of these trajectories actually reach the origin. Moreover, $(h, h') = (0, 0)$ is a trajectory (satisfies Eq. (17)). Thus again the trajectories are separated on TQ, and ψ_q is injective.

That ψ_q might not always be injective is seen from the following example.

Example 7.3. Constant Friction.

Let $Q = \mathbb{R}$ and the trajectories be given by

$$h_{q_0\dot{q}_0}(t) = \begin{cases} -\dot{q}_0\mu t^2/|2\dot{q}_0| + \dot{q}_0 t + q_0 & \text{for } \dot{q}_0 \neq 0, t < |\dot{q}_0|/\mu, \\ \dot{q}_0|\dot{q}_0|/2\mu + q_0 & \text{for } \dot{q}_0 \neq 0, t > |\dot{q}_0|/\mu, \\ 0 & \text{for } \dot{q}_0 = 0, \end{cases} \tag{18}$$

with $\mu > 0$. Note that this produces a sort of constant frictional force in the direction opposite to the velocity, and gives zero force for velocity zero. The second of Eqs. (4) now becomes

$$dh'/dt = \begin{cases} -\mu h'/|h'| = -\mu \operatorname{sign} h' & \text{for } h' \neq 0, \\ 0 & \text{for } h' = 0. \end{cases} \tag{19}$$

The canonical vector field is now

$$\Delta(q, \dot{q}) = \begin{cases} (\dot{q}, -\mu \operatorname{sign} \dot{q}) & \text{for } \dot{q} \neq 0, \\ (\dot{q}, 0) = (0, 0) & \text{for } \dot{q} = 0. \end{cases} \tag{20}$$

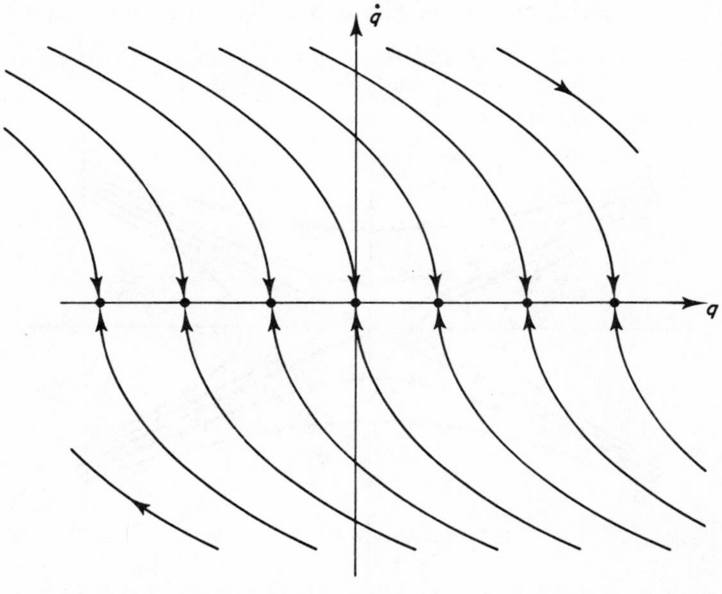

Fig. 7.4. Phase portrait for Example 7.3, constant friction

Fig. 7.4 is the phase portrait of this dynamical system. All the trajectories either approach *and reach* $\dot{q}=0$ *in a finite time* or are zero-dimensional, points on the $q=0$ axis. Thus (recall again that in the canonical lifting $\dot{q}=dq/dt$ is the velocity) for each value of q there are three trajectories at $\dot{q}=0$: the one which starts at $q_0<q$ with $\dot{q}>0$ and comes to rest at q after a time $t=\dot{q}_0/\mu$, the one that starts at $q_0>q$ with $\dot{q}_0<0$ and comes to rest at q after a time $t=-\dot{q}_0/\mu$, and the one with $\dot{q}_0=0$, $q_0=q$. Thus the orbits are not separated on TQ, and the canonical lifting is not injective.

Example 7.4. Relativistic free particle

Let $Q=\mathbb{R}$ and the trajectories be given by

$$h_{q_0\dot{q}_0}(t)=\dot{q}_0 t+q_0, \quad \text{where } \dot{q}_0\in\,]-1,1[. \tag{21}$$

Note the important restriction on the velocity. In this case the second of Eqs. (4) becomes

$$dh'/dt=0. \tag{22}$$

Fig. 7.5 is the rather simple phase portrait for this system. The restriction on the velocity causes the entire phase portrait to lie inside the strip between the lines $\dot{q}=1$ and $\dot{q}=-1$. Nevertheless, within this strip the orbits are separated (including one-dimensional orbits on the $\dot{q}=0$ axis), and ψ_q, though it is not surjective (it does not map $\{h_q\}$ onto all of T_qQ), is injective.

We have now exhibited several examples of systems lifted from Q to TQ, and are ready to discuss some restrictions that will be made in the sequel. The first

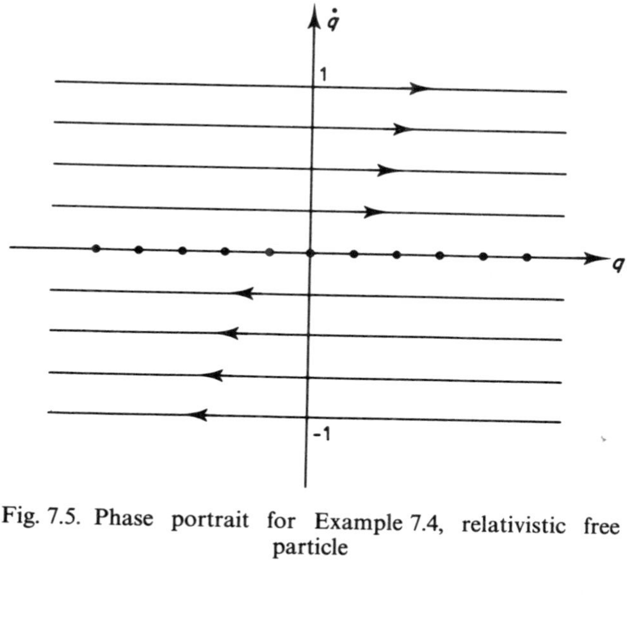

Fig. 7.5. Phase portrait for Example 7.4, relativistic free particle

restriction is the following. Henceforth we shall deal only with dynamical systems that are separable on TQ (unless otherwise stated for some special cases), and it will therefore be *assumed that* ψ_q *is injective for all* $q \in Q$.

A topology will now be chosen on $\psi_q\{h_q\}$ in an obviously convenient way, namely the topology induced on $\psi_q\{h_q\} \subset T_qQ$ by the topology of T_qQ, which will be called the *canonical topology*. This makes $\psi_q\{h_q\}$ a topological space, and with the corresponding topology on $\{h_q\}$ the lifting ψ_q is a homeomorphism. This means that two elements $h_{q,\dot{q}}$ and $h_{q,\dot{q}'}$ are near each other if $\dot{q}$ and $\dot{q}'$ are near each other.

The assumption that TQ is a suitable carrier space requires that the structure of TQ be appropriate for describing the dynamics. In the present context, as will now be explained, this leads to the assumption that ψ_q is *regular* at each point $q \in Q$, or in other words that $\psi_q\{h_q\} \subset T_qQ$ is a *regular submanifold* of T_qQ. We make some remarks about this assumption here, but details will be found in the next chapter.

There are two aspects to this assumption. First, it requires that $\psi_q\{h_q\}$ be a manifold and therefore that it be possible to establish a differentiable structure on it. Second, it requires that it be a regular submanifold of T_qQ, i.e. that the differentiable structure be derived from the differentiable structure on T_qQ. Equivalently, the inclusion map is required to be a *regular imbedding*. Evidently, all of the examples exhibited so far satisfy this technical assumption. In Example 7.4, for instance, $\psi_q\{h_q\}$ is a *proper* submanifold of T_qQ (that is, it is not T_qQ itself), and the assumption is still satisfied.

One pathology whose elimination is guaranteed by (but does not require) this assumption is a differentiable structure that changes from fiber to fiber. The assumption thus relates to the global properties of ψ. It then becomes appropriate to pass from a discussion of the properties of ψ_q at each point q of Q to a discussion of the entire map $\psi:\{h\} \rightarrow TQ$ on all of Q. Consider therefore the following example.

Example 7.5. The quadratic star

We shall call vector fields *stars* if (but not necessarily only if) they are associated with differential equations of the form $dh'/dt = nh'^2/h$, where n is a positive integer, called the *degree* of the star. Much of the pathology of the present example is exhibited by stars of other degrees.

Let $Q = \mathbb{R}$ and the trajectories be given by

$$h_{q_0\dot{q}_0}(t) = \frac{q_0^2}{q_0 - \dot{q}_0 t}. \tag{23}$$

The second of Eqs. (4) now becomes

$$dh'/dt = 2h'^2/h, \tag{24}$$

which is seen to yield a quadratic star. To the extent that this system has a physical interpretation, it involves a force which depends quadratically on the velocity and inversely on the distance from the origin $q=0$. Equation (24), which is singular, can also be written in the nonstandard form

$$hdh'/dt = 2h'^2. \tag{25}$$

The phase portrait for this system can be found, for instance, by dividing both sides of (24) by h' and integrating. When the initial conditions are inserted, this results in

$$h' = \dot{q}_0 h^2/q_0^2,$$

and the resulting phase portrait is shown in Fig. 7.6. All of the trajectories which are not of dimension zero (points on the $\dot{q}=0$ axis) are parabolas which approach the origin of TQ as $t \to \pm\infty$, but never reach it. Each of these trajectories also reaches infinity (both in q and in $\dot{q}$) in a finite time depending on the initial conditions. Such a vector field is said to be not complete (see Chapter 5). Moreover, the only trajectory which reaches the $q=0$ axis in a finite time is the zero-dimensional one at the origin $(0,0)$ of TQ. Thus ψ_q is injective for each q, but surjective only for $q \neq 0$. In fact although $\psi_q\{h_q\}$ is a submanifold of T_qQ for each q, $\{h\}$ is not a submanifold of TQ, for the dimension pinches off to zero at $q=0$.

In order to avoid the kind of pathology seen in this example, we shall make the following additional assumption: for every $q \in Q$ there is a neighborhood $U \subset Q$ containing q such that if $q' \in U$, then $\psi_{q'}\{h_{q'}\}$ is diffeomorphic to $\psi_q\{h_q\}$. In other words, it will be assumed that the union of $\psi_q \forall q \in Q$ is a bundle (see Chapter 5) $W_0 \subset TQ$, which will be called the *canonical bundle* associated with Q and $\{h\}$. It is seen that $q=0$ in Example 7.5 does not satisfy this assumption, for $\psi_q\{h_q\}$ is of dimension zero for $q=0$ and of dimension one for every other q' in any U

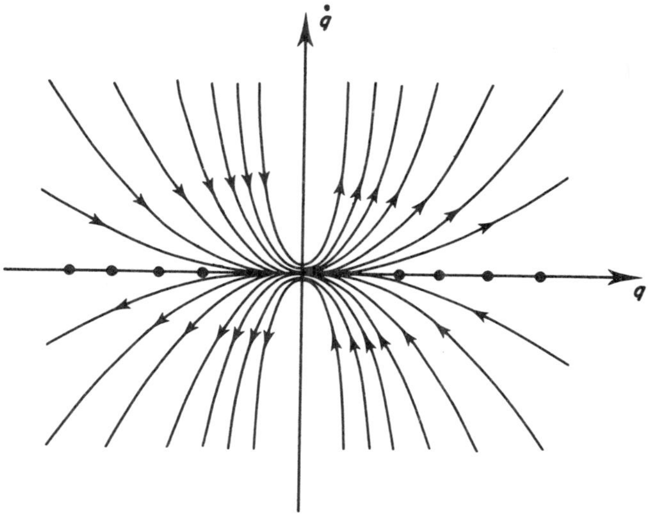

Fig. 7.6. Phase portrait for Example 7.5, quadratic star

containing $q = 0$. In this connection see the rather more complicated dynamical system of Eqs. (19) and (20) of Chapter 6.

Example 7.6. Langue de Chat (a parameter-dependent family of systems)

Let $Q = \mathbb{R}$ and the trajectories be given by

$$h_{rC}(t) = \alpha r \cot(\alpha\sqrt{r^2+\alpha^2 t} + C)[r^2 + \alpha^2 csc^2(\alpha\sqrt{r^2+\alpha^2 t} + C)]^{-1/2}. \quad (26)$$

where $\alpha \in \mathbb{R}$ is a parameter. Note that in order to avoid notational complications we have not labelled h_{rC} by its initial conditions $q_0, \dot{q}_0$. On the other hand, the constant C can be dropped, for it merely tells where on each trajectory in TQ the particle starts at $t = 0$. If C is set equal to zero the particle is constrained to start at $(q_0, \dot{q}_0) = (r, 0)$, and then when h_{rC} is written h_{r0} the subscript takes on the usual meaning. It is perhaps easier to work with the implicit expression for h_{r0} in the form

$$\frac{h_{r0}}{\sqrt{r^2 - h_{r0}^2}} = \frac{\alpha}{\sqrt{r^2+\alpha^2}} \cot(\alpha\sqrt{\alpha^2 + r^2}\, t). \quad (27)$$

First consider $\alpha > 0$. Then the second of Eqs. (4) becomes

$$\frac{dh'}{dt} = 2\frac{hh'^2}{\alpha^2 + h^2} - h(\alpha^2 + h^2)^2. \quad (28)$$

The force depends, as in the previous example, in a complicated way on the position and velocity.

The phase portrait for this system can be obtained by taking the derivative of (27) and solving for h'. Since C has already been omitted, the result appears in terms of r alone, namely

$$h' = \pm(h^2 + \alpha^2)\sqrt{r^2 - h^2}. \quad (29)$$

Fig. 7.7 is the resulting phase portrait. The velocity along each orbit has several extrema as a function of position, and since the phase portrait is symmetric about both the $q = 0$ and $\dot{q} = 0$ axes, these extrema can be studied in the first quadrant only (including the positive $\dot{q}$ axis). The extrema occur at the points $(q, \dot{q})$ with coordinates

$$(0, \alpha^2 r) \quad \text{and} \quad \left(\left[\frac{2r^2 - \alpha^2}{3}\right]^{1/2}, 2\left[\frac{r^2+\alpha^2}{3}\right]^{3/2}\right). \quad (30)$$

The second of these is a maximum and occurs only for $r \geqslant \alpha/\sqrt{2}$, whereas the first is a maximum for r below this critical value and a minimum otherwise. Note that $-r$ and r are the limit points of each orbit. (Recall that C has been eliminated from consideration so that all the orbits start at $q_0 = r$.) Each orbit is periodic in time with period $2\pi(\alpha\sqrt{\alpha^2 + r^2})^{-1}$. This system meets all of our assumptions for any $\alpha > 0$.

Now consider the limit as $\alpha \to 0$. In this limit Eq. (28) becomes singular at $h = 0$,

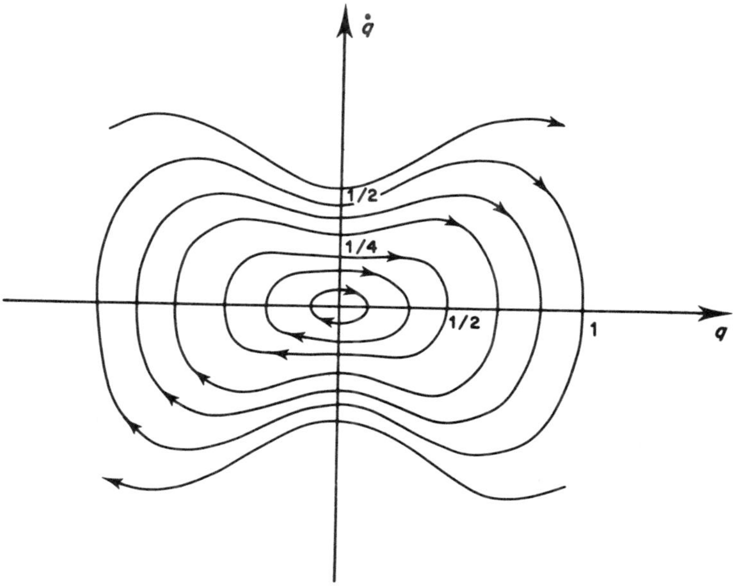

Fig. 7.7. Phase portrait for Example 7.6. $\alpha = \sqrt{2}$

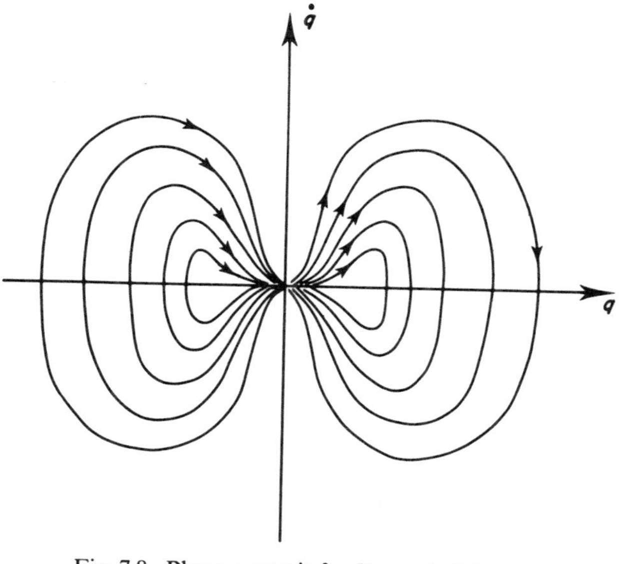

Fig. 7.8. Phase portrait for Example 7.6. $\alpha = 0$

and the system changes character in a radical way. The critical value $\alpha/\sqrt{2}$ of r becomes zero, so that every one of the orbits now exhibits the second extremum (maximum), and for all orbits the first extremum (minimum) occurs at (0, 0). In the limit the equation of the trajectory becomes

$$h_{r0} = r[t^2r^4 + 1]^{-1/2}, \tag{31}$$

and it is seen that the trajectories are no longer periodic. In fact, they approach the origin of TQ as $t \to \pm\infty$. The sign of q is now fixed by the sign of r (essentially the sign of q_0), and each orbit is pinched off at the origin, much like those of Example 7.5, into one to the right and one to the left of the $q = 0$ axis in the phase portrait of Fig. 7.8. In addition, again as in Example 7.5, there is a point trajectory at the origin. At $q = 0$ this system, although its vector field is complete, exhibits the pathology of Example 7.5.

Example 7.7. The plane pendulum

It is not always easiest to start with an analytic expression for h or even for $h_{q_0\dot{q}_0}$ as in most of the examples so far. This may be because the measurements do not yield results that are easily put in analytic form or because the measurements are in fact of other aspects of a physical system than those required for h. Partly as an example of such a case and partly in order to give an example on some manifold other than $\mathbb{R}^2$, we now turn to the simple plane pendulum. In this example Q is the circle S^1, so that TQ is the cylinder $S^1 \times \mathbb{R}$. No single chart can form an atlas on this Q, and thus none on TQ either. Consider therefore the two-chart atlas on S^1 given by

$$\begin{aligned} q&:]0, 2\pi[\subset S^1 \to \mathbb{R}, \\ q'&:]-\pi, \pi[\subset S^1 \to \mathbb{R}. \end{aligned} \tag{32}$$

In the first of these charts the equations for the plane pendulum which corresponds to (4) are

$$h'(t) = dh(t)/dt, \quad -\sin h(t) = dh'(t)/dt. \tag{33}$$

Equivalently, the vector field could be given as

$$\Delta(q, \dot{q}) = (\dot{q}, -\sin q). \tag{34}$$

It is left to the reader to find the equations for h or Δ in the other chart of the atlas. Of course the second term in Δ is what is called the (generalized) force.

Now as in the other examples, Eq. (33) can be integrated to obtain

$$h' = \sqrt{2(E + 1 + \cos h)}, \tag{35}$$

where the constant of integration E has been chosen in this form for convenience, as will be seen in what follows. Note that the minimum value of E is -2.

Fig. 7.9 shows the phase portrait of this dynamical system on the cylinder; it is given in two views, and the chart used is the first of Eqs. (32). When $E = -2$ the trajectory consists of one point, the origin $q = 0$, $\dot{q} = 0$. For $-2 < E < 0$ the

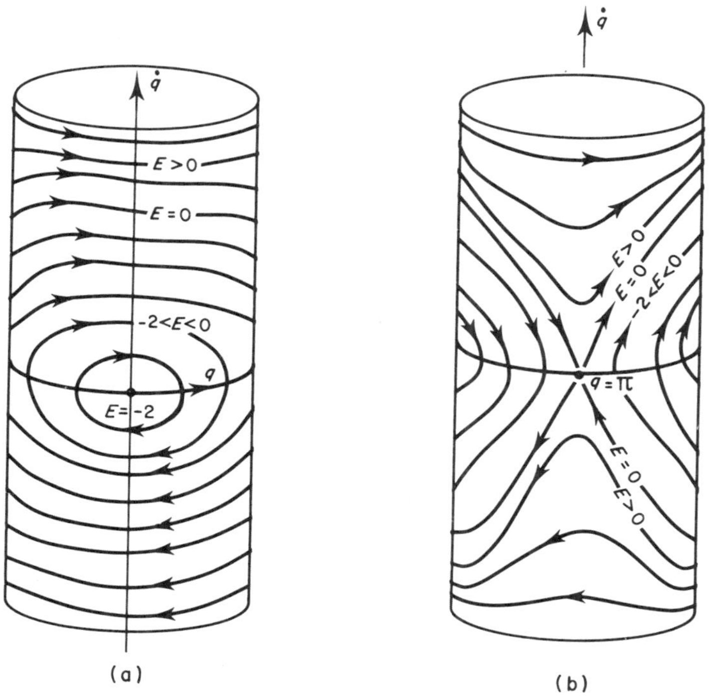

Fig. 7.9. Phase portrait for Example 7.2. (a) Front view. (b) Rear view

trajectories are closed curves which are covered periodically as the pendulum swings back and forth through the equilibrium point at $q = 0$. For $E > 0$ (we shall return to $E = 0$ in a moment) there are two trajectories for each value of E, one for which $\dot{q}$ is always positive, and one negative, the first going to the right on the cylinder, the other to the left. These trajectories correspond to rotation of the pendulum which now passes through the two equilibrium points, at $q = 0$ and at $q = \pi$. For $E = 0$ there are three trajectories. One is the equilibrium point at $(\pi,0)$, and the other two correspond to motion which reaches this equilibrium point at $t = \infty$ or leaves it at $t = -\infty$ (or both), one trajectory to the right and the other to the left on the cylinder. Thus the curve on the cylinder that looks like an orbit with a cross-over point at $(\pi,0)$ consists in fact of three trajectories. The situation is somewhat similar to Example 7.2. It is seen that ψ_q is injective in this example.

This chapter has exhibited some of the pathologies it is desirable to avoid in vector fields which describe dynamical systems. Some of the assumptions that have been made here to avoid them will be applied when possible to vector fields in the sequel, but unfortunately this is not always possible. In such cases exceptions will have to be dealt with explicitly.

After the next chapter, a digression on submanifolds and smooth maps, we shall return to TQ and study general diffeomorphisms on it.

8

A digression on submanifolds and smooth maps

It often turns out in the evolution of a dynamical system on a manifold M that a subset of M takes on a special importance or interest. For example, integral curves are important subsets of M. Surfaces of constant energy are another familiar example. For one reason or another, as will be seen, it is often convenient or even necessary to restrict the analysis to such a subset, and then it is important to know to what extend the formalism used on manifolds can be carried over to the subset. The subset might itself end up being a manifold, but often it may not, and even when it does, its differentiable structure may differ in important ways from that of M itself. Thus two questions arise immediately in this context. Can the subset be endowed with a differentiable structure which is relevant in some way to the dynamical system? What is the relation, if any, between this differentiable structure and the original one on M? This digression is concerned with these and similar questions. It will explain and enlarge on some terminology already introduced in Chapter 7 (e.g. regular imbedding).

Most of the examples in this chapter will involve one-dimensional subsets of two-dimensional manifolds, often of $\mathbb{R}^2$. It is convenient, however, to bear in mind another two-dimensional manifold, namely the torus T^2, which is perhaps richer in interesting topological properties. Particular subsets on it which we will consider later (§8.1(3)) are winding lines, both of rational and irrational slope. The torus T^2 is of dynamical interest, for example, because it is the constant-energy surface of the harmonic oscillator in two degrees of freedom, and the winding lines correspond to trajectories, their slopes being the ratios of the frequencies.

Let M be a (differentiable, Hausdorff, second countable) manifold of dimension μ with a differentiable structure consisting of charts $(U_\alpha, \varphi_\alpha)$. Roughly speaking, a *submanifold* N of M is a subset of M which is itself a manifold, but difficulties arise, as mentioned above, with the differentiable structures on N and M. Two procedures will be discussed for arriving at submanifolds: *imbeddings* and *submersions*, which will be defined in this chapter.

8.1 SUBMANIFOLDS FROM IMBEDDINGS

1. An *open submanifold* N of M is a subset $N \subset M$ which is open in the manifold topology of M, with differentiable structure consisting of charts $(N \cap U_\alpha, \varphi_\alpha)$(and therefore with the subset topology); $\dim N = \mu$. Open submanifolds are essentially open chunks of M. They are seldom of dynamical interest.

A *regular submanifold* is roughly what one thinks of when one tries to visualize a smooth submanifold. For example, the circle of radius r in the plane is a regular submanifold of $\mathbb{R}^2$.

More precisely let $N \subset M$ be a subset of M, and let i be the *natural injection* $i:N \to M:n \mapsto n$, $n \in N$. Then N is a *ν-dimensional regular submanifold* of M if there exists an atlas $(\bar{U}_\alpha, \bar{\varphi}_\alpha)$ in the differentiable structure on M and an integer ν such that the maps $(N \cap \bar{U}_\alpha, \psi_\alpha)$ defined by

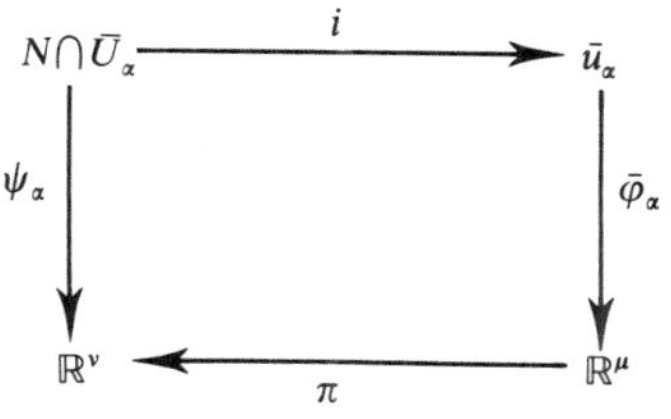

provide a smooth atlas, and hence a differentiable structure, for N. Here π is the projection $\pi:(x^1, \ldots, x^\mu) \mapsto (x^1, \ldots, x^\nu)$. For a regular submanifold so defined, the manifold topology and subset topology on N coincide. (An open submanifold is also regular.)

2. The idea of a regular submanifold is not general enough for some purposes; it is useful to generalize the concept of a *submanifold* to a subset whose topology might be different from the subset topology. An example is the winding line of irrational slope on the torus $T^2 = S \times S$ (see Figs. 8.1, and 8.2).

Let each circle S of the torus be thought of as the unit circle in the complex plane, so that the points of S can be described by $e^{i\theta}$, and those of T^2 by $(e^{i\theta_1}, e^{i\theta_2})$, $(\theta_1, \theta_2) \in \mathbb{R}^2$. The winding line passing through the origin $\theta_1 = \theta_2 = 0$ can then be written in the form

$$c: I \subset \mathbb{R} \to T^2 : t \mapsto (e^{i\alpha t}, e^{i\beta t}), \tag{1}$$

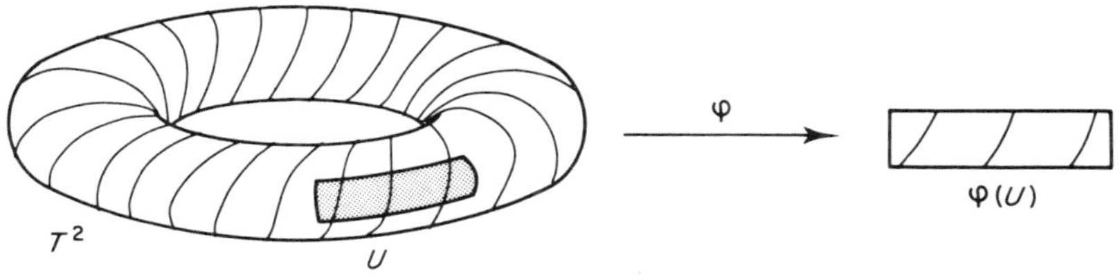

Fig. 8.1. The torus with a winding line of rational slope. There exist subneighborhoods of U through which the line passes just once

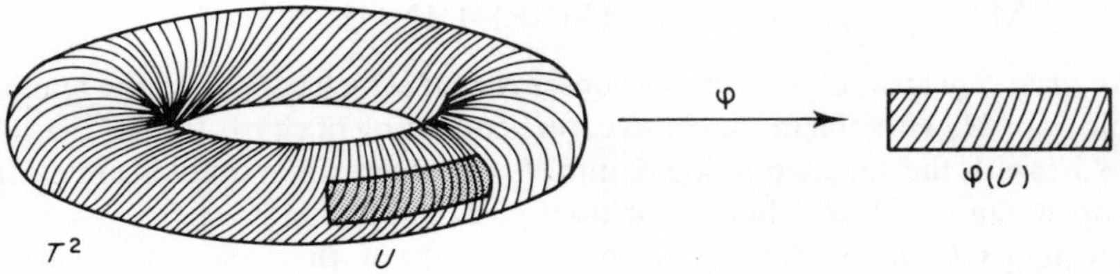

Fig. 8.2. An attempt to represent the torus with a winding line of irrational slope. The line passes an infinite number of times through each neighborhood, no matter how small

where I (the range of t) is determined by $\alpha, \beta \in \mathbb{R}$. Then if the slope α/β is rational, the line comes back on itself after winding a finite number of times around the torus, the range of t is finite, and at each point $m \in T^2$ on the line there are neighborhoods U such that the line passes only once through U (or $\{c\} \cap U$ is connected). If α/β is irrational, the line never comes back on itself, winding an infinite number of times through every neighborhood U of any point $m \in T^2$ (or $\{c\} \cap U$ is not connected) and $I = \mathbb{R}$.

Now, the winding line of irrational slope is clearly a subset N of the torus (though a strange one: its closure is T^2), and it is diffeomorphic to $\mathbb{R}$. There are points on the line, however, which are far away in the topology which comes from $\mathbb{R}$, but are very close in the subset topology which comes from T^2. That is, in every neighborhood U of any point $m \in T^2$ on the line there are other points m' which lie on other components (segments) of c passing through U. These other points can be very far away from m if one moves only along the line, i.e. very far away in the topology which comes from $\mathbb{R}$. Thus this winding line is not a *regular* submanifold of T^2, and it is useful to define a *submanifold* so as to include it.

In order then to extend the definition, it helps to think of a subset of M as the image of a mapping $\varphi: K \to M$, where K is a manifold of dimension κ (in the example of the winding line, M is T^2 and K is $\mathbb{R}$). It can be shown that if φ is an injection it defines a differentiable structure on $N = \varphi(K)$, which thus becomes a manifold (in the example, N is the winding line). If, moreover, $T\varphi$ is also an injection (i.e. if $T_k\varphi$ is an injection for each $k \in K$, as is the case in the example, and then it is clear that $\kappa \leqslant \mu$), then N is diffeomorphic to K, and therefore $\dim N \equiv \nu = \kappa$ (Brickell and Clark, 1970, p. 74). Although in this way N becomes a manifold with a well-defined differentiable structure *induced* by $\varphi(K)$, and hence with a certain specific *induced* topology, this topology may not be the subset topology obtained from M. An injective mapping for which $T\varphi$ is also injective is called an *imbedding*, and thus one arrives at the following definition.

A *submanifold* of M is the image $N = \varphi(K)$ of an imbedding $\varphi: K \to M$, where K is also a (differentiable) manifold. It follows that $\nu = \kappa \leqslant \mu$.

Remark. A *regular imbedding* is one for which the topology induced by $\varphi(K)$ is the subset topology on M. Thus a regular submanifold is the image of a regular imbedding. Examples of regular submanifolds are thus the winding line of rational slope α/β on T^2 and the circle S in $\mathbb{R}^2$.

There are several results on submanifolds and regular submanifolds that may prove useful. If a subset N of M can be given the structure of a regular submanifold of M, there are no other submanifold structures of the same dimension that N can be given. Even if a submanifold N is not regular, an open neighborhood in the subset topology is still open in the induced topology, and this implies that submanifolds are Hausdorff. Imbeddings which are not regular, however, may not give rise to second countability, so that submanifolds as defined here may turn out not to be manifolds! On the other hand, compact or connected submanifolds are always second countable, so just to be safe one could deal only with such submanifolds. But this, like the restriction to regular submanifolds, is also too restrictive. In fact recall that in the case of the plane pendulum (Example 7.7) there are two disconnected trajectories for each energy that is sufficiently high: the submanifolds defined by fixed energy are not all connected. Thus it is useful to generalize even further.

3. As an example of an *immersed submanifold*, consider the figure-eight ($F8$) in $\mathbb{R}^2$, illustrated in Fig. 8.3. The peculiarities of $F8$ arise from the crossing at the origin. Clearly no chart in the differentiable structure of $\mathbb{R}^2$ can project a neighborhood

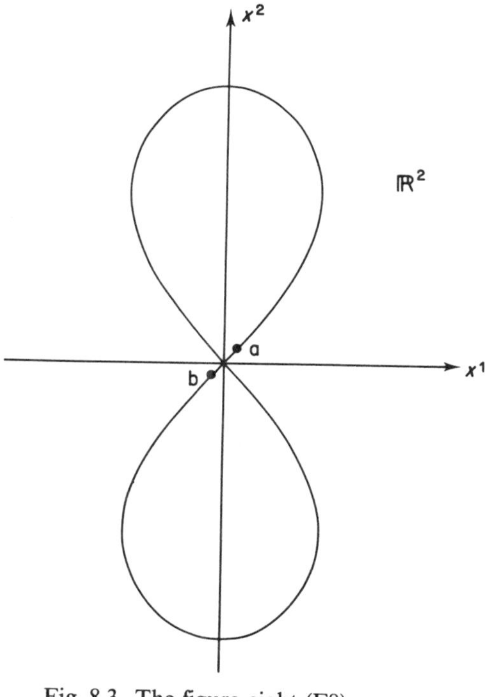

Fig. 8.3. The figure-eight (F8)

of the origin $(0, 0)$ into $\mathbb{R}$: the subset topology cannot possibly be the manifold topology on $F8$.

One can think of $F8$ as the circle S^1 twisted once and laid onto the plane. Then the mapping from S^1 to $\mathbb{R}^2$ may be written in the form

$$\varphi: S^1 \to \mathbb{R}^2: e^{i\theta} \mapsto (\sin 2\theta, \sin \theta), \tag{2}$$

and as a subset $F8$ is $\varphi(S^1)$. But this mapping is not an imbedding, for it is not an injection at the origin $(0, 0) \in \mathbb{R}^2$. In fact two points on the circle, $\theta = 0$ and $\theta = \pi$, map onto the origin. Therefore φ does not endow $F8$ with a submanifold structure in our definition, and it is then called an immersed submanifold (an *immersion* φ is a map for which $T\varphi$ is injective, but φ itself may be only locally injective).

On the other hand, $F8$ can be viewed as the image of an open interval $I \subset \mathbb{R}$ in accordance with

$$\varphi': I \to \mathbb{R}^2: t \mapsto (\sin 2t, \sin t), 0 < t < 2\pi. \tag{3}$$

The mapping φ' does endow $F8$ with a submanifold structure, but not that of a regular submanifold. The points a and b, which in the subset topology lie close to the origin in $\mathbb{R}^2$, lie about as far away from it as they can in the topology induced by φ', for $\varphi'(\pi) = (0, 0)$, whereas $a = \varphi'(\varepsilon)$ and $b = \varphi'(2\pi - \varepsilon)$, with $\varepsilon > 0$ small.

Formally an *immersed submanifold* N of M is the image of an immersion $\varphi: K \to M$, where K is a manifold of dimension $\kappa \leqslant \mu$. With this definition an immersed submanifold is not necessarily a submanifold of M; it is if the immersion is an imbedding (i.e. an injection).

Recall Example 7.7, in which there were three trajectories that lay on a curve with a cross-over point. It will be shown in Chapter 10, if it is not already clear, that the curve on which they lie is essentially $F8$, and thus it is evident that this kind of submanifold can arise even in a simple dynamical system. It is interesting that, as was pointed out in Chapter 7, the physics breaks up this rather unpleasant submanifold into three more reasonable ones: the cross-over point at $(\pi, 0)$ and the two regular submanifolds which correspond to the two loops obtained from $F8$ when the cross-over point is removed.

4. There are, moreover, other possibilities depending on the mapping $\varphi: K \to M$, but then $\varphi(K)$ is not called a submanifold. One of these occurs if φ is only locally

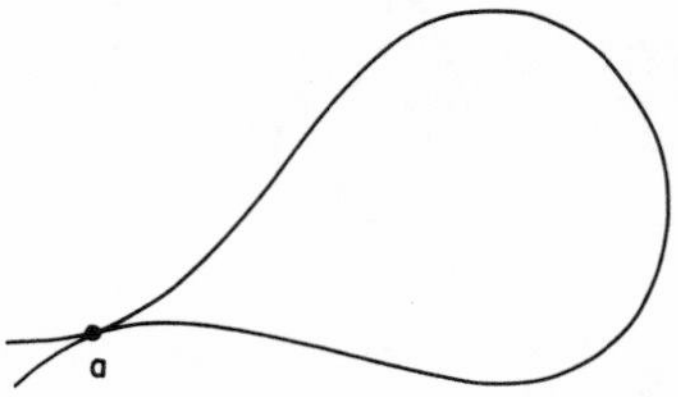

Fig. 8.4. A point of self-tangency

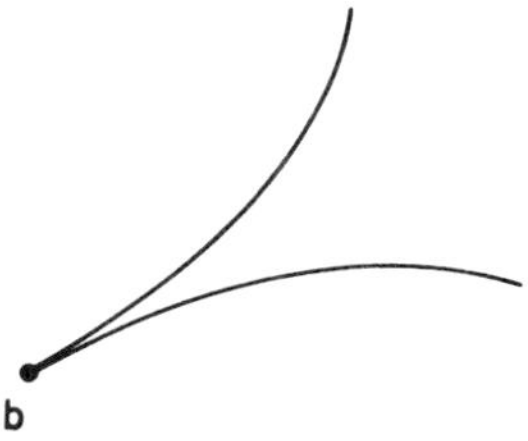

Fig. 8.5. A cusp where $T\varphi$ is not injective

injective but $T\varphi$ is not injective. For example $\varphi:\mathbb{R}\to\mathbb{R}^2$ may have a point such as the one illustrated at a in Fig. 8.4, where the curve is tangent to itself. Another example is illustrated by the cusp at b in Fig. 8.5, where φ is injective, but $T\varphi$ is not.

There is no reason to go into all the possible ways in which a subset can or cannot be endowed with a differentiable structure. The point has been to illustrate some of the more important ways and to induce a certain degree of caution which will be important later, especially in dealing with foliations (Chapter 18).

8.2 SUBMANIFOLDS FROM SUBMERSIONS

Consider a mapping $\varphi:M\to N$, where $\mu>\nu\equiv\dim N$, in which sets of points in M get mapped into one point in N. Under certain conditions these sets of points $\varphi^{-1}(n)\subset M$, $n\in N$, may be submanifolds. For example, let $M=\mathbb{R}^2-\{0,0\}$ and $N=\mathbb{R}^+$, and let φ be defined by $\varphi\colon(x,y)\mapsto r=\sqrt{x^2+y^2}$. Then in an obvious way $\varphi^{-1}(r)$, the circle of radius r, is a regular submanifold of $\mathbb{R}^2-\{0,0\}$.

Let $\varphi:M\to N$ be a *submersion*, which means that $T_m\varphi$ is surjective for all $m\in M$. This implies, incidentally, that $\mu\geqslant\nu$. Then it can be shown as a consequence of the Implicit Mapping Theorem (see AM, p. 49) that a differentiable structure on $\varphi^{-1}(n)$ can be given for each $n\in N$, which makes $\varphi^{-1}(n)$ a submanifold of M whose *codimension* is ν (i.e. whose dimension is $\mu-\nu$). In fact it turns out (Brickell and Clark, 1970, p. 88) that the imbedding $i:\varphi^{-1}(n)\to M$ is regular, so that $\varphi^{-1}(n)$ is a regular submanifold of M, and then, as has been mentioned, this represents the only submanifold structure of dimension $\mu-\nu$ which is possible for $\varphi^{-1}(n)$.

As n is allowed to run through N, many regular submanifolds are obtained in this way, one passing through each $m\in M$. The resulting set of submanifolds is called a *foliation* of M, and will be discussed in some detail in Chapter 18. In the example above, for instance, the circles of radius r foliate $\mathbb{R}^2-\{0,0\}$ as r runs through $\mathbb{R}^+$. Actually, the requirement that $T_m\varphi$ be surjective $\forall m\in M$ globalizes another possibility, namely that $T_m\varphi$ be surjective only for some $m\in M$. Then $\varphi^{-1}(n)$, with $n=\varphi(m)$, would be a manifold, but this would not necessarily lead to a foliation.

We conclude this chapter with some remarks about the dynamical significance of submanifolds. Submanifolds and imbeddings have already arisen in Chapter 7, and they will arise later when not an entire dynamical system is considered, but a part of one, restricted to a subset of the carrier space. If that subset can be given a submanifold structure, then the apparatus of vector fields on differentiable manifolds can be put to use. If the submanifold is regular, then it follows that it is Hausdorff and second countable, but if it is not regular, these properties must be checked. Thus regularity is not an academic matter only. Often submanifolds are defined by functions, through submersion (like the example of the circle in $\mathbb{R}^2 - \{0,0\}$), and the regularity is guaranteed. This will be discussed in Chapter 18. Even later, in Section B of Part II, Lie algebras and Lie groups will play an important role. Then it will become important to know (Brickell and Clark, 1970, p. 215) that a regular submanifold of a Lie group is a Lie subgroup.

BIBLIOGRAPHY TO CHAPTER 8

As a general reference, see Brickell and Clark (1970, Ch. 5). A slightly less technical approach, with several examples, can be found in Chillingworth (1976, Ch. 3).

9

Transformations on TQ

The manifold TQ was constructed in Chapter 5 by putting a vector-bundle structure on the set $\bigcup_q T_q Q$ by means of the natural atlases. But a *differentiable* structure on TQ is something more: it is a *maximal* atlas, in which the transition functions from chart to chart (the *overlap maps*) are required to be C^∞ rather than to be *vector-bundle isomorphisms* (which preserve the vector-bundle structure just as tangent-bundle isomorphisms preserve the tangent-bundle structure; see Chapter 5), like the ones obtained from the natural atlas. In other words, the differentiable structure on TQ is larger than the vector-bundle structure. In fact there are three such structures on TQ that are relevant for present purposes: (a) the differentiable structure just mentioned, in which the overlap maps are diffeomorphisms; (b) a fiber-bundle structure, in which the overlap maps are fiber-preserving diffeomorphisms; (c) a vector-bundle structure, that is, the tangent-bundle structure defined by the natural atlases. These structures, together with their obvious inclusion relations, are possessed also by any other vector bundle.

What is important for tangent bundles is that there is a standard procedure, the algorithm of Chapter 5, which allows one to construct the tangent bundle of any differentiable manifold. In terms of diagrams this can be put as follows: there exists a *tangent functor* T which associates the diagram

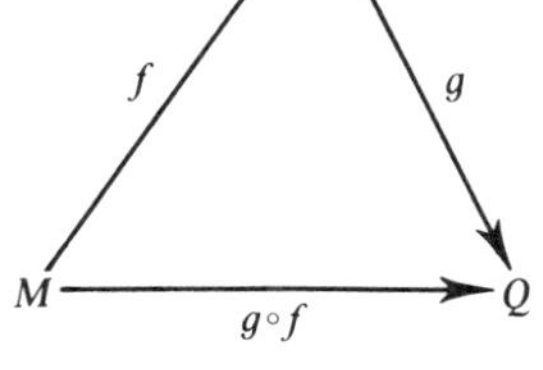

with the diagram

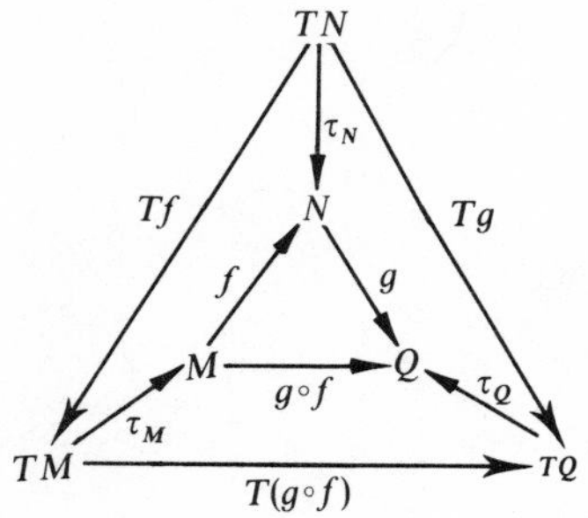

Moreover, $T\mathbb{1}_M = \mathbb{1}_{TM}$. These properties characterize the tangent functor from the *category* of manifolds and smooth mappings to the category of tangent bundles and tangent-bundle mappings. Such a functor is *covariant* (essentially this means that the arrows between two manifolds and those between their tangent bundles point in the same direction in both levels of the diagram).

We proceed now to a discussion of possible transformations on TQ, and of their relation to dynamics, in terms of the three kinds of structure on TQ.

1. Recall Chapter 2, which discussed differentiable manifolds in general and diffeomorphisms on them in particular. The considerations of that chapter apply also to TQ, which is now to be treated as differentiable manifold, and it will be found that general diffeomorphisms of the kind discussed there, which do not take into account the fibered nature of TQ, can be useful for dealing with general symmetry properties of vector fields, including symmetry properties which do not arise from symmetries of the dynamical system on Q. For instance, the Runge–Lenz vector for the Kepler problem is related to such a general symmetry property.

An important example of diffeomorphisms which do not preserve the fibered nature of TQ is the time-development of the system itself. Roughly speaking, one may think of the time evolution of the system in terms of mappings $F_t: TQ \to TQ$ which pick up any point $m_0 \in TQ$ at time $t = 0$ and move it along the integral curves to where it will be at time t later. This view is complicated by the fact that the dynamical vector field Δ may not be complete; that is, its maximal integral curves may not have all of $\mathbb{R}$ as their domain in time (see Example 7.5 and the end of Chapter 5). Thus we introduce the local notion of a *flow box*. (As elsewhere in this chapter, the discussion will be restricted to TQ, but it applies as well to other manifolds. See the first Remark in Chapter 10.)

A flow box for Δ at $m \in TQ$ is a triple (U, a, F), composed of a neighborhood $U \subset TQ$ which is open and contains m, an interval $I_a = (-a, a) \subset \mathbb{R}$ which is defined by $a \in \mathbb{R}$ with $a > 0$ or $a = +\infty$, and a map $F: U \times I_a \to TQ$ which is C^∞ and such that $F_{m'} \equiv F|\{m'\} \times I_a: I_a \to TQ$ is an integral curve of Δ at $m' \in U$. It is also required that $F_t = F|U \times \{t\}: U \to TQ$, where $t \in I_a$, be a diffeomorphism onto the image. A theorem (AM, p. 67) tells us that for each $\Delta \in \mathfrak{X}(TQ)$ there exists a unique flow box at every $m \in TQ$.

This may seem like an unnecessarily complicated definition, but the point is that locally the theorem provides the desired mapping $F_t: TQ \to TQ$ and guarantees that it is a diffeomorphism where defined. The most interesting case is when Δ is complete and F can be extended to all of $TQ \times \mathbb{R}$. Then the map is called the *flow* $F_\Delta: TQ \times \mathbb{R} \to TQ$. It would be nice if it were possible in general to characterize the conditions under which such a flow exists (under which the vector field Δ is complete), but unfortunately no such characterizations exist.

An important property of the flow box is its *group property*, namely

$$F_t F_{t'} = F_{t'} F_t = F_{t+t'}, \quad t, t', t+t' \in I_a. \tag{1}$$

It is seen also that F_0 is the identity. This means that if m, n, p are three points in TQ, going along the flow from m to n and then along the flow from n to p is the same as going along the flow directly from m to p (passing through n). This group property is essentially what was established for the mapping H at Eq. (8), Chapter 6.

When the vector field is complete, what is obtained in this way is a parameter-dependent family of diffeomorphisms from TQ to TQ. That they are not fiber-bundle diffeomorphisms (see below) is evident: two points in TQ in the same fiber over a point $q \in Q$ have different velocities, and thus in general after a time t will arrive over different points in the base space, i.e. be in different fibers.

2. We now turn to the structure of TQ as a fiber bundle. Suppose that (M, π, Q') is a fiber bundle, and consider a mapping $\varphi: TQ \to M$. Then φ is called a C^∞ *fiber-bundle mapping* if there exists a C^∞ mapping $\varphi_0: Q \to Q'$ which enters the diagram

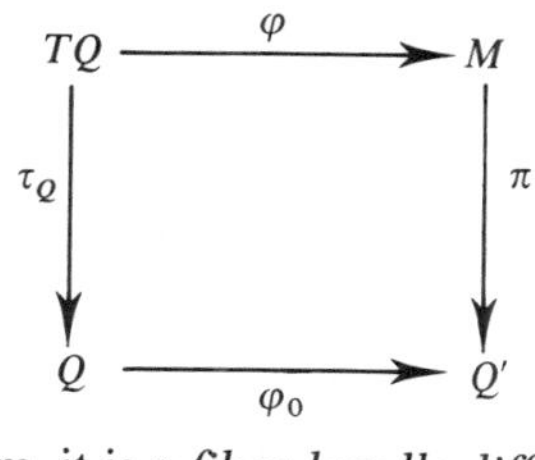

If φ is also a diffeomorphism, it is a *fiber-bundle diffeomorphism*. These mappings are said to *preserve* or *respect* fibers. Finally, for the case in which $Q = Q'$, a fiber-preserving diffeomorphism is called *base invariant* iff it enters into the diagram

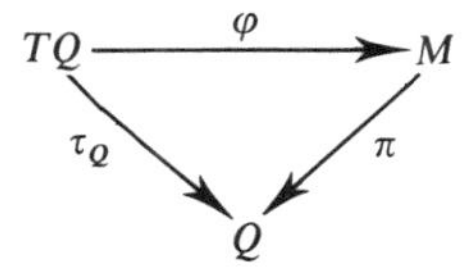

Since M and TQ are differentiable manifolds, it follows from Chapter 2 that the pullback through φ of any atlas for M is an atlas for TQ. If φ is a fiber-bundle mapping, a fiber-bundle atlas for M will be pulled back in this way to a fiber-bundle atlas for TQ.

Often we shall be concerned not with Q so much as with TQ and with the tangent bundle of TQ, namely $T(TQ) \equiv T^2Q$, so it is of interest to know what can be said in addition when the manifold Q is itself a tangent bundle or, more generally, a fiber bundle. Thus consider a fiber bundle (Q, π, N) and its tangent bundle $(TQ, T\pi, TN)$. Then TQ keeps track of the fiber nature of its base manifold Q, and the tangent space T_qQ at each $q \in Q$ has a privileged vector subspace consisting of those vectors which are projected onto zero by $T\pi$. This subspace $\ker(T_q\pi)$ is said to consist of the *vertical* vectors. It can be shown that $T_qQ/\ker(T_q\pi)$ is diffeomorphic to $T_{\pi(q)}N$.

Remark. Recall that the vector fields on TQ, as defined in Chapter 5, are the sections of the vector bundle $(T(TQ), \tau_{TQ}, TQ)$. By applying the tangent functor to (TQ, τ_Q, Q), one can also get the vector bundle $(T(TQ), T\tau_Q, TQ)$, but the sections of this vector bundle are not vector fields on TQ, with the sole exception of those that are of second order.

3. We now turn to the structure of TQ as a tangent bundle. Vector-bundle isomorphisms have been mentioned at the beginning of this chapter. More formally, their definition is the following. If (M, π, Q') is a vector bundle, $\varphi: TQ \to M$ is a C^∞ *vector-bundle mapping* if a φ_0 exists as in the definition of a fiber-bundle mapping, but now φ must be linear on fibers. If, moreover, φ is a diffeomorphism, it is then an isomorphism on fibers, and it is called a *vector-bundle isomorphism.* The map $T\varphi_0: TQ \to TQ'$ is then a tangent-bundle mapping. Every fiber-preserving diffeomorphism $\varphi: TQ \to TQ'$ can be decomposed uniquely into a tangent-bundle part and a fiber-preserving base-invariant part, in accordance with the diagram

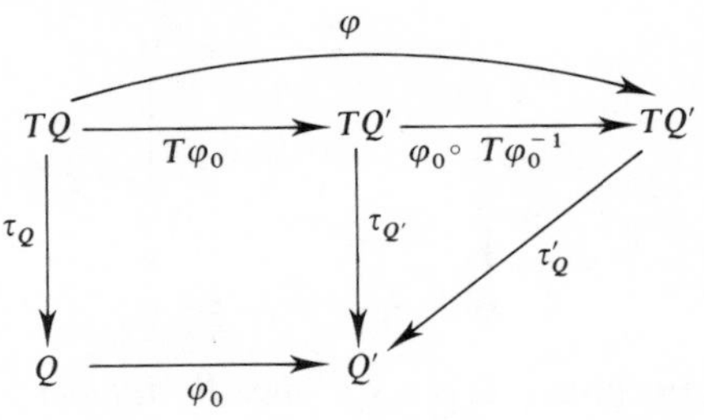

That is,

$$\varphi = (\varphi_0 \circ T\varphi_0^{-1}) \circ T\varphi_0, \tag{2}$$

where φ_0 is the unique base-space diffeomorphism associated with φ. Then $T\varphi_0$ is, of course, the tangent-bundle part, and $\varphi_0 \circ T\varphi_0^{-1}$ is base invariant.

Finally, in view of the special role of second-order vector fields in dynamics, it is convenient to define one more type of diffeomorphism. A diffeomorphism on TQ is called *Newtonian* iff it carries second-order vector fields into second-order vector fields. A link connecting these diffeomorphisms with the kind we have been discussing is given by the following.

Theorem. A diffeomorphism on TQ is Newtonian iff it is the tangent lifting $T\varphi$ of a diffeomorphism $\varphi: Q \to Q$.

Sufficiency is obvious: if $\varphi: Q \to Q$, then $T\varphi$ is Newtonian.

To prove the necessity, observe first that a fiber-preserving base-invariant diffeomorphism is Newtonian iff it is the identity. This is quite clear because such a transformation is necessarily the identity on the base space, and if it is to preserve any relations between base and fiber, it must be the identity also on the fiber. Then in view of Eq. (2), all that needs proof is that a Newtonian diffeomorphism is fiber-preserving. We shall prove this in a natural chart, but the result has global validity. Let the transformed vector field be $X_\varphi = T\varphi \circ X \circ \varphi^{-1}$, and calculate X_φ at the point $\varphi(q, \dot{q}) \equiv (\varphi_0(q, \dot{q}), \varphi_1(q, \dot{q}))$. One obtains

$$X_\varphi(\varphi(q, \dot{q})) = T\varphi \circ X(q, \dot{q}) = T\varphi(q, \dot{q}; X_1(q, \dot{q}), X_2(q, \dot{q}))$$

$$= \left(\varphi_0(q, \dot{q}), \varphi_1(q, \dot{q}); \frac{\partial \varphi_0}{\partial q^j} \dot{q}^j + \frac{\partial \varphi_0}{\partial \dot{q}^j} X_2^j, \frac{\partial \varphi_1}{\partial q^j} \dot{q}^j + \frac{\partial \varphi_1}{\partial \dot{q}^j} X_2^j \right),$$

where use has been made of the fact that X is of second order, i.e. that $X_1^k = \dot{q}^k$. Now, since X_φ is also to be of second order, its first component must be φ_1:

$$\frac{\partial \varphi_0}{\partial q^j} \dot{q}^j + \frac{\partial \varphi_0}{\partial \dot{q}^j} X_2^j = \varphi_1(q, \dot{q}).$$

The transformation, and therefore also φ_1, must be independent of the vector field X, so that

$$\partial \varphi_0 / \partial \dot{q}^k = 0,$$

and φ is thus fiber-preserving. This completes the proof.

Finally, we give an example only of the unique decomposition of a fiber-preserving diffeomorphism into its tangent-bundle and base-invariant parts. Consider $TQ = \mathbb{R}^2$ in the usual natural coordinates, and the diffeomorphism

$$\varphi: \mathbb{R}^2 \to \mathbb{R}^+ \times \mathbb{R}: (q, \dot{q}) \mapsto (e^q, e^{2q} \dot{q}).$$

This decomposes into $\varphi = \chi \circ \eta$, where

$$\eta: \mathbb{R}^2 \to \mathbb{R}^+ \times \mathbb{R}: (q, \dot{q}) \mapsto (e^q, e^q \dot{q})$$

is a tangent-bundle map, the lift of $\varphi_0: \mathbb{R} \to \mathbb{R}: q \mapsto e^q$, and

$$\chi: \mathbb{R}^+ \times \mathbb{R} \to \mathbb{R}^+ \times \mathbb{R}: (q, \dot{q}) \mapsto (q, q\dot{q})$$

is base invariant. In this example $e^{2q}\dot{q}$ could have been replaced more generally by $f(q, \dot{q})$ and $q\dot{q}$ by $f(\log q, \dot{q}/q)$.

10

Integrating the dynamics on TQ. Hamiltonian and Lagrangian formalisms

10.1 INTEGRATING THE DYNAMICS

1. This chapter begins a discussion of some aspects of what will be called *integrating the dynamics*, a subject with which we shall deal in much more detail in Part II. It is assumed now that the vector field which describes the dynamics on TQ is known (we shall often refer to it as the dynamics or the *dynamical vector field*) and this knowledge is now to be used to answer essentially all reasonable questions about the dynamical system. The only way one can be sure that every such question can be answered is to know the integral curves or trajectories on TQ or, equivalently, the base integral curves or trajectories on Q. Thus the dynamical system will be said to have been *solved* or *integrated* when these integral curves have been found.

This does not mean, however, that a complete solution involving the enumeration of essentially all the integral curves is a necessary or even a reasonable procedure for answering any given question. In fact conservation, stability, closed orbits and many other aspects of dynamical systems are much more easily discussed in altogether different terms, which one might call synthetic as opposed to analytic. For example, a study of the qualitative aspects of the vector field or the symmetries of the potential function, if one exists, leads directly to useful results. Moreover, the answers one obtains in such synthetic ways are often more informative and, if not already general enough, more easily generalized than detailed analytical answers. In fact they often suggest new questions leading to important classifications of dynamical systems (e.g. systems with central potentials) which would hardly have been found from the analytical solutions obtained from integration as defined here. This chapter, as will be seen, is largely concerned with such synthetic results, but they are viewed here as steps that can be taken toward integration. It may seem strange that after Δ has been constructed from the trajectory germs, essentially from the integral curves, we are now about to launch into a discussion of how to find the integral curves from Δ. But this procedure becomes reasonable when one realizes that the theoretical analysis starts with the vector field, as has been discussed elsewhere (in the Introduction and in Chapter 6). It is the vector field (in Newtonian terms, the force) which characterizes any physical system; the discussion of Chapter 6 was of how to obtain this characterization from the observed data.

Remark. The discussion here will be restricted to TQ, but many of the results apply to manifolds in general. This means, of course, that they can be used for other carrier spaces (see, for example, Chapters 12 and 14), but also that they can be used on submanifolds of such carrier spaces. This is of particular interest in this chapter, where certain submanifolds play an important role nearly from the start.

The equations for obtaining integral curves from Δ are ordinary differential equations, and any standard book on the subject gives existence and uniqueness theorems which tell one that locally they can always be solved: a local solution passes through each point of TQ, and such solutions can in general be extended to maximal integral curves. Local solutions can also be defined by constants of the motion other than the initial conditions, but in general such definitions will not be valid for the maximal curves, or as is said, not valid globally. To assert more than this is difficult. There are no general rules for calculating global solutions of dynamical systems, and methods are usually ad hoc, though with reasonably wide ranges of applicability. For this reason it is important to discuss types of systems for which some general statements can made, and we now turn to the subclass of dynamical systems admitting symmetries, which is of particular interest for this book.

2. The evolution of a dynamical system can be seen as a family of transformations, parametrized by the time t, from some *initial state* or *condition* $m_0 \in TQ$ to states m_t at other times. Consider a set of initial conditions all of which lie on some submanifold $M \subset TQ$. It may turn out that all of the m_ts arising from such m_0s also lie on M. That is, the integral curve passing through each m_0 lies on M and, as time varies, the system, moving along this curve, remains on M. If such an M can be found, then for the states lying on M the analysis can be restricted to this submanifold, and because it is a submanifold and therefore has differential properties, one can continue to apply the techniques developed for dynamical systems on manifolds. But now the dimension, previously determined by the number of degrees of freedom, is reduced to that of M. This is illustrated in Fig. 10.1.

Such an M will be called an *invariant submanifold*. Formally $M \subset TQ$ is an invariant submanifold for Δ iff

$$\Delta(m) \in T_m M \subset T_m(TQ) \, \forall m \in M. \tag{1}$$

This means, clearly, that the dynamical field Δ is everywhere tangent to M, or that those integral curves which pass through points in M lie in M.

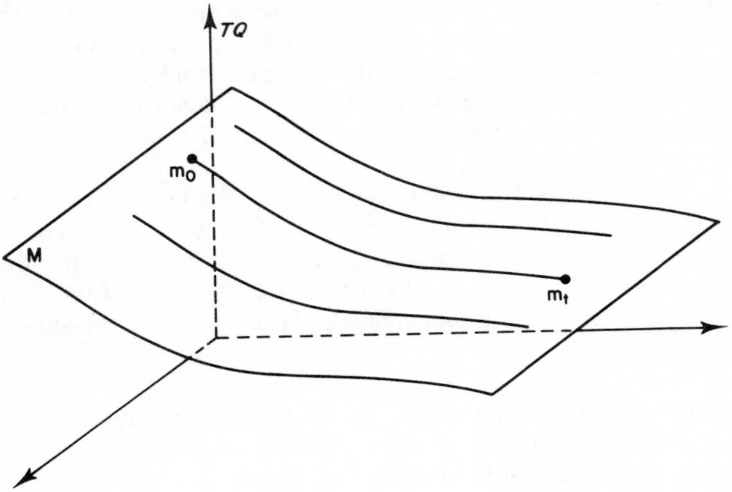

Fig. 10.1. An invariant submanifold, showing some trajectories

In this chapter such manifolds, or rather sets of them, will be defined by functions. To see how a function may define a submanifold, consider the following example. Let $TQ = \mathbb{R}^2$, as in many of the examples of Chapter 7, and consider the function

$$f: TQ \to \mathbb{R}: (q, \dot{q}) \mapsto q^2 + \dot{q}^2 \equiv r. \tag{2}$$

Then $f^{-1}(r)$ is a submanifold for $r > 0$, namely the circle of Cartesian radius $r^{1/2}$. That it is a submanifold follows from the definition (see Chapter 8) because there is obviously a chart in which it is a coordinate surface. For $r = 0$ as well, $f^{-1}(r)$ is a submanifold (the point at the origin), but one which is different from all the others in that it is of dimension zero. We shall return to this point later (see Fig. 10.2).

More generally, then, when can a function be used in this way to define a submanifold of some TQ? The answer is provided by the following. Suppose $f: TQ \to \mathbb{R}$ is of class C^n, $n > 0$. A point $r \in \mathbb{R}$ is called a *regular value* of f iff $T_m f$ is surjective for each $m \in f^{-1}(r)$. Let R_f be the set of regular values of f. Then the inverse function theorem says that if f is of class C^∞ and $r \in R_f$, then $f^{-1}(r)$ is a submanifold of TQ whose codimension is one, i.e. whose dimension is one less that that of TQ. (Here we have specialized to C^∞ because we are concerned with C^∞ functions.)

For instance, in the example of Eq. (2), with $m = (q, \dot{q})$, one obtains

$$T_{q\dot{q}} f = 2q\,dq + 2\dot{q}\,d\dot{q}$$

(there is slight abuse of notation here, for this is what AM, in their definition 8.1, call the *second component*). Then $r > 0$ implies that $(q, \dot{q}) \neq (0, 0)$ and thus that $T_m f: T_m(TQ) \to T_r\mathbb{R}$ is surjective. Therefore $r > 0$ is in R_f, and the theorem tells us

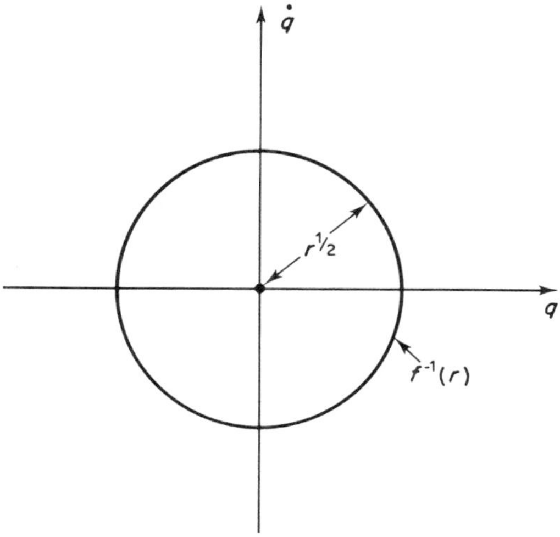

Fig. 10.2. The submanifold $f^{-1}(r)$ for
$f(q, \dot{q}) = q^2 + \dot{q}^2 = r$.

that the circle is indeed a submanifold, as of course we knew already. Concerning $r = 0$, however, the theorem says nothing, for $r = 0$ is not in R_f. Nevertheless, the origin $(q, \dot{q}) = (0, 0)$ is a submanifold, as mentioned above. It is thus seen that the theorem is only a sufficiency test. That $f^{-1}(0)$ is a single point is a reflection of the next observation.

The usefulness of the above technique is seen from the following theorem due to Sard (AM, p. 50). Let M and N be manifolds, and let $\chi: M \to N$ be of class C^∞. Then $\mathbb{R}_\chi$ is dense in N. For instance, if M is TQ and N is $\mathbb{R}^k$, consider

$$\chi \equiv (f_1, \ldots, f_k): TQ \to \mathbb{R}^k.$$

The set of regular values of χ is dense in $\mathbb{R}^k$, and each regular value yields submanifold of TQ. Matters of this kind will be discussed more fully in Chapter 18.

The functions used to define invariant submanifolds are invariant under Δ-evolution, i.e. are *constants of the motion.* A constant of the motion is a function $f \in \mathscr{F}(TQ)$ which does not vary along the integral curves; that is, the integral curves lie in subsets of constant f, of the form $f^{-1}(r)$, where r is a constant. Then it is obvious that if such a subset is a submanifold, it is an invariant one. More formally, $f \in \mathscr{F}(TQ)$ is a constant of the motion iff for $g: I \to TQ$ an integral curve of Δ,

$$d[f \circ g]/dt \equiv df(g(t))/dt = 0, \quad t \in I. \tag{3}$$

This last equation can also be written in the form

$$L_\Delta f \equiv i_\Delta df = 0, \tag{4}$$

and therefore f is a constant of the motion iff Eq. (4) is satisfied.

It may be of some interest to exhibit the equivalence of (3) and (4). Let Δ be written in accordance with Eq. (18) of Chapter 5 in the form

$$\Delta = \Delta_1^k \partial/\partial q^k + \Delta_2^k \partial/\partial \dot{q}^k,$$

and let $w \in TQ$. Then

$$L_\Delta f(w) = \Delta_1^k \partial f/\partial q^k|_w + \Delta_2^k \partial f/\partial \dot{q}^k|_w.$$

Now, there is a unique integral curve g such that $g(t) = w$, and the fact that it is an integral curve means that $\Delta_1^k(w) = dh^k/dt$ and $\Delta_2^k(w) = dh'^k/dt$, where $g(t) = Th(t, 1)$, or $h(t)$ is the base integral curve corresponding to $g(t)$. Thus

$$L_\Delta f(w) = \left[\frac{dh^k}{dt}\frac{\partial f}{\partial q^k} + \frac{dh'^k}{dt}\frac{\partial f}{\partial \dot{q}^k}\right]_w$$

at every point $w \in TQ$ through which there passes an integral curve. But the right-hand side is what is actually meant by $(d/dt)f(g(t))$. Thus (3) and (4) are equivalent.

In this way the problem of finding invariant submanifolds becomes one of finding constants of the motion. We know of no general method for finding such constants, but for certain special kinds of problems it can be shown that certain functions (energy, angular momentum, the canonical conjugates of ignorable coordinates) are constants. Often it is a matter of insight and experience. Sometimes, on the other hand, it is easier to find objects other than functions whose Lie derivatives along Δ vanish, like vector fields or forms. Such objects will be called *invariant structures*, or simply *invariants*, of the dynamics (and then a constant of the motion is an invariant structure which is a function). When they are not functions, such invariant structures can often be contracted with others to form functions, and then these in turn are constants of the motion and can be used to obtain invariant submanifolds.

Consider first vector-field and one-form invariants. For $X \in \mathfrak{X}(TQ)$ and $\theta \in \mathfrak{X}^*(TQ)$, the Lie derivative of their contracted product is

$$L_\Delta(i_X\theta) = i_{[\Delta, X]}\theta + i_X L_\Delta \theta, \tag{5}$$

so that if both $[\Delta, X] \equiv L_\Delta X = 0$ and $L_\Delta \theta = 0$, it is found that $i_X\theta \in \mathscr{F}(TQ)$ is a constant of the motion. In particular, one may choose $X = \Delta$, the dynamical vector field, and then it is seen that if $L_\Delta\theta = 0$, that is, if θ is invariant, so is the function $i_\Delta\theta$. Thus one-form invariants can lead directly to constants of the motion.

If $X, Y \in \mathfrak{X}(TQ)$ and $\omega \in \Omega^2(TQ)$ are all invariants, so is the function $i_X i_Y \omega \equiv -\omega(X, Y)$, but there is nothing more we know of to say about this combination. That is, nothing new is added if one of the vector fields is Δ, and certainly nothing if they both are. Nor do we know of any further advantages to be gained in this way of studying p-forms for any $p > 2$. This is not to say that there is no inherent interest in studying invariant structures other than functions, as will soon be seen, but for the present, contracting invariant structures in this way to obtain constants of the motion seems to provide no further advantage.

3. There is, on the other hand, another way in which a two-form ω can yield a constant of the motion, specifically, when it has the property that

$$i_\Delta \omega = df \tag{6}$$

for some function $f \in \mathscr{F}(TQ)$. Then

$$L_\Delta f \equiv i_\Delta df = \omega(\Delta, \Delta) = 0,$$

and f is a constant of the motion. Note that if ω is *nondegenerate* (this means that $i_X \omega = 0$ implies $X = 0$), then Δ is uniquely determined by f. However, this 'pairing' between f and Δ does not itself help integrate the dynamics. As usual, the constant of the motion can be used only to reduce the dimension of the problem by restricting it to an invariant submanifold. If, however, ω is not only nondegenerate, but closed, i.e. if $d\omega = 0$, even more is obtained: essentially the Hamiltonian formalism on TQ. Even if ω is not closed, moreover, it is sometimes possible to find an $F(f)$, where f satisfies Eq. (6), such that $F(f)\omega \equiv \omega'$ is closed; ω is then said to admit an integration factor (see Flanders, 1969). Then

$$i_\Delta \omega' = i_\Delta[F(f)\omega] = F(f)df \equiv dG(f),$$

and again the Hamiltonian formalism is obtained on TQ.

Remark. It is not clear to us at this writing that a reasonable quasi-Hamiltonian dynamics could not be constructed without the requirement that ω be closed.

An ω which is closed and nondegenerate is called a *symplectic form.* As will now be explained, the existence of such a symplectic form adds a Lie-algebra structure to the differential dynamics under discussion. The discussion which follows refers to TQ, but it is valid for any even-dimensional manifold whatsoever. The essential ingredient is the symplectic form (see AM, p. 165).

Let ω be a nondegenerate two-form on TQ, and $X_g \in \mathfrak{X}(TQ)$ be the (unique) vector field associated with the function $g \in \mathscr{F}(TQ)$ according to

$$i_{X_g} \omega = dg. \tag{7}$$

Then for every pair (g, h) of functions, the *Poisson bracket* of g and h may be defined by

$$\{g, h\} \equiv \omega(X_g, X_h) = -\{h, g\}. \tag{8}$$

It can be shown (Marmo and Simoni, 1976) that if Δ satisfies (6), then

$$L_\Delta\{h, g\} = \{L_\Delta h, g\} + \{h, L_\Delta g\} \tag{9}$$

if and only if ω is closed, and thus for such ω if g and h are constants of the motion, so is $\{g, h\}$. Moreover, the Poisson bracket then satisfies the *Jacobi identity*

$$\{g, \{h, f\}\} + \{h, \{f, g\}\} + \{f, \{g, h\}\} = 0 \ \forall f, g, h \in \mathscr{F}(TQ). \tag{10}$$

Remark. There are other ways to define the Poisson bracket, and for completeness we present one of them here. This method does not require vector fields, and its generalizations will therefore differ from generalizations based upon the first definition (Marmo, Saletan, Simoni and Zaccaria, 1981). Let the dimension of TQ be $\mu = 2\nu$. Both $\Omega^{(\nu)} = \omega \wedge \omega \wedge \ldots \wedge \omega$ (ν factors) and $\Omega^{(\nu-1)} \wedge dg \wedge dh$ are 2ν-forms on TQ. Since the set of 2ν-forms on TQ is of dimension one and since $\Omega^{(\nu)} \neq 0$, there exists a function, depending of course on g and h, which connects the two 2ν-forms. This function is the Poisson bracket:

$$\Omega^{(\nu-1)} \wedge dg \wedge dh = \{g, h\}\Omega^{(\nu)}. \tag{11}$$

4. An example of all this is the simple harmonic oscillator in one degree of freedom (with angular frequency $\omega = 1$). As is well known (twice) the total energy $f = q^2 + \dot{q}^2$ is a constant of the motion, and the invariant submanifolds associated with this constant have been discussed at Eq. (2). The dynamical vector field is

$$\Delta = \dot{q}\partial/\partial q - q\partial/\partial\dot{q}.$$

That is, the force function is $-q$. Now consider the restriction of this vector field to the invariant manifold $f^{-1}(r)$, $r > 0$. To do so, one may choose an atlas whose charts have the invariant submanifolds as one set of coordinate surfaces; in this case polar coordinates will serve. Then one may write (this is one of the charts; the simple task of completing the atlas is left to the reader)

$$\begin{aligned} q &= r^{1/2}\cos\varphi, \quad r = q^2 + \dot{q}^2, \\ \dot{q} &= r^{1/2}\sin\varphi, \quad \varphi = \arctan(\dot{q}/q). \end{aligned} \tag{12}$$

To calculate Δ one needs

$$\frac{\partial}{\partial q} = \frac{\partial r}{\partial q}\frac{\partial}{\partial r} + \frac{\partial \varphi}{\partial q}\frac{\partial}{\partial \varphi},$$

etc., which yields

$$\Delta = -\partial/\partial\varphi. \tag{13}$$

Then to integrate, let $g(t)$ be an integral curve with coordinates $\{r(t), \varphi(t)\}$, and calculate $\Delta(g(t)) = Tg(t, 1)$. For this purpose, recall that although $Tg(t, 1) = (g(t), g'(t))$, the $g(t)$ on the right-hand side is generally suppressed in the writing. In other words, the two components, dr/dt and $d\varphi/dt$, of $dg(t)/dt$ need only be set equal to the two components of Δ which are exhibited in Eq. (13) (the first, that is the r-component, is zero). This yields

$$dr/dt = 0, \quad d\varphi/dt = -1.$$

Thus there is only one integration to perform, which gives $\varphi = -t + \delta$, where δ is an arbitrary constant. The result can also be written in the form

$$q = r^{1/2}\cos(t - \delta), \quad \dot{q} = -r^{1/2}\sin(t - \delta).$$

This shows how the knowledge of a constant of the motion and its associated invariant submanifold can be used to simplify the integration.

This harmonic oscillator can be used to illustrate also how an invariant one-form can lead to a constant of the motion. Consider the one-form

$$\theta = q d\dot{q} - \dot{q} dq.$$

A simple calculation shows that $L_\Delta \theta = 0$, that is, that θ is an invariant. Then $i_\Delta \theta$ should be a constant of the motion according to Eq. (5), and indeed

$$i_\Delta \theta = -q^2 - \dot{q}^2 = -f.$$

An invariant vector field other than Δ is

$$X = q\partial/\partial q + \dot{q}\partial/\partial \dot{q},$$

and then $i_X \theta$ should be a constant of the motion. Another simple calculation shows that $i_X \theta = 0$, which is trivially a constant of the motion, so trivially, in fact, that it is of no help in integrating the dynamics.

It is not surprising, incidentally, that in this simple example one keeps coming up with f as essentially the only nontrivial constant of the motion. In our definition a constant of the motion is in $\mathscr{F}(TQ)$ and therefore can not depend explicitly on the time, and it follows that a dynamical system with one degree of freedom can have at most one independent constant of the motion. In this example that constant is the energy.

10.2 SIMPLE REDUCTION OF HAMILTONIAN DYNAMICS

An invariant submanifold, found (as in §10.2) with the help of a constant of the motion, serves as a carrier manifold of lower dimension for the dynamical system: if the initial manifold TQ (or, more generally, M) has dimension $\mu = 2\nu$, the submanifold has dimension $2\nu - 1$. This transition from TQ to the submanifold represents the first step, the first example, of what is to become one of the enduring themes of this book, the *reduction* of dynamical systems to what one may hope are simpler ones of lower dimension. Various properties of the initial system, such as (but not only) symmetries, add power to even the most elementary reduction process, allowing reduction by more than one dimension at a time. One such property is that of being *Hamiltonian.* This section is devoted, then, to Hamiltonian systems on TQ, and it includes an introduction to the *Hamiltonian formalism* on TQ. Later sections develop this theme further, and this will eventually lead to a discussion of the *Lagrangian formalism.* It should be clear that this is not the usual order of presentation, which starts with the Lagrangian formalism on TQ and arrives at the Hamiltonian one, but on T^*Q. We, too, will eventually arrive there, but that will be in the next chapter, Chapter 11.

1. Consider a dynamical system Δ for which there exists a nondegenerate (antisymmetric) two-form ω on TQ and a function $E \in \mathscr{F}(TQ)$ such that

$$i_\Delta \omega = dE. \tag{14}$$

Such a dynamics is said to be *Hamiltonian with respect to* ω (briefly ω-

Hamiltonian), and E is called its *Hamiltonian function* (briefly its *Hamiltonian*). For reasons that will become evident after Eq. (18) (and were hinted at around Eq. (9)), ω will be assumed closed. Now suppose that a constant of the motion is known for this dynamical system, so that it can be *reduced* to an invariant submanifold. There is, as will be seen, more that can be done with this constant because Δ is ω-Hamiltonian. Let (U, φ) be a local chart for TQ. Since $\dim(TQ) = 2\nu$ is even, $\varphi(m)$ for $m \in U$ can be written in the form $\varphi(m) = (y^k, z^k)$, $k \in \{1, 2, \ldots, \nu\}$. In this way the local coordinates are divided into two subsets of ν coordinates each, but no assumption is made that these coordinates are natural in any sense. Local expressions for Δ and ω are then

$$\Delta = g^k \partial/\partial y^k + h^k \partial/\partial z^k \tag{15}$$

and

$$\omega = a_{jk} dy^j \wedge dz^k + \tfrac{1}{2}(b_{jk} dy^j \wedge dy^k + c_{jk} dz^j \wedge dz^k), \tag{16}$$

where the $g^k, h^k, a_{jk}, b_{jk} = -b_{kj}, c_{jk} = -c_{kj}$ are all in $\mathscr{F}(TQ)$. Then Eq. (14) can be written, in U, in the form

$$\begin{aligned} a_{kj}h^j + b_{kj}g^j &= -\partial E/\partial y^k, \\ a_{jk}g^j - c_{kj}h^j &= \partial E/\partial z^k. \end{aligned} \tag{17}$$

Recall that the integral curves of Δ are given by $dy^k/dt = g^k$ and $dz^k/dt = h^k$. Thus one may write

$$\begin{aligned} a_{kj}dz^j/dt + b_{kj}dy^j/dt &= -\partial E/\partial y^k, \\ a_{jk}dy^j/dt - c_{kj}dz^j/dt &= \partial E/\partial z^k. \end{aligned} \tag{18}$$

Eqs. (18) may be considered the equations for the integral curves of Δ in terms of ω and E: if this two-form and this function are known *a priori*, then Eqs. (18) can be used as the starting point for integrating the dynamics. These equations are similar to those of the usual Hamiltonian formalism, for if the b_{jk} and c_{jk} all vanish and if $a_{jk} = \delta_{jk}$, they result in Hamilton's canonical equations, but on TQ rather than on phase space T^*Q (the cotangent bundle; recall Chapter 5). In fact this similarity is quite real, for there is an important theorem by Darboux (1882) which says that if ω is closed, there exist *symplectic charts* in which (18) take on the simple form of the canonical equations. More precisely, if ω is, as assumed, a closed nondegenerate two-form (it is then called a *symplectic form*), there exist local charts in which the integral curves of Δ satisfy the differential equations

$$\begin{aligned} dz_k/dt &= -\partial E/\partial y^k, \\ dy^k/dt &= \partial E/\partial z_k, \end{aligned} \tag{19}$$

where $z_k = \delta_{kj} z^j$.

Remark. Note that from Eq. (14) it follows that $L_\Delta \omega = 0$ when ω is closed. The converse, however, is not true. As Δ is called ω-Hamiltonian when (14) is satisfied, it is called *locally ω-Hamiltonian* when $L_\Delta \omega = 0$. It can be shown that the locally ω-Hamiltonian vector fields at any point m of a manifold M (generalizing here from TQ to M) span T_mM.

2. It will now be shown that if Δ is ω-Hamiltonian (we emphasize again that ω is a symplectic form, i.e. closed and nondegenerate), a constant of the motion has a sort of double value: it can be used to reduce the dynamics not by one, but by two dimensions.

Let f be a constant of the motion for Δ, and let a be a regular value of f. Let $N_a = f^{-1}(a)$ be the invariant submanifold defined by the value a and let X be the vector field uniquely defined by the equation

$$i_X\omega = df. \tag{20}$$

Note that then X is also ω-Hamiltonian.

Since f is a constant of the motion, $L_\Delta f = 0$. It follows in addition from (20) that

$$L_X f \equiv i_X df = 0, \tag{21}$$

which means that the integral curves not only of Δ, but also of X lie on the N_a submanifolds, as in Fig. 10.3. Moreover,

$$[\Delta, X] = 0. \tag{22}$$

Indeed,

$$i_{[\Delta,X]}\omega = L_\Delta i_X\omega - i_X L_\Delta\omega = dL_\Delta f - i_X L_\Delta\omega = -i_X L_\Delta\omega.$$

But $L_\Delta\omega = i_\Delta d\omega + di_\Delta\omega = d(dE) = 0$ because ω is closed. Equation (22) follows because ω is nondegenerate.

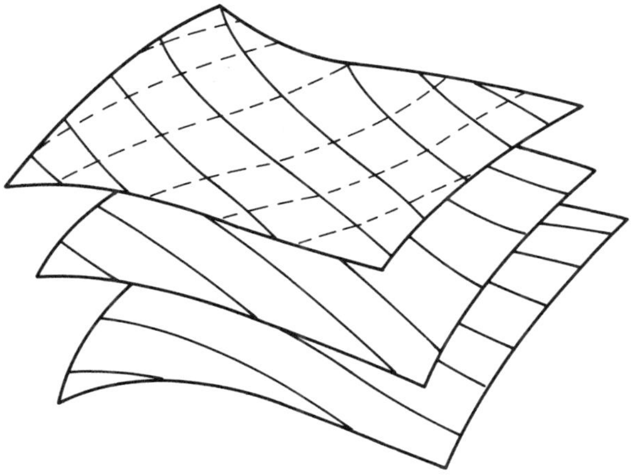

Fig. 10.3. Invariant submanifolds N_a for three different values of a. The solid curves on the N_a are integral curves of Δ. The dotted curves one the upper one of the N_a are integral curves of X

Now N_a is of the odd dimension $2\nu - 1$, which means that any two-form on it must be degenerate. (To see this, let α be any two-form on N_a, and consider its antisymmetric matrix in any local chart; the determinant of this matrix must necessarily vanish.) This means that there can be no Hamiltonian vector fields in $\mathfrak{X}(N_a)$, because there are no symplectic forms on N_a. The vector field Δ is everywhere tangent to N_a and can therefore be thought of as a vector field in $\mathfrak{X}(N_a)$, and it is seen that on N_a it is therefore not Hamiltonian. For the fixed value a of f, then, although one can think of the problem of integrating the dynamics as a problem no longer on TQ, but rather on N_a, the new reduced problem is not Hamiltonian. On the other hand, from N_a one can construct another manifold one dimension lower, and on this manifold Δ, or rather a part of Δ, turns out to be Hamiltonian. It is in this sense that a constant of the motion can be used to reduce a Hamiltonian dynamical system by two dimensions, instead of only one; moreover, the reduced dynamics is also Hamiltonian.

The construction proceeds as follows. Consider the equivalence relationship on N_a defined by the integral curves of X: two points $n_1, n_2 \in N_a$ are equivalent if they lie on the same maximal integral curve of X. It is assumed that at each point of N_a where $X \neq 0$ there exists a neighborhood U such that no integral curve of X passes more than once through U, and that the quotient of N_a with respect to the equivalence is a manifold. These and other similar technical assumptions will be discussed in some detail in Part II, where f will be replaced by a more general mapping and use will be made of the fact that the equivalence relationship *foliates* N_a (see Chapter 18). The dimension of the quotient manifold is then $2\nu - 2$, and we now assert that Δ, considered as a vector field on N_a, *projects* with respect to the equivalence relationship onto a Hamiltonian vector field on the quotient manifold.

The explanation and proof of this assertion will proceed in two steps. First it will be shown that Δ defines a unique vector field, called its projection, on the quotient manifold, and second it will be shown that the projected vector field is Hamiltonian with respect to a symplectic form suggested in a natural way by ω.

Let π be the projection with respect to the equivalence relationship described above. That is $\pi n = [n], n \in N_a$, where $[n]$ is the equivalence class of points in N_a all of which lie on the integral curve of X which passes through n. Let the image of this projection be called $M_a \equiv \pi N_a$. Then the vector field $Y \in \mathfrak{X}(N_a)$ is said to be *projectable with respect to π onto the vector field $\tilde{Y} \in \mathfrak{X}(M_a)$* if a vector field $\tilde{Y} \in \mathfrak{X}(M_a)$ exists such that the following diagram commutes:

$$\begin{array}{ccc} TN_a & \xrightarrow{T\pi} & TM_a \\ \uparrow{\scriptstyle Y} & & \uparrow{\scriptstyle \tilde{Y}} \\ N_a & \xrightarrow[\pi]{} & M_a \end{array} \tag{23}$$

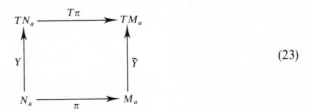

When $\tilde{Y}$ exists it is called the *projection* of Y. This diagram can be used to define projectability of a vector field also more generally: π may be the projection onto the quotient manifold with respect to an equivalence relationship other than that obtained from a vector field $X \in \mathfrak{X}(N_a)$. In the discussion which follows it is so interpreted.

Another definition of projectability, equivalent to that of (23), is the following. Let $\tilde{g} \in \mathscr{F}(M_a)$, and define

$$g = \tilde{g} \circ \pi \in \mathscr{F}(N_a). \tag{24}$$

Then g is a function which is constant on equivalence classes (in our particular case, constant along the integral curves of X). Moreover, any function $g \in \mathscr{F}(N_a)$ which is constant on equivalence classes can be written in the form of (24) with some $\tilde{g} \in \mathscr{F}(M_a)$. Then $Y \in \mathfrak{X}(N_a)$ is said to be projectable with respect to π if for every $\tilde{g} \in \mathscr{F}(M_a)$ there exists a $\tilde{g}' \in \mathscr{F}(M_a)$ such that

$$g' \equiv L_Y g \equiv L_Y(\tilde{g} \circ \pi) = \tilde{g}' \circ \pi. \tag{25}$$

The equivalence of (23) and (25) can be demonstrated by noting first that (25) defines a derivation and hence a vector field $Y' \in \mathfrak{X}(M_a)$ [see the Remark which follows Eq. (30), Chapter 5]. Indeed, let $\tilde{k} = \tilde{g}\tilde{h}$, with $\tilde{g}, \tilde{h}, \tilde{k} \in \mathscr{F}(M_a)$. Then

$$\begin{aligned} \tilde{k}' \circ \pi \equiv k' = L_Y k = L_Y(gh) &= (L_Y g)h + g(L_Y h) \\ &= g'h + gh' = (\tilde{g}'\tilde{h} + \tilde{g}\tilde{h}') \circ \pi. \end{aligned}$$

Thus there exists a Y' as asserted, such that $\tilde{g}' = L_{Y'}\tilde{g}$ for all $\tilde{g} \in \mathscr{F}(M_a)$. What remains to show is that $Y' = \tilde{Y}$. This follows immediately:

$$L_{Y'}\tilde{g} \circ \pi = \tilde{g}' \circ \pi = g' = L_Y g = i_{\tilde{Y}} dg \circ \pi = L_{\tilde{Y}}\tilde{g} \circ \pi, \forall \tilde{g} \in \mathscr{F}(M_a).$$

Summarizing, a vector field $Y \in \mathfrak{X}(N_a)$ is projectable with respect to $\pi : N_a \to M_a$ (or with respect to the equivalence relationship for which π is the projection onto the quotient manifold; or even, as we shall say for this particular case, with respect to the vector field X) iff the Lie derivative with respect to Y carries functions constant on the equivalence classes into other functions constant on the equivalence classes, these equivalence classes being $\pi^{-1}(m), m \in M_a$.

Now consider our particular case. Let Y be projectable with respect to X and let $g \in \mathscr{F}(N_a)$ be any function constant along integral curves of X, so that $L_X g = 0$. Then $g' = L_Y g$ is also constant along integral curves of X, or

$$0 = L_X g' = L_X L_Y g = L_{[X,Y]} g - L_Y L_X g = L_{[X,Y]} g. \tag{26}$$

Therefore a necessary and sufficient condition that Y be projectable with respect to X (recall that g is an arbitrary function such that $L_X g = 0$) is that

$$[X, Y] = hX, h \in \mathscr{F}(N_a). \tag{27}$$

In particular, a sufficient condition is that X and Y commute, and thus it follows from (22) that Δ is projectable. This completes the first step in proving the assertion.

The second step requires showing that the projected field is Hamiltonian with respect to an appropriately chosen symplectic form on M_a. This symplectic form will be obtained from the form ω on TQ by passing through N_a: let ω_a be the degenerate two-form on N_a defined as follows. Consider any two vector fields $Z_1, Z_2 \in \mathfrak{X}(N_a)$. Every vector field on N_a can be thought of as coming from at least one vector field on TQ: if Z is a vector field on TQ which is tangent to N_a, then at any point $n \in TQ$ which lies on N_a the vector $Z(n)$ is in $T_n N_a$. In this way every vector field on N_a can be identified with at least one vector field on TQ which is tangent to N_a. Then ω_a can be defined by

$$\omega_a(Z_1, Z_2) = \omega(Z_1, Z_2)|_{N_a}, \tag{28}$$

where we have committed an abuse of notation, for on the left-hand side the vector fields are on N_a and on the right-hand side they are on TQ. Now, as has been pointed out, ω_a is degenerate. Recall that the kernel of a two-form α on a manifold M is defined (see Chapter 5, §5.3(1)) by

$$\ker \alpha = \{X \in \mathfrak{X}(M) \vdash \alpha(X, Y) = 0 \ \forall \ Y \in \mathscr{F}(M)\}.$$

The kernel of ω_a is the set of all vector fields parallel to X. Indeed,

$$i_Y i_X \omega_a = (i_Y df)|_{N_a} = i_Y(df|_{N_a}) = 0 \ \forall \ Y \in \mathfrak{X}(N_a),$$

for N_a is characterized by $df = 0$. Moreover, since N_a is of dimension $2\nu - 1$ and ω itself is nondegenerate, $\ker \omega_a$ must be of dimension one, and thus

$$\ker \omega_a = \{hX \vdash h \in \mathscr{F}(N_a)\}.$$

Thus ω_a has been defined and its degeneracy exhibited.

A symplectic form Ω_a will now be defined on M_a by using ω_a and by projecting vector fields from N_a onto M_a. Let $\tilde{Y}, \tilde{Z} \in \mathfrak{X}(M_a)$; then Ω_a is defined by

$$\pi^*[\Omega_a(\tilde{Y}, \tilde{Z})] = \omega_a(Y, Z), \tag{29}$$

where $Y, Z \in \mathfrak{X}(N_a)$ project onto $\tilde{Y}, \tilde{Z}$. This is a valid definition of a symplectic form on M_a only if it can be shown (i) that every vector field on M_a can be obtained by projection from a vector field on N_a, and (ii) that Ω_a is in fact a symplectic form (i.e. is closed and nondegenerate).

First we turn to a proof of (i). Consider the set of vector fields $Y \in \mathfrak{X}(N_a)$ such that $L_Y \omega_a = 0$. It can be shown (this is related to the last Remark of §10.2(1)) that such vector fields span $T_n N_a$ at each point $n \in N_a$. Moreover, as will now be shown, they are all projectable, and then dimensional considerations can be used to establish that they project at every point onto the relevant tangent space in M_a. Suppose that Y is such a vector field. Then

$$i_{[X,Y]}\omega_a = L_Y i_X \omega_a - i_X L_Y \omega_a = 0,$$

so that $[X, Y] = hX$, $h \in \mathscr{F}(N_a)$, and therefore Y is projectable. Since these fields span $T_n N_a$ at each point $n \in N_a$, their dimension is $2\nu - 1$. The kernel of the projection at each point n is just the set of vectors parallel to $X(n)$ and hence the dimension of the projected set is $2\nu - 2$, which is just the dimension of $T_m M_a$, where $m = \pi(n) \in M_a$. Thus the set of projected vectors spans $T_m M_a$ at each point

m, and hence every vector field $\tilde{Y} \in \mathfrak{X}(M_a)$ can be obtained by projection from a vector field $Y \in \mathfrak{X}(N_a)$.

Remark. A more general result is obtained by Brickell and Clark (1970, p. 126). They show that every vector field on a quotient manifold of a paracompact manifold can be obtained in this way by projection. In other words, this completeness property with respect to projection does not depend on the existence of a symplectic form.

To prove (ii) we note first that closure of Ω_a follows immediately from its definition (29). Nondegeneracy may be seen as follows. If $\Omega_a(\tilde{Y}, \tilde{Z}) = 0$, then $\omega_a(Y, Z) = 0$, and hence Y or Z (or both) are in $\ker \omega_a$. But $\ker \omega_a = \{hX\}$, and hence Y or Z (or both) project to the null vector field: $\tilde{Z} = 0$ or $\tilde{Y} = 0$, as follows from the second definition of projectability (that of Eqs. (24) and (25)).

What remains is to show that $\tilde{\Delta}$ is Hamiltonian:

$$\begin{aligned} \pi^*(i_{\tilde{\Delta}}\Omega_a) &= i_\Delta \pi^* \Omega_a = i_\Delta \omega_a = i_\Delta \omega|_{N_a} \\ &= (dE)|_{N_a} = d(E|_{N_a}) = \pi^* d\tilde{E}, \end{aligned} \tag{30}$$

where $E|_{N_a} = \tilde{E} \circ \pi$ because $E|_{N_a}$ is constant along the integral curves of X. Thus it follows that

$$i_\Delta \Omega_a = d\tilde{E}. \tag{31}$$

3. It is now seen that a constant of the motion f for a Hamiltonian dynamical system does double work: it can be used to reduce the dimension not by one, but by two, from $\mu = 2\nu$ to $\mu - 2$, and the reduced dynamical system is again Hamiltonian, and with respect to a symplectic form obtained from the original one. In the process, however, a piece of Δ has been lost. That is, $\tilde{\Delta}$ is not the dynamical vector field Δ on N_a (and certainly not on TQ), but only the projection of it onto M_a, and even when the new, reduced dynamical system is integrated, only part of the problem is solved. This can be seen if f is taken as one of the coordinates in a chart, so that locally

$$dE = (\partial E/\partial f)df + (\partial E/\partial \xi^k)d\xi^k, \tag{32}$$

where $\xi^1, \ldots, \xi^{2\nu-1}$ are the other coordinates. Then $dE|_{N_a}$ is just the second term in (32), the sum over k, and thus not all of dE appears in (31).

We now describe this double reduction process briefly in local coordinates. Consider a local chart (U, φ), and assume that $df \neq 0$ in U, where f is still the constant of the motion under discussion. Then as in Eq. (32), f can be chosen as the first coordinate, this time of a symplectic chart (whose properties were described above Eq. (19); it follows from a detailed proof of Darboux's theorem that the first coordinate is arbitrary). One can then write

$$\omega = -df \wedge dw + \Omega. \tag{33}$$

A further consequence of the theorem is that the coordinate w can be chosen such that $d(L_\Delta w) \wedge df = 0$, and such that Ω depends in no way on f, w, df, or dw, is (of course) closed, and is nondegenerate when restricted to the rest of the local coordinates.

For the vector field X defined in Eq. (20), write

$$X = X_f \partial/\partial f + X_w \partial/\partial w.$$

It then follows that $X_f = \tilde{X} = 0$ and $X_w = 1$, so that w is the variable which parametrizes the integral curves of X. Eq. (22) then implies that Δ is independent of w, as is E (for just as $L_\Delta f = 0$, so $L_X E = 0$). Then a simple calculation shows that Δ can be written in the form $\Delta = \Delta_f \partial/\partial f + \tilde{\Delta}$, where

$$i_{\tilde{\Delta}}\Omega = dE|_{f=\text{const.}} \tag{34}$$

This may be recognized as a restatement in local coordinates of Eq. (31); the $\tilde{\Delta}$ that appears there is essentially the same as the one that appears here, Ω can be thought of as a two-form on the (local) space of the remaining $2\nu - 2$ coordinates (it is then what was previously called Ω_a), and $E|_{f=\text{const.}}$ is $E\restriction_{N_a}$.

4. This concludes the discussion of simple reduction of a Hamiltonian system using a single constant of the motion. As has been mentioned, this is the first treatment of reduction, one of the main themes of this book. The 'double value' possessed by constants of the motion of Hamiltonian systems will play a role later. For a detailed illustration of the reduction procedure described above, the reader is referred to the example at the beginning of Chapter 17.

10.3 EXISTENCE OF SYMPLECTIC FORMS ON TQ

1. If there exists a symplectic form ω on TQ and if Δ satisfies Eq. (14) for some function $E \in \mathscr{F}(TQ)$, then it can be said that the Hamiltonian formalism has been established for Δ on TQ. There are, on the other hand, vector fields on TQ which, though they may not satisfy (14), nevertheless satisfy the condition $d(i_\Delta \omega) = 0$. Such vector fields are what was called locally Hamiltonian in the Remark at the end of §10.2(1), for the closure of ω implies that

$$L_\Delta \omega = 0. \tag{35}$$

For an example of a locally, but not globally Hamiltonian vector field, let $Q = S$ be the circle, so that $TQ = S \times \mathbb{R}$ is the cylinder. The coordinates on TQ can be chosen as in Fig. 10.4, and then $\omega = d\varphi \wedge dz$ is a symplectic form on TQ. Although φ is not globally defined, $d\varphi$ and of course dz are, and therefore so is ω. Now consider the two vector fields $X_1 = f(z)\partial/\partial\varphi$ and $X_2 = \partial/\partial z$. These are both locally Hamiltonian, for

$$i_{X_1}\omega = f(z)dz = d\left[\int f(z)dz\right],$$

and

$$i_{X_2}\omega = -d\varphi.$$

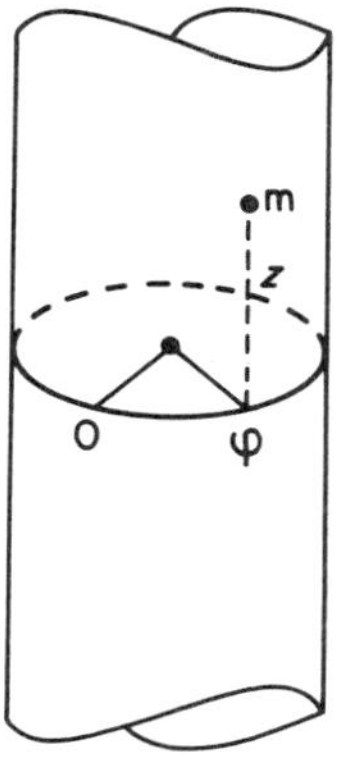

Fig. 10.4. A symplectic chart on the cylinder $S \times \mathbb{R}$. The coordinates of m are (z, φ)

Both of these fields satisfy (35), but X_2 does not satisfy (14) because φ is not globally defined. Thus X_1 is globally Hamiltonian, while X_2 is only locally Hamiltonian.

2. So far our discussion of the Hamiltonian formalism has proceeded as though it had been established that symplectic forms must necessarily exist on TQ. This is not obvious, however, and we now go on to a constructive proof of their existence.

Symplectic forms will be constructed on TQ by starting with functions. It is easy to construct a one-form out of a function on any manifold M by applying the exterior derivative

$$d:\Omega^k(M) \to \Omega^{k+1}(M): \alpha \mapsto d\alpha$$

with $k = 0$, so that $\Omega^k(M) = \mathscr{F}(M)$. But this construction cannot be applied twice, for $d^2 = 0$, and thus another way must be found to construct a two-form from a function. One approach would be to look for another homomorphism from Ω^k to Ω^{k+1} which could be used twice or perhaps once with the exterior derivative. Of course there could be other approaches entirely, but the construction presented below is of this kind. It is specific to $M = TQ$ and lies at the very basis of the usual Lagrangian formalism.

The C^∞ mapping $d_v:\Omega^k(TQ) \to \Omega^{k+1}(TQ)$ that is used for this purpose is called the *vertical derivative*. It will not be defined here globally or in all generality; that is done in the Appendix to this chapter. Here, because it will be applied only to functions, we give its action only on functions, and locally:

$$d_v f = (\partial f / \partial \dot{q}^k) dq^k. \tag{36}$$

Then d and d_v are used to define a one-form and a two-form in terms of f through

$$\theta_f = d_v f, \quad \omega_f = -d\theta_f. \tag{37}$$

In local coordinates ω_f is then given by

$$\omega_f = \frac{\partial^2 f}{\partial \dot{q}^k \partial \dot{q}^j} dq^k \wedge d\dot{q}^j + \frac{\partial^2 f}{\partial \dot{q}^k \partial q^j} dq^k \wedge dq^j. \tag{38}$$

The minus sign in Eq. (37) follows the convention adopted by AM.

It is clear that ω_f is closed, but in order to be a symplectic form it must also be nondegenerate. When it is, f is called a *regular* function; from (38) it is seen that in local coordinates f is regular iff

$$\det|\partial^2 f/\partial\dot{q}^k\partial\dot{q}^j| \neq 0; \tag{39}$$

this determinant is called the *Hessian* of f. Thus a symplectic form can always be constructed from a regular function $f\in\mathscr{F}(TQ)$.

Remark. Since our manifolds are paracompact, it is always possible to define a partition of unity on them and therefore a Riemannian metric on Q which can be used to construct a regular (in fact hyperregular; see Chapter 11) function on TQ. For more details, see AM, especially pp. 127, 223.

Two special cases of this construction of ω_f are worth mentioning. First, if Q is of dimension one, the second term in Eq. (38) vanishes. Second, although the natural atlas is not in general symplectic for ω_f, it is seen that it will be symplectic if f is of the form

$$f(q,\dot{q}) = \delta_{kj}\dot{q}^k\dot{q}^j + V(q),$$

which is a familiar form for Lagrangian functions.

3. Suppose now that a dynamical vector field Δ is given and a function f has been used to construct a symplectic form ω_f on TQ. Then in accordance with the definition, Δ is ω_f-Hamiltonian iff there exists a function $E_f\in\mathscr{F}(TQ)$ such that

$$i_\Delta\omega_f = dE_f, \tag{40}$$

or in local coordinates, with

$$\Delta = \Delta_q^k\partial/\partial q^k + \Delta_{\dot{q}}^k\partial/\partial\dot{q}^k,$$

iff the equations

$$\frac{\partial E_f}{\partial q^k} = -\frac{\partial^2 f}{\partial\dot{q}^k\partial\dot{q}^j}\Delta_{\dot{q}}^j + \left(\frac{\partial^2 f}{\partial\dot{q}^j\partial q^k} - \frac{\partial^2 f}{\partial\dot{q}^k\partial q^j}\right)\Delta_q^j,$$
$$\frac{\partial E_f}{\partial\dot{q}^k} = \frac{\partial^2 f}{\partial\dot{q}^k\partial\dot{q}^j}\Delta_q^j \tag{41}$$

have solutions for E_f.

Consider now an ω_f-Hamiltonian vector field Δ, and assume that Eqs. (41) have been solved for E_f. Then the usual differential equations for the integral curves $c(t)$ of Δ, that is the equations $\dot{q}^k = \Delta_q^k$, $\ddot{q}^k = \Delta_{\dot{q}}^k$, can be cast in the local form

$$\frac{\partial E_f}{\partial q^k} = -\frac{\partial^2 f}{\partial\dot{q}^k\partial\dot{q}^j}\ddot{q}^j + \left(\frac{\partial^2 f}{\partial\dot{q}^j\partial q^k} - \frac{\partial^2 f}{\partial\dot{q}^k\partial q^j}\right)\dot{q}^j,$$
$$\frac{\partial E_f}{\partial\dot{q}^k} = \frac{\partial^2 f}{\partial\dot{q}^k\partial\dot{q}^j}\dot{q}^j. \tag{42}$$

It can be shown, incidentally, that nonsingularity of the Hessian is the necessary condition for reducing this set of differential equations to standard form, that is, to explicit equations for the $\dot{q}^k$ and $\ddot{q}^k$.

Thus if E_f has been found, the local equations for the integral curves can be written in terms of this function alone, rather than in terms of the 2ν components of Δ. But of course E_f must be found from Eqs. (41) and thus depends in general both on the particular function f for which Δ has been found to be ω_f-Hamiltonian and also explicitly on Δ. It turns out, however, that *if* Δ *is of second order*, there is a relation between f and E_f alone. That is, the relation between f and E_f is the same for all second-order ω_f-Hamiltonian vector fields Δ, independent of Δ, so that for them E_f can be obtained from f alone. This fact lies at the basis of the Lagrangian formalism and will now be demonstrated.

10.4 THE LAGRANGIAN FORMALISM

1. The Lagrangian formalism can be introduced by rewriting Eq. (40) so as to define a new function, the *Lagrangian* $\mathscr{L}\in\mathscr{F}(TQ)$. In order to remain quite general at first, consider not TQ, but some *symplectic manifold* (that is, a differentiable manifold endowed with a symplectic form) (M,ω) whose symplectic form ω is not only closed, but exact, i.e. such that there exists a one-form θ satisfying $\omega = -d\theta$. It will be seen that this is particularly applicable to the case in which $M = TQ$ and $\theta = \theta_f$.

Let Δ be ω-Hamiltonian on this symplectic manifold. Then if E is the Hamiltonian function, it follows that

$$dE = i_\Delta\omega = -i_\Delta d\theta = -L_\Delta\theta + di_\Delta\theta,$$

or

$$L_\Delta\theta = d\mathscr{L}, \quad \text{where} \quad \mathscr{L} = i_\Delta\theta - E \tag{43}$$

is the *Lagrangian* function for Δ (with respect to ω). So far all that has been done is to rewrite Eq. (14) under the condition that there exists a θ such that $\omega = -d\theta$, and thus Eq. (43) is equivalent to (14) with this special condition. If Eq. (14) is called the equation of the Hamiltonian formalism, Eq. (43) can now be called the equation of the Lagrangian formalism.

In particular, if $M = TQ$ and $\theta = \theta_f \equiv d_v f$, Eq. (43) becomes

$$L_\Delta\theta_f = d\mathscr{L}, \quad \text{where} \quad \mathscr{L} = i_\Delta\theta_f - E_f, \tag{44}$$

and it is seen that $\mathscr{L}$, like E_f, depends on both f and Δ. It is interesting, however, that the relation between E_f and $\mathscr{L}$ does not involve all of Δ, but only its first component. Indeed, in local coordinates Eq. (43) yields

$$\mathscr{L} = \Delta_q^k \partial f/\partial\dot{q}^k - E_f \equiv A_f - E_f, \tag{45}$$

where $A_f = i_\Delta\theta_f$ is called the *action*.

The situation is now the following. A vector field $\Delta\in\mathfrak{X}(TQ)$ is given and a function f is found such that Δ is Hamiltonian with respect to ω_f. The

Hamiltonian function is then called E_f and is obtained from Δ and ω_f in accordance with Eq. (40). This equation of the Hamiltonian formalism is recast in the form of (44), so that now the dynamical system is described by the two functions f and $\mathscr{L}$. Suppose now that f and $\mathscr{L}$ had been given at the start. Then the vector field could have been recovered from (44) or from its equivalent (40) by working back through the definition of E_f in (44). But what is obtained in this way is not the usual Lagrangian formalism, in which the vector field is recovered from a single function, the Lagrangian $\mathscr{L}$ alone.

2. It turns out that the two functions can be reduced to one in the special case in which Δ is a second-order vector field. A hint of this is given by the local-coordinate form of (45), for recall that for a second-order vector field $\Delta_q^k = \dot{q}^k$, so that for all such vector fields

$$\mathscr{L} = (\partial f/\partial q^k)\dot{q}^k - E_f, \tag{46}$$

which is independent of Δ. In fact it can be shown that if Δ is of second order, then $\theta_f = \theta_{\mathscr{L}}$. This is shown intrinsically in the Appendix to this chapter. The local demonstration is quite simple. According to (45) and (41)

$$\begin{aligned}\theta_{\mathscr{L}} &= \frac{\partial \mathscr{L}}{\partial \dot{q}^k} dq^k = \left[\frac{\partial^2 f}{\partial \dot{q}^k \partial q^j}\dot{q}^j + \frac{\partial f}{\partial q^k} - \frac{\partial^2 f}{\partial \dot{q}^k \partial q^j}\Delta_q^j\right] dq^k \\ &= \frac{\partial f}{\partial q^k} dq^k = \theta_f.\end{aligned}$$

For Δ of second order, therefore, Eq. (44) becomes

$$L_\Delta \theta_{\mathscr{L}} = d\mathscr{L}, \tag{47}$$

and Eq. (40) becomes

$$i_\Delta \omega_{\mathscr{L}} = dE_{\mathscr{L}}, \tag{48}$$

where $\mathscr{L}$ and $E_{\mathscr{L}}$ are related by

$$E_{\mathscr{L}} = i_\Delta \theta_{\mathscr{L}} - \mathscr{L}. \tag{49}$$

The Lagrangian function alone can therefore be used to recover the dynamical vector field Δ, so long as Δ is both of second order and Hamiltonian on the symplectic manifold $(TQ, \omega_{\mathscr{L}})$.

In local coordinates the $d\dot{q}$ part of (45) is an identity, and the dq part reads

$$\frac{\partial \mathscr{L}}{\partial q^k} = \frac{\partial^2 \mathscr{L}}{\partial \dot{q}^k \partial \dot{q}^j}\frac{d\dot{q}^j}{dt} + \frac{\partial^2 \mathscr{L}}{\partial \dot{q}^k \partial q^j}\dot{q}^j,$$

where the integral-curve equations $\Delta_{\dot{q}}^k = d\dot{q}^k/dt$ have been inserted. These are the usual Euler–Lagrange equations.

It should be emphasized that two conditions were needed to arrive at this result: Δ had to be of second order and it had to be Hamiltonian with respect to a symplectic form ω obtained from a function in accordance with Eq. (37).

An example of the case in which ω is not obtainable from a function is the following. Consider $Q = \mathbb{R}^2$, so $TQ = \mathbb{R}^2 \times \mathbb{R}^2$, and let the dynamics be

$$\Delta = \dot{q}^1 \frac{\partial}{\partial q^1} + \dot{q}^2 \frac{\partial}{\partial q^2} - q^1 \frac{\partial}{\partial \dot{q}^1} - q^2 \frac{\partial}{\partial \dot{q}^2}, \tag{50}$$

which is the vector field of the isotropic harmonic oscilator (as often, the superscripts here are indices, not powers). Then with ω given by

$$\omega = d\dot{q}^1 \wedge d\dot{q}^2 + dq^1 \wedge dq^2,$$

Δ is Hamiltonian and the Hamiltonian function is

$$E = q^2 \dot{q}^1 - q^1 \dot{q}^2,$$

as is easily verified. It should be clear from (38) that ω cannot be generated by some function in accordance with (37), for such an ω could never include a term of the form $d\dot{q}^1 \wedge d\dot{q}^2$. This is thus an example on TQ of the general situation: Eq. (14) is satisfied, but (44) and (47) are not, and Δ enters into the Hamiltonian formalism, but not into the usual Lagrangian one with this E. It is interesting that although there is no Lagrangian function $\mathscr{L}$ such that the symplectic form of this example is $\omega_{\mathscr{L}}$, there is an $\mathscr{L}$ such that the E of this example is $E_{\mathscr{L}}$, namely

$$\mathscr{L} = q^2(\dot{q}^1 \ln \dot{q}^1 - \dot{q}^1) - q^1(\dot{q}^2 \ln \dot{q}^2 - \dot{q}^2).$$

The Lagrangian formalism has been important in physics because a standard way has been found to write down Lagrangian functions of very many physical systems. Thus dynamical vector fields of such systems are easily recovered, and more importantly, the equations of motion for the integral curves of each system can be written down in terms of the single function $\mathscr{L}$. The formalism has also been of importance, through the Noether theorem (see Chapter 15), for the growth of ideas about symmetry and invariance and for developing methods of passing to quantum descriptions of dynamical systems.

10.5 OTHER CLASSES OF DYNAMICAL SYSTEMS

One should not be left with the impression that the only classes of dynamical systems that can be of importance are the three that have so far been treated (second order, Hamiltonian, and Lagrangian). For this reason we pass on to a very brief discussion of two other classes, *Liouville* and *gradient* dynamics.

1. Often in systems which possess ergodic properties or sastisfy the usual kind of conservation laws (say, of charge or mass), or in situations that arise in statistical mechanics, it is impossible to find differentiable constants of the motion which are not trivial, that is, not identically constant on the manifold. An example of such a system with ergodic properties is the irrational winding line on the torus T^2. The torus can be thought of as a finite cylinder whose ends are identified, and then a two-chart atlas can be constructed by using more or less the coordinates of

Fig. 10.2:

$$(z, \varphi): \{]0, 2\pi[\times]0, 2\pi[\} \subset T^2 \to \mathbb{R}^2,$$
$$(z', \varphi'): \{]-\pi, \pi[\times]-\pi, \pi[\} \subset T^2 \to \mathbb{R}^2.$$

Consider the vector field which, in this atlas, is given by

$$\Delta = \alpha\partial/\partial z + \beta\partial/\partial\varphi = \alpha\partial/\partial z' + \beta\partial/\partial\varphi',$$

where α and β are real numbers. If α/β is irrational, the integral curves of Δ are irrational winding lines on the torus (see Chapter 8, §8.1(2)), and any trajectory is an immersed submanifold whose closure is the entire torus. This means that any constant of the motion f must be a constant on the entire torus and thus is trivial. (If α/β is rational, the closure of a trajectory is not the entire torus, and a nontrivial constant of the motion can exist and in fact does.)

In ergodic statistical cases of this kind one deals with a measure on the carrier manifold to define such things as probabilities. This can be done in terms of a volume element (by using the Riesz representation theorem (AM, p. 135)). In many other situations it is also possible to define an invariant volume element, and dynamical systems with such invariant volume elements are called *Liouville dynamics*.

Liouville's theorem (AM, p. 188) states that a Hamiltonian dynamical system preserves the volume element in phase space. The theorem is easily proved when one realizes that the nth exterior power of a symplectic form can be used to define a volume element, and Liouville's theorem follows immediately from Eq. (35). Thus the Liouville property is of some interest in Hamiltonian systems, but we shall nevertheless not go any further into the description of Liouville dynamics.

2. In many areas of physics a vector field is written as the gradient of a potential function (the obvious example is electrostatics). Such descriptions are of particular interest when the equilibrium points of the dynamical system, that is, the zeros of the vector field, are important. For instance if the potential function is like the one in Fig. 10.5 for a one-dimensional system, it is immediately clear that there are three equilibrium points (q_1, q_2, q_3) and that two of them are stable (q_1, q_3) and the other unstable. A more detailed analysis requires much more, of course, but these relevant properties have been caught by inspection.

There is a certain similarity between Hamiltonian and gradient dynamics. The role played by the symplectic form in Hamiltonian dynamics is played in gradient dynamics by a symmetric two-form, and the analogue of the Hamiltonian function is the potential. More formally, let $G \in \tau_2^0(M)$ be symmetric and positive definite (G is then called a *Riemannian metric*). Then the vector field $X \in \mathfrak{X}(M)$ is a *gradient vector field* for G iff there exists a function $f \in \mathcal{F}(M)$ such that

$$G_\flat(X) \equiv G(X, \quad) = df. \tag{51}$$

Since G is positive definite, the pairing $G_\flat \in \mathrm{Lin}\,(\mathfrak{X}, \mathfrak{X}^*)$ is an isomorphism, and one writes the inverse of (51) in the form

$$X = \operatorname{grad} f = G_\natural(df).$$

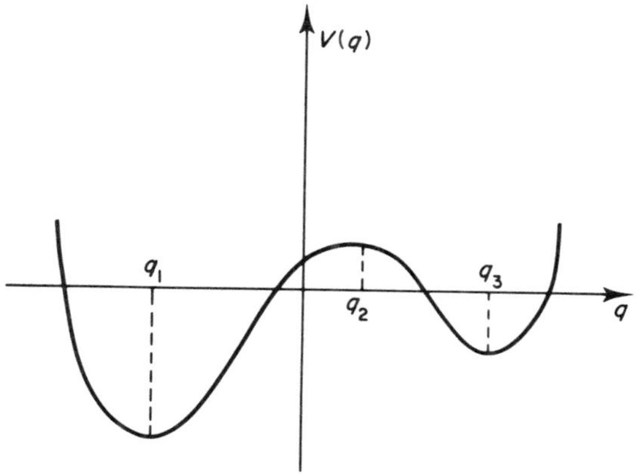

Fig. 10.5. A potential function $V(q)$

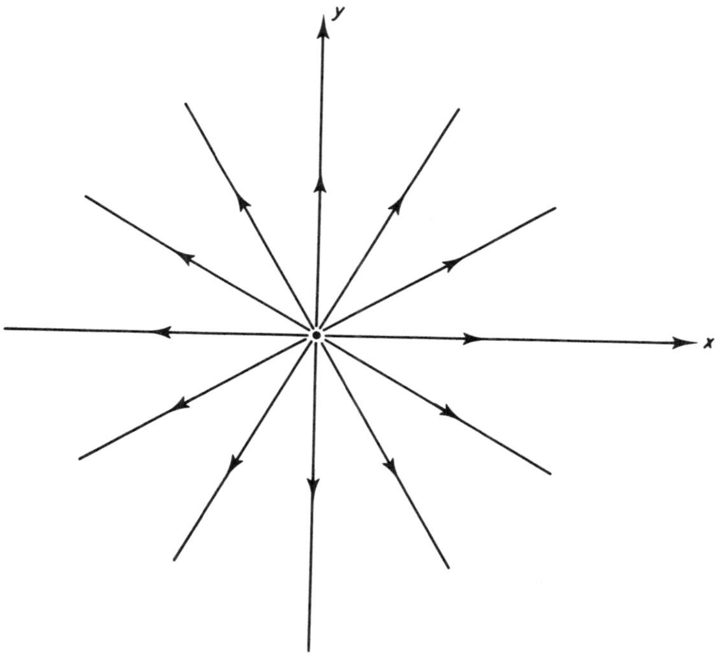

Fig. 10.6. Phase portrait of the vector field $X = (x, y; x, y)$

An example of a gradient vector field which is not globally Hamiltonian for any symplectic structure is given by the following. Let $M = \mathbb{R}^2$ and let X be given by

$$X = x\partial/\partial x + y\partial/\partial y,$$

and consider the positive-definite symmetric tensor

$$G = dx \otimes dx + dy \otimes dy.$$

This means that

$$G_\flat(X) = x\,dx + y\,dy = \tfrac{1}{2}d(x^2 + y^2),$$

so that X is a gradient vector field for G, and the potential function is $f = \frac{1}{2}(x^2 + y^2)$.

Fig. 10.6 is the simple phase portrait of this system. Any function which is constant along trajectories must have the same value on each trajectory, for they all meet at the origin. Thus any constant of the motion is trivial. Now, if the system were Hamiltonian, the Hamiltonian function E would be a nontrivial constant of the motion, so it follows that the system is not globally Hamiltonian for any ω.

Note that in this example the submanifolds of constant f (circles, in this case) are normal to the trajectories. This is characteristic of gradient dynamics. This is in contrast to Hamiltonian dynamics, whose trajectories characteristically lie within the submanifolds of constant E: the dynamical vector field is tangent to those submanifolds.

Although we shall occasionally introduce a metric G for computational purposes, we shall not in this book make use of gradient dynamics.

APPENDIX

The purpose of this Appendix is to demonstrate intrinsically that if Δ is of second order, the Lagrangian function $\mathscr{L}$ defined by (44) has the property that

$$\theta_{\mathscr{L}} = \theta_f. \tag{A1}$$

This will be done in the intrinsic notation developed in Chapter 5 for TQ.

Consider the *vertical endomorphism* $v: TTQ \to TTQ$ defined by

$$v: (q, \dot{q}; q', \dot{q}') \mapsto (q, \dot{q}; 0, q'). \tag{A2}$$

Note that a similar mapping would have been impossible to define on some carrier manifold other than TQ, for it depends on the possibility of identifying the vectors q', $\dot{q}$, and $\dot{q}'$, even though they are in different (but in this case isomorphic) vector spaces, as has already been explained in Chapter 5.

Now consider the vertical derivative d_v, which was represented locally in its action on functions by Eq. (36). This C^∞ mapping from $\Omega^k(TM)$ to $\Omega^{k+1}(TM)$ can be defined intrinsically in three steps by using the v defined in (A2), and then Eq. (37) will constitute an intrinsic definition of θ_f. In the first step one defines $v_*: \mathfrak{X}(TQ) \to \mathfrak{X}(TQ)$ from v by thinking of vector fields on TQ as sections of $T^2Q = TTQ$. Let $X \in \mathfrak{X}(TQ)$ be given by

$$X(q, \dot{q}) = (q, \dot{q}; X_q, X_{\dot{q}}),$$

where X_q and $X_{\dot{q}}$ are vectors in the appropriate tangent spaces and depend, of course, on the point $(q, \dot{q}) \in TQ$. Then $v_* X \in \mathfrak{X}(TQ)$ is defined by

$$v_* X(q, \dot{q}) \equiv X_v(q, \dot{q}) \equiv (q, \dot{q}; 0, X_q). \tag{A3}$$

Later (for instance in Chapter 15, §15.2) we shall often omit the asterisk from this operation, writing $vX \equiv v_* X \equiv X_v$. In the second step the *vertical insertion* $i_v : \Omega^k(TQ) \to \Omega^k(TQ)$ is defined by

$$i_v f = 0, \quad f \in \mathcal{F}(TQ),$$
$$(i_v \alpha)(Y_1, Y_2, \ldots, Y_k) = \sum_j \alpha(Y_1, \ldots, Y_{jv}, \ldots, Y_k) \tag{A4}$$

for $\alpha \in \Omega^k(TQ)$ and for all $Y_j \in \mathfrak{X}(TQ)$. Finally, d_v is defined in the third step by

$$d_v \equiv [i_v, d] \equiv i_v d - d i_v. \tag{A5}$$

It is immediately evident that

$$d \cdot d_v = d i_v d = -d_v \cdot d,$$

or that d_v anticommutes with d. According to the definition at (37),

$$\theta_{\mathscr{L}} = d_v \mathscr{L} = i_v d\mathscr{L} = i_v L_\Delta \theta_f = i_v L_\Delta i_v df. \tag{A6}$$

Now, it can be shown in general that

$$i_v L_X i_v df + L_{X_v} i_v df = i_v d i_{X_v} df. \tag{A7}$$

If X is of second order, so that $X(q, \dot{q}) = (q, \dot{q}; \dot{q}, X_{\dot{q}})$, it follows that $X_v(q, \dot{q}) = (q, \dot{q}; 0, \dot{q})$ defines a vector field $V \equiv X_v \in \mathfrak{X}(TQ)$ which is independent of how $X_{\dot{q}}$ depends on the point $(q, \dot{q}) \in TQ$. This vector field V obtained by v_* from each and every second-order vector field is called the *Liouville* vector field. For a second-order vector field, then, Eq. (A7) becomes

$$i_v L_X df + L_V i_v df = i_v d i_V df.$$

Since Δ is of second order, (A6) reads

$$\theta_{\mathscr{L}} = i_v d i_V df - L_V i_v df,$$

which is already seen to be independent of $\mathscr{L}$. Finally, it can be shown that

$$i_v L_V = L_V i_v + i_v, \tag{A8}$$

so that

$$\begin{aligned}\theta_{\mathscr{L}} &= i_v L_V df - i_v i_V d^2 f - L_V i_v df \\ &= i_v df = \theta_f.\end{aligned}$$

This completes the proof. Note that two unproved statements have been used in the course of it, namely (A7) and (A8). They can be verified in Godbillon (1969, Chapter IX).

We add to this Appendix the definition of one more vertical operation, $v^* : \Omega^k(TQ) \to \Omega^k(TQ)$. It is defined by

$$v^* f = f, \quad f \in \mathcal{F}(TQ),$$
$$v^* \alpha(Y_1, \ldots, Y_k) = \alpha(Y_{1v}, \ldots, Y_{kv}), \quad \alpha \in \Omega^k(TQ).$$

BIBLIOGRAPHY TO CHAPTER 10

General references to dynamical systems as differential equations are Hirsch and Smale (1974), Arnol'd (1978), and Brickell and Clark (1970, Chapter 8).

The use of constants of the motion to reduce the number of degrees of freedom is discussed in Whittaker (1904, Chapter 3).

A general setting for Hamiltonian and Lagrangian dynamics is laid down by AM, Chapter 3.

The use of constants of the motion to reduce the dimension of a symplectic carrier space is discussed for instance in Marmo, Saletan and Simoni (1979).

Modern approaches to Darboux's theorem can be found in Sternberg (1963), Weinstein (1977), and Arnol'd (1978).

Projectability of vector fields is discussed in Brickell and Clark (1970, Section 7.5).

Our presentation of the transition from the Hamiltonian to the Lagrangian formalism is, as far as we know, original. A different approach which follows AM more closely can be found in Caratù, Marmo, Simoni, Vitale and Zaccaria (1976).

Liouville dynamics in the context of ergodic systems is discussed in Arnol'd and Avez (1967); see also Marmo, Saletan, Simoni and Zaccaria (1981).

The intrinsic formalism on TQ and the detailed definition and properties of the relevant operators can be found in Godbillon (1969, Chapters IX and X).

11

From the tangent bundle to the cotangent bundle

In the preceding chapter we discussed the Hamiltonian formalism on TQ; yet most physicists know this formalism and use it only on T^*Q, that is, on phase space. In this chapter we discuss the transition from the description of a dynamical system on the tangent bundle TQ to its description on the cotangent bundle T^*Q, and in the next we shall turn to the usual canonical Hamiltonian formalism on T^*Q.

We start by saying a few words to motivate this transition. If the Hamiltonian formalism exists already on TQ, why go over to T^*Q? This question is best answered *a posteriori*, when the advantages and usefulness of the new description become evident. In fact, it is difficult to discuss its advantages before the new terminology has been established and before the new objects one deals with have been studied. Nevertheless some things can be said already, especially in terms of symplectic forms, which have already been introduced in the discussion of the Hamiltonian formalism on TQ.

11.1 THE FIBER DERIVATIVE

1. Recall the definition of a symplectic chart on TQ. Consider a Lagrangian function $\mathscr{L}$ and assume that its associated symplectic form $\omega_{\mathscr{L}}$ is nondegenerate. Let $\{x^1, \ldots, x^\nu, y^1, \ldots, y^\nu\}$ be local coordinates in a chart symplectic for $\omega_{\mathscr{L}}$, so that locally one may write

$$\omega_{\mathscr{L}} = -\delta_{jk} dx^j \wedge dy^k. \tag{1}$$

Then x_k can be defined by $x_k = \delta_{kj} x^j$, and $\omega_{\mathscr{L}}$ can be written in the form

$$\omega_{\mathscr{L}} = -dx_j \wedge dy^j. \tag{2}$$

Symplectic charts obtained in this way depend, of course, on $\mathscr{L}$; a chart which is symplectic for one Lagrangian will not be so for another. In fact, the $\mathscr{L}$-dependence can be exhibited explicitly in local coordinates by using Eqs. (37) and

(38) of Chapter 10: a symplectic chart is obtained when one sets

$$x_j \equiv p_j = \partial \mathscr{L} / \partial \dot{q}^j, \quad y^j = q^j. \tag{3}$$

It is clear then that the Hamiltonian formalism on TQ leads to charts which depend on Lagrangians, charts which are, in addition, not the natural ones in which almost all of the preceding analysis took place. Moreover, the p_j (or the x_j) are components of covariant vectors (or *covectors*), which are duals of contravariant ones. These lead immediately to one-forms, and this in turn suggests strongly that the analysis be carried over to T^*Q. It will turn out that when this is done $\omega_{\mathscr{L}}$ will be replaced always by the same symplectic form ω_0 on T^*Q and that the charts which are in a sense natural are also symplectic with respect to ω_0. These factors form a large part of the motivation for going to T^*Q.

2. Eq. (3) is essentially the local form of the map which defines the transition from TQ to T^*Q. The next task is to globalize this map and to analyze its actions on the vector fields and forms that have been discussed on TQ.

We have been using TQ as a carrier manifold for the dynamics, but it is evident that any fiber bundle (J, π, Q) constructed over Q will serve as well, so long as the integral curves in J project down to the same trajectories on Q. In other words, J is a suitable carrier manifold if it is diffeomorphic to TQ and the diffeomorphism satisfies the diagram

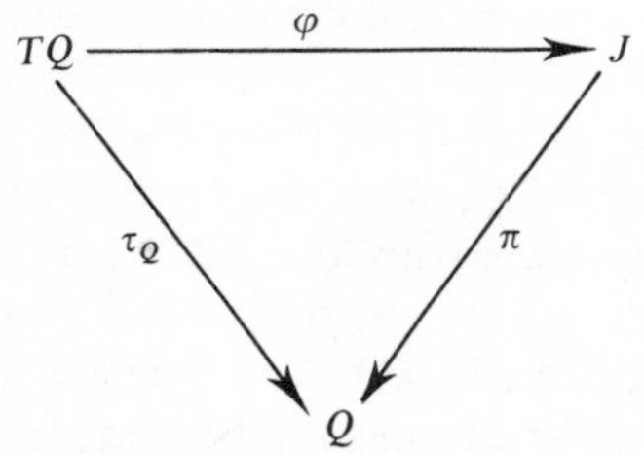

The diffeomorphism from TQ to T^*Q which will now be defined satisfies this diagram.

Let $f \in \mathscr{F}(TQ)$ be regular (see the definition of regularity in Chapter 10), and define $f_q: T_qQ \to \mathbb{R}$ by

$$f_q = f \mid T_qQ, \quad q \in Q. \tag{4}$$

Then the *fiber derivative* Ff of f is defined by its action on any element $v_q \in T_qQ$, namely by

$$Ff: v_q \mapsto Df_q(v_q) \in \mathrm{Lin}\,(T_qQ, \mathbb{R}) \equiv T_q^*Q. \tag{5}$$

In words, since $Df_q(v_q)$ is by definition a linear map from T_qQ to $\mathbb{R}$ (its definition in local coordinates is exhibited in Eq. (9)), Ff maps each element $v_q \in T_qQ$ to an element of $\mathrm{Lin}\,(T_qQ, \mathbb{R})$, or to an element of T_q^*Q. If f is regular, $Ff: TQ \to T^*Q$ is a local fiber-preserving diffeomorphism (but not a vector-bundle mapping; AM, p. 209). If in addition Ff is a global diffeomorphism, f is said to be *hyperregular*. Henceforth f will be assumed hyperregular.

Remark. One may ask whether hyperregular functions always exist on TQ. It can be shown that if Q is paracompact, they do. This is because there then exists a partition of unity on Q and a metric tensor, which can be written locally as g_{jk}, and then $g_{jk}\dot{q}^j\dot{q}^k$ is a hyperregular function on TQ.

The application of all this to dynamics will involve fiber derivatives $F\mathscr{L}$ of Lagrangian functions. Under such a map a vector field $X\in\mathfrak{X}(TQ)$ will be carried over to a vector field on T^*Q which will sometimes be written in the form $(F\mathscr{L})^*X \equiv X_{\mathscr{L}}\in\mathfrak{X}(T^*Q)$. When the vector field under discussion is the dynamics Δ, we shall call $\Delta_{\mathscr{L}}$ the dynamical vector field on T^*Q.

3. An important property of the fiber derivative is that there exists a unique symplectic form ω_0 on T^*Q such that

$$(Ff)^*\omega_0 = \omega_f \;\forall f\in\mathscr{F}(TQ). \tag{6}$$

That is, independent of which function $f\in\mathscr{F}(TQ)$ is used, ω_f is the pullback with respect to Ff of this fixed symplectic form ω_0 on T^*Q. Recall that according to the definition at Eq. (38), Chapter 10, $\omega_f = -d\theta_f$, so the assertion of Eq. (6) can be proved by showing that there exists a unique $\theta_0\in\mathfrak{X}^*(T^*Q)$ such that

$$(Ff)_*\theta_0 = \theta_f, \tag{7}$$

for then $\omega_0 = -d\theta_0$. To establish this, we shall first define θ_0 and then show in local coordinates that (7) is indeed satisfied.

The one-form θ_0 can be defined by its local action on vector fields $X\in\mathfrak{X}(T^*Q)$. Let $\alpha_q\in T_q^*Q$. Then the action of θ_0 is given by

$$\theta_0(\alpha_q)\cdot X(\alpha_q) = \alpha_q\cdot T\tau_Q^* X(\alpha_q). \tag{8}$$

To help understand this definition, recall that τ_Q^* is the projection from T^*Q to Q according to the diagram

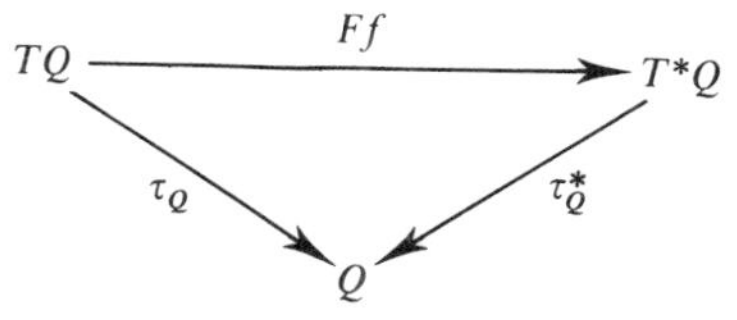

(see the diagram just above Eq. (4)), and that $T\tau_Q^*: T(T^*Q)\to TQ$ is obtained by applying the tangent functor.

Now, starting with Eq. (5), we put these definitions into local coordinates. Let $\{q, \dot{q}\}$ be natural local coordinates on TQ, and let the coordinates of $v_q\in T_qQ$ be $\dot{q}^j = v^j$. Then $Df_q(v_q)$ has component, which will be called p_j and which are equal to the derivatives $\partial f/\partial\dot{q}^j$ evaluated at v^j. That is, if w_q with components w^j is the general point in T_qQ, then $Df_q(v_q)$ maps it according to

$$Df_q(v_q): w_q \mapsto p_j w^j = w^j(\partial f/\partial\dot{q}^j)(v_q). \tag{9}$$

For an arbitrary point in TQ this can be written symbolically in the form

$$Ff:(q,\dot{q})\mapsto(q,\partial f/\partial\dot{q}).$$

We emphasize that the components of what is called here $\partial q/\partial\dot{q}$ are $\partial f/\partial\dot{q}^j=p_j$, which correspond to the definition of the momentum as given in most of the classical texts.

The definition of Eq. (8) can be used to establish the form of θ_0 in local coordinates on T^*Q. Let the coordinates of α_q be p^j, so that $\{q^j,p_j\}$ are coordinates in a chart. Then in analogy with Eq. (18) Chapter 5, the general vector field $X\in\mathfrak{X}(T^*Q)$ can be written in the form

$$X=a^j\partial/\partial q^j+b_j\partial/\partial p_j,$$

where the a^j and b_j are functions on T^*Q. Then on the right-hand side of Eq. (8), $T\tau_Q^*X(\alpha_q)$ is a vector in T_qQ whose components are the a^j evaluated at the point $(q,p)\in T^*Q$, or evaluated for the vector $\alpha_q\in T_q^*Q$. Thus the right-hand side of (8) becomes p_ja^j. Then by writing θ_0 as a linear combination of the dq^j and dp_j and applying it to X in order to evaluate the left-hand side of (8), one finds that in local coordinates

$$\theta_0=p_jdq^j. \tag{10}$$

The components of θ_0 are thus seen in local coordinates to be independent of the q^j, depending only on the p_j, and in this sense θ_0 is independent of the point in the base space. Eq. (10) is the local form of Eq. (8). It tells us that for any vector field $X\in\mathfrak{X}(T^*Q)$, the function $\theta_0X\in\mathscr{F}(T^*Q)$ is given locally by a^jp_j.

The next step is to find the one-form on TQ to which θ_0 is pulled back by the fiber derivative. Note first that $(Ff)_*X$ is, of course, a vector field on TQ, and that it can be written in local coordinates in the form

$$(Ff)_*X=a^j\partial/\partial q^j+B^j\partial/\partial\dot{q}^j,$$

where the precise form of the B^j will not concern us here. Now consider the function $a^jp_j\in\mathscr{F}(T^*Q)$. According to (9) its pullback under the fiber derivative is just $a^j\partial f/\partial\dot{q}^j$, or

$$a^j\partial f/\partial\dot{q}^j=(Ff)_*\theta_0\cdot(Ff)_*X.$$

This implies that in local coordinates

$$(Ff)_*\theta_0=(\partial f/\partial\dot{q}^j)dq^j\in\mathfrak{X}^*(TQ). \tag{11}$$

Comparison with (10.38) shows that Eq. (7) is satisfied locally.

This establishes also what was asserted at Eq. (6), namely that the fiber derivative of f maps ω_f into a fixed symplectic form ω_0 on T^*Q, independent of f. The advantage of this is the following. A Lagrangian dynamical system on TQ can be mapped over to T^*Q to yield a dynamical system which is Hamiltonian always with respect to the same symplectic structure ω_0 on T^*Q, given in local coordinates by

$$\omega_0=-d\theta_0=dq^j\wedge dp_j. \tag{12}$$

There is thus a natural symplectic structure on T^*Q, and the natural local coordinates are automatically symplectic. That is, the p_j are the coefficients of the dq^j in the natural one-form θ_0 of Eqs. (8) and (10) (the dq^j are the duals of the $\partial/\partial q^j$).

Remark. Note that Eq. (8) defines θ_0 intrinsically, so that ω_0 is also defined intrinsically through (12).

4. This does not mean, however, that there are no other symplectic charts on T^*Q. Consider, for example, the simple case of two dimensions (the base space $Q = \mathbb{R}$ is one-dimensional) and the new chart

$$p' = \sqrt{1/2}\,(p - q), \quad q' = \sqrt{1/2}\,(p + q). \tag{13}$$

Then it is easy to see that

$$\omega_0 = -\,dp \wedge dq = -\,dp' \wedge dq',$$

so that the new chart is also symplectic. One may think of (13) as a diffeomorphism rather than a change of coordinates, and it is then seen to leave ω_0 invariant.

Eq. (13) is an example of a large class of diffeomorphisms on T^*Q which leave the natural symplectic form ω_0 invariant, called *canonical diffeomorphisms*, or *symplectomorphisms*. Recall the theorem of Chapter 9 concerning Newtonian diffeomorphisms on TQ. What it says essentially is that if there is some reason to require diffeomorphisms on TQ to preserve the second-order nature of vector fields, then (in local terms) the diffeomorphisms are not allowed to mix the q^j and the $\dot{q}^j$. It is seen from the above that symplectomorphisms, required only to preserve ω_0, form a much larger class, for they mix the q^j and the p_j. Thus, to the extent that these requirements are important, one on TQ and the other on T^*Q, many more diffeomorphisms are allowed on the contangent bundle than on the tangent bundle. On the other hand, to the extent that these requirements are not important, there would seem to be no particular advantage in T^*Q over TQ, although the availability of a larger class of diffeomorphisms is often cited as one of the reasons for making the transition. Actually, the availability of the class of symplectomorphisms does become significant in certain applications, in particular in canonical perturbation theory and some results following from it (e.g. the KAM theorem; see AM (1978, p. 405)).

5. The canonical Hamiltonian formalism on T^*Q will be discussed in detail in the next chapter; here we limit ourselves to some remarks on how the equations of motion are carried over to T^*Q by the fiber derivative $F\mathscr{L}$ of $\mathscr{L}$. Suppose that the Lagrangian function $\mathscr{L}$ is given. From it can be constructed the Hamiltonian on TQ

$$E_{\mathscr{L}} = \dot{q}^k \partial\mathscr{L}/\partial\dot{q}^k - \mathscr{L}.$$

Then the *Hamiltonian function* $H_{\mathscr{L}}$ *on* T^*Q is defined by

$$H_{\mathscr{L}} = E_{\mathscr{L}} \circ (F\mathscr{L})^{-1} \equiv (F\mathscr{L})^* E_{\mathscr{L}}. \tag{14}$$

This can be written in local coordinates ($\mathscr{L}$ now plays the role of what was called f around Eq. (9)) in the form

$$H_{\mathscr{L}} = (\dot{q}^k \circ (F\mathscr{L})^{-1}) p_k - \mathscr{L} \circ (F\mathscr{L})^{-1}, \tag{15}$$

which is the well-known expression for the Hamiltonian, except that it is usually written without the $(F\mathscr{L})^{-1}$.

The Hamiltonian form (Chapter 10, Eq. (6)) of the equations of motion on TQ is then carried over to

$$i_{\Delta_{\mathscr{L}}} \omega_0 = H_{\mathscr{L}}, \tag{16}$$

where $\Delta_{\mathscr{L}}$ is the dynamical vector field on T^*Q. In local coordinates, if one writes

$$\Delta_{\mathscr{L}} = \dot{q}^k \partial/\partial q^k + \dot{p}_k \partial/\partial p_k,$$

Eq. (16) becomes

$$-\dot{p}_k dq^k + \dot{q}^k dp_k = \frac{\partial H}{\partial q^k} dq^k + \frac{\partial H}{\partial p_k} dp_k, \tag{17}$$

which are Hamilton's canonical equations in their familiar form.

In the next chapter the last remarks will be discussed in more detail.

11.2 LIFTING VECTOR FIELDS FROM Q TO T^*Q

1. Just as vector fields on Q need sometimes to be lifted to the tangent bundle TQ, they must also sometimes be lifted to T^*Q, in particular in connection with the Hamiltonian formalism on the cotangent bundle. We therefore show briefly how this is done, in clear analogy with the lifting to TQ discussed in Chapter 5, §5.5(2).

In outline the procedure is the following. From the given vector field $X \in \mathfrak{X}(Q)$ one forms its Lie group action φ^X. The cotangent $T^*\varphi^X$ of this action is then constructed in accordance with Eqs. (7) and (8) of Chapter 5. The infinitesimal generator of this cotangent is then defined as the lifted vector field, written X^{T^*}, or simply X^*. The analogy with the lifting to TQ is then obvious.

Assume, therefore, that $X \in \mathfrak{X}(Q)$ is given and that its action φ_t^X is known for each $t \in \mathbb{R}$. For simplicity, we shall suppress the superscript X and subscript t, writing simply φ. Now form $T^*\varphi$ in accordance with Eqs. (7) and (8) of Chapter 5. Let $w_q \in T_qQ$ and $\alpha_{\varphi(q)} \in T^*{}_{\varphi(q)}Q$. Then what Eq. (7) of Chapter 5 says is that

$$w_q \cdot (T^*\varphi)\alpha_{\varphi(q)} = \alpha_{\varphi(q)} \cdot (T\varphi) w_q. \tag{18}$$

Here we have left some subscripts off the maps $T\varphi$ and $T^*\varphi$ (compare Eq. (7), Chapter 5). Now, if the local coordinates of w_q are $\dot{q}^k$, then the local coordinates of $(T\varphi)w_q$ are $\dot{q}^j \partial\varphi^k/\partial q^j$. Then if the local coordinates of $\alpha_{\varphi(q)}$ are p_k, the right-hand side of (18) can be written in local coordinates in the form

$$\alpha_{\varphi(q)}\cdot(T\varphi)w_q = p_k(\partial\varphi^k/\partial q^j)\dot{q}^j,$$

and it then follows that in local coordinates

$$[(T^*\varphi)\alpha_{\varphi(q)}]_j = p_k\partial\varphi^k/\partial q^j. \tag{19}$$

Eq. (18) defines $T^*\varphi_t^X$, and (19) exhibits its action in local coordinates. Its infinitesimal generator $X^*\in\mathfrak{X}(T^*Q)$ is then defined as the *lift* of $X\in\mathfrak{X}(Q)$. With the aid of (19), the analogs of Eqs. (36) and (37) of Chapter 5 can be derived. If X is written in the form $X=(q, X(q))$, then

$$X^*(q, p) = (q, p; X(q), \tilde{X}'(q, p)), \tag{20}$$

where $\tilde{X}'(q, p)$ is given by

$$\tilde{X}'_k(q, p) = -(\partial X^j/\partial q^k)p_j. \tag{21}$$

It is easily checked that the following diagram commutes:

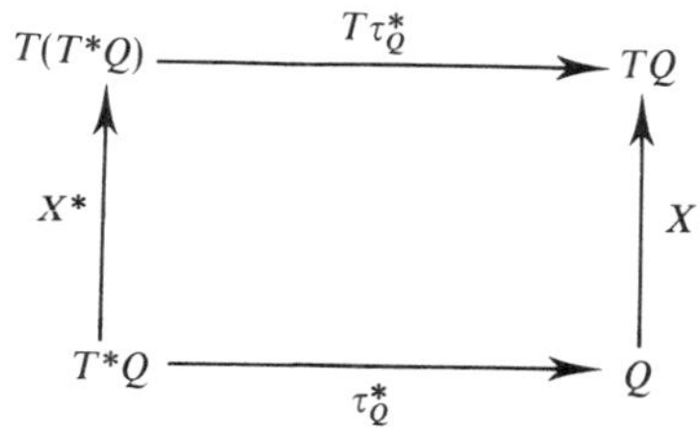

2. The above procedure for lifting vector fields from Q to T^*Q depends on the definition at Eqs. (7) and (8) of Chapter 5 of the cotangent map or, as can now be said, on lifting diffeomorphisms of Q to diffeomorphisms of T^*Q. There is in addition another way to carry diffeomorphisms from Q to T^*Q, this one based on the dynamics Δ, or rather on the Lagrangian $\mathscr{L}$. One first lifts the diffeomorphism φ of Q to its tangent diffeomorphism $T\varphi$ of TQ, and then uses the fiber derivative $F\mathscr{L}$ to carry $T\varphi$ from TQ to T^*Q. What is obtained in this way can in general be quite different from $T^*\varphi$; for instance, it need not be linear on fibers, whereas $T^*\varphi$ is, as is seen from (19). Moreover, the lifting procedure described in § 11.2(1) is independent of dynamics, which this one is obviously not. The two procedures lead to the same diffeomorphism on T^*Q only when $\mathscr{L}$ and φ are related in a special way. We shall not go into that at this time.

12

The canonical Hamiltonian formalism on T^*Q

This chapter discusses in more detail the last part of Chapter 11 as well as some of the properties of the cotangent bundle T^*Q and diffeomorphisms on it. We shall give some examples of the Hamiltonian formalism on T^*Q and demonstrate, among other things, that it is not entirely equivalent to the Lagrangian formalism on TQ.

12.1 DYNAMICS ON T^*Q

1. Like the tangent bundle TQ, the contangent bundle T^*Q has three relevant structures, together with their obvious inclusion relations. These are (a) the differentiable structure, (b) the fiber-bundle structure, and (c) the vector-bundle structure. Also as in the case of the tangent bundle, associated with each structure is a set of diffeomorphisms. These are (a), the full set $\mathrm{Diff}(T^*Q)$, (b) the fiber-bundle diffeomorphisms, and (c) the vector-bundle diffeomorphisms. Among the last two there are also the base-invariant ones. As was mentioned at the end of §11.1, the existence of the natural, intrinsically defined symplectic structure ω_0 on T^*Q is in some sense analogous to the existence of second-order vector fields on TQ (see the discussion following Eq. (5.12)), and associated with these are, respectively, the canonical transformations (or symplectomorphisms) on T^*Q and the Newtonian ones on TQ.

As was described in Chapter 11, $\varphi \in \mathrm{Diff}(T^*Q)$ is called canonical (or a symplectomorphism) iff

$$\varphi^*\omega_0 \equiv \varphi_*\omega_0 = \omega_0. \tag{1}$$

Since $\omega_0 = -d\theta_0$, a diffeomorphism which preserves θ_0 also preserves ω_0 and is therefore canonical. The converse, however, is not true. The example of Eq. (13) of Chapter 11 shows, for instance, that a symplectomorphism need not preserve θ_0.

Remarks. 1. Canonical transformations or diffeomorphisms are sometimes defined as those which preserve ω_0 up to a constant, those which satisfy $\varphi_*\omega_0 = c\omega_0$ for some $c \in \mathbb{R}$. With this definition, those we call canonical are usually called *homogeneous* canonical.

2. It can be shown that a diffeomorphism preserves θ_0 iff it is the lifting of a diffeomorphism on the base. The proof of this assertion is similar to the proof in Chapter 9 of the similar assertion that a diffeomorphism on TQ is Newtonian iff it is the lifting of one on the base: every fiber-preserving diffeomorphism can be decomposed into a *cotangent* one and another acting only on the fibers, and all that needs proof is that a diffeomorphism preserving θ_0 is fiber-preserving.

The relevance of the canonical diffeomorphisms is seen when they are applied to dynamical systems. Let $\Delta \in \mathfrak{X}(T^*Q)$ be a Hamiltonian dynamical vector field. It satisfies the equation (we drop the subscript 0 from ω_0 since for a while we shall be dealing only with the natural symplectic form)

$$i_\Delta \omega = dH \tag{2}$$

for some function $H \in \mathscr{F}(T^*Q)$. Now let φ be any symplectomorphism, and apply φ^* to (2). This yields

$$i_{\varphi^*\Delta}\varphi^*\omega = d\varphi^* H,$$

and then it follows from Eq. (1) that

$$i_{\varphi^*\Delta}\omega = d(H \circ \varphi^{-1}) \equiv dK. \tag{3}$$

Thus if Δ is ω-Hamiltonian, so is $\varphi^*\Delta$, but with the new Hamiltonian $K = H \circ \varphi^{-1}$.

2. One-parameter groups of symplectomorphisms can be generated in the following way. If $X \in \mathfrak{X}(T^*Q)$ is a complete vector field such that $L_X\omega = 0$, its flow F^X determines a one-parameter group of diffeomorphisms $\{F_t^X, t \in \mathbb{R}\}$ (AM, p. 60) which leave ω invariant. On the other hand $L_X\omega = 0$ implies that X is locally Hamiltonian, which in turn implies that locally there exists a function f such that $i_X\omega = df$. Conversely, given a function $f \in \mathscr{F}(T^*Q)$, one can construct the unique vector field X which satisfies $i_X\omega = df$; this vector field will also be such that $L_X\omega = 0$. If X turns out to be complete, f is thus associated with a unique one-parameter group of symplectomorphisms. One might expect that a 'generating function' f can be defined in this way for any one-parameter group of symplectomorphisms, and in a certain sense this is true. It will be discussed more fully in Chapter 14.

The classical form of these considerations is obtained in local coordinates. Let $\{q^k, p_k\}$ be a symplectic chart for ω, so that $\omega = dq^k \wedge dp_k$. Then as has been seen at Eq. (17) of Chapter 11, if the general vector field on T^*Q is written in the form

$$X = a^k \partial/\partial q^k + b_k \partial/\partial p_k,$$

the equation

$$i_X\omega = df \tag{4}$$

becomes

$$a^k = \partial f/\partial p_k, \quad b_k = -\partial f/\partial q^k, \tag{5}$$

and this defines the vector field X in terms of the function f. That Eq. (4) can also be used to define f in terms of X depends on the fact that ω is closed, and hence so is $i_X\omega$. To see this in local coordinates, write $d(i_X\omega)$ locally:

$$\begin{aligned} 0 = d(a^k dp_k - b_k dq^k) &= da^k \wedge dp_k - db_k \wedge dq^k \\ &= (\partial a^k/\partial q^j)dq^j \wedge dp_k + (\partial a^k/\partial p_j)dp_j \wedge dp_k \\ &\quad - (\partial b_k/\partial q^j)dq^j \wedge dq^k - (\partial b_k/\partial p_j)dp_j \wedge dq^k, \end{aligned}$$

from which it follows that

$$\begin{aligned} &\partial a^k/\partial p_j = \partial a^j/\partial p_k, \quad \partial b_k/\partial q^j = \partial b_j/\partial q^k, \\ &\partial a^k/\partial q^j = -\partial b_j/\partial p_k. \end{aligned}$$

These are the local integrability conditions for the existence of a function $f(q, p)$ satisfying (5). If φ is a symplectomorphism and X satisfies (4), then so does φ^*X, as has been seen, except that when X is replaced by φ^*X in (4), f must be replaced by $g = f \circ \varphi^{-1}$. In local coordinates this means that if $\varphi:(q, p) \mapsto (Q, P)$, then φ^*X is of the form

$$\varphi^*X = (\partial g/\partial P_k)\partial/\partial Q^k - (\partial g/\partial Q^k)\partial/\partial P_k.$$

3. To obtain Hamilton's equations in the usual form, as at the end of Chapter 11, X must be replaced by the dynamical vector field Δ on T^*Q and f by the Hamiltonian function H. Let $c: I \to T^*Q$ be an integral curve of Δ, and let it be represented in local coordinates as $\{q^k(c(t)), p_k(c(t))\}$, or as is more commonly written by physicists, $(q^k(t), p_k(t))$. Then according to Eq. (11) of Chapter 5, $c(t)$ is a solution of the equation $\Delta(c(t)) = Tc(t, 1)$. In local coordinates this may be written

$$(q, p; a, b) = (q(t), p(t); \dot{q}(t), \dot{p}(t)),$$

where the dot represents the time derivative, as usual. Eq. (5) with f replaced by H yields the desired expressions, namely Eq. (17) of Chapter 11.

12.2 EXAMPLES

Of the three examples given below, the first two are examples on T^*Q which are pathological in the sense that they cannot be pulled back in the usual way to TQ. In both of them, as will be seen, the carrier manifold for the dynamics is something less that T^*Q. Hamiltonians like the ones in these examples are classically called linear in the momenta and stand in contrast to the kind of Hamiltonians usually treated, which are quadratic in the momenta. The third example also involves linearity, this time of the Lagrangian. In it T^*Q is treated as a base space, and its tangent bundle is formed, a procedure sometimes used to establish a variational principle for obtaining Hamilton's canonical equations. This example is similar to the first two in that the carrier manifold is less than the tangent bundle. That is, as will be seen in that example, the carrier manifold is T^*Q, not $T(T^*Q)$.

Example 12.1. Hamiltonian linear in the p_k

This is an example of a Hamiltonian vector field on T^*Q which cannot be pulled back to a second-order vector field on TQ. Let $X \in \mathfrak{X}(Q)$ and consider the function $P_X \in \mathscr{F}(T^*Q)$ defined by

$$P_X(\alpha_q) = \alpha_q(X(q)), \tag{6}$$

where $\alpha_q \in T_q^*Q$. Now consider the Hamiltonian system for which $H = P_X$. If in a chart on Q the vector field $X = a^k \partial/\partial q^k$, so that in a natural chart on T^*Q the function P_X is given by $P_X = a^k p_k$, the dynamical vector field on T^*Q is then given by

$$\Delta = a^k \partial/\partial q^k - (\partial a^j/\partial q^k) p_j \partial/\partial p_k.$$

This vector field cannot be mapped into a second-order vector field on TQ by any fiber-preserving diffeomorphism, for the a^k are functions only on Q. As was mentioned following Eq. (5) of Chapter 11, the fiber derivative is a fiber-preserving diffeomorphism, and therefore Δ corresponds to no second-order vector field on TQ. This means also that reversing the procedure of Eq. (15) of Chapter 11 will not provide a nontrivial Lagrangian on TQ. In fact Q itself is a carrier manifold for this dynamics, for the base integral curves are already separated.

Example 12.2. First degree homogeneous Hamiltonian

This example has a pathology similar to the previous one. Let $Q = \mathbb{R}^2$, so that $T^*Q = \mathbb{R}^4$, and in a natural chart define the Hamiltonian function by

$$H(q, p) = (p_1^2 + p_2^2)^{1/2}. \tag{7}$$

It is true that this H is not C^∞, but the point $p_1 = p_2 = 0$ can be removed from each fiber (the zero section can be removed) and the system treated in what remains of T^*Q. Be that as it may, the vector field on T^*Q is then

$$\Delta = p_k(p_1^2 + p_2^2)^{-1/2} \partial/\partial q^k, \tag{8}$$

and its base integral curves are given by

$$c_1 = p_1 t(p_1^2 + p_2^2)^{-1/2} + c_{10},$$
$$c_2 = p_2 t(p_1^2 + p_2^2)^{-1/2} + c_{20}.$$

When these are lifted to TQ (or simply when the velocities are calculated) it is found that

$$\dot{q}^k = p_k(p_1^2 + p_2^2)^{-1/2}, \quad k = 1,2.$$

These velocities, however, are not independent, for

$$(\dot{q}^1)^2 + (\dot{q}^2)^2 = 1. \tag{9}$$

This relation between the velocities is a reflection of the fact that again there is no nontrivial Lagrangian associated with H by reversing the procedure of Eq. (15) of Chapter 11. It is as though Eq. (9) were a constraint, forcing the carrier space for the integral curves to be not all of TQ, but $\mathbb{R}^2 \times S$.

Example 12.3. Lagrangian linear in the velocities

The procedure of this example, involving an irregular Lagrangian (one that is not a regular function on TQ; see Chapter 10), is often used in obtaining the equations of motion or in applying the variational principle (that is, in deriving the Euler–Lagrange equations) in phase space. It will be discussed first in local terms and only then in global terms.

Consider a manifold $M = \mathbb{R}^2$ (in the applications this is usually T^*Q) with a symplectic structure, and let $\{q^k, p_k\}$ be a symplectic chart for this structure. Consider a function $H \in \mathscr{F}(M)$. The tangent bundle of M is $= \mathbb{R}^4$, and locally it will be treated in a natural chart of the form $\{q^k, p_k; \dot{q}^k, \dot{p}_k\}$. Let the Lagrangian function $\mathscr{L} \in \mathscr{F}(TM)$ be

$$\mathscr{L}(q, p; \dot{q}, \dot{p}) = p_k \dot{q}^k - H(q, p). \tag{10}$$

Then with the standard procedure of Chapter 10 (see Eq. (39) of Chapter 10) $\theta_{\mathscr{L}}$ and $\omega_{\mathscr{L}}$ can be constructed:

$$\theta_{\mathscr{L}} = p_k dq^k, \quad \omega_{\mathscr{L}} = -dp_k \wedge dq^k. \tag{11}$$

The energy function is now given in the same chart by

$$E_{\mathscr{L}} = \dot{q}^k \partial \mathscr{L} / \partial \dot{q}^k + \dot{p}_k \partial \mathscr{L} / \partial \dot{p}_k - \mathscr{L},$$

which, because $\partial \mathscr{L} / \partial \dot{p}_k = 0$, is just

$$E_{\mathscr{L}}(q, p; \dot{q}, \dot{p}) = H(q, p). \tag{12}$$

The dynamical vector field which satisfies $i_\Delta \omega_{\mathscr{L}} = dE_{\mathscr{L}}$ is then

$$\Delta = -(\partial H / \partial p_k) \partial / \partial q^k + (\partial H / \partial q^k) \partial / \partial p_k + a^k \partial / \partial \dot{q}^k + b_k \partial / \partial \dot{p}_k, \tag{13}$$

where the a^k and b_k are arbitrary. This arbitrariness is not suprising, for although $\omega_{\mathscr{L}}$ is a symplectic form on M, it is degenerate on TM and thus does not give rise to a unique vector field in $\mathfrak{X}(TM)$: the resulting Δ is arbitrary up to a vector in the kernel of $\omega_{\mathscr{L}}$. Nevertheless, its projection by $T\tau_M$ is a unique vector field in $\mathfrak{X}(M)$ (see Eqs. (17)–(19) of Chapter 5). By requiring that Δ in Eq. (13) be of second order, one obtains Hamilton's canonical equations on M.

We now globalize this example: consider any one-form $\alpha \in \mathfrak{X}^*(M)$, and define the function $f_\alpha \in \mathscr{F}(TM)$ through the restrictions of α to $T_m M$ for all $m \in M$:

$$f : v_m \mapsto \alpha_m(v_m), \quad v_m \in T_m M.$$

In addition, for any (Hamiltonian) function $H \in \mathscr{F}(M)$, define the Lagrangian function $\mathscr{L} \in \mathscr{F}(TM)$ by

$$\mathscr{L} = f_\alpha - H \circ \tau_M. \tag{14}$$

Then it turns out that $\theta_{\mathscr{L}} = \tau_{M*}\alpha$ and $\omega_{\mathscr{L}} = \tau_{M*}d\alpha$, which is clearly degenerate. As above, one finds that $E_{\mathscr{L}} = H \circ \tau_M$ and that the dynamical vector field is not unique.

12.3 FUNCTIONS IN INVOLUTION

Recall the definition of the Poisson bracket at Eq. (7) of Chapter 10, which is applicable to any manifold with a symplectic form. Two functions $f, g \in \mathscr{F}(T^*Q)$ are said to be *in involution* with respect to ω iff their Poisson bracket vanishes, i.e. iff

$$\{f, g\} = 0,$$

where the Poisson bracket is the one defined in terms of ω. For the purposes of dynamics, one is interested in sets of functions which can become coordinate systems on T^*Q, so the definition of involution will be further restricted by requiring what is classically called functional independence of the two functions. A set $\{f_1, \ldots, f_\lambda\}$ of λ functions on a symplectic manifold (M, ω) is *involutive* (the functions are in involution) iff (a) $df_1, \ldots, df_\lambda$ are $\mathscr{F}(M)$-independent in $\mathfrak{X}^*(M)$ and (b) $\{f_j, f_k\} = 0$, $j, k \in \{1, \ldots, \lambda\}$.

It can be shown that if M is 2ν-dimensional, then no more than ν functions can be in involution. There is, moreover, a theorem due to Carathéodory and Jacobi which says that if $f_1, \ldots, f_\lambda$ are in involution, $\lambda < \nu$, then about each point of M there is a neighborhood in which it is possible to find functions $f_{\lambda+1}, \ldots, f_\nu$ such that the set $f_1, \ldots, f_\nu$ is involutive. These results will become important in Part II, in the discussion of reduction of dynamical systems, in particular on symplectic manifolds.

We quote also a strong version of Darboux's theorem. Let (M, ω) be a 2ν-dimensional symplectic manifold with ν functions $f_1, \ldots, f_\nu$ in involution. Then about each point of M there is a neighborhood U in which it is possible to find ν other functions $g^1, \ldots, g^\nu$ which are in involution and have the property that

$$\omega|_U = df_j \wedge dg^j.$$

From this it follows that $\{f_j, g^k\} = \delta_j^k$. Clearly, then, the 2ν functions f_k, g^k provide a symplectic chart in U, sometimes called a Heisenberg coordinate system.

Note again that the remarks concerning functions in involution apply equally to any symplectic manifold, not only to T^*Q.

13

Equivalent Lagrangians and Hamiltonians

13. EQUIVALENT LAGRANGIANS AND HAMILATIONIANS

The example of Eq. (50) of Chapter 10 shows that a vector field can be Hamiltonian with respect to two different symplectic forms. More thorough investigation would show also that each symplectic form induces a different Hamiltonian function and, if one exists, a different Lagrangian function. It is thus possible that of two theorists, each could be using what he would call *the* Lagrangian description of a certain dynamical system, and yet the two descriptions would be different and might even seem unrelated. This chapter discusses the relation between two such descriptions, making some comments also on the uniqueness (or lack of it) in associating a Lagrangian or Hamiltonian function with a given vector field. We will find in the process that when different Lagrangian or Hamiltonian functions exist, they are related to different constants of the motion and hence to different invariances, so that one choice of Lagrangian may be better suited than another to exhibit certain invariances of the dynamical system. The discussion starts on TQ and passes later to T^*Q.

13.1 AMBIGUITY ON TQ

1. Recall how a Lagrangian function is associated with a second-order vector field Δ. Let $f \in \mathscr{F}(TQ)$ be regular, and calculate ω_f. If Δ is ω_f-Hamiltonian, that is if there exists a function $E \in \mathscr{F}(TQ)$ such that

$$dE = i_\Delta \omega_f, \tag{1}$$

then f is a suitable Lagrangian function for Δ (and we shall call it $\mathscr{L}$ instead of f) and it is related to E by

$$E = i_\Delta \theta_{\mathscr{L}} - \mathscr{L}. \tag{2}$$

This brief summary avoids the problem of how, given a second-order Δ, to find a

Lagrangian function. This is a complicated problem, sometimes called the inverse problem of Lagrangian dynamics; we shall not go into it at all.

Instead, we turn first to the lack of uniqueness of the function $\mathscr{L}$ which yields a given symplectic structure $\omega_{\mathscr{L}}$. Recall that $\omega_{\mathscr{L}}$ is defined in two steps. The first step is the derivation

$$\mathscr{L} \mapsto \theta_{\mathscr{L}} = v^* d\mathscr{L} \tag{3}$$

(see the end of the Appendix to Chapter 10 for a definition of v^*), and the second is the definition of $\omega_{\mathscr{L}}$ in terms of $\theta_{\mathscr{L}}$:

$$\omega_{\mathscr{L}} = -d\theta_{\mathscr{L}}. \tag{4}$$

If two Lagrangians $\mathscr{L}$ and $\mathscr{L}'$ lead to the same symplectic form, then

$$0 = d\theta_{\mathscr{L}} - d\theta_{\mathscr{L}'} = d\theta_{\mathscr{L}-\mathscr{L}'} = -\omega_{\mathscr{L}-\mathscr{L}'}. \tag{5}$$

In looking for Lagrangian functions which lead to the same symplectic form, therefore, one looks for functions $f \in \mathscr{F}(TQ)$ such that $\omega_f = 0$.

One such set of functions is immediate. The kernel of the derivation v^*d defined in Eq. (3) consists of functions on TQ which are pullbacks under $(\tau_Q)_*$ of functions on Q; i.e. it consists of functions on TQ which in natural local coordinates do not depend on the $\dot{q}^k$. Indeed, let $v^* df = 0$. Then

$$0 = (v^* df)X = df(X_v) = L_{vX} f \;\forall\; X \in \mathfrak{X}(TQ),$$

(recall that $vX \equiv X_v$) which means that f is constant along the integral curves of all vector fields that are tangent to fibers, or is constant in each fiber. Thus f depends only on which fiber it is calculated in, hence only on the point $q \in Q$ which defines the fiber. The converse is obvious (it will be demonstrated in passing in natural local coordinates below). It then follows that if $\mathscr{L}' = \mathscr{L} + (\tau_Q)_* f$ for any $f \in \mathscr{F}(Q)$, Eq. (5) is satisfied. More even than that: $\theta_{\mathscr{L}'} = \theta_{\mathscr{L}}$.

Additional ambiguity in the Lagrangian comes from the exterior derivative in Eq. (4), for it annihilates any part of $\theta_{\mathscr{L}}$ which is closed. A first step in understanding this can perhaps be obtained in local coordinates. Let $g \in \mathscr{F}(TQ)$. Then in local coordinates

$$\theta_g = (\partial g/\partial \dot{q}^k) dq^k$$

(and now it is clear why functions independent of the $\dot{q}^k$ are in the kernel of v^*d), and what are needed are functions g such that

$$\omega_g = -d\theta_g = -d[(\partial g/\partial \dot{q}^k) dq^k] = 0.$$

This means for one thing that the coefficients of the $d\dot{q}^j \wedge dq^k$ in this expression must vanish, and hence that g must be linear in the $\dot{q}^k$. In Example 12.3 of §12.2 it is shown that any such function can be written in the form $g \equiv g_\alpha = i_\Delta (\tau_Q)_* \alpha$, where Δ is any second-order vector field, and $\alpha \in \mathfrak{X}^*(Q)$ determines g_α. For such functions it is then found that $\theta_g = (\tau_Q)_* \alpha$, and thus $d\theta_{g_\alpha} = 0$ iff $d\alpha = 0$, that is iff α is closed. (In local coordinates this reflects the symmetry of the coefficients of the $dq^k \wedge dq^j$ in $d\theta_g$.)

One therefore concludes that Eq. (5) is satisfied, or $\omega_{\mathscr{L}} = \omega_{\mathscr{L}'}$, iff

$$\mathscr{L}' = \mathscr{L} + (\tau_Q)_* f + g_\alpha, \tag{6}$$

where $f \in \mathscr{F}(Q)$ and $g_\alpha = i_\Delta(\tau_Q)_*\alpha$ for some closed $\alpha \in \mathfrak{X}^*(Q)$ and for arbitrary second-order vector field Δ. The $(\tau_Q)_* f$ term has to do with the definition of the Lagrangian already discussed around Eq. (44) of Chapter 10: two different functions f will lead to two different energy functions E in Eq. (2). The g_α term, on the other hand, will not affect E, for $i_\Delta\theta_{g_\alpha} - g_\alpha = 0$ by definition.

Remark. If in local coordinates α is written $\alpha = \alpha_k(q)dq^k$, then $g_\alpha(q, \dot{q}) = \alpha_k(q)\dot{q}^k$. In particular, since α is closed, locally there exists an $s(q)$ such that $\alpha = ds(q)$, and then $g_\alpha(q, \dot{q}) = \dot{q}^k \partial s/\partial q^k$, which can also be written as ds/dt. Thus the g_α term corresponds to the well-known ambiguity in the Lagrangian up to the total time derivative of a function on Q. But closure of α does not mean that globally there exists an $s \in \mathscr{F}(Q)$ such that $\alpha = ds$, and thus from the global point of view the g_α term is not quite the same thing.

2. A different kind of ambiguity is achieved if one can find different Lagrangians yielding different symplectic forms, but whose Euler–Lagrange equations are satisfied by the same vector field Δ. If two such different Lagrangians can be found, the same dynamics is associated with both of them: they yield the same integral curves and trajectories, both in TQ and in Q.

To attempt to find this kind of ambiguity, consider Eq. (47) of Chapter 10 again:

$$L_\Delta\theta_{\mathscr{L}} - d\mathscr{L} = 0. \tag{7}$$

This may be treated as an equation in which the vector field Δ is given, but the Lagrangian function $\mathscr{L}$ is to be found, and the focus now is on the uniqueness of the solution. As the operations involved in (7) are linear, the set of solutions, which may be called the set of *admissible Lagrangians for* Δ, has a linear structure. Among these admissible Lagrangians are those which determine Δ uniquely by Eq. (7) or, equivalently, by (1) and (2). Such uniquely Δ-determining Lagrangians will be called *equivalent* for Δ (some authors call them s-equivalent, s for 'solution'). Note, for example, that every g_α function of §13.1(1) is an admissible Lagrangian for every second-order vector field Δ, since $L_\Delta\theta_{g_\alpha} = dg_\alpha$ for any such Δ. No two of these functions are equivalent Lagrangians, however, for clearly no such function even begins to determine Δ (this follows also from the fact that ω_{g_α} vanishes). What are to be found, then, are regular functions in the set of admissible Lagrangians for a given vector field Δ. That the set of such equivalent Lagrangians can contain more than the functions related as in Eq. (6) (related, as we shall say, trivially) is seen from the following example.

Remark. If two Lagrangians $\mathscr{L}$ and $\mathscr{L}'$ are admissible and $\mathscr{L}$ determines Δ uniquely, whereas $\mathscr{L}'$ does not, $\mathscr{L}'$ is said to be *subordinate* to $\mathscr{L}$.

Example 13.1. Harmonic oscillator in one dimension

Let the carrier manifold be $TQ = T\mathbb{R} = \mathbb{R}^2$ and consider the Lagrangians

$$\mathscr{L} = (q^2 - \dot{q}^2)/2 \tag{8}$$

and

$$\mathscr{L}' = (\dot{q}/q)\arctan(\dot{q}/q) - (1/2)\log(\dot{q}^2 + q^2). \tag{9}$$

These are equivalent locally where both are defined. To see this, one writes the general second-order vector field in the form

$$\Delta = \dot{q}\partial/\partial q + f(q, \dot{q})\partial/\partial\dot{q}, \tag{10}$$

and calculates the functions $f_{\mathscr{L}}$ and $f_{\mathscr{L}'}$ obtained from the two Lagrangians through Eq. (7). The two one-forms $\theta_{\mathscr{L}}$ and $\theta_{\mathscr{L}'}$ are found to be

$$\theta_{\mathscr{L}} = \dot{q}dq \quad \text{and} \quad \theta_{\mathscr{L}'} = (1/q)\arctan(\dot{q}/q)dq,$$

and the functions obtained are

$$f_{\mathscr{L}}(q, \dot{q}) = f_{\mathscr{L}'}(q, \dot{q}) = -q.$$

Consequently, when they are treated by the established procedure, the two Lagrangians yield the same vector field Δ, the vector field of the harmonic oscillator in one dimension with (circular) frequency one. Nevertheless, they yield different symplectic forms, as can be observed by applying the exterior derivative to $\theta_{\mathscr{L}'}$ and $\theta_{\mathscr{L}}$: one obtains

$$\omega_{\mathscr{L}} = -d\dot{q} \wedge dq \quad \text{and} \quad \omega_{\mathscr{L}'} = -(q^2 + \dot{q}^2)^{-1} d\dot{q} \wedge dq.$$

The energy functions obtained by means of Eq. (2) are

$$E_{\mathscr{L}} = (q^2 + \dot{q}^2)/2 \quad \text{and} \quad E_{\mathscr{L}'} = (1/2)\log(2E_{\mathscr{L}}).$$

Example 13.2. A one-dimensional 'relativistic' dynamics.

As in Example 13.1, let $Q = \mathbb{R}$, so $TQ = \mathbb{R}^2$. Consider the two Lagrangians

$$\mathscr{L} = \dot{q}\arcsin\dot{q} + (1 - \dot{q}^2)^{1/2} - V(q) \tag{11}$$

and

$$\mathscr{L}' = \dot{q}F(\dot{q})\exp[-V(q)], \tag{12}$$

where $V(q) \in \mathscr{F}(Q)$ is the 'potential function' and $F(q)$ is defined by

$$F(q) = \int^q (1/x^2)\exp\sqrt{1 - x^2}\, dx.$$

It is left to the reader to show that these two Lagrangians are equivalent where they are defined and that they give rise to different symplectic forms. The second-order vector field obtained from them is found to be

$$\Delta = \dot{q}\partial/\partial q - (1 - \dot{q}^2)^{1/2} V'(q)\partial/\partial\dot{q}.$$

The coefficient of $\partial/\partial\dot{q}$, which plays the role of the force function, clearly goes to zero as the velocity $\dot{q}$ approaches one, and the vector field Δ is confined to the strip between $\dot{q} = 1$ and $\dot{q} = -1$. It is in this sense that this dynamical system is relativistic. The Hamiltonian functions on TQ obtained from these Lagrangians are

$$E_{\mathscr{L}} = V(q) - (1 - \dot{q}^2)^{1/2} \quad \text{and} \quad E_{\mathscr{L}'} = \exp[-E_{\mathscr{L}}],$$

the first of which looks at least a bit like the sum of a potential and a kinetic part, which might cause one to prefer the Lagrangian of Eq. (11) to that of (12).

Remark. It should not be surprising that in both of the above examples the energy functions are not independent. These are one-dimensional systems and can have only one constant of the motion.

Example 13.3. The isotropic harmonic oscillator in three dimensions

Let $Q = \mathbb{R}^3$, so that $TQ = \mathbb{R}^6$, and for each nonsingular symmetric matrix $A = \{a_{jk}\} \in GL(3, \mathbb{R})$ consider the Lagrangian function

$$\mathscr{L}_A = a_{jk}(\dot{q}^j\dot{q}^k - q^j q^k)/2. \tag{13}$$

These are all equivalent.

This assertion is proved by calculating the θ_A obtained from each of $\mathscr{L}_A$ and inserting them into Eq. (7) with arbitrary second-order vector field Δ, which can be written in the form

$$\Delta = \dot{q}^j\partial/\partial q^j + f^j(q, \dot{q})\partial/\partial\dot{q}^j. \tag{14}$$

It is a simple matter to find the f^j: they are

$$f^j(q, \dot{q}) = -q^j$$

independent of which $\mathscr{L}_A$ is used. Thus all of the $\mathscr{L}_A$ yield the same vector field and are therefore equivalent. The symplectic forms obtained from them are

$$\omega_A = -a_{jk}d\dot{q}^j \wedge dq^k,$$

a different one for each $A \in GL(2, \mathbb{R})$.

Associated with each Lagrangian in each of the above examples is its own Hamiltonian function on TQ (we have not demonstrated this in the last example), each a constant of the motion. One may wonder whether this relationship is reflexive, whether every constant of the motion can serve as a Hamiltonian function with its own (equivalent) Lagrangian. This question and others involving the relation between equivalent Lagrangians and constants of the motion has been treated by several authors; the reader is referred to the bibliography at the end of this chapter.

13.2 AMBIGUITY ON T^*Q

1. The fiber derivative of any one of the Lagrangians in the set of equivalent ones can be used to make the transition to T^*Q. Since in general the mapping from the tangent bundle to the cotangent bundle will depend on which of the equivalent Lagrangians is used, the vector field Δ on TQ will not always be mapped to the same vector field on T^*Q, but the different symplectic forms obtained from different Lagrangians will always be mapped, each by its own fiber derivative, onto the same symplectic form ω_0 on T^*Q.

It can be put this way: let $F\mathscr{L}_1$ and $F\mathscr{L}_2$ be the fiber derivatives of two equivalent Lagrangians $\mathscr{L}_1$ and $\mathscr{L}_2$. Then the two mappings symbolized by

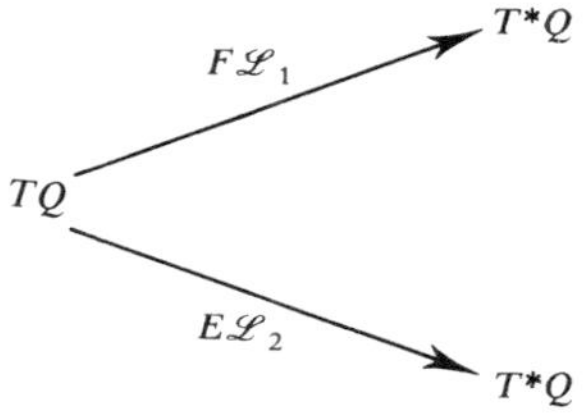

carry the two equations of the Hamiltonian formalism on TQ onto T^*Q according to

$$i_\Delta \omega_k = dE_k \xrightarrow{F\mathscr{L}_k} i_{\Delta_k}\omega_0 = dH_k, \tag{15}$$

where $k \in \{1, 2\}$, and ω_k is the symplectic form on TQ obtained from $\mathscr{L}_k$. The two vector fields Δ_1 and Δ_2 in $\mathfrak{X}(T^*Q)$ are now different; the only way they can be seen to be equivalent without passing back through TQ is by observing that their integral curves project down to the same base integral curves on Q. Clearly they are related by

$$\Delta_2 = (F\mathscr{L}_2)^*(F\mathscr{L}_1)^{*-1}\Delta_1.$$

Then the map

$$\varphi = F\mathscr{L}_2 \cdot (F\mathscr{L}_1)^{-1} \tag{16}$$

is a fiber-preserving base-invariant diffeomorphism from T^*Q to T^*Q. It is not surprising then that the two Δ_k project down to the same base integral curves, for all vector fields related by a fiber-preserving base-invariant diffeomorphism do so.

2. Eq. (15) uses two different fiber derivatives to carry the equations of the Hamiltonian systems from TQ to T^*Q, but both systems could just as well have been carried over by the same fiber derivative. Suppose that $F\mathscr{L}_1$ is chosen for this purpose. Then (15) is replaced by

$$\left.\begin{aligned} i_\Delta\omega_1 &= dE_1 \\ i_\Delta\omega_2 &= dE_2 \end{aligned}\right\} \xrightarrow{F\mathscr{L}_1} \left\{\begin{aligned} i_{\Delta_1}\omega_0 &= dH_1 \\ i_{\Delta_1}\omega_{12} &= dH_{12} \end{aligned}\right. \tag{17}$$

Since the same diffeomorphism is used for both systems, and since the dynamical vector field $\Delta \in \mathfrak{X}(TQ)$ is the same one in both systems, the same vector field $\Delta_1 \in \mathfrak{X}(T^*Q)$ is obtained in both cases (this is the Δ_1 of Eq. (15)), but now the two symplectic forms on T^*Q are different and so are the Hamiltonians. It should be clear that

$$\omega_{12} = \varphi_*^{-1}\omega_0 = \varphi^*\omega_0. \tag{18}$$

It is seen that Δ_1 is Hamiltonian with respect to both ω_0 and ω_{12}, though with different Hamiltonians. Another way to say this is that φ is a diffeomorphism such that Δ_1 is Hamiltonian with respect to both ω_0 and $\varphi^*\omega_0$. In local coordinates this means that there are two charts, one in which ω_0 takes on the Darboux canonical form and the other in which $\varphi^*\omega_0$ does, in which the differential equations for the integral curves of Δ_1 are the classical form of Hamilton's canonical equations even though φ is not a canonical transformation, not a symplectomorphism (for it does not leave ω_0 invariant). We shall then say that φ is *canonoid* with respect to Δ_1. A diffeomorphism φ is canonical (is a symplectomorphism) iff every vector field $X \in \mathfrak{X}(T^*Q)$ which is Hamiltonian with respect to ω_0 is also Hamiltonian with respect to $\varphi^*\omega_0$ (and then there is a theorem which says that it is canonical iff $\varphi^*\omega_0 = c\omega_0$ for some $c \in \mathbb{R}$). This can now be stated by saying that φ is canonical iff it is canonoid with respect to every $X \in \mathfrak{X}(T^*Q)$. Notice that the diffeomorphism φ of Eq. (16) is not only canonoid, but fiber-preserving and base-invariant.

Unlike most other classes of diffeomorphisms that are often treated, the canonoids with respect to a given vector field Δ do not form a group, and it is perhaps for this reason that they have not been discussed extensively in the classical literature. They are of interest, however, in connection with the last point of §13.1: they allow one to consider the dynamics on T^*Q as generated by dynamical variables other than what is usually called the energy. Their counterparts exist also on TQ, but they are not fiber-preserving, and it is more common to ignore the fiber structure on T^*Q then on TQ.

As an example, consider $T^*Q = \mathbb{R}^4$ in a symplectic chart and the diffeomorphism

$$\eta:(q^1, q^2, p_1, p_2) \mapsto (q^1, p_1, q^2, p_2).$$

The vector field

$$\Delta = p_k \partial/\partial q^k - q^k \partial/\partial p_k$$

of the harmonic oscillator, Hamiltonian with respect to $\omega_0 = dq^k \wedge dp_k$, is also Hamiltonian with respect to

$$\eta^*\omega_0 = dq^1 \wedge dq^2 + dp_1 \wedge dp_2.$$

Indeed, a simple calculation shows that

$$i_\Delta \eta^*\omega_0 = -d(q^1 p_2 - q^2 p_1),$$

so that apart from a constant the Hamiltonian with this symplectic form is

$H = -(q^1 p_2 - q^2 p_1)$, which is commonly called the angular momentum. This entire example can be pulled back by η, and then it is found that

$$\eta_* \Delta = q^2 \partial/\partial q^1 - q^1 \partial/\partial q^2 + p_2 \partial/\partial p_1 - p_1 \partial/\partial p_2$$

is Hamiltonian with respect to ω_0 and that the Hamiltonian function remains the same: $\eta_* K = K$. The reader may verify that in the symplectic chart in which ω_0 is in Darboux canonical form, the differential equations for the integral curves of Δ are Hamilton's canonical equations in their classical form with Hamiltonian function K. In this example η is canonoid, but not fiber-preserving, for it mixes the qs and ps.

Let us return to two of the examples of §13.1. In Example 13.1 the transition to T^*Q is familiar for the unprimed Lagrangian $\mathscr{L}$. For $\mathscr{L}'$ the transition is given by

$$F\mathscr{L}':(q, \dot{q}) \mapsto (q, p) = (q, [1/q] \arctan [\dot{q}/q]).$$

Then the canonoid diffeomorphism we have called φ is defined by

$$\varphi:(q, p) \mapsto (q, q \tan pq).$$

It is now easily checked that φ is canonoid with respect to the vector field $(F\mathscr{L})^* \Delta \in \mathfrak{X}(T^*Q)$, where $\Delta \in \mathfrak{X}(TQ)$ is the vector field of Example 13.1.

Now consider Example 13.3. The vector field $\Delta_A \in \mathfrak{X}(T^*Q)$, obtained by mapping $\Delta \in \mathfrak{X}(TQ)$ by the fiber derivative

$$F\mathscr{L}_A:(q, \dot{q}) \mapsto (q, p) = (q, A\dot{q}),$$

is

$$\Delta_A \equiv (F\mathscr{L}_A)^* \Delta = a^{jk} p_j \partial/\partial q^k - a_{jk} q^j \partial/\partial p_k,$$

where $A^{-1} = \{a^{jk}\}$ is also symmetric. Then

$$i_{\Delta_A} \omega_0 = d(a^{jk} p_j p_k + a_{jk} q^j q^k).$$

where $\omega_0 = -dp_j \wedge dq^j$. For any two matrices $A, B \in GL(3, \mathbb{R})$, the diffeomorphism φ can be defined by

$$\varphi = (F\mathscr{L}_B)(F\mathscr{L}_A)^{-1}:(q,p) \mapsto (q, BA^{-1}p),$$

from which it follows that

$$\omega_{AB} \equiv \varphi^* \omega_0 = -b_{jk} a^{ki} dp_i \wedge dq^j$$

and

$$i_\Delta \omega_{AB} = d(b_{kr} a^{ki} p_i a^{rj} p_j + b_{kr} q^k q^r).$$

It is thus seen that Δ_A is Hamiltonian with respect to both ω_0 and ω_{AB}.

In both of these examples, φ is not only canonoid, but fiber-preserving and base-invariant.

We shall not go further into the question of ambiguity in the Lagrangian and Hamiltonian formalisms. In recent years the literature on this subject has grown, in particular with regard to the connection between ambiguity and constants of the motion; the reader is referred to the bibliography.

BIBLIOGRAPHY TO CHAPTER 13

Some papers concerning ambiguity and equivalence, with particular emphasis on the relation to symmetry and constants of the motion, are Henneaux (1982), Hojman and Urrutia (1981), Lutzky (1981), Marmo and Saletan (1977), Marmo and Simoni (1976), Sarlet and Cantrijn (1983). These papers contain many other relevant references.

14

Other carrier spaces: action-angle variables and the Hamilton–Jacobi method

So far in this book we have been dealing with only two carrier manifolds, TQ and T^*Q. Now, so long as Q is paracompact these two manifolds are diffeomorphic, for in that case it is possible to construct a hyperregular Lagrangian (see the Remark following Eq. (5) of Chapter 11), whose fiber derivative is a fiber-bundle diffeomorphism. This explains why the classical local formalism is able to treat the two manifolds as though they were a single one and to consider the fiber derivative, often called the Legendre map, as simply a change of coordinates. Many other such 'changes of coordinates' are performed in the classical formalism, going by such names as Routh's procedure, action-angle variables, Hamilton–Jacobi transformations. But in the global or intrinsic formalism, these are thought of, like the transition from the tangent bundle to the contangent bundle, as transitions to new carrier manifolds.

What really makes TQ and T^*Q different carrier manifolds is that the charts on them are restricted in different ways, or that the charts and the overlap maps are required to have different properties. On TQ the overlap maps are required to be linear on fibers, and on T^*Q they are required to respect the symplectic structure. Since, moreover, the treatment is global, the charts are required to cover the entire manifold, and each and every one of the overlap maps is required to respect the relevant structure. In this sense, then, in looking for a new carrier manifold, one is in fact looking for an atlas with certain desirable properties, what is desirable depending on the particular dynamical system being treated. For example, a desirable property might be that the dynamical vector field locally look like $\partial/\partial x$. In general what one tries to find is not a minimal atlas, but one that contains many others.

When dealing with second-order vector fields, one is led in this way to TQ; when dealing with Hamiltonian vector fields, to T^*Q. Other properties of vector fields lead to other carrier spaces, but few are of wide applicability or interest in classical mechanics, with the notable exceptions of action-angle variables and the

Hamilton–Jacobi method, which are useful for calculational purposes, for perturbation theory, and for the transition to quantum mechanics.

One of the aims of Part II of this book is to develop some general points of view from which to look for structures defining carrier manifolds in the sense described above, structures which can be adapted to particular dynamical systems. For the present, however, we deal only with action-angle variables and with the Hamilton–Jacobi approach.

There are other ways, incidentally, in which one can be led to new carrier manifolds. On occasion we have suggested that points be removed from manifolds because the vector fields which are being considered on them have some singular behavior at these points. When points or perhaps even regions are removed from a carrier manifold for such reasons, what remains is a new carrier manifold, usually quite different from the first. A carrier manifold obtained in this way, although motivated by a certain particular vector field, is often appropriate for many other dynamical systems.

14.1 ACTION-ANGLE VARIABLES

1. Action-angle variables are used to handle Hamiltonian dynamical systems which are *completely integrable*, that is systems which have v constants of the motion in involution, where v is the number of degrees of freedom. A symplectic chart is then found in which these constants of the motion form half of the coordinates. Let these constants of the motion be called the *action variables* J_k, and their conjugates (that is, those which complete the symplectic chart) be called the *angle variables* $w^k, k \in \{1, \ldots, v\}$. Written out in terms of these variables, the Hamiltonian function will depend only on the J_k, whose time-derivatives, the partials of H with respect to the w^k, all vanish. In fact, in terms of these symplectic variables, the canonical equations become

$$\dot{J}_k = -\partial H/\partial w^k = 0,$$
$$\dot{w}^k = \partial H/\partial J_k \equiv v^k(J). \tag{1}$$

The first set of these equations, from which it is clear why H cannot depend on the w^k, merely restates the constancy of the J_k, and the second is then trivially solved. In the classical treatment, then, the problem at this point would be to find a canonical transformation from the $\{q, p\}$ chart to the $\{w, J\}$ chart.

Actually there is more to action-angle variables than has been stated so far. The w^k are assumed to be cyclic or periodic variables, as the name 'angle' implies. That is, the dynamical system must be one that returns to its initial state when all of the J_k remain fixed and just one of the w^k changes by some finite value, often taken to be 2π. Therefore, although we are about to define action-angle variables in terms of the properties of differentiable manifolds, if they are to be applicable to any given dynamical system, that system must have properties which are compatible with this definition. A sufficient condition (Arnol'd (1978, §49)) is essentially that the level sets of the J_k functions be compact connected submanifolds (see §14.2).

Consider the cotangent bundle (this construction can be performed just as well on the tangent bundle). For any fiber bundle, a *model neighborhood* is a contractible neighborhood of the base space crossed with a fiber. Model neighborhoods so far have been of the form $U \times \mathbb{R}^\nu$, where U is open in (its own) $\mathbb{R}^\nu$, and the cotangent bundle has been constructed from such model neighborhoods. To describe action-angle variables, one chooses a different kind of model neighborhood, and this will lead to a different manifold, which can serve as a carrier manifold for systems which are compatible with it. In the fiber-bundle structure finally arrived at, the base manifold B will be parametrized by the J_k and the fibers by the w^k, and the model neighborhoods will look like $U \times T^\nu$, rather than $U \times \mathbb{R}^\nu$, where T^ν is the ν-torus.

Let $M = \mathbb{R}^\nu \times \mathbb{R}^\nu$ be a symplectic manifold, and let $\{x, y\}$ be a symplectic chart on M. Consider the equivalence relationship on M defined by

$$(x, y) \approx (x', y') \Leftrightarrow x - x' \in \mathbb{Z}, y = y'. \tag{2}$$

It is easily established that this is indeed an equivalence relationship. If $T^\nu(y)$ is the equivalence class for fixed y, then it follows immediately that $T^\nu(y)$ is diffeomorphic to $T^\nu(y')$ for any y and y' and that they are all diffeomorphic to the ν-torus T^ν. It then follows that $\bigcup_{y \in \mathbb{R}^\nu} T^\nu(y)$ is a fiber bundle.

Now for any open ball $B^\nu \subset \mathbb{R}^\nu$ define $T^\nu \times B^\nu$, and on this the symplectic chart with coordinates $\{w^1, \ldots, w^\nu, J_1, \ldots, J_\nu\}$, where $w^k = x^k (\text{mod } 1)$ is in T^ν, and $J_k = y_k$ is in B^ν. This manifold $T^\nu \times B^\nu$ is called an action-angle variable model neighborhood, and $\{w, J\}$ are called action-angle variables.

Another way to build an action-angle model, necessarily diffeomorphic, is to define something like polar coordinates on the same manifold M using the same set of symplectic charts as before. One writes

$$\rho_k = (x^k)^2 + (y_k)^2,$$
$$\varphi^k = \arctan(x^k / y_k).$$

The chart obtained in this way is suitable only for a region of M which is diffeomorphic to $T^\nu \times B^\nu$, but it is no longer symplectic. This is easily fixed up, however, and the new chart

$$J_k = \pi \rho_k, \quad w^k = \varphi^k / 2$$

is symplectic. Not only is the chart symplectic, but this model is equivalent to the first one, for they are diffeomorphic.

The fibers of the action-angle carrier manifold are toruses, whereas in the tangent and cotangent bundles they were vector spaces. In principle other fiber bundles could also be used, whose fibers were neither toruses nor vector spaces ($U \times S^\nu$ comes to mind), but the usefulness of such models would depend on whether there were physical systems to which they could apply. Eventually, in Part II of the book, we go beyond even fibered structures to foliated ones; the resulting more general models will be applicable to many dynamical systems and will involve projectability of vector fields.

Returning to action-angle variables, a 2ν-dimensional manifold M is said to admit action-angle variables if there exists a symplectomorphism $\varphi : M \to T^\nu \times B^\nu$; it admits action-angle variables locally in an open submanifold $U \subset M$ if similarly there exists a symplectomorphism $\varphi : U \to T^\nu \times B^\nu$. A dynamical vector field Δ is compatible with the action-angle variables if each of its integral curves lies on the fibers, i.e. on the toruses. If Δ is ω-Hamiltonian with Hamiltonian function $H \in \mathscr{F}(M)$, then H admits action-angle variables in the neighborhood $U \subset M$ if M admits local action-angle variables on U and Hamilton's canonical equations for $\varphi^* \Delta | \varphi(U)$ are

$$\begin{aligned} \dot{J}_k &= -\partial K/\partial w^k = 0, \\ \dot{w}^k &= \partial K/\partial J_k \equiv \nu^k(J), \end{aligned} \tag{3}$$

where $K = \varphi_* H$ is the transformed Hamiltonian. Then $\{w, J\}$ are called action-angle variables for H in U.

2. Under what conditions does a 2ν-dimensional symplectic manifold (M, ω) admit action-angle variables globally or locally? Some sufficient conditions are provided by the following theorem due to Liouville (see Arnol'd (1978, § 49 and AM § 16). Let functions $f_1, \ldots, f_\nu \in \mathscr{F}(M)$ in involution be given, and consider the map $F : M \to \mathbb{R}^\nu : m \mapsto \{f_1(m), \ldots, f_\nu(m)\}$. Then the set $N_a = \{m \in M \vdash F(m) = a\}$ is a submanifold of dimension ν, and there exists a contractible neighborhood $V \subset \mathbb{R}^\nu$ about $a \in \mathbb{R}^\nu$ such that

$$F^{-1}(V) = N_a \times V. \tag{4}$$

Let X_k be the ω-Hamiltonian vector field whose Hamiltonian function is f_k, so that $i_{X_k}\omega = df_k$. Then N_a is an invariant submanifold for each of the X_k, and if the X_k are complete, each N_a is diffeomorphic to a disjoint union of cylinders of the form $T^\lambda \times \mathbb{R}^\sigma, \lambda + \rho = \nu$. Action-angle variables are obtained when $\sigma = 0$, i.e. when these cylinders reduce to the torus parts alone, for then each component of N_a is diffeomorphic to T^ν, and V, being contractible, is diffeomorphic to B^ν. Locally then a diffeomorphism is obtained from M to $T^\nu \times B^\nu$. Note that a sufficient condition for the cylinders to reduce to the torus parts alone is that they be compact, and if in addition they are connected, they reduce to a single torus.

So far no conditions have been put on the f_k functions, but now assume that they are all constants of the motion for an ω-Hamiltonian dynamical vector field Δ, so that the N_a are all invariant submanifolds. Then a chart $\{w, J\}$ can be chosen in $F^{-1}(V)$ such that the J_k are coordinates for V and the w^k are coordinate for N_a, defined mod 1, with the dynamics in terms of these coordinates being given locally by

$$\begin{aligned} \dot{J}_k &= 0, \\ \dot{w}^k &= \nu^k(J). \end{aligned} \tag{5}$$

If the w^k can be chosen to satisfy Eq. (3), essentially so that ω can be written in the form $dJ_k \wedge dw^k$, action-angle variables are obtained for the dynamical system.

Our aim here has been to show how the transformation to action-angle variables becomes, in our intrinsic treatment, a transition to a new carrier manifold, whose fibers are no longer vector spaces, but toruses, and whose zero section need no longer be the base space Q.

3. The spherical pendulum, a mass point moving on a sphere of radius r under the influence of gravity, will now be presented to give an example of the use of action-angle variables. The configuration space is then $Q = S^2$, the 2-sphere, and the carrier manifold is TS^2 or T^*S^2. We shall present this example in the Hamiltonian formalism on TQ, rather than on T^*Q.

The Lagrangian function $\mathscr{L} \in \mathscr{F}(TS^2)$ is given in spherical polar coordinates ($\theta = 0$ is taken along the positive z axis) by

$$\mathscr{L} = mr^2(\dot{\theta}^2 + \dot{\varphi}^2 \sin^2 \theta)/2 - mgr \cos \theta. \tag{6}$$

Note, incidentally, that θ and φ are well-defined everywhere in Q except at the north and south poles (actually on the z axis), where $\theta = \pm\pi$ and φ is completely ambiguous. Now, one might think that $d\theta$ and $d\varphi$ are well-defined all over the sphere, but even this is not so, for, if they were, TS^2 would be trivializable, which it is not. Thus special care will have to be taken at the north and south poles.

Proceeding according to the standard Lagrangian formalism as described in Chapter 10, one constructs the necessary geometrical objects (forms, functions, and vector fields):

$$\theta_{\mathscr{L}} = mr^2(\dot{\theta}\, d\theta + \dot{\varphi} \sin^2 \theta\, d\varphi); \tag{7}$$

$$\omega_{\mathscr{L}} \equiv -d\theta_{\mathscr{L}} = -mr^2(d\dot{\theta} \wedge d\theta + \sin^2 \theta d\dot{\varphi} \wedge d\varphi + 2\dot{\varphi} \sin\theta \cos\theta d\theta \wedge d\varphi); \tag{8}$$

$$E_{\mathscr{L}} = mr^2(\dot{\theta}^2 + \dot{\varphi}^2 \sin^2 \theta)/2 + mgr \cos \theta; \tag{9}$$

$$\Delta = \dot{\theta}\partial/\partial\theta + \dot{\varphi}\partial/\partial\varphi + (\dot{\varphi}^2 \sin\theta \cos\theta + gr^{-1} \sin\theta)\partial/\partial\dot{\theta}. \tag{10}$$

The angular momentum about the z axis

$$p_\varphi \equiv \partial\mathscr{L}/\partial\dot{\varphi} = mr^2 \dot{\varphi} \sin^2 \theta$$

is an immediate constant of the motion. Another constant of the motion is the energy E (we omit the subscript $\mathscr{L}$), from which $\dot{\varphi}$ can be eliminated in favor of p_φ:

$$E = \tfrac{1}{2}mr^2\dot{\theta}^2 + p_\varphi^2/(2mr^2 \sin^2 \theta) + mgr \cos \theta. \tag{11}$$

It follows from the fact that p_φ is a constant of the motion that $\{E, p_\varphi\} = 0$, or that E and p_φ are in involution, so that one may attempt to arrive at action-angle variables by using the energy, the angular momentum, and the variables which are conjugate to them to parametrize TQ. Actually to find the conjugate variables is a nontrivial task, so that we shall limit ourselves to a qualitative discussion of the map

$$\mu: TS^2 \to \mathbb{R}^2 : m \mapsto (E(m), p_\varphi(m)) \tag{12}$$

in order to show to what extent $\mu^{-1}(c)$ is a torus for $c \in \mathbb{R}^2$.

That it is a torus for $\sin\theta \neq 0$ and $dE \wedge dp_\varphi \neq 0$ follows from the fact that it is compact (θ and p_φ are limited for each fixed E) and from Liouville's theorem described above. It will be argued further that $\mu(M)$ looks like the shaded area in

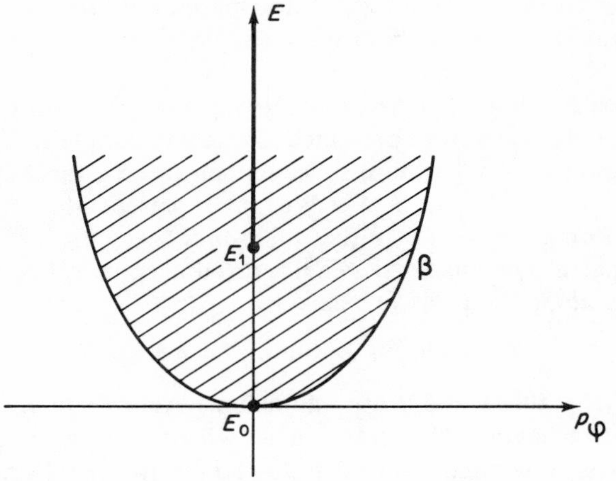

Fig. 14.1. A representation of $\mu(m)$, possible values of E and p_φ, for the spherical pendulum. The equilibrium points are E_0 (stable) and E_1 (unstable)

Fig. 14.1, in which the points E_0 and E_1 represent equilibrium points (stable and unstable), and the boundary β represents those points in TS^2 where $\dot{\theta} = 0$ and $\theta = \text{const.} > \pi/2$.

First the boundary β. When $\dot{\theta} = 0$ the pendulum is undergoing circular motion in a horizontal orbit, and Eq. (11) becomes

$$E = p_\varphi^2/(2mr^2 \sin^2 \theta) + mgr\cos\theta,$$

so that for each fixed value of θ the energy E and the angular momentum p_φ are related quadratically: p_φ determines E. The boundary β is not a parabola, however, for each value of E corresponds to a different fixed value of θ, determined through the equations of motion (we shall not go into that). The inverse image of each point on β is one-dimensional, a circle, whose points may represent for example the initial values of φ.

The equilibrium point E_0 is clearly the motionless stable one, and E_1 is the unstable one in which the pendulum is at rest at $\theta = 0$. On the E axis, where $p_\varphi = 0$, Eq. (11) becomes

$$E = \tfrac{1}{2}mr^2\dot{\theta}^2 + mgr\cos\theta,$$

the same equation as for the simple plane pendulum, whose phase portrait was discussed in Example 7.7. So long as E lies between E_0 and E_1 the inverse image of a point on the E axis is a torus: in its motion, the pendulum oscillates through a finite range of θ, sweeping out a closed curve in the $(\theta, \dot{\theta})$ 'plane'. Within this range of E values, therefore, as one moves from the E axis toward the boundary β the torus gets thinner and thinner until on β it becomes a circle. At $E = E_1$ the phase portrait of the plane pendulum becomes the figure-eight. It is instructive to redraw Fig. 7.9 as in Fig. 14.2 here, in order to show more clearly the relation

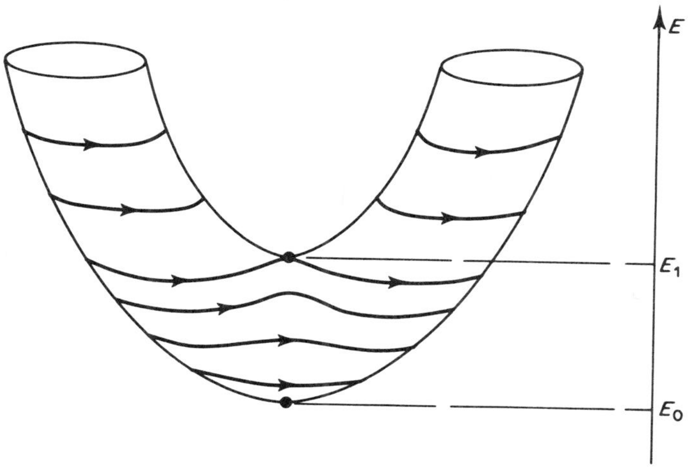

Fig. 14.2. A redrawing of Fig. 7.9. The cylinder of Fig. 7.9 is laid on its side and bent so that the stable equilibrium point (called $E = -2$ in Fig. 7.9, but E_0 here) is on the bottom. The point called E_1 here was called $E = 0$ in Fig. 7.9

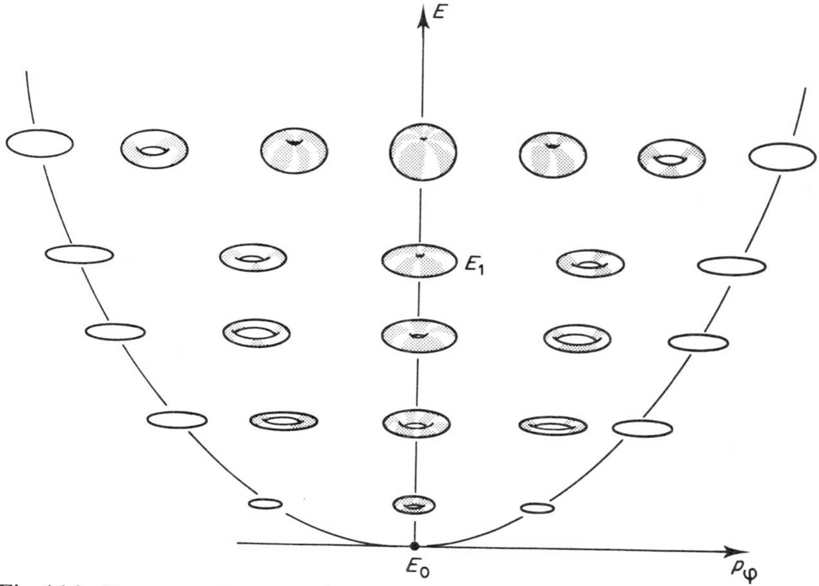

Fig. 14.3. Representations of $\mu^{-1}(E, p_\varphi)$ for some values of E and p_φ, drawn on the background of Fig. 14.1. As one moves up the E axis, the hole in the torus gets smaller and smaller, disappearing at E_1. At this point the image is a sphere with the north and south poles identified. Above E_1, the image on the E axis is a sphere, drawn here with an identation to show its connection both with the torus on the lower part of the E axis and with the torus at this value of E but with $p_\varphi \neq 0$. On the boundary β the image is always a circle (except at $E = E_0$)

between the energy and the kind of orbit. At this point the hole in the torus has disappeared, and it becomes a figure-eight rotated about a line perpendicular to the common axis of the two loops (as illustrated in Fig. 14.3). At energies higher than E_1, still on the E axis, the inverse image becomes the sphere S^2, for the two disconnected orbits of the plane pendulum at high energy are connected in the circular one: a rotation by π in φ will bring a clockwise orbit into one which is counterclockwise.

Off the E axis, when $p_\varphi \neq 0$, the so-called centrifugal barrier will not allow θ to get to 0 or to π, so that again θ oscillates, and again the inverse image of μ is a torus. As one approaches the E axis and p_φ becomes smaller, the hole of the torus created by the centrifugal barrier also becomes smaller, disappearing at the E axis, where the inverse image of μ becomes a sphere.

Fig. 14.3 is an illustration of $\mu^{-1}(c)$ for various values of $c \in \mu(M)$. It is seen to be a torus everywhere except on β and on the E axis above E_1. At all those points where it is a torus, action-angle variables can be used.

14.2 THE HAMILTON–JACOBI METHOD

1. Finding a different carrier manifold for a Hamiltonian dynamical system is equivalent, as has been said already, to finding a different symplectic atlas for ω. Such an atlas is particularly useful when the coordinates in it include constants of the motion, for then some of the canonical equations become trivial. From this point of view, the more constants of the motion among the coordinates, the better. The Hamilton–Jacobi method is then probably the most ambitious such program, in which one tries to maximize the number of constant coordinates, in fact tries to find $2\nu - 1$ of them. As in the action-angle approach, the Hamiltonian will not depend on the canonical conjugates of the constants of the motion (this was observed at the beginning of §14.1), and then if the Hamiltonian is chosen as one of the constant coordinates, the equations of motion become marvelously simple.

Remark. In the classical local treatment one says that all of the Hamilton–Jacobi variables are constants of the motion. But, with our definition, a constant of the motion cannot depend on the time explicitly, and therefore there cannot exist 2ν of them. In fact it is clear that if the Hamiltonian is one of the variables, its conjugate is essentially the time up to an additive constant. In the classical treatment it is the difference between this variable and the time which is sometimes called the 2ν-th constant of the motion.

Roughly speaking, then, the intention is to find a symplectic chart in which

$$\omega = dH \wedge dt + \omega_1, \tag{13}$$

where $i_\Delta \omega_1 = 0$, so that $i_\Delta \omega = dH$ implies that $L_\Delta t = -1$. This can be visualized on $\mathbb{R}^{2\nu}$ with symplectic coordinates $\{x^1, \ldots, x^\nu, y_1, \ldots, y_\nu\}$, so that $\omega = dy_j \wedge dx^j$, if the Hamiltonian is taken to be $H = x^1$, the dynamical vector field is $\Delta = -\partial/\partial y_1$,

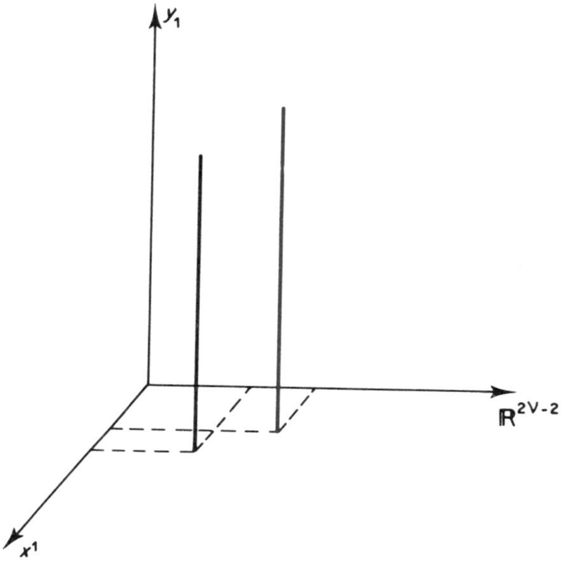

Fig. 14.4. A simple example of Hamilton–Jacobi variables

and all of the $\{x, y\}$ variables other than y_1 are constants of the motion. Moreover $L_\Delta y_1 = -1$, and ω is of the form of (13). Fig. 14.4 is an illustration of the situation. The vertical lines are integral curves, and along them the values of all the x^k remain constant, as do those of all the y_k other than y_1.

It has been mentioned that symplectomorphisms in general lose track of the fact that the carrier manifold $M = T^*Q$ is a cotangent bundle over the base space Q. This was true also on going to action-angle variables, and is again true for the Hamilton–Jacobi method. The zero section obtained by setting $y_k = 0$ in the example above for all $k \in \{1, \ldots, v\}$ need not be Q or any subset of Q. In the Hamilton–Jacobi method, one tries to make M a fiber space in a different way. One writes $m_1 \approx m_2$ for $m_1, m_2 \in M$ iff m_1 and m_2 lie on the same integral curve of the dynamical vector field Δ. Since this clearly is an equivalence relation, it yields a set M_Δ of equivalence classes and a projection $\pi_\Delta : M \to M_\Delta$. But in general the $(M, \pi_\Delta, M_\Delta)$ constructed in this way is not a fiber bundle, for M_Δ will not always be Hausdorff and two 'fibers' need not be diffeomorphic (for instance, some orbits may be closed and others open). If this were a fiber bundle, it would be possible to find not only a chart, but an atlas for M with the properties desired, and the problem of integrating the dynamics would essentially be solved. In general it is only locally, in regions of the carrier manifold, that one can replace the integration problem by the problem of finding a symplectomorphism to Hamilton–Jacobi coordinates. Thus one is often forced to work in local coordinates when using the Hamilton–Jacobi method.

2. An interesting example on which one can construct this kind of fibration is

the isotropic harmonic oscillator in two dimensions. Let

$$H = \tfrac{1}{2}(p_x^2 + x^2) + \tfrac{1}{2}(p_y^2 + y^2). \tag{14}$$

With the usual symplectic form

$$\omega = dx \wedge dp_x + dy \wedge dp_y \tag{15}$$

this Hamiltonian yields the well-known equations of motion for the vector field Δ of the harmonic oscillator. We wish to establish the form of M_Δ is this example (Hopf, 1931).

To do so we transform first to the complex coordinates

$$z_1 = p_x - \mathrm{i}x, \quad z_2 = p_y - \mathrm{i}y, \tag{16}$$

in which the Hamiltonian becomes

$$H = [|z_1|^2 + |z_2|^2]/2, \tag{17}$$

the symplectic form becomes

$$\omega = (\mathrm{i}/2)(dz_1 \wedge d\bar{z}_1 + dz_2 \wedge d\bar{z}_2), \tag{18}$$

and the equations of motion become

$$\dot{z}_k = -\mathrm{i}z_k. \tag{19}$$

(Incidentally, it follows that the four functions $f_{jk} = z_j\bar{z}_k$ are constants of the motion; they contain four real constants of the motion, only three of which are independent.) It is perhaps a little easier to analyze the form of M_Δ from the vantage point of these complex variables. In any case, whether in the real or complex variables, it is easily seen that the level sets of H, which are of course invariant submanifolds, are all diffeomorphic to the 3-sphere S^3. Consider one of these constant-H spheres. From the equations of motion it is seen that

$$z_j(t) = \lambda_j \mathrm{e}^{-\mathrm{i}t}, \tag{20}$$

where the λ_j are complex constants, so that the value of H on the sphere is $\frac{1}{2}\sum|\lambda_j|^2$. If for simplicity the sphere is chosen with $H = 1/2$, the system point $z \equiv \{z_1, z_2\}$ is a point in $\mathbb{C}^2$ of modulus one.

We now assert that the quotient of S^3 with respect to the motion, or as we shall say, with respect to Δ, is

$$S^3/\Delta \equiv \mathbb{C}P^1, \tag{21}$$

where $\mathbb{C}P^1$ is the complex projective plane, diffeomorphic to S^2, which is defined as follows. Let $z, w \in [\mathbb{C}^2] \equiv \mathbb{C}^2 - \{0\}$, and define an equivalence relation on $[\mathbb{C}^2]$ by $z \approx w$ iff there exists a complex number $\lambda \in \mathbb{C}$ such that $z = \lambda w$. Then the projection of $[\mathbb{C}^2]$ with respect to this equivalence is defined as $\mathbb{C}P^1$; it consists essentially of the 'rays' in $\mathbb{C}^2$. It should be noted that as the integral curves, or rather the trajectories, are circles, Eq. (21) defines a certain fibration of the 3-sphere into a base manifold which is the 2-sphere and fibers which are circles Hopf (1931).

For the proof of (21), note first that according to (20) two points lie on a trajectory iff there is some value of t such that one is e^{-it} times the other, i.e. iff they differ only in phase. Let us call two such points dynamically equivalent, and then the left-hand side of (21) is just the projection with respect to this dynamical equivalence. Now in the definition above of $\mathbb{C}P^1$ let $\lambda = re^{-i\varphi}$. Fix λ by fixing first r and then φ. When r is fixed, $S^3 \subset [\mathbb{C}^2]$ is obtained, diffeomorphic to the level sets of H. Then fixing φ is exactly fixing the phase on S^3, as in the dynamical equivalence. It is thus seen that in the definition of $\mathbb{C}P^1$ the manifold $[\mathbb{C}^2]$ is projected onto a manifold diffeomorphic to the one obtained by projecting any level set of H with respect to the dynamical equivalence. This establishes (21). Since $\mathbb{C}P^1$ is diffeomorphic to S^2, it follows that $S^3/\Delta = S^2$.

All that is left to characterize M_Δ in this example is to include the possible values of H, which parametrize the set of nontrivial 3-spheres. It follows almost immediately that $M_\Delta = S^2 \times \mathbb{R}^+$.

3. One way to approach the problem of finding a Hamilton–Jacobi symplectomorphism, i.e. one that leads to Hamilton–Jacobi coordinates on the carrier manifold, is to parametrize the full set of symplectomorphisms and then to restate the requirement for a Hamilton–Jacobi symplectomorphism as an equation involving the parameters. This will require a restriction to a subclass of the symplectomorphisms, called the *free transformations*.

Let φ be a symplectomorphism. Then φ uniquely determines a closed one-form α_φ by

$$\alpha_\varphi = \varphi_* \theta_0 - \theta_0, \tag{22}$$

where θ_0 is the natural one-form defined by Eq. (8) of Chapter 11, with the property that $\omega = -d\theta_0$ (recall that we omit the subscript 0 on ω_0). That α_φ is closed follows from the fact that φ is symplectic:

$$d\alpha_\varphi = \omega - \varphi_* \omega = 0.$$

The association $\varphi \mapsto \alpha_\varphi$ is not one-to-one, however, for $\alpha_\varphi = \alpha_\psi$ implies that $\varphi_* \theta_0 = \psi_* \theta_0$, or that

$$(\varphi \circ \psi^{-1})_* \theta_0 = \theta_0. \tag{23}$$

Then according to Remark 2 following Eq. (1) of Chapter 12, this means not that $\varphi \circ \psi^{-1}$ is the identity, but only that it is the lift of a diffeomorphism on the base. The symplectomorphisms on $M = T^*Q$ can therefore not be parametrized by the one-forms.

It is instructive to see this also from the local point of view. Since α_φ is closed, about every $m \in M$ there is a neighborhood U and a local function $F \in \mathscr{F}(U)$ such that

$$\alpha_\varphi|_U = dF. \tag{24}$$

Note that F depends on φ. Let U and $\varphi(U)$ be the domains of local coordinates $\{q, p\}$ and $\{Q, P\}$, respectively, so that as in §12.1 one may write $\varphi:(q, p) \mapsto (Q, P)$.

The statement that φ is symplectic can now be written locally in the form

$$dq^k \wedge dp_k = dQ^k(q,p) \wedge dP_k(q,p), \tag{25}$$

and Eq. (22) becomes

$$dF(q,p) = P_k(q,p)dQ^k(q,p) - p_k dq^k. \tag{26}$$

From this one obtains a set of differential equations for F:

$$\begin{aligned} \partial F/\partial q^k &= P_j \partial Q^j/\partial q^k - p_k, \\ \partial F/\partial p_k &= P_j \partial Q^j/\partial p_k. \end{aligned} \tag{27}$$

Suppose now that φ is given, which means locally that the functions $Q^k(q,p)$ and $P_k(q,p)$ are given. Then (27) determines $F(q,p)$ uniquely up to an additive constant, and therefore α_φ uniquely (locally). This reflects the fact that also globally the one-form is determined uniquely by the symplectomorphism. On the other hand, given a function $F(q,p)$ defined on U, Eqs. (27) do not determine the $Q^k(q,p)$ and $P_k(q,p)$ functions uniquely. This also reflects the global situation, in which the symplectomorphism is not determined uniquely by the one-form.

In general, therefore, one-forms (or, for that matter, functions) cannot be used to parametrize the symplectomorphisms. Suppose, however, that only those symplectomorphisms are considered for which the set $\{q, Q\}$ provides local coordinates, that is, for which the Jacobian

$$\det|\partial(q,Q)/\partial(q,p)| = \det|\partial Q/\partial p|$$

fails to vanish in U. Such a symplectomorphism is sometimes called a free transformation. One may now choose $\{q, Q\}$ rather than $\{q, p\}$ as the independent variables in U, and if $f'(q,Q)$ is defined by $f'(q,Q) = f(q, p'(q,Q))$ for any function $f \in \mathscr{F}(U)$, Eq. (26) can be rewritten in the form

$$dF'(q,Q) = P'_k(q,Q)dQ^k - p'_k dq^k \tag{26'}$$

and (27) in the form

$$\begin{aligned} p'_k(q,Q) &= -\partial F'/\partial q^k, \\ P'_k(q,Q) &= \partial F'/\partial Q^k. \end{aligned} \tag{27'}$$

In analogy with the previous result, this defines $F'(q,Q)$ uniquely up to an additive constant, but now it can be shown that if a function $F'(q,Q)$ is given such that

$$\det|\partial^2 F'/\partial q \partial Q| \neq 0, \tag{28}$$

then Eqs. (27′) uniquely define a local symplectomorphism, and therefore that such functions can be used to parametrize the local free transformations (see Arnol'd, 1978, §49).

Another way to arrive at the result of (27′) is to treat F not as a function defined on a neighborhood U of M, but on a neighborhood of $U \times \varphi(U)$. In fact, Eq. (26) can be treated in this way: all of the forms and functions in that equation may be considered as defined not on U but on $U \times \varphi(U)$, where U is the range of the $\{q, p\}$ and $\varphi(U)$ is the range of the $\{Q, P\}$. Then Eq. (26) reads

$$dF = p_k dq^k - P_k dQ^k. \tag{29}$$

This shows that in fact F depends only on the q^k and Q^k and therefore that p_k and P_k can be written as functions of $\{q, Q\}$.

In the same spirit of doubling the dimension of the manifold, a global version of the above parametrization can be obtained by starting with functions $f \in \mathscr{F}(Q \times Q)$. Then two one-forms $d_1 f$ and $d_2 f$ are defined by

$$d_1 f \equiv d[f | Q \times \{q\}], \quad d_2 f \equiv d[f | \{q\} \times Q] \; \forall q \in Q.$$

In this way $d_1 f(q_1, q_2)$ defines a mapping which carries each q_2 in the second Q of $Q \times Q$ into an element of $T^*_{q_1} Q$, where q_1 is in the first Q. The roles of q_1 and q_2 can be interchanged in this prescription, as well, and then if the mappings are invertible (this is a restriction on f, corresponding to Eq. (28)), one arrives in this way at a map $\varphi_f : T^*Q \to T^*Q$ which is symplectic and reduces in local coordinates to (27′) with F' replaced by f. We do not go into details.

Remark. The functions here called f and F' are known in the classical literature as generating functions of the free canonical transformations. They are defined not on $M = T^*Q$ or submanifolds of M, but on objects like $Q \times Q$ or even, as in the approach of Eq. (29), on $M \times M$ (although it turns out that for free transformations they reduce in this last case also to functions on $Q \times Q$).

4. How is all of this related to the well-known Hamilton–Jacobi equation? Let $U \subset Q$, and consider T^*U with coordinates $\{q, p\}$ (recall that one is often forced into local coordinates in the Hamilton–Jacobi method). The first object is to find a free transformation $\{q, p\} \to \{Q, P\}$ such that in terms of the new coordinates the Hamiltonian is a function only of the Q^k, i.e. such that

$$H(q, p) = K(Q). \tag{30}$$

It follows then that $H'(q, Q)$ is independent of the q^k and therefore equal to $K(Q)$. In other words,

$$H'(q, Q) \equiv H(q, p'(q, Q)) = H(q, -\partial f / \partial q) = K(Q), \tag{31}$$

where f replaces the F' of Eq. (27′). This equation may now be viewed as a differential equation for the generating function f which defines the free transformation in accordance with Eq. (27′). Now, whether or not K depends on all of the Q^k or only on some of them, all of them are constants of the motion, for $\partial K / \partial P_k = 0$ for all k. The similarity with action-angle variables is striking at this point: if K depends on all of the Q^k this is the action-angle case, although here the constants of the motion are thought of as position variables, whereas the J_k are usually thought of as generalized momenta. If, however, for some j the Hamiltonian is independent of Q^j, then not only Q^j, but also P_j is a constant of the motion.

Suppose now that one of the Q^k, say Q^1, is the Hamiltonian, so that $K(Q) = Q^1$. Then all of the Q^k and all of the P_k except P_1 are constants of the motion, and the equation of motion for P_1 is simply

$$\dot{P}_1 = -1. \tag{32}$$

Eq. (31) then becomes

$$H(q, -\partial f/\partial q) = Q^1. \tag{33}$$

This is the Hamilton–Jacobi equation. Its solution is the generating function for a local symplectomorphism such that all but one of the new coordinates in U are constants of the motion, and one of these is the Hamiltonian itself. This is what was to be found.

Among the solutions to this equation are those that are called complete. These depend on ν parameters, which may be called the Q^k, in such a way that Eq. (28) is fulfilled, and then (27′) yields the desired local symplectomorphism.

Both of the techniques discussed in this chapter depend on finding many constants of the motion. Since dynamical systems do not in general have many global constants of the motion, this approach usually forces the analysis to be local. To develop global techniques takes considerably more work and cannot rely so greatly on finding constants of the motion. The approaches that will be discussed later will depend on such geometric objects as distributions and foliations and will rely eventually on Lie algebras and groups and on their actions on manifolds. Nevertheless, before proceeding with those other approaches, we turn in the next chapter to the Noether theorem, which connects constants of the motion with symmetries of dynamical systems.

15

The Noether theorem

Because constants of the motion can be used as in Chapter 10 to reduce the dimension of the dynamical system, a general method for finding them, when they exist, would be useful. Unfortunately, no reliable recipe exists for finding them, though certain classes of systems always lead to certain constants of the motion. For example, if the system is Hamiltonian, the Hamiltonian function is necessarily a constant of the motion. Another example: in central force problems the angular momentum is always conserved. Other similar examples can be found.

One method for finding constants of the motion, related to these examples, is the one provided by the Noether theorem. This method has proved quite useful in classical and quantum mechanics and in field theory, and it has therefore received great attention in the literature. It associates a *symmetry* of the dynamical system, broadly speaking, with a constant of the motion, and thus if one can easily recognize a symmetry, one is led simply to a constant. Precisely what is meant here by 'dynamical system' has varied over the years, but now it is almost always taken to be the Lagrangian function $\mathscr{L} \in \mathscr{F}(TQ)$. What the Noether theorem says is something like this: if the Lagrangian is invariant under a one-parameter group of transformations (diffeomorphisms) φ_τ on TQ, then there is a constant of the motion which can immediately be associated with this invariance, or rather with the *symmetry group* φ_τ. Over the last fifty years or so the theorem has been discussed by many authors and many attempts have been made to generalize it and to obtain its converse. The generalizations have usually involved extending the meaning of symmetry (or relaxing the requirement of strict invariance of the Lagrangian). The converse would state that with every constant of the motion can be associated a one-parameter symmetry group for the Lagrangian $\mathscr{L}$.

15.1 THE HAMILTONIAN FORMALISM

1. A Noether type of theorem, one connecting a symmetry to a constant of the motion, can be stated in the Hamiltonian formalism as well as in the Lagrangian, perhaps even more easily. Let $Y, Z \in \mathfrak{X}_{\mathscr{H}}(M)$ be Hamiltonian vector fields (the

subscript $\mathscr{H}$ stands for 'Hamiltonian') on a symplectic manifold (M, ω):

$$i_Y\omega = dF_Y, \quad i_Z\omega = dF_Z, \quad F_Y, F_Z \in \mathscr{F}(M).$$

Then it follows that

$$L_Y F_Z = i_Y dF_Z = i_Y i_Z \omega = -L_Z F_Y, \tag{1}$$

and in particular that $L_Y F_Z = 0$ iff $L_Z F_Y = 0$.

Now let Δ be a Hamiltonian dynamical system with Hamiltonian function H, and let $X \in \mathfrak{X}_{\mathscr{H}}(M)$ be such that $L_X H = 0$. The vector field X defines its Hamiltonian function $F \in \mathscr{F}(M)$ up to an additive constant by the relation $i_X\omega = dF$, and then it follows immediately from (1) that F is a constant of the motion: $L_\Delta F = 0$. Conversely, let F be a constant of the motion for Δ, and let X be the vector field for which $i_X\omega = dF$. Then it follows from (1) that $L_X H = 0$.

These two statements may be taken as the *Hamiltonian Noether theorem* and its converse. When X is complete, they can be interpreted as follows. The vector field X defines by its flow (see §5.5) the one-parameter group of diffeomorphisms φ_τ^X of which it is said to be the *infinitesimal generator* (in the literature this term is sometimes applied to F). Then to say that $L_X H = 0$ is to say that $(\varphi_\tau^X)^* H = H$, or that H is invariant under the one-parameter group φ_τ^X of symplectomorphisms, which itself is said to be a symmetry for H. Thus F is a constant of the motion iff it gives rise in this way to (generates) a one-parameter group of symplectomorphisms which leave H invariant.

If X is locally Hamiltonian (so that $L_X\omega = 0$) and $L_X H = 0$, then it follows that

$$0 = dL_X H = L_X i_\Delta \omega = i_{[X,\Delta]}\omega, \tag{2}$$

and thus

$$L_X H = 0 \Rightarrow [X, \Delta] \equiv L_X \Delta = 0: \tag{3}$$

a symmetry of the Hamiltonian is a symmetry of Δ. The converse is not true, however, for from $dL_X H = 0$ follows only that $L_X H = \text{const}$. From (2) it also follows that if X is Hamiltonian more than locally, then $L_\Delta F = \text{const.}$, where $i_X\omega = dF$, as before.

2. In general a Noether type of theorem can be understood as a prescription for associating a symmetry (like that of H) with an invariant structure, that is with some object (like F) whose Lie derivative along Δ vanishes, but there is no reason why such associations should not be obtained by procedures other than the Noether theorem described above. For example, let $\varphi \in \mathrm{Diff}(M)$ be a symmetry for the Hamiltonian dynamical vector field Δ, so that $\varphi_* \Delta = \Delta$. It is well known (this is also called Liouville's theorem) that $L_\Delta \omega^\nu = 0$, where $2\nu = \mu = \dim(M)$ and ω^ν is the νth power with respect to the exterior product. Now, because ω^ν is a μ-form on M and the set of μ-forms is a one-dimensional $\mathscr{F}$-module, every μ-form α can be written $\alpha = f\omega^\nu$ with some $f \in \mathscr{F}(M)$. Thus the equation $\varphi^*\omega^\nu \equiv (\det \varphi)\omega^\nu$ defines a function $\det \varphi$, and this function, as will now

be shown, is a constant of the motion. Indeed,

$$0 = \varphi^* L_\Delta \omega^v = L_{\varphi_* \Delta} \varphi^* \omega^v = L_\Delta(\det \varphi \cdot \omega^v) = (L_\Delta \det \varphi)\omega^v,$$

so that

$$L_\Delta \det \varphi = 0.$$

For another example, again with Hamiltonian Δ, let $X \in \mathfrak{X}(M)$, $[X, \Delta] = 0$, $g = i_X i_\Delta \omega$. Then

$$L_\Delta g = i_{[X,\Delta]} i_\Delta \omega + i_X i_\Delta L_\Delta \omega = 0,$$

so that g is a constant of the motion.

On the other hand such associations of symmetries with invariances are not inevitably fruitful. It is easily seen, for instance, that in the above example

$$dg = L_X i_\Delta \omega - i_X L_\Delta \omega = L_{[X,\Delta]} \omega + i_\Delta L_X \omega = i_\Delta L_X \omega,$$

so that if X is locally Hamiltonian, $dg = 0$ and $g = \text{const.} \equiv C$, which is only trivially a constant of the motion.

All of the information for the case of a Hamiltonian dynamical vector field Δ is contained in the following diagram, where $\alpha \equiv i_\Delta \omega$, $\beta \equiv i_X \omega$, and $X \in \mathfrak{X}_{\mathscr{H}}(M)$ is the infinitesimal generator of a symmetry for H.

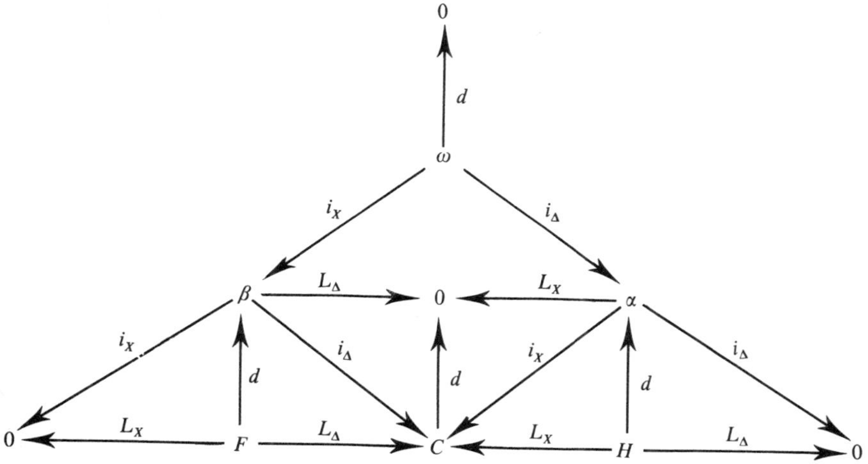

15.2 THE LAGRANGIAN FORMALISM

1. We now turn to the Noether theorem in its usual setting, the Lagrangian formalism. Let $\mathscr{L} \in \mathscr{F}(TQ)$ be the Lagrangian function, and let Δ be the

dynamical vector field, so that with

$$\theta_{\mathscr{L}} \equiv d_{\mathrm{v}}\mathscr{L} \equiv i_{\mathrm{v}}d\mathscr{L}$$

(recall Chapter 10) the equation of motion is

$$L_{\Delta}\theta_{\mathscr{L}} = d\mathscr{L}, \tag{4}$$

or equivalently

$$i_{\Delta}\omega_{\mathscr{L}} = dE_{\mathscr{L}}; \quad E_{\mathscr{L}} = i_{\Delta}\theta_{\mathscr{L}} - \mathscr{L}, \quad \omega_{\mathscr{L}} = -d\theta_{\mathscr{L}}. \tag{5}$$

Consider now another complete vector field $X \in \mathfrak{X}(TQ)$, the infinitesimal generator of a one-parameter group φ_τ of diffeomorphisms on TQ. Let us calculate the *infinitesimal transformation* which X induces on $\mathscr{L}$, which it is traditional to write in the form $\delta\mathscr{L} \equiv L_X\mathscr{L}$. It is easily shown that

$$\delta\mathscr{L} = i_Z\theta_{\mathscr{L}} + L_{\Delta}(i_X\theta_{\mathscr{L}}), \tag{6}$$

where $Z = [X, \Delta]$. (Note that $Z = L_X\Delta$, which becomes $Z = \delta\Delta$ in the traditional notation.) Indeed,

$$L_{\Delta}i_X\theta_{\mathscr{L}} = -i_Z\theta_{\mathscr{L}} + i_XL_{\Delta}\theta_{\mathscr{L}} = -i_Z\theta_{\mathscr{L}} + i_Xd\mathscr{L},$$

and Eq. (6) follows immediately.

Eq. (6) is the identity which is taken as the starting point for the usual classical Noether theorem. Roughly speaking, in that theorem it is shown that the first term on the right-hand side of (6) vanishes, and then if X is a symmetry (an infinitesimal symmetry) for $\mathscr{L}$, i.e. if $\delta\mathscr{L} = 0$, then $i_X\theta_{\mathscr{L}}$ is clearly a constant of the motion.

Remark. Prompted by this description of the Noether theorem, one might ask under what condition $i_X\theta_{\mathscr{L}}$ will be a constant of the motion. Clearly all that is necessary is that $i_Z\theta_{\mathscr{L}} - \delta\mathscr{L}$ vanish, which is less stringent than the separate vanishing of each of the terms.

We start now by showing why the first term on the right-hand side of (6) vanishes in the usual classical treatment. That treatment involves diffeomorphisms which are called *point transformations*; in our terms, a point transformation φ is the lift to TQ of a diffeomorphism ψ on Q, or $\varphi = T\psi, \psi \in \mathrm{Diff}(Q)$. It was shown in Chapter 9 that a diffeomorphism is of this type iff it is Newtonian, i.e. iff it carries second-order vector fields into second-order vector fields. The infinitesimal generator X^T of a one-parameter group of such diffeomorphisms was described in §5.5(2); it is also called Newtonian. Assume, therefore, that φ_τ is a one-parameter group of Newtonian diffeomorphisms, so that X in (6) is replaced by X^T, and so that $Z = [X^T, \Delta]$. Then, as will be shown, Z is a vertical vector field, and this causes the first term on the right-hand side of (6) to vanish. Indeed, since Δ is of second order and φ_τ is Newtonian, $\varphi_{\tau *}\Delta \equiv (\varphi_\tau)_*\Delta$ is of second order and can therefore be written (in one of the notations of Chapter 5) in the form

$$\varphi_*\Delta = (q, \dot{q}; \dot{q}, Y(\tau)).$$

Then

$$Z(q,\dot{q}) = L_{X^T}\Delta(q,\dot{q}) = \lim_{\tau\to 0}[(\varphi_{\tau*}\Delta)(q,\dot{q}) - (\varphi_{0*}\Delta)(q,\dot{q})]/\tau$$

$$= \lim_{\tau\to 0}[(q,\dot{q};\dot{q},Y(\tau)) - (q,\dot{q};\dot{q},Y(0))]/\tau$$

$$= \lim_{\tau\to 0}(q,\dot{q};0,Y(\tau) - Y(0))/\tau$$

$$= (q,\dot{q};0,Z_{\dot{q}}).$$

This means precisely that Z is *vertical*, i.e. that $vZ = 0$ (see Eq. (A3) of Chapter 10 and the sentence following it). But then

$$i_Z\theta_{\mathscr{L}} = i_Z d_v\mathscr{L} = i_Z i_v d\mathscr{L} = i_{vZ}d\mathscr{L} = 0. \tag{7}$$

Thus for X^T Newtonian, Eq. (6) becomes

$$\delta\mathscr{L} = L_\Delta(i_{X^T}\theta_{\mathscr{L}}). \tag{8}$$

Noether's theorem now follows in its usual classical form: Let φ_τ be a one-parameter group of Newtonian diffeomorphisms and X^T be its infinitesimal generator. If $\mathscr{L}$ is invariant under φ_τ, i.e. if $\delta\mathscr{L} = 0$, then $i_{X^T}\theta_{\mathscr{L}}$ is a constant of the motion for the dynamical field Δ obtained from $\mathscr{L}$.

With these conditions, moreover, X^T is $\omega_{\mathscr{L}}$-Hamiltonian (recall that a similar statement had to be imposed in the Hamiltonian version of the Noether theorem) and φ_τ is a symmetry for Δ and for $E_{\mathscr{L}}$. The proof of these statements requires showing first that $L_{X^T}\theta_{\mathscr{L}} = \theta_{L_{X^T}\mathscr{L}}$, or that $\delta\theta_{\mathscr{L}} = \theta_{\delta\mathscr{L}}$. This result does not itself depend on $\delta\mathscr{L}$ vanishing; it may be seen as follows. Let $X^T \in \mathfrak{X}(TQ)$ be Newtonian and $f \in \mathscr{F}(TQ)$, and consider $\theta_f \equiv v^*df \equiv i_v df$ (again, refer to the Appendix of Chapter 10 for the vertical operations). Let $Y \in \mathfrak{X}(TQ)$ be arbitrary, and form

$$i_Y\delta\theta_f \equiv i_Y L_{X^T}v^*df = L_{X^T}i_Y v^*df - i_{[X^T,Y]}v^*df.$$

If α is a one-form, $i_X v^*\alpha = i_{X_v}\alpha \equiv i_{vX}\alpha$; therefore

$$i_Y\delta\theta_f = L_{X^T}i_{vY}df - i_{v[X^T,Y]}df. \tag{9}$$

Now form

$$i_Y\theta_{\delta f} \equiv i_Y v^* L_{X^T}df \equiv i_{vY}L_{X^T}df. \tag{10}$$

Eq. (9) and (10) together yield

$$\begin{aligned} i_Y(\delta\theta_f - \theta_{\delta f}) &= L_{X^T}i_{vY}df - i_{v[X^T,Y]}df - i_{vY}L_{X^T}df \\ &= (i_{[X^T,vY]} - i_{v[X^T,Y]})df \\ &= i_{([X^T,vY]-v[X^T,Y])}df. \end{aligned} \tag{11}$$

We shall show in local coordinates that this expression vanishes for any vector field $X \in \mathfrak{X}(Q)$. In one of the notations of Chapter 5 one may write, for $X \in \mathfrak{X}(TQ)$,

$$X = X_q^j\partial/\partial q^j + X_{\dot{q}}^j\partial/\partial\dot{q}^j$$

and

$$X_v \equiv vX = X_q^j \partial/\partial \dot{q}^j.$$

In this notation

$$[X, Y]_r^k = X_q^j \partial Y_r^k/\partial q^j + X_{\dot{q}}^j \partial Y_r^k/\partial q^j - Y_q^j \partial X_r^k/\partial q^j - Y_{\dot{q}}^j \partial X_r^k/\partial q^j,$$

where r stands for either q or $\dot{q}$. Then

$$[X, vY]_q^k = -Y_q^j \partial X_q^k/\partial \dot{q}^j,$$
$$[X, vY]_{\dot{q}}^k = X_q^j \partial Y_q^k/\partial q^j + X_{\dot{q}}^j \partial Y_q^k/\partial \dot{q}^j - Y_q^j \partial X_{\dot{q}}^k/\partial \dot{q}^j,$$
$$v[X, Y]_q^k = 0,$$
$$v[X, Y]_{\dot{q}}^k = X_q^j \partial Y_q^k/\partial q^j + X_{\dot{q}}^j \partial Y_q^k/\partial \dot{q}^j - Y_q^j \partial X_q^k/\partial q^j - Y_{\dot{q}}^j \partial X_q^k/\partial \dot{q}^j.$$

If $X = X^T$ is Newtonian, $\partial X_q^k/\partial \dot{q}^j = 0$ and

$$\partial X_{\dot{q}}^k/\partial \dot{q}^j = \partial X_q^k/\partial q^j,$$

and therefore $v[X^T, Y] = [X^T, vY]$, and it follows from (11) that for Newtonian vector fields X^T

$$\delta\theta_f = \theta_{\delta f}. \tag{12}$$

It can now be shown that X^T is $\omega_{\mathscr{L}}$-Hamiltonian. In general for any $X \in \mathfrak{X}(TQ)$,

$$i_X \omega_{\mathscr{L}} = -i_X d\theta_{\mathscr{L}} = -L_X \theta_{\mathscr{L}} + d(i_X \theta_{\mathscr{L}}).$$

But for Newtonian $X = X^T$ it has already been shown that $L_{X^T}\theta_{\mathscr{L}} \equiv \delta\theta_{\mathscr{L}} = \theta_{\delta\mathscr{L}}$, and if $\mathscr{L}$ is moreover invariant under X^T, then $\theta_{\delta\mathscr{L}} = 0$. Thus

$$i_{X^T}\omega_{\mathscr{L}} = d(i_{X^T}\theta_{\mathscr{L}}),$$

and X^T is $\omega_{\mathscr{L}}$-Hamiltonian.

That X^T is a symmetry for $E_{\mathscr{L}}$ can be seen as follows. In general from (5) one has

$$\delta E_{\mathscr{L}} = L_X i_\Delta \theta_{\mathscr{L}} - \delta\mathscr{L} = i_Z \theta_{\mathscr{L}} + i_\Delta \delta\theta_{\mathscr{L}} - \delta\mathscr{L}.$$

If $X = X^T$ is Newtonian and a symmetry for $\mathscr{L}$, it then follows from (7) and (12) that

$$\delta E_{\mathscr{L}} = i_\Delta \theta_{\delta\mathscr{L}} = 0.$$

That X^T is a symmetry for Δ can be seen as follows. In general

$$i_Z \omega_{\mathscr{L}} \equiv i_{[X,\Delta]}\omega_{\mathscr{L}} = L_X i_\Delta \omega_{\mathscr{L}} - i_\Delta L_X \omega_{\mathscr{L}} = L_X dE_{\mathscr{L}} - i_\Delta L_X \omega_{\mathscr{L}}.$$

Because L_X commutes with d, if X is $\omega_{\mathscr{L}}$-Hamiltonian and a symmetry for $E_{\mathscr{L}}$, as has already been established, then $Z \equiv [X, \Delta] = 0$ (of course $\omega_{\mathscr{L}}$ is regular).

A simple example of the classical Noether theorem in action is the following.

Let $Q = \mathbb{R}^3$ and let φ_τ be the one-parameter group of Newtonian diffeomorphisms on $TQ = T\mathbb{R}^3$ which is obtained by lifting the one-parameter group ψ_τ of diffeomorphisms on Q given by

$$\psi_\tau : \mathbb{R}^3 \to \mathbb{R}^3 : x^k \mapsto \psi^k(x, \tau), \quad \psi^k(x, 0) = x^k.$$

This group lifts to

$$\varphi_\tau : T\mathbb{R}^3 \to T\mathbb{R}^3 : (x^k, \dot{x}^j) \mapsto \left(\psi^k(x, \tau), \frac{\partial \psi^k}{\partial x^j} \dot{x}^j \right).$$

Then

$$X^T = (\partial \psi^k / \partial \tau) \partial / \partial x^k + (\partial^2 \psi^k / \partial \tau \partial x^j) \dot{x}^j \partial / \partial \dot{x}^k$$

(where the derivatives with respect to τ are evaluated at $\tau = 0$) and

$$\theta_{\mathscr{L}} = (\partial \mathscr{L} / \partial \dot{x}^k) dx^k. \tag{13}$$

Then

$$i_{X^T} \theta_{\mathscr{L}} = (\partial \psi^k / \partial \tau)(\partial \mathscr{L} / \partial \dot{x}^k)$$

is conserved if $\mathscr{L}$ is invariant under φ_τ. If $\mathscr{L}$ depends on the $\dot{x}^k$ only through the kinetic energy $(1/2)\dot{x}^k \dot{x}^j \delta_{jk}$ and if ψ_τ represents rotation through an angle τ about a fixed axis, this conserved quantity is the component of angular momentum about that axis.

2. A slight generalization of the Noether theorem, often included in the classical treatments, is to relax the condition of invariance for $\mathscr{L}$. Consider a function $f \in \mathscr{F}(Q)$ and define $g \in \mathscr{F}(TQ)$ as its pullback $g = \tau_Q^* f$. Suppose there exists an f such that

$$\delta \mathscr{L} = L_\Delta \tau_Q^* f \equiv L_\Delta g. \tag{14}$$

The Lagrangian may then be said to be *quasi-invariant* under X. Then with Newtonian $X = X^T$, so that Eq. (7) is valid, Eq. (6) becomes

$$L_\Delta (i_{X^T} \theta_{\mathscr{L}} - g) = 0, \tag{15}$$

and thus $i_{X^T} \theta_{\mathscr{L}} - g$ is a constant of the motion. Moreover, the function g, so long as it is the pullback of some $f \in \mathscr{F}(Q)$, is unique up to an additive constant, as will now be shown in local coordinates. Assume $\mathscr{L}$ to be quasi-invariant, i.e. assume (14) to hold. Then

$$\begin{aligned} \delta \mathscr{L} \equiv L_X \mathscr{L} &= X_q^k \partial \mathscr{L} / \partial q^k + X_{\dot{q}}^k \partial \mathscr{L} / \partial \dot{q}^k \\ &= \dot{q}^k \partial f / \partial q^k. \end{aligned} \tag{16}$$

This equation puts restrictions on $\mathscr{L}$ (and in general on X), for the right-hand side is linear in the $\dot{q}^k$. But if a function f exists to satisfy the differential equation (16), it is clear that f is determined up to an additive constant, and therefore the same is true of its pullback g. Note that this calculation does not depend on X being Newtonian.

It also follows from the condition of Eq. (14) that for Newtonian X

$$\delta\theta_{\mathscr{L}} = d(\tau_Q^* f) \equiv dg. \tag{17}$$

Again, this will be demonstrated in local coordinates (the intrinsic proof depends on the detailed properties of the vertical operators). From Eqs. (12), (16), and the definition of θ_f one obtains

$$\delta\theta_{\mathscr{L}} = [\partial(\dot{q}^k \partial f/\partial q^k)/\partial \dot{q}^j]dq^j = (\partial f/\partial q^j)dq^j = dg,$$

as asserted. With these results it is relatively straightforward to show that X^T is $\omega_{\mathscr{L}}$-Hamiltonian, that $\delta E_{\mathscr{L}} = 0$, and that $Z = 0$. In other words, even if $\delta\mathscr{L} \neq 0$, but (14) is satisfied, all the same results are obtained as in the $\delta\mathscr{L} = 0$ case, except that the associated constant of the motion is now the one in (15). As has been mentioned, this generalization to quasi-invariance is often included in the classical treatments: $\delta\mathscr{L}$, it is said, need not be zero, but may be the total time derivative of some function only of the q^k. See the Remark following Eq. (6) of Chapter 13.

Remark. It can be shown that with Newtonian X^T a sort of converse statement can be made: If $\delta\theta_{\mathscr{L}} = df$ and $\delta\mathscr{L} = L_\Delta f$ for some $f \in \mathscr{F}(TQ)$, then f is the pullback of some function on Q. We do not prove this statement.

3. We now turn to further attempts to generalize the Noether theorem by removing the restriction that φ_τ, and hence also X, be Newtonian. Nevertheless, because (6) is an identity true for all vector fields X, it cannot tell us very much unless X is restricted somehow. For example, even if $\delta\mathscr{L} = 0$, it is not generally true that $i_X\theta_{\mathscr{L}}$ is a constant of the motion.

One might try to restrict X as little as possible by making a statement modeled roughly on the generalization of the preceding section: assume that X is such that for some $g \neq i_X\theta_{\mathscr{L}} \in \mathscr{F}(TQ)$ it turns out that $\delta\mathscr{L} = i_Z\theta_{\mathscr{L}} + L_\Delta g$. Then it follows immediately from (6) that $i_X\theta_{\mathscr{L}} - g$ is a constant of the motion. But this is in fact no restriction on X, for if any constant of the motion exists (recall that there may actually be no global ones), every $X \in \mathfrak{X}(TQ)$ satisfies this requirement with g defined by

$$g = i_X\theta_{\mathscr{L}} + F, \tag{18}$$

where $F \in \mathscr{F}(TQ)$ is any arbitrary constant of the motion. This is because with $L_\Delta F = 0$, Eq. (6) implies that

$$\delta\mathscr{L} = i_Z\theta_{\mathscr{L}} + L_\Delta(i_X\theta_{\mathscr{L}} + F) = i_Z\theta_{\mathscr{L}} + L_\Delta g$$

for arbitrary X. Thus the generalization of the Noether theorem obtained in this way would be so general as to be empty, for it would associate every vector field X with any constant of the motion F. Clearly what is needed is some restriction.

Remark. Of course it may occur that someone, in calculating $L_X\mathscr{L}$, may discover that it is equal to $L_\Delta g$ for some $g \neq i_X\theta_{\mathscr{L}}$. This would show that a constant of the motion exists and would exhibit it, but it is hardly a systematic prescription for finding such constants.

Recall that among the first advantages obtained from Newtonian X was Eq. (7), namely the vanishing of the first term $i_Z\theta_{\mathscr{L}}$ on the right-hand side of (6). As a first restriction on X, therefore, weaker than that it be Newtonian, assume that

$$i_Z\theta_{\mathscr{L}}=0, \tag{19}$$

where $Z=[X,\Delta]$ as before. It turns out, however, that this restriction alone does little to improve the situation, for now (6) becomes

$$L_X\mathscr{L}\equiv\delta\mathscr{L}=L_\Delta(i_X\theta_{\mathscr{L}}), \tag{20}$$

an identity for every X satisfying (19). Again, every such X can be associated with any constant of the motion F (if any exists) in accordance with (18), except that now $\delta\mathscr{L}=L_\Delta g$. Restriction in addition to (19) is needed if a meaningful theorem is to result.

Consider then the situation if, in addition, X is a symmetry for $\mathscr{L}$, i.e. if $\delta\mathscr{L}=0$. This is essentially the same as in the classical Noether theorem, except that X is no longer Newtonian. It follows immediately from (20) that $i_X\theta_{\mathscr{L}}$ is a constant of the motion, but it does not follow that $Z=0$, that $\delta E_{\mathscr{L}}=0$, or that X is $\omega_{\mathscr{L}}$-Hamiltonian. These last three results required Eq. (12), which is valid only for Newtonian X.

Is it possible now to relax the requirement that X be a symmetry for $\mathscr{L}$? It was shown above that to require that $\delta\mathscr{L}=L_\Delta g$ provides essentially no restriction: it is too general. It is possible, however, to restrict the form of g in this equation, thereby obtaining meaningful results. We demonstrate two ways.

First, let $g=\tau_Q^*f$, $f\in\mathscr{F}(Q)$, which yields Eq. (14). Then as was shown, g is unique up to an additive constant, and $i_X\theta_{\mathscr{L}}-g$ is a constant of the motion. In this non-Newtonian case, however, (17) does not follow, for that requires (12).

Second, instead of (14), let g satisfy (17); that is, let g be determined (up to an additive constant) by

$$\delta\theta_{\mathscr{L}}=dg. \tag{21}$$

As usual, with g still satisfying $\delta\mathscr{L}=L_\Delta g$, it follows that $i_X\theta_{\mathscr{L}}-g$ is then a constant of the motion. In this version of a generalized Noether theorem (see Candotti Palmieri and Vitale (1972)) it turns out that X is $\omega_{\mathscr{L}}$-Hamiltonian, that $\delta E_{\mathscr{L}}=0$, and that $Z=0$. Indeed,

$$i_X\omega_{\mathscr{L}}=-i_Xd\theta_{\mathscr{L}}=-L_X\theta_{\mathscr{L}}+di_X\theta_{\mathscr{L}}=-d(g-i_X\theta_{\mathscr{L}}),$$

showing that X is $\omega_{\mathscr{L}}$-Hamiltonian and that its Hamiltonian function is the associated constant of the motion. From this and the Hamiltonian–Noether theorem it follows immediately that $L_XE_{\mathscr{L}}=\delta E_{\mathscr{L}}=0$. Finally,

$$\begin{aligned}-i_Z\omega_{\mathscr{L}}&=i_Zd\theta_{\mathscr{L}}=L_Z\theta_{\mathscr{L}}-di_Z\theta_{\mathscr{L}}=L_Z\theta_{\mathscr{L}}\\&=L_XL_\Delta\theta_{\mathscr{L}}-L_\Delta L_X\theta_{\mathscr{L}}=L_Xd\mathscr{L}-L_\Delta dg=d(\delta\mathscr{L}-L_\Delta g)=0,\end{aligned}$$

so that $Z=0$.

Remark. Requiring that $L_\Delta g = \delta\mathscr{L}$, that $dg = \delta\theta_{\mathscr{L}}$ and that $V[X,\Delta]=0$ is equivalent to requiring that X be $\omega_{\mathscr{L}}$-Hamiltonian and that its Hamiltonian function be a constant of the motion.

We now present two examples. First as a trivial example of a non-Newtonian symmetry, consider the Lagrangian

$$\mathscr{L} = (1/2)a_{jk}\dot{q}^j\dot{q}^k$$

with constant coefficients $a_{jk} = a_{kj}$, and the vector field

$$X = f^j(\dot{q})\partial/\partial q^j.$$

The Lagrangian one-form in this example is $\theta_{\mathscr{L}} = a_{jk}\dot{q}^k dq^j$, and the dynamical vector field is $\Delta = \dot{q}^k\partial/\partial q^k$. Note that X is not Newtonian, and that $L_X\theta_{\mathscr{L}} \equiv \delta\theta_{\mathscr{L}} = a_{jk}\dot{q}^j df^k$, whereas $\delta\mathscr{L} = 0$ so that $\theta_{\delta\mathscr{L}} = 0$ (and if g is to be the pullback of a function on Q it also vanishes). It then follows that $i_X\theta_{\mathscr{L}} - g = a_{jk}\dot{q}^j f^k$ is a constant of the motion. This is almost obvious, for the dynamical system represents the free particle, and any function of the velocity is a constant of the motion. Notice that $[X,\Delta]=0$, but X is not $\omega_{\mathscr{L}}$-Hamiltonian for general f.

For the next example, we turn to the Kepler problem. Consider the Lagrangian

$$\mathscr{L} = (1/2)m\dot{q}^2 + c/q,$$

where $q = (q^j q^k \delta_{jk})^{1/2}$ and similarly for $\dot{q}$, and the indices range from 1 to 3. Then the dynamical vector field is given by

$$\Delta = \dot{q}^k\partial/\partial q^k - [cq^k/(mq^3)]\partial/\partial\dot{q}^k$$

and the Lagrangian one-form by

$$\theta_{\mathscr{L}} = m\dot{q}^j\delta_{jk}dq^k.$$

There will be three vector fields corresponding to the X of the Noether theorem; let us call them $X^{(p)}$, with the index p ranging also from 1 to 3. They are given by

$$\begin{aligned} X^{(p)} = &-m(2\dot{q}^j q^p - q^j\dot{q}^p - \dot{\mathbf{q}}\cdot\mathbf{q}\,\delta^{pj})\partial/\partial q^j \\ &- [c(q^2\delta^{pj} - q^p q^j)/q^3 - m\dot{q}^2\delta^{pj} + m\dot{q}^p\dot{q}^j]\partial/\partial\dot{q}^j \end{aligned}$$

where $\dot{\mathbf{q}}\cdot\mathbf{q} = q^r q^s\delta_{rs}$. Then some calculation will show that $Z^{(p)} \equiv [X^{(p)},\Delta] = 0$, so that the condition of Eq. (19) is automatically fulfilled. One can now calculate $\delta_p\mathscr{L} \equiv L_{X(p)}\mathscr{L}$:

$$\delta_p\mathscr{L} = 2mc[(\dot{\mathbf{q}}\cdot\mathbf{q})\dot{q}^p - q^2\dot{q}^p]/q^3. \tag{22}$$

If this is set equal to $L_\Delta g^p$, $g^p \in \mathscr{F}(TQ)$, the function g^p will be found to exist, but will not be uniquely defined. Now condition (21) may be added; it is found that

$$\delta_p\theta_{\mathscr{L}} = -md[(m\dot{q}^2 + c/q)q^p - m(\dot{\mathbf{q}}\cdot\mathbf{q})\dot{q}^p],$$

so that (up to an additive constant)

$$g^p = m[m(\dot{\mathbf{q}}\cdot\mathbf{q})\dot{q}^p - (m\dot{q}^2 + c/q)q^p].$$

It is easily verified that this g^p satisfies (21) and it then follows that

$$A^p \equiv g^p - i_{X(p)}\theta_{\mathscr{L}} = m[(m\dot{q}^2 - c/q)q^p - m(\dot{\mathbf{q}}\cdot\mathbf{q})\dot{q}^p]$$

is a constant of the motion. In fact A^p is the pth component of the Runge–Lenz vector.

15.3 ON THE CONVERSE OF THE NOETHER THEOREM

1. The Noether theorem in any one of its variants succeeds in associating a unique constant of the motion with a given vector field $X \in \mathfrak{X}(TQ)$, with the stipulation that the vector field may not be chosen arbitrarily, but must satisfy certain requirements (e.g. it must be a symmetry for $\mathscr{L}$). An *inverse Noether theorem* would of course do the opposite: with a given constant of the motion it would associate a unique vector field, again with certain stipulations about the vector field, and it would do this through the identity (6). One may therefore attempt to proceed in a simple and straightforward way: given a constant of the motion F, find a vector field $X \in \mathscr{F}(TQ)$ which is both a symmetry for $\mathscr{L}$ and satisfies

$$i_X\theta_{\mathscr{L}} = F. \tag{23}$$

This approach, however, will not yield a unique X, for to any solution one can add any vector field in the kernel of $\theta_{\mathscr{L}}$, and this kernel has a basis of dimension $2\nu - 1$ at each point of TQ, where $\nu = \dim Q$. The fact that X must also be a symmetry for $\mathscr{L}$ does not help very much, for the set of such symmetries is the kernel of $d\mathscr{L}$, and this set is just as large as $\ker\theta_{\mathscr{L}}$. In this straightforward sense, then, it is seen that there is no converse Noether theorem within the Lagrangian formalism.

On the other hand, the Hamiltonian Noether theorem does have a converse, and this converse can be used to associate a unique vector field $X \in \mathfrak{X}(TQ)$ with each constant of the motion $F \in \mathscr{F}(TQ)$ in the following way. The Lagrangian function $\mathscr{L}$ determines the symplectic form $\omega_{\mathscr{L}}$, and then $\omega_{\mathscr{L}}$ can be used to determine the vector field which satisfies $i_X\omega_{\mathscr{L}} = dF$. This X is uniquely determined by F, but it is no longer necessarily a symmetry for $\mathscr{L}$, although as has been seen in the Hamiltonian theorem it is a symmetry for $E_{\mathscr{L}}$ and for Δ.

2. The last observation can be thought of as a Noether *type* of theorem within the Lagrangian formalism, in the same sense that this term was used in §15.1(1), for it associates a symmetry with a constant of the motion. There are also other ways to do this within the Lagrangian formalism.

For instance let X^T be Newtonian and a symmetry for Δ (i.e. satisfying $[X^T, \Delta] = 0$), yet not a symmetry for $\mathscr{L}$. Assume, in fact, not only that $\delta\mathscr{L} \neq 0$, but that (14) is also not satisfied, so that $\delta\mathscr{L} \neq L_\Delta\tau_Q^* f$ for any $f \in \mathscr{F}(Q)$.

Then $\delta E_{\mathscr{L}}$ is a constant of the motion because $\delta\mathscr{L}$ is subordinate (see Chapter 13) to $\mathscr{L}$.

As in the Hamiltonian case, these associations are not always fruitful. It can be shown that $\delta E_{\mathscr{L}} = E_{\delta\mathscr{L}}$. If (14) is in fact satisfied then $\delta\mathscr{L}$ is linear in the $\dot{q}$s and the new constant of the motion is identically zero.

3. One way to try to obtain a converse of the Noether theorem might be to alter the theorem slightly in such a way that the resulting theorem has a converse. For this purpose, let us reconsider Eq. (14). In that equation the $\delta\mathscr{L} \equiv L_X\mathscr{L}$ on the left-hand side was consistently taken to be the Lie derivative with respect to a Newtonian vector field $X = X^T$, one which is the infinitesimal generator of a one-parameter group of diffeomophisms which carries each and every second-order vector field into a second-order one. There exist, however, diffeomorphisms on TQ which carry not every second-order vector field, but just a given one into a second-order vector field. If $\varphi\in\text{Diff}(TQ)$ carries a given second-order vector field W into a second-order vector field, we shall call φ *Newtonoid with respect to W* (recall the definition of canonoid transformation just after Eq. (18) of Chapter 13). As a vector field X is Newtonian iff $v[X, Y] = 0$ for all second-order vector fields Y, so it is Newtonoid with respect to a given second-order vector field W iff $v[X, W] = 0$.

Remark. Like the canonoid diffeomorphisms on T^*Q, the Newtonoids on TQ do not form groups, so it is impossible to define Newtonoid vector fields in terms of infinitesimal generators of one-parameter groups of Newtonoid diffeomorphisms.

In altering, or generalizing, the Noether theorem, the first thing we shall do will be to change the requirement on the vector field X with respect to which the Lie derivative is taken on the left-hand side of (14): now it is not to be Newtonian, but Newtonoid with respect to some second-order vector field W. The right-hand side will also be altered, the Lie derivative with respect to Δ changing to the Lie derivative with respect to W. We delay writing down the resulting equation, for its discussion requires the following procedure for associating a Newtonoid vector field with every $X\in\mathfrak{X}(TQ)$.

Let $X\in\mathfrak{X}(TQ)$ and let $W\in\mathfrak{X}(TQ)$ be of second order. Then X_W, a vector field Newtonoid with respect to W and associated with X, will be defined by

$$L_{X_W}L_W g \equiv L_W L_X g, \tag{24a}$$

$$L_{X_W} g \equiv L_X g, \tag{24b}$$

for every g which is the pullback of some function on Q, i.e. for $g = \tau_Q^* f$ and for every $f\in\mathscr{F}(Q)$. In local coordinates Eq. (24a) defines the action of L_X on the $\dot{q}$s and (24b) defines its action on the qs. In fact if in a natural chart one writes

$$X = A^j\partial/\partial q^j + B^j\partial/\partial\dot{q}^j, \tag{25}$$

then from (24) one obtains

$$L_{X_W}\dot{q}^j = L_W A^j,$$
$$L_{X_W}q^j = A^j = L_X q^j.$$

This result can also be written in the form

$$X_W = A^j\partial/\partial q^j + (L_W A^j)\partial/\partial \dot{q}^j. \tag{26}$$

This can be made even more explicit by writing out the Lie derivative in local coordinates, but we shall not do that. In any case, it is seen that X_W depends only on the first component of X, so that the association $X \to X_W$ is many-to-one, and it is easily shown from (26) that if the A^j are functions only of the qs, then X_W is the lift of a vector field in $\mathfrak{X}(Q)$ and therefore Newtonian.

It is now a simple matter to show, by using the fact that any two second-order vector fields differ by a vertical one, that if W and V are both of second order, then locally $X_V - X_W = L_{(V-W)}A^j\partial/\partial q^j$ or more generally that $X_V - X_W = X_{(V-W)}$.

Eq. (24b) allows (24a) to be written in the form

$$L_{[W,X_W]}g = 0 \tag{27}$$

for g, as above, the pullback of some function on Q, or

$$v[W, X_W] = 0,$$

which shows that X_W is indeed Newtonoid with respect to W.

Now the altered Eq. (14) can be written: a vector field $X \in \mathfrak{X}(TQ)$ will now be defined as an infinitesimal symmetry for $\mathscr{L}$ if there exists a function $F \in \mathscr{F}(TQ)$ such that

$$L_{X_W}\mathscr{L} = L_W F \tag{28}$$

for all second-order vector fields W. Then $\mathscr{L}$ will be called quasi-invariant in this altered sense with respect to X, and the altered Noether theorem will be stated in terms of such quasi-invariant Lagrangians. But before proceeding to the theorem itself, we illustrate the meaning of (28) in local coordinates.

When Eq. (28) is written out in local coordinates and account is taken of the fact that it is to hold for all second-order vector fields W, it splits up into the two equations

$$A^j\partial\mathscr{L}/\partial q^j + \dot{q}^k(\partial A^j/\partial q^k)(\partial\mathscr{L}/\partial\dot{q}^j) = q^k\partial F/\partial q^k,$$
$$W^k(\partial A^j/\partial\dot{q}^k)(\partial\mathscr{L}/\partial\dot{q}^j) = W^k\partial F/\partial\dot{q}^k, \tag{29}$$

where the W^k are the coefficients of the $\partial/\partial\dot{q}^k$ in the local expression for W. From these equations the auxiliary role of W becomes apparent: W allows one to avoid introducing the accelerations $\ddot{q}^j$, which would carry the discussion from TQ to T^2Q.

We now state the Noether theorem, generalized in the way we have been describing.

Let $\mathscr{L} \in \mathscr{F}(TQ)$ be a Lagrangian, Δ its dynamical vector field, and $\omega_{\mathscr{L}}$ its (nondegenerate) symplectic form, and let $X \in \mathfrak{X}(TQ)$ be an infinitesimal symmetry for $\mathscr{L}$ in the sense of Eq. (28). Then

(a) $G = F - i_X\theta_{\mathscr{L}}$ is a constant of the motion;
(b) X_Δ is a symmetry for Δ, i.e. $[X_\Delta, \Delta] = 0$.

Proposition (a) is proved immediately. Since

$$L_\Delta\theta_{\mathscr{L}} - d\mathscr{L} = 0,$$

one obtains

$$L_\Delta G \equiv L_\Delta(F - i_X\theta_{\mathscr{L}}) = L_{X_\Delta}\mathscr{L} - i_{X_\Delta}d\mathscr{L} = 0. \tag{30}$$

Proposition (b) is also easily proved. Indeed,

$$\begin{aligned} -i_{X_\Delta}\omega_{\mathscr{L}} \equiv i_{X_\Delta}\mathrm{d}\theta_{\mathscr{L}} &= L_{X_\Delta}\theta_{\mathscr{L}} - di_{X_\Delta}\theta_{\mathscr{L}} \\ &= dF - di_{X_\Delta}\theta_{\mathscr{L}} = dG, \end{aligned}$$

where (28) has been used. Then

$$i_{[\Delta, X_\Delta]}\omega_{\mathscr{L}} = L_\Delta i_X\omega_{\mathscr{L}} - i_{X_\Delta}L_\Delta\omega_{\mathscr{L}} = -L_\Delta dG = 0,$$

for both $\omega_{\mathscr{L}}$ and G are invariant under Δ. The result then follows from nondegeneracy of $\omega_{\mathscr{L}}$.

Remark. It can be shown that $L_{X_\Delta}\omega_{\mathscr{L}} = 0$.

Now the converse theorem can be stated.

Let $\mathscr{L} \in \mathscr{F}(TQ)$ be a Lagrangian, Δ its dynamical vector field, and $\omega_{\mathscr{L}}$ its (nondegenerate) symplectic form, and let $G \in \mathscr{F}(TQ)$ be any constant of the motion for Δ. Then associated with G is a unique vector field $X \in \mathfrak{X}(TQ)$ which is an infinitesimal symmetry for $\mathscr{L}$ in the sense of Eq. (28).

Consider the vector field X defined by

$$i_X\omega_{\mathscr{L}} = dG. \tag{31}$$

This implies immediately, as has been demonstrated more than once, that $[X, \Delta] = 0$, and this implies trivially that

$$v[X, \Delta] = 0,$$

or that X is already Newtonoid with respect to Δ. In other words, $X = X_\Delta$.

What must be shown is that there exists a function F such that Eq. (28) is satisfied for every second-order vector field W. We assert that $F = G + i_X\theta_{\mathscr{L}}$ is such a function. For each second-order W define the vertical vector field Y_W by $Y_W = W - \Delta$, and consider

$$L_W F = L_\Delta F + L_{Y_W} F.$$

One obtains

$$\begin{aligned} L_W F &= L_\Delta(i_X\theta_{\mathscr{L}} + G) + L_{Y_W}(i_X\theta_{\mathscr{L}}) + L_{Y_W}G \\ &= i_X d\mathscr{L} + i_{[Y_W, X]}\theta_{\mathscr{L}} + i_X L_{Y_W}\theta_{\mathscr{L}} + i_{Y_W}dG, \end{aligned} \tag{32}$$

where we have used $L_\Delta\theta_{\mathscr{L}} = d\mathscr{L}$.

Now consider the second term in the last line of (32). In general $i_X\theta_{\mathscr{L}} = i_{vX}d\mathscr{L}$ for any $X \in \mathfrak{X}(TQ)$, so that this term can be written $d\mathscr{L}(v[Y_W, X])$. If Y is vertical and X_W Newtonoid, as in this case (for $X = X_\Delta$), then $v[Y, X_W] = X_{W+Y} - X_W$. This is easily shown in local coordinates: the expressions for both sides are identical. Thus the second term becomes $d\mathscr{L}(X_{\Delta+Y_W} - X) \equiv d\mathscr{L}(X_W - X)$. What now enters is X_W, no longer X_Δ (see the discussion around Eq. (27)).

The last two terms of (32) cancel, for Y_W being vertical, $i_X L_{Y_W}\theta_{\mathscr{L}} = i_X i_{Y_W} d\theta_{\mathscr{L}} = -i_{Y_W} i_X = -i_{Y_W} dG$. Putting it all together, one arrives at

$$L_W F = L_{X_W}\mathscr{L}$$

for all second-order vector fields W. This completes the proof.

BIBLIOGRAPHY TO CHAPTER 15

Only a brief sketch can be given here of the large bibliography on the Noether theorem. Most of the classical papers are listed in Candotti Palmieri and Vitale (1972), which also contains a discussion of some possible generalizations and the possibility of inverting the classical theorem.

A detailed analysis within the context of differential geometry can be found in Takens (1977) and in Palis and do Carmo (1977).

For a recent list of relevant papers on the theorem and its inversion, and for discussion of time-dependent generalizations, see Sarlet and Cantrijn (1981).

Appendix to Part I

Time-dependent Hamiltonian dynamics

The purpose of this appendix is to show how the formalism we have been discussing, mostly Hamiltonian, but also by extension Lagrangian, can be generalized so as to include dynamical systems which depend on the time. So far we have been assuming that the vector field Δ is independent of the time. In terms of the Hamiltonian or Lagrangian functions, this means that neither $\mathscr{L}$ nor H have any time dependence. In this appendix we no longer make this assumption. The formalism so far has been developed for a carrier manifold whose dimension is even, but when the time t is added as a new variable, the dimension becomes odd. The generalization to take account of this will be in terms of what is called reduction to *the autonomous case*. We shall start in a local chart to illustrate the procedure, gradually introducing more formal intrinsic consideration. Most details will be omitted; for them the reader is referred to AM and to Arnol'd (1978).

1. Let M be a manifold of dimension μ and consider a local chart (U, φ) with coordinates x^j. A given dynamical vector field $\Delta \in \mathfrak{X}(M)$, when restricted to U, can be specified by its (local) components Δ^j, which means that

$$i_\Delta dx^j = \Delta^j. \tag{A1}$$

The equations of motion can be written in the form $L_\Delta x^j \equiv dx^j/dt = \Delta^j$, where the x^j are now interpreted as the local coordinates of an integral curve of Δ. It may be noted that if M is TQ, of dimension 2ν, the first ν coordinates can be taken as the q^k, and the last ν as the $\dot{q}^k$; then when the first ν of the Δ^j are the $\dot{q}^k$ the last ν are the F^k, the components of the force. Physically this shows that when one adds time dependence to Δ one is considering time-dependent forces.

We now add the other variable t to U, replacing U by $V = U \times I$, $I \in \mathbb{R}$ (we shall sometimes write simply $V = U \times \mathbb{R}$), and define a basis $\{\alpha^j, dt\}$ of one-forms in $V, j \in \{1, \ldots, \mu\}$, by

$$\alpha^j = dx^j - \Delta^j dt. \tag{A2}$$

The equations of motion can now be written in the form

$$i_{\tilde{\Delta}}\alpha^j = 0, \tag{A3}$$

where $\tilde{\Delta}\in\mathfrak{X}(V)$ is given by

$$\tilde{\Delta} = \Delta + \partial/\partial t. \tag{A4}$$

In addition to (A3), $\tilde{\Delta}$ satisfies $i_{\tilde{\Delta}}dt = 1$.

All other differential forms on V can be constructed by using the basis of one-forms consisting of the α^j and dt; for instance all two-forms ω such that $i_{\tilde{\Delta}}\omega = 0$ can be expressed by $\omega = \omega_{jk}\alpha^j \wedge \alpha^k$ with $\omega_{jk}\in\mathscr{F}(V)$. To find a constant of the motion, one looks for *integrating factors* $a_k\in\mathscr{F}(V)$ such that $a_k\alpha^k$ is an exact one-form, for if it is, then $a_k\alpha^k = df, f\in\mathscr{F}(V)$, and

$$i_{\tilde{\Delta}}a_k\alpha^k = 0 = i_{\tilde{\Delta}}df \equiv L_{\tilde{\Delta}}f.$$

2. In applying this approach to the Hamiltonian formalism one runs immediately into the difficulty that if $M = T^*Q$, then $\dim M = 2\nu$ and $\dim V = 2\nu + 1$ is odd. Therefore V cannot support a symplectic structure: any two-form on an odd-dimensional manifold is necessarily degenerate.

On a symplectic manifold (M, ω) the symplectic structure ω provides a pairing between Hamiltonian vector fields and functions (up to additive constants). Hamilton's canonical equations $i_\Delta\omega = dH$ are just such a pairing: the Hamiltonian function $H\in\mathscr{F}(M)$ determines $\Delta\in\mathfrak{X}(M)$ uniquely. A somewhat weaker association is provided on a manifold of odd dimension $2\nu + 1$ by a two-form ω' of the maximum rank 2ν. Again, consider a neighborhood V like the one at Eq. (A2). In that case the equation $i_{\tilde{\Delta}}\omega' = dH, H\in\mathscr{F}(V)$, determines $\tilde{\Delta}\in\mathfrak{X}(V)$ only up to the addition of a solution X of

$$i_X\omega' = 0. \tag{A5}$$

Any two solutions X, Y of (A5) are related by $Y = fX$, $f\in\mathscr{F}(V)$: the full set of solutions forms a one-dimensional distribution. To fix a particular solution from this set, another condition is needed, for instance $i_X dt = 1$.

To obtain something closer to a symplectic structure one requires that ω' be closed, or that at each point there be a neighborhood in which there exists a one-form θ such that $\omega' = -d\theta$. Such a pair (M, ω'), consisting of a manifold of dimension $2\nu + 1$ and a closed two-form ω' of rank 2ν, is called a *contact manifold*, and ω' is called a *contact structure*. Let π be the projection $\pi: V \to U:(u,t)\mapsto u$, $u\in U, t\in\mathbb{R}$, and let ω' be the contact form on V defined by $\omega' = \pi^*\omega$, where ω is the symplectic form on M. If H is a function in $\mathscr{F}(V)$ which is the pullback of a function in $\mathscr{F}(U)$, then the equation

$$i_{\tilde{\Delta}}\omega' = dH, \tag{A6}$$

together with (A4), uniquely defines a vector field $\tilde{\Delta}\in\mathfrak{X}(V)$ which can be shown to project (see § 19.1) onto the vector field Δ which is locally Hamiltonian on M (in U) and whose Hamiltonian function in U is H. This suggests a way of generalizing the Hamiltonian formalism to time-dependent vector fields and Hamiltonian functions: make H any function on V (more generally on $M = T^*Q \times \mathbb{R}$) and proceed as with Eq. (A6). It is clear that then if H is independent of t, the time-independent Hamiltonian formalism is retrieved. We shall return to this point soon.

Remarks. 1. If ω' is exact, the one-form θ can be used in place of dt to 'normalize' a solution of (A5) by $i_X\theta = 1$. Moreover, if the vector field X is so normalized and, like $\tilde{\Delta}$, has component 1 along $\partial/\partial t$ and satisfies (A3) in V, the one-form can be written $\theta = dt + \theta_k\alpha^k$, for $\{\alpha^k, dt\}$ is a basis for one-forms on V. In that case $\omega' = d\theta_k \wedge \alpha^k + \theta_k d\alpha^k$.

2. If, in addition, it were possible in general to replace the α^k by exact one-forms da^k, $a^k \in \mathscr{F}(V)$, ω' could be written $\omega' = d\theta_k \wedge da^k$. Then (A5) would imply that $L_X\theta_k = 0$, and the 'Darboux' basis which put ω' into this canonical form would provide 2ν constants of the motion for X, independent because ω' is of maximum rank.

A convenient way to write (A6) is to define a new contact form $\tilde{\omega}$ by

$$\tilde{\omega} = \omega' + dH \wedge dt, \tag{A7}$$

for then (A6) with condition (A4) becomes

$$i_{\tilde{\Delta}}\tilde{\omega} = 0. \tag{A8}$$

Now the Hamiltonian function and hence the dynamics is sort of built into the contact structure. Note that $\tilde{\omega}$ is closed because the term added to ω' is closed (in fact exact), and ω' is closed by definition.

3. In order to obtain something even more closely resembling the Hamiltonian formalism, one can add still another dimension to the manifold, raising the dimensionality to $2(\nu + 1)$. This procedure is called *reduction to the autonomous case.* Let the additional variable so added be called s, and on $V' = V \times \mathbb{R}$ (more generally on $T^*Q \times \mathbb{R} \times \mathbb{R}$) define the one-form

$$\tilde{\theta} = -s\,dt + \theta', \tag{A9}$$

where θ' is the one-form on V (more generally, on $T^*Q \times \mathbb{R}$) whose exterior derivative is $-\omega'$. Then through $\tilde{\theta}$ define the symplectic form $\Omega = -d\tilde{\theta}$ on V'. Now let H be a generalized Hamiltonian defined initially on V and pulled back to V', and assume that $\tilde{\Delta}$ satisfies (A4) (or that $i_{\tilde{\Delta}}dt = 1$); then because $i_{\tilde{\Delta}}ds = 0$, the equation

$$i_{\tilde{\Delta}}\Omega = 0 \tag{A10}$$

is satisfied where $H = -s$, for a simple calculation shows that $i_{\tilde{\Delta}}\Omega = d(H + s)$.

In this way one arrives at something resembling the Hamiltonian formalism for time-dependent dynamical systems. The configuration space has been extended to include the time, and the energy, or the Hamiltonian, plays the role of th variable canonically conjugate to t That this formalism reduces to the usual one for time-independent systems should be clear from the way the last step was taken from V to V' and from the fact, mentioned above, that the contact dynamics on V reduces to the usual dynamics in the time-independent case. Such reduction of the time-independent case thus takes place in two steps, first to V (more generally to $T^*Q \times \mathbb{R}$) and then to U (or T^*Q). We do not go into details, but briefly describe the first step. Because even in the time-dependent case H is assumed to be the pullback of a function on $T^*Q \times \mathbb{R}$, the full manifold $T^*Q \times \mathbb{R} \times \mathbb{R}$ can be treated as a fibered manifold with fiber $\mathbb{R}$ (consisting of the variable s) and base manifold $T^*Q \times \mathbb{R}$. The Hamiltonian function H on the base space can be used as a section, writing

$$H: T^*Q \times \mathbb{R} \to (T^*Q \times \mathbb{R}) \times \mathbb{R}: m \mapsto (m, H(m)),$$

and pulling back all functions, forms and vector fields through this mapping. Actually, a vector field cannot be pulled back, so $\tilde{\Delta}$ must be projectable (see § 19.1), which it always is if, for example, $L_{\tilde{\Delta}}H = 0$.

4. We now give an overview of the whole picture.

One starts with $Q \times \mathbb{R}$ for the configuration manifold, rather than Q itself, where $\mathbb{R}$ is what may be called the time axis. Then one passes to the tangent or cotangent bundle

$$T(Q \times \mathbb{R}) \approx TQ \times T\mathbb{R} \approx TQ \times \mathbb{R} \times \mathbb{R},$$
$$T^*(Q \times \mathbb{R}) \approx T^*Q \times T^*\mathbb{R} \approx T^*Q \times \mathbb{R} \times \mathbb{R}.$$

We are restricting our discussion to the second of these options: the cotangent bundle. The last factor $\mathbb{R}$ consists of the variable s conjugate to the time t. A symplectic structure Ω on such a manifold can be obtained by adding the pullbacks of symplectic structures on T^*Q and on $T^*\mathbb{R}$. It is clear that $T^*(Q \times \mathbb{R})$ is a trivial bundle over $T^*Q \times \mathbb{R}$; consider sections σ of this bundle,

$$\sigma: T^*Q \times \mathbb{R} \to T^*(Q \times \mathbb{R}).$$

If everything in the dynamics is to be independent of the additional parameter s, the treatment can be restricted to the image of $T^*Q \times \mathbb{R}$ under σ, and then one obtains a familiar formula for the pullback of Ω to such a submanifold:

$$\tilde{\omega} = \omega_0 + dH \wedge dt. \tag{A11}$$

This is essentially a generalization of (A7); it is on a more general section of $T^*(Q \times \mathbb{R})$. We write ω_0 rather than ω' to emphasize that it is obtained from the symplectic form on T^*Q, as described at (A7).

Remark. One need not start from the natural symplectic structure on $T^*Q \times T^*\mathbb{R}$; many other symplectic structures would do. It must, however, be possible to restrict the resulting formalism to a submanifold of codimension one in $T^*(Q \times \mathbb{R})$, and it must be possible to write such a submanifold as a section of the fiber bundle. This is necessary in order to arrive at a time-dependent type of dynamical system. Such a more general approach would replace $s + H(t, x)$ by a more general function $K(s, t, x)$ which has the property that $\partial K/\partial s \neq 0$.

In using this formalism, many qualifications are necessary in order to avoid mixing s with other coordinates, and this requires some care in defining symmetries and invariance. Thus although the purpose of this appendix is to show how the Hamiltonian formalism can be generalized to include time-dependent dynamical systems, it is important to remember that some freedom is lost in the kind of transformations or diffeomorphisms allowed in this generalization, for the auxiliary variable s has a special role.

Concerning such transformations, we make the following remarks. If time is to be treated in a Newtonian way, that is as absolute time, it is useful to treat $Q' = Q \times \mathbb{R}$ as a fiber space over $\mathbb{R}$ whose fibers are the configurations manifold Q at different times. Then only those transformations φ are allowed on Q' which are fiber-preserving:

$$\begin{array}{ccc} Q' & \xrightarrow{\quad\varphi\quad} & Q \\ \downarrow & & \downarrow \\ \mathbb{R} & \xrightarrow[\quad\varphi_0\quad]{} & \mathbb{R} \end{array}$$

Moreover, φ_0 is required, as usual, to be linear (cf. Chapter 4). In applications to relativistic dynamics, however, the fiber structure of Q' would be disregarded.

Remark. Contact structures on manifolds of odd dimension are useful in other ways, as on energy manifolds. The reader is referred to the literature.

PART II

Reduction, actions of algebras and groups

16

Introduction to Part II

The second part of this book is concerned with invariant structures of various kinds and the ways in which they aid in integrating the dynamics. Invariant structures have already been introduced in Part I (constants of the motion, invariant differential forms), but in neither the depth nor the variety with which they will be treated now.

It was shown in Chapter 10 how a constant of the motion can be used to provide invariant submanifolds of the carrier manifold M of the dynamical system Δ, that is submanifolds $N \subset M$ with the property that if an integral curve of Δ passes through a point of N, then it lies entirely in N. In the way described there, a function $f: M \to \mathbb{R}$ provides a submanifold N whose dimension is $\nu = \mu - 1$, where $\mu = \dim M$, and the problem of integrating the dynamics is *reduced* from that of finding the integral curves of Δ on a manifold of dimension μ to one of finding the integral curves of a vector field (no longer Δ, but Δ restricted to N) on a manifold of dimension ν.

Actually, reduction changes the problem not simply to integrating a vector field on a single manifold of one lower dimension, but of integrating many vector fields, one on each of a whole set of such lower-dimensional manifolds, one for each value of $c \in f(M)$. One may assume for the moment that through each point of M passes one such submanifold $N_c = f^{-1}(c)$, $c \in \mathbb{R}$, that these submanifolds are all isomorphic, and that they are stacked nicely so that one can think of them as providing a laminated structure for M. Such a slicing up of M by submanifolds of dimension less than μ (but not necessarily of dimension $\mu - 1$), is called a *foliation*, described in some detail in Chapter 18. Each N_c is called a *leaf* of the foliation, and if f is a constant of the motion, each leaf is invariant under Δ. In this way, constants of the motion lead to foliations of M with invariant leaves.

Chapter 10, §10.2, describes not only reduction by a constant of the motion in the sense used here, but reduction also of Hamiltonian systems, in which the constant plays a 'double role'. That reduction is accomplished in two steps, the first of which is just what has been described here, but the second of which is different. The second reduction also involves a foliation, this time not of M but of N_c, and this foliation involves no function which is invariant under (i.e. which is a constant of the motion for) the vector field in $\mathfrak{X}(N_c)$ which plays the role of $\Delta \in \mathfrak{X}(M)$ in the description just presented. This vector field, which could be called the dynamical system on N_c, is Δ restricted to N_c (recall that Δ is tangent to N_c and can therefore be thought of as in $\mathfrak{X}(N_c)$). The leaves of the foliation in § 10.2 were the integral curves of another vector field $X \in \mathfrak{X}(N_c)$, and these leaves are not invariant under Δ. Nevertheless, the foliation itself is invariant, for the motion carries leaves into leaves. The foliation is thus an invariant structure which aids in integrating the dynamical system by reducing its dimension.

The foliation just described differs from the kind discussed above in that there is no constant of the motion underlying it. There is, however, another invariant geometric object underlying it, and this is what will be discussed and generalized in Chapter 19. Section A of Part II is devoted to reduction through foliations. It will show how invariant foliations can be obtained from conformally invariant κ-forms through the distributions defined by their kernels (these terms are defined in Chapters 18 and 19), and then how dynamical systems can be reduced by projection with respect to such invariant foliations. All of this, together with the simple example of Chapter 17, generalizes the discussion of § 10.2.

Section B of Part II deals with similar material, but when the objects involved have additional algebraic structures. It will be concerned, for example, with group manifolds, distributions (sets of vector fields; see Chapter 18) which form finite-dimensional Lie algebras, sets of functions which form what are called function groups under the Poisson bracket, and other such matters. It will show how such additional structure provides additional information which aids in integrating the dynamics.

Section A
Reduction

17

Linear dynamical systems: a prelude to reduction

This chapter starts with an example the point of which is to illustrate two kinds of reduction of dynamical systems. The first is the reduction of Hamiltonian systems (expanding the discussion of §10.2.) by means of a *foliation* related to a constant of the motion. The second is the reduction of a system, not necessarily Hamiltonian, by means of a foliation not related to a constant of the motion. This will be followed by a discussion of reduction of linear systems as a prelude to the same kind of procedure in more general cases, to be taken up in later chapters.

17.1 REDUCTION BY A CONSTANT OF THE MOTION (HAMILTONIAN SYSTEM)

Consider the dynamical system on $\mathbb{R}^4 - \{0\}$ given by

$$\Delta = (p_1 + \lambda q^2)\partial/\partial q^1 - (q^1 - \lambda p_2)\partial/\partial p_1 + p_2\partial/\partial q^2 - q^2\partial/\partial p_2. \qquad (1)$$

When $\lambda = 0$ this becomes the isotropic harmonic oscillator in two degrees of freedom. The integral curves of Δ are obtained from the differential equation

$$\begin{aligned} \dot{q}^1 &= p_1 + \lambda q^2, & \dot{p}_1 &= -q^1 + \lambda p_2, \\ \dot{q}^2 &= p_2, & \dot{p}_2 &= -q^2 \end{aligned} \qquad (2)$$

or from

$$dc/dt = Ac,$$

where $c(t) = (q^1, p_1, q^2, p_2) \in \mathbb{R}^4$ and A is the linear operator whose matrix is

$$A = \begin{vmatrix} 0 & 1 & \lambda & 0 \\ -1 & 0 & 0 & \lambda \\ 0 & 0 & 0 & 1 \\ 0 & 0 & -1 & 0 \end{vmatrix}.$$

Any of several techniques can be used to solve this rather simple linear system.

However, consider the following one, which, though not particularly straightforward, illustrates one of the methods of §10.2 and of this Section A. Later, another, simpler method will be discussed.

A constant of the motion f for Δ is given by

$$f(q, p) = [(q^2)^2 + (p_2)^2]^{1/2},$$

and Δ is Hamiltonian with respect to

$$\omega = dq^2 \wedge dq^1 + dp_2 \wedge dp_1 - dp_2 \wedge dq^2, \tag{3}$$

for $i_\Delta \omega = dH$, where $H = q^1 p_2 - p_1 q^2$. Thus one may proceed as in §10.2. The submanifold $N_c \subset \mathbb{R}^4$ given by $N_c = f^{-1}(c), c > 0$, has the topology of $S \times \mathbb{R}^2$, and a suitable local chart on it has coordinates $\{q^1, p_1, \theta\}$, where $\theta = \arctan(p_2/q^2)$. In accordance with §10.2, the usefulness of the constant of the motion is doubled when one more dimension is projected out with respect to the vector field X which is defined by

$$i_X \omega = df. \tag{4}$$

Note that not only Δ, but also X is tangent to the N_c. Let w parametrize the integral curves of X. Then it is easily found that

$$X = -\cos\theta\, \partial/\partial q^1 - \sin\theta\, \partial/\partial p_1,$$

and that the integral curves are given by

$$\begin{aligned} q^1 &= -w\cos\theta + q_0^1, \\ p_1 &= -w\sin\theta + p_{10}, \end{aligned} \tag{5}$$

and, of course, $\theta = \text{const.}$, $f = \text{const.} \equiv c$. We choose the parametrization so that $w = 0$ when $p_1 = 0$, as illustrated in Fig. 17.1.

Fig. 17.1 represents a small neighborhood of N_c (recall that $\dim N_c = 3$) around the origin of the local $\{q^1, p_1, \theta\}$ chart. Two sets of the integral curves of Eq. (5) are drawn in, one for $\theta = 0$ and one for $\theta = \theta_0$, $0 < \theta_0 < \pi/2$.

In accordance with §10.2, the dimension of N_c is reduced by one through projection with respect to the integral curves of X: for $\theta > 0$ this may be done by taking (θ, Q) as the coordinates, where Q is the value of q^1 at the intersection of the integral curve with the (θ, q^1) plane. The result of all this is that the chart whose coordinates are $\{q^1, p_1, q^2, p_2\}$ has been replaced by another, valid now only in a neighborhood U of $\mathbb{R}^4$ (for example, θ must be greater than zero). The coordinates in the new chart are $\{f, \theta, w, Q\}$, and the transformation equations in U are

$$\begin{aligned} f &= [(q^2)^2 + (p_2)^2]^{1/2}, & q^1 &= Q - w\cos\theta, \\ \theta &= \arctan(p_2/q^2), & p_1 &= -w\sin\theta, \\ w &= p_1[(q^2/p_2)^2 + 1]^{1/2}, & q^2 &= f\cos\theta, \\ Q &= q^1 - p_1 q^2/p_2, & p_2 &= f\sin\theta. \end{aligned} \tag{6}$$

Note that $w = f p_1/p_2$. (To form an atlas for all of $\mathbb{R}^4 - \{0\}$ with similar charts is

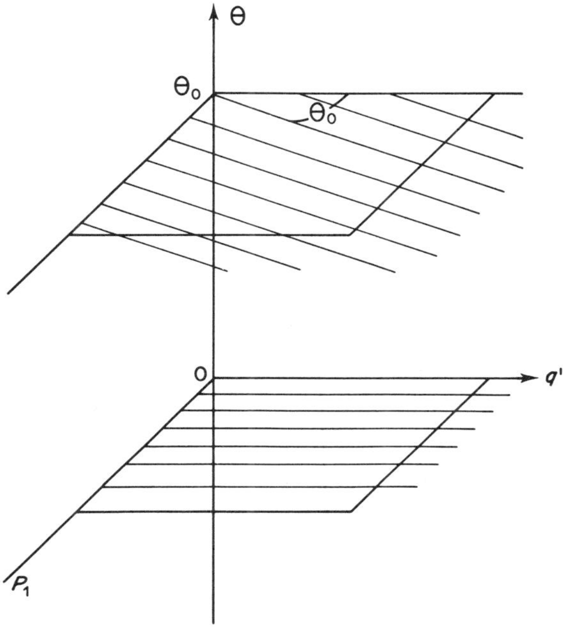

Fig. 17.1. Integral curves of X

not difficult.) Then H, ω, Δ, and X can be written in the new coordinates:

$$H = fQ\sin\theta,$$

$$\omega = -df\wedge dw + \lambda f\,df\wedge d\theta + \cos\theta\,df\wedge dQ - f\sin\theta\,d\theta\wedge dQ,$$

$$\Delta = -\frac{\partial}{\partial\theta} + Q\cot\theta\frac{\partial}{\partial Q} + \left[\frac{Q}{\sin\theta} - \lambda f\right]\frac{\partial}{\partial w},$$

$$X = \partial/\partial w.$$

Note that the equation for ω contains a term of the form $-df\wedge dw$, as asserted at Eq. (33) of Chapter 10. Also, the equation for X merely restates the fact that w parametrizes its integral curves. Finally, it is clear from the expressions for Δ and X that these two fields commute, or that $L_\Delta X = 0$, as was also asserted in Chapter 10. The equations of motion are easily written down in these new coordinates:

$$\dot{f} = 0, \qquad \dot{\theta} = -1,$$
$$\dot{Q} = Q\cot\theta, \quad \dot{\omega} = -\lambda f + Q/(\sin\theta). \tag{7}$$

The vector field Δ is tangent to N_c (that is, to the $f = c$ submanifold), which is of the odd dimension three; considered as a vector field on N_c it can therefore not be Hamiltonian. But Δ can be *projected with respect to* X, in terminology established in Chapter 10, onto the (θ, Q) plane in U, which is the (θ, q^1) plane of Fig. 17.1. In other words, it can be written in the form $\Delta = \Delta_f + \tilde{\Delta}$, where

$$\tilde{\Delta} = -\partial/\partial\theta + Q \cot\theta\, \partial/\partial Q$$

is tangent to the (θ, Q) plane and Δ_f is what remains. Note that $[\tilde{\Delta}, \Delta_f] = 0$. If one writes

$$\Omega_c = -c \sin\theta\, d\theta \wedge dQ$$

for ω restricted to the (θ, Q) plane and

$$H_c = cQ \sin\theta$$

for H restricted to the same plane, it is easily verified that

$$i_{\tilde{\Delta}}\Omega_c = dH_c.$$

Thus the resulting two-dimensional system is in fact (locally) Hamiltonian; it can be integrated in the usual way to obtain the integral curves of $\tilde{\Delta}$. These integral curves are essentially the solutions of the two middle equations of (7), and they can be inserted into the other two equations to solve completely and obtain the integral curves of Δ. As was discussed in Chapter 10, the solution of the reduced problem for the integral curves of $\tilde{\Delta}$ is not the same as the solution of the entire dynamical system. The last step metioned here, i.e. inserting the solution of the middle two equations of (7) into the last of those equations (the first is, of course, trivial), is necessary before the task is completed. When all of this is done, the solution in U may be written down. It is

$$\begin{aligned} f(t) &= c, & Q(t) &= K/\sin(t+\alpha), \\ \theta(t) &= -(t+\alpha), & w(t) &= -\lambda ct + B + K\cot(t+\alpha). \end{aligned} \tag{8}$$

This rather complicated procedure for integrating the dynamics of Eq. (1) involves two foliations of $\mathbb{R}^4$ (foliations will be described in detail in the next chapter). The leaves of the first foliation are the $N_c = f^{-1}(c)$, and those of the second are the integral curves of $X = \partial/\partial w$. Now, X has two interesting properties: it commutes with Δ (so that Δ maps integral curves of X into integral curves of X or, in terms of foliations, leaves into leaves), and it is in the kernel of ω_c, the pullback of ω to N_c. It will eventually be seen that these properties are related and that they are what makes it possible to split Δ into Δ_f and $\tilde{\Delta}$.

On the other hand, it is somewhat surprising that the seemingly most straightforward kind of reduction, namely by a constant of the motion, of a linear dynamical system leads almost immediately to intrinsically nonlinear structures such as $S \times \mathbb{R}^2$ and the non-Cartesian local chart of this example. Clearly this is due to the fact that a constant of the motion is seldom linear. In fact it can be shown that a system on $\mathbb{R}^n$ which is linear in the sense of Eq. (2) and the equation that follows it admits linear constants of the motion only if the operator corresponding to A is singular. If a constant of the motion is not a linear function, the submanifolds it leads to will not be subspaces, but true complicated submanifolds. In some cases, moreover, it will not even be possible to identify the manifold obtained in this way from a foliation of N_c with any submanifold of N_c.

17.2 SIMPLER REDUCTION OF THE SAME SYSTEM

There are, of course, simpler ways to deal with the dynamical system of Eq. (1), applicable more generally to linear systems. Among these ways is the one we shall now present, which is analogous to the usual treatment of small vibrations. A glance at Eq. (2) shows that the second two equations can be solved independently of the first two (they yield the oscillator in one degree of freedom). This solution can then be inserted into the first two equations, whose solution then completes the problem. (This second step is similar to the last step of § 17.1, the step in which the differential equation for $w(t)$ was obtained.) In other words, Δ can be written in the form $\Delta = \Delta_1 + \Delta_2$, where

$$\Delta_1 = (p_1 + \lambda q^2)\partial/\partial q^1 - (q^1 - \lambda p_2)\partial/\partial p_1,$$
$$\Delta_2 = p_2\partial/\partial q^1 - q^2\partial/\partial p_2.$$

Note that $[\Delta_1, \Delta_2] \neq 0$.

Clearly Δ_2 can be thought of as a vector field on $\mathbb{R}^2$ with the one-chart atlas whose coordinates are $\{q^2, p_2\}$. For reasons that will become clear in Chapter 18, it is convenient to think of this $\mathbb{R}^2$ as the quotient of $\mathbb{R}^4$ with respect to the foliation Φ provided by the planes (submanifolds) $q^2 = \text{const.}$, $p_2 = \text{const.}$ Thus Δ projects *with respect to this foliation* onto $\Delta_2 \in \mathfrak{X}(\mathbb{R}^2)$, a dynamical vector field which can be integrated independently. The rest of the integration, that of Δ_1, is more complicated and will not be discussed here in any generality. For completeness, however, we write down the whole solution (b and β are related to B and K of Eq. (8)):

$$q^1 = b\cos(t + \beta) + \lambda at\cos(t + \alpha), \qquad q^2 = a\cos(t + \alpha),$$
$$p_1 = -b\sin(t + \beta) - \lambda at\sin(t + \alpha), \quad p_2 = -a\sin(t + \alpha). \tag{9}$$

It can be verified with the aid of (6) that this solution agrees in U with Eq. (8).

In this second approach Δ has been reduced to the dynamical system represented by Δ_2 on $\mathbb{R}^2$, but this second reduction was associated with no constant of the motion. This demonstrates that useful reductions of dynamical systems can take place not only without the involvement of constants of the motion, but even without making use of invariant submanifolds. Indeed, not only the planes $q^2 = \text{const.}$, $p_2 = \text{const.}$ fail to be invariant, so do the planes $q^1 = \text{const.}$, $p_1 = \text{const.}$ There are no simple invariant submanifolds in this decomposition of Δ into $\Delta_1 + \Delta_2$. Although each leaf of Φ is not itself invariant under Δ (except the one given by $q^2 = p_2 = 0$), the entire foliation Φ is invariant: each leaf is carried by the flow of Δ into another leaf. It turns out that this invariance can be stated in terms of the Lie derivative along Δ of an appropriate geometric object. The two-form $\sigma = dq^2 \wedge dp_2$ is invariant under Δ: that is, $L_\Delta \sigma = 0$. From σ one obtains the foliation Φ which reduces Δ in the following way. The two-dimensional kernel of

σ is spanned by the vector fields $\partial/\partial q^1$ and $\partial/\partial p_1$, and these can be used, as will eventually be described, to construct the leaves of Φ. At any rate, it is easily seen that these two vector fields are everywhere tangent to those leaves. This general view of reduction of dynamical systems will be described in some detail in the next chapter.

17.3 REDUCTION OF LINEAR SYSTEMS

1. The linear example of §17.1 and §17.2 has been reduced in two ways. First it was reduced as a Hamiltonian system and with the use of a constant of the motion. Then it was reduced by means of a rather obvious foliation that is not associated with a constant of the motion. This section discusses the generalization of this second approach from the example given to linear systems in general.

Let W be a linear space of dimension μ, and consider the differential equation

$$\dot{x} = Bx, \tag{10}$$

where $x \in W$ and $B \in \mathrm{Lin}(W, W)$. Let $W_1 \subset W$ be an eigenspace of B, so that $B(W_1) \subset W_1$. Another subspace W_2 can then be found such that $W = W_1 \times W_2$ (W_2 is not unique), and every $x \in W$ decomposes uniquely into $x = x_1 + x_2$, with $x_1 \in W_1$ and $x_2 \in W_2$. Thus $Bx_1 \in W_1$, and Bx_2 can be decomposed uniquely, like any other vector, into $Bx_2 = y_1 + y_2$. Eq. (10) then becomes

$$\begin{aligned} \dot{x}_1 &= Bx_1 + y_1 \equiv Bx_1 + (Bx_2)|W_1, \\ \dot{x}_2 &= y_2 \equiv (Bx_2)|W_2, \end{aligned} \tag{11}$$

and is thus reduced on W_2. If, in addition, W_2 is an eigenspace of B, then $y_1 = 0$, and (11) *completely reduces* (10); one says that the dynamical system is composed of two noninteracting systems, one one W_1 and the other on W_2.

Remarks. 1. What we are calling the direct product $W_1 \times W_2$ of two vector spaces is often written as the direct *sum* $W_1 \oplus W_2$. We choose the notation which is in keeping with the notation for manifolds. An element $x \in W_1 \times W_2$ is the ordered pair (x_1, x_2) with $x_j \in W_j$, and then $(x_1, x_2) + (y_1, y_2) = (x_1 + y_1, x_2 + y_2)$.

2. Although this section is not concerned primarily with Hamiltonian systems, a few words may be said about them. It may be desirable to reduce a Hamiltonian Δ in ways that preserve its Hamiltonicity. The dynamics of (10) is Hamiltonian if, but not only if, there exists an antisymmetric nonsingular bilinear form ω on W (possible only if μ is even) such that $\alpha(x, y) \equiv \omega(Bx, y)$ is a symmetric bilinear form. Then $\alpha(x, x) = H(x)$ is the Hamiltonian function of the dynamics. (There are other kinds of linear dynamical systems which are Hamiltonian, for symplectic forms ω exist which are not bilinear forms, which cannot be represented by matrices with constant coefficients. Here, however, we are considering only this restricted class.) The proof that the system is then Hamiltonian and of the form of the Hamiltonian, omitted here, is perhaps

simplest to perform in a canonical chart (Arnol'd, 1978, Section 41 and Appendix 6). The requirement that the reduced dynamics also be Hamiltonian places restrictions on acceptable eigenspaces of B. It forces W_1 and W_2 to be even-dimensional and the restriction of ω and α to W_1 and W_2 to have appropriate singularity and symmetry properties.

2. In order for this discussion of linear systems, now no longer necessarily Hamiltonian, to serve as a prelude to more general considerations, it should be cast in terms of vector fields. At each point $x \in W$ the tangent space $T_x W$ can be identified with W itself and in this way Bx can be considered a vector in $T_x W$. Then the dynamical vector field which gives rise to Eq. (10) may be written $\Delta = (x, Bx)$ or, in Cartesian local coordinates on W, equivalently as $\Delta = B_j^k x^j \partial / \partial x^k$ (we are using two of the notations of Chapter 5, around Eqs. (15) to (20) of that chapter), which we shall sometimes abbreviate in the form

$$\Delta = \langle Bx | \partial \rangle. \tag{12}$$

Now splitting W into W_1 and W_2 can be thought of in terms of foliations, as has been mentioned several times and as will be explained in more detail in the following chapters. There are, moreover, many foliations possible, and of them we choose one in which x and y are on the same leaf iff $x - y \in W_1$, i.e. iff $x_2 = y_2$. The foliation is then a set of stacked W_1 subspaces, and the quotient of W with respect to the foliation can be identified with W_2. The dynamical vector field can now be written $\Delta = \Delta_1 + \Delta_2$, where, in the notation of Eqs. (11) and (12),

$$\begin{aligned} \Delta_1 &= \langle Bx_1 + y_1 | \partial_1 \rangle, \\ \Delta_2 &= \langle y_2 | \partial_2 \rangle. \end{aligned} \tag{13}$$

Here ∂_j contains derivatives only with respect to the coordinates in W_j. It is then seen that Δ_2 is a vector field on the quotient space W_2, but that Δ_1 is not a vector field on W_1, for $y_1 \equiv (Bx_2)| W_1$ depends on variables in W_2.

The vector field Δ_2, viewed as living on W_2 rather than on W, is the *projection* of Δ onto the quotient space with respect to the foliation. This follows essentially from the definition of projectability in the diagram of (23) of Chapter 10 and from the subsequent discussion. That definition can be extended in an obvious way from projection π with respect to a vector field to other projections, in our case to projection from W to W_2. More specifically, even though every vector field $X \in \mathfrak{X}(W)$ can be written in the form $X = X_1 + X_2$, with $X_1 = \langle u | \partial_1 \rangle$, $X_2 = \langle v | \partial_2 \rangle$, $u, v \in W$, the vector field X_2 is not necessarily on W_2. It is, however, if $v \in W_2$, for then X_2 depends only on variables in W_2 and as far as X_2 is concerned, it is as if the rest of W did not exist. If this is the case, X is projectable with respect to the foliation.

This explanation of the projectability for linear systems agrees with that part of Chapter 10 in which projectability was discussed in terms of functions defined over the image space of the projection, i.e. over the quotient. Indeed, let $g = g_2 \circ \pi \in \mathscr{F}(W)$, with $g_2 \in \mathscr{F}(W_2)$ and π the projection with respect to the foliation (i.e. $\pi x = x_2$). Then g is constant along the leaves (i.e. on W_1) and is therefore a function *of* the leaves. If $L_X g = g'$ is also constant on the leaves, that is if there exists a $g'_2 \in \mathscr{F}(W_2)$ such that $g' = g'_2 \circ \pi$, then X is projectable in the sense just described. To see this, consider

$$L_X g = \langle u | \partial_1 g \rangle + \langle v | \partial_2 g \rangle = \langle v | \partial_2 g \rangle = \langle v | \partial_2 g_2 \circ \pi \rangle.$$

This is of the form $g'_2 \circ \pi$ iff $\langle v | \partial_2 g_2 \rangle \in \mathscr{F}(W_2)$, which holds iff $v \in W_2$. Clearly, then Δ is projectable, for $y_2 \in W_2$.

As in the example at the beginning of this chapter, the reduction from Δ to Δ_2 is associated with the invariance under Δ of a certain differential form. Let $\mu_k = \dim W_k$, $k = 1, 2$, and consider the closed (in fact exact) μ_2-form

$$\theta = dx^{\mu_1 + 1} \wedge \ldots \wedge dx^{\mu},$$

where $\{x^{\mu_1 + 1}, \ldots, x^{\mu}\}$ are coordinates on W_2. Then θ (or any constant multiple of it) has the following two properties:

(a) θ is *conformally invariant* under Δ (which means that $L_\Delta \theta = A\theta$, where A is some constant);
(b) every vector field Y in $\ker \theta$ (which means every one for which $i_Y \theta = 0$) is tangent to W_1.

If W is identified with its tangent space at any point, Property (b) states that $\ker \theta$ spans W_1.

Property (b) is easy to prove. Here we prove only (a). First, calculate $i_\Delta \theta$ using $\Delta = B^k_j x^j \partial / \partial x^k$. The calculation yields

$$\begin{aligned} i_\Delta \theta = {} & B^{\mu_1 + 1}_j x^j dx^{\mu_1 + 2} \wedge \ldots \wedge dx^{\mu} \\ & - B^{\mu_1 + 2}_j x^j dx^{\mu_1 + 1} \wedge dx^{\mu_1 + 3} \wedge \ldots \wedge dx^{\mu} + \\ & + (-1)^{\mu - \mu_1 + 1} B^{\mu}_j x^j dx^{\mu_1 + 1} \wedge \ldots \wedge dx^{\mu - 1}. \end{aligned}$$

Second, use $L_\Delta \theta = i_\Delta d\theta + d i_\Delta \theta = d i_\Delta \theta$, so that

$$\begin{aligned} L_\Delta \theta = {} & B^{\mu_1 + 1}_j dx^j \wedge dx^{\mu_1 + 1} \wedge \ldots \wedge dx^{\mu} + \ldots \\ & + B^{\mu}_j dx^{\mu_1 + 1} \wedge \ldots \wedge dx^j. \end{aligned}$$

Third, recall that $B(W_1) \subset W_1$, which implies that all matrix elements of the form $B^{\mu_1 + k}_j$ are zero unless $\mu_1 + 1 \leqslant j \leqslant \mu$. It follows that

$$L_\Delta \theta = (\mathrm{Tr}_2 B)\theta,$$

where $\mathrm{Tr}_2 B$ is the trace (the sum of the diagonal elements) of B restricted to W_2. Note that if $\mathrm{Tr}_2 B = 0$, then $L_\Delta \theta = 0$, or θ is invariant, not only conformally, under Δ.

The same technique can be used to prove the following converse result: let σ be a conformally invariant κ-form, $\kappa \leqslant \mu$, with constant coefficients, and let $V \subset W$ be the subspace of W spanned by $\ker \sigma$. Then V is an eigenspace of B (that is, $BV \subset V$), and therefore Δ is projectable with respect to the foliation whose leaves are the subspaces of W which are parallel to V and of the same dimension.

In conclusion, then, a method for obtaining a reduced dynamics such as Δ_2 is contained in these four steps: (1) find a conformally invariant closed κ-form; (2) find the vector field, in its kernel; (3) find the subspaces spanned by these vector fields and form the foliation which they then define; (4) project the dynamical vector field with respect to this foliation to obtain the reduced dynamical system.

In subsequent chapters this procedure will be generalized from linear spaces to differential manifolds. The major difference arises from the impossibility of identifying the manifold M with the tangent space at a point, so that in the usual way many charts have to be used. Another difference is that even if M is a linear space, a dynamical system on it which is not also linear will not lead to foliations whose leaves are linear subspaces, and this changes the whole discussion. Thus what has been treated in this chapter is a special case, and it will be seen that it is a special case of a more general procedure for reducing dynamical systems in four very similar steps. But before going on to reduction in general, we digress to discuss the concept of foliation.

BIBLIOGRAPHY TO CHAPTER 17

The linear case is discussed in AM, p. 161ff., as a preparation for Hamiltonian and Lagrangian dynamics on manifolds.

A more detailed and complete analysis can be found in Hirsch and Smale (1974).

For a special example (the harmonic oscillator) see Cushman (1974).

18

A digression on foliations and distributions

The linear case of the preceding chapter shows that foliations can play an important role in the integration of dynamical systems. In fact, foliations have been used already in several places in this book without always having been mentioned explicitly and certainly without careful definition. The time has come to discuss them and some other related geometrical objects with some attention to details.

18.1 FOLIATIONS

The idea of a foliation may be roughly understood from some examples in previous chapters. It is a slicing of the manifold M into a stack of closely packed submanifolds, called leaves, which fill up M. As the leaves will be required if not always to carry dynamical systems, at least to be intimately bound up with them, they will be endowed with desirable properties (e.g. second countability, Hausdorff), properties which they will obtain automatically if the definition of leaves is restricted in convenient ways. We will, therefore, usually consider only those foliations whose leaves are closed or compact, often regular submanifolds, and of the many foliations possible on any given manifold, those will of course be of interest which are related in relevant ways to the vector fields which are the dynamical systems.

1. For us a *foliation* Φ of a manifold M will be defined as follows: it is a family $\{\hat{\ell}_\alpha\}$ of disjoint subsets $\hat{\ell}_\alpha$ (the *leaves* of the foliation), one passing through each $m \in M$, on which differentiable structures can be given so that the natural injection $i:\hat{\ell}_\alpha \to M$ is an imbedding (recall that an imbedding is an injective mapping φ from a connected manifold, for which $T\varphi$ is also injective) and $i(\hat{\ell}_\alpha)$ is connected in the induced topology.

It follows that $i(\hat{\ell}_\alpha) \equiv \ell_\alpha$ is a submanifold (and we shall henceforth write ℓ_α for $\hat{\ell}_\alpha$), so that Φ consists of a family of submanifolds, each of dimension less than $\mu = \dim M$, which stack together to fill up M. The leaves need not all be of the same dimension, and even when they are, they need not be diffeomorphic to each other.

It is not necessary either that the imbedding i be regular, for the differentiable structure on ℓ_α might be such that the induced topology on the leaves is different from the subset topology (recall the usual example of the irrational winding line on the torus; see below). The submanifolds ℓ_α will therefore not necessarily be regular, but their connectedness assures one that they will be Hausdorff and second countable and therefore suitable carrier manifolds for the reduced dynamical systems (see Brickell and Clark, 1970, p. 180).

The following example illustrates some of the properties of foliations. Let

$$M \equiv \mathbb{R}^2 - \{(-1,0) \bigcup (1,0)\}$$

and let each leaf ℓ_b of Φ be the set of points given in the usual Cartesian chart by

$$f(x, y) \equiv [(x-1)^2 + y^2]^{-1/2} - [(x+1)^2 + y^2]^{-1/2} = b. \tag{1}$$

Fig. 18.1 shows the leaves of Φ in the plane. The leaves are regular submanifolds (the subset topology makes f a regular imbedding) and are obviously connected. All of the ℓ_b are of dimension one, compact for $b \neq 0$, but not compact for $b = 0$, since ℓ_0 is the entire y axis. This is an example of a foliation whose leaves are not all diffeomorphic.

Although the points $(-1, 0)$ and $(1, 0)$ have been removed from $\mathbb{R}^2$, they can easily be put back and Φ can be extended to include each of them as an additional leaf (a submanifold of dimension zero). This then becomes an example of what we shall call a *singular* foliation, one whose leaves are not all of the same dimension.

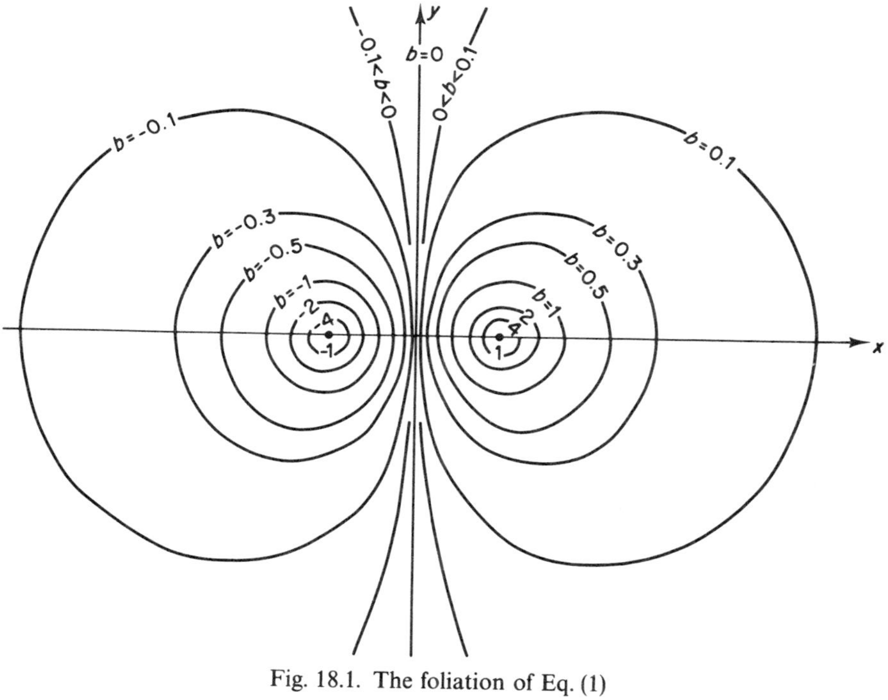

Fig. 18.1. The foliation of Eq. (1)

Another example of a singular foliation is the one discussed in connection with Fig. 10.2, the foliation of $\mathbb{R}^2$ by circles around the origin. Each leaf in that foliation is given by

$$f(x, y) \equiv x^2 + y^2 = \rho^2, \qquad (2)$$

where $\rho \geqslant 0 \in \mathbb{R}$. Again, all of the $\ell_\rho, \rho \neq 0$, are diffeomorphic and of the same dimension, but ℓ_0, the point at the origin, is of dimension zero, different from all the others.

A nonsingular foliation, one whose leaves are all of the same dimension $\mu - \kappa$ is said to have *codimension* κ. From the definition it follows that locally each leaf of a (nonsingular) foliation whose codimension is κ looks like a linear space, a plane of dimension $\mu - \kappa$ in $\mathbb{R}^\mu$, and then the local set of leaves looks like a κ-dimensional set of such parallel planes. Locally these planes can be taken as the coordinate planes of a chart, and the existence of such local coordinates, called coordinates *distinguished* by the foliation, is sometimes taken as the defining property of a foliation. In the present context the existence of such distinguished coordinates needs no proof: it follows from the ability to flatten an immersion locally (but see Brickell and Clark (1970, p. 70); also Bourbaki (1967, p. 46)).

Consider, for instance, the foliation of $M = \mathbb{R}^2 - \{0\}$ defined by Eq. (2), from which the origin has been removed to make the leaves all have the same dimension (i.e. to make the foliation nonsingular). Note, incidentally, that this foliation is like that of Eq. (1) in that both are nonsingular, but unlike it in that the leaves of this one are all diffeomorphic to each other. Each leaf ℓ_ρ is a circle of radius $r = \rho$. A neighborhood U can be chosen about an arbitrary $m \in M$ such that polar coordinates can be taken as a local chart φ in U. Let the coordinates of m be r_0, θ_0 in this local chart, and then $\ell_r|U$ is the inverse image of $r = r_0$. In other words, $\varphi(\ell_{r_0})$ is just the line segment in $\varphi(U)$ given by $r = r_0$. In fact each $\ell_\rho|U$ is

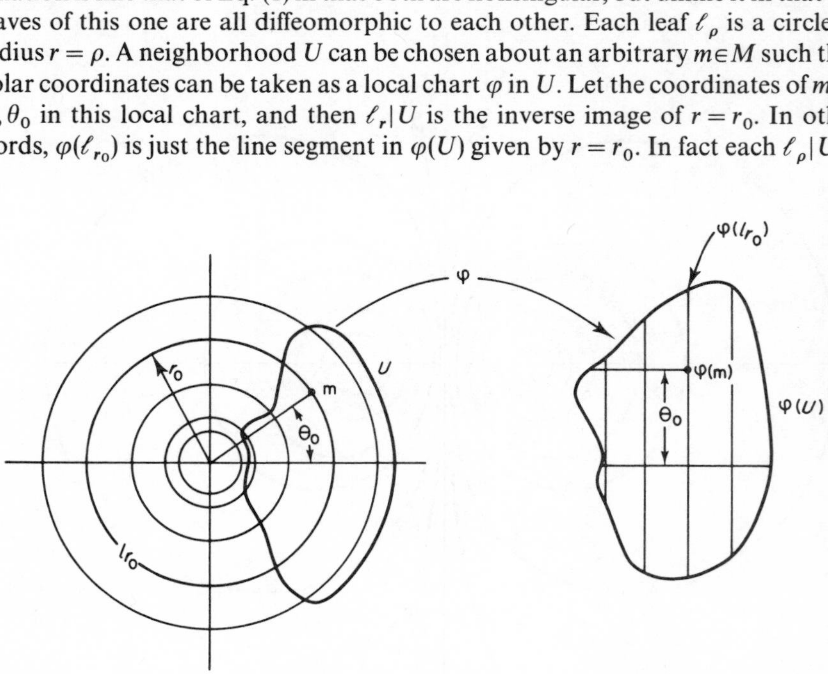

Fig. 18.2. An example of a distinguished chart

just the inverse image of a similar line segment given by $r = \rho$, and thus φ straightens out the foliation locally, as illustrated in Fig. 18.2. Clearly this foliation can be straightened out in this way only locally.

Suppose that (U, φ) is a distinguished chart for a foliation Φ at a point $m \in U \subset M$, and let ℓ_α be any leaf which intersects U. If there exists a $U' \subset U$ such that $m \in U'$ and ℓ_α intersects U' only once (i.e. such that $\ell_\alpha \cap U'$ is connected), then (U, φ) is said to be a *regular chart for* Φ *at m*. If there exists a regular chart at every $m \in M$, then Φ is said to be a *regular foliation*. An example of an *irregular foliation* (one that is not regular) is that of the torus by irrational winding lines all of the same slope, for each such winding lines intersects every neighborhood an infinite number of times. Note that an irregular foliation is not the same as a singular one (for instance, all of the leaves in the above example of the torus have the same dimension). In a regular foliation the set of indices $\{\alpha\}_U$ in each regular chart (U, φ) is contained in a obvious way in $\mathbb{R}^\kappa$, where $\mu - \kappa$ is still the dimension of the leaves in U, and thus the entire set can itself form a manifold, although it may be impossible to identify this manifold with a submanifold of M. This aspect of foliations, important from the dynamical point of view, will be discussed more fully later.

2. The leaves of foliations are submanifolds, and their dimensions are necessarily lower than that of M. For this reason submersions, discussed in §8.2, are helpful in understanding foliations.

Recall that a submersion is a mapping $\varphi: M \to N$, with $\nu \equiv \dim N \leqslant \mu \equiv \dim M$, such that T_φ is surjective. As has been seen, it follows that $\varphi^{-1}(n)$, $n \in N$, is a submanifold of M with dimension $\mu - \nu$. According to the definition of a foliation, the family of these submanifolds, as n runs through N, is a foliation of M whose codimension is ν, and this foliation is said to be *generated* by the submersion φ. Such (global) submersions generate regular foliations, for each leaf is a regular submanifold of M (see Chapter 8 for this result and for the definition of a regular submanifold; the regularity of the foliation follows immediately from the definition).

An example of this is the submersion

$$\varphi: M = \mathbb{R}^3 - \{0\} \to S^2$$

which maps $m \in M$ with polar coordinates (r, θ, φ) into the point (θ, φ) on the sphere S^2 imbedded in $\mathbb{R}^3$. In this example $\varphi^{-1}(n)$ is the ray from the origin which passes through the point $n \in S^2$, and these rays provide a regular foliation of M. We shall refer to this foliation of M again in several examples.

When $\varphi: M \to N$ is a submersion, all points $n \in N$ are regular values for φ, or the set R_φ is all of N (recall the definitions from §10.1: $n = \varphi(m)$ is a regular value for φ iff $T_m\varphi$ is surjective for all $m \in \varphi^{-1}(n)$; R_φ is the set of regular values). Foliations generated by submersions, necessarily regular, can be divided into two classes, those whose leaves are all diffeomorphic to each other and those some of whose leaves are not diffeomorphic to some others. The example at Eq. (2) with the origin of $\mathbb{R}^2$ removed is of the first type, and that of Eq. (1) is of the second. It is easily verified that in both examples $df \neq 0$ everywhere, so that $T\varphi$ is indeed surjective in both examples.

3. Dynamical systems often involve foliations that are not regular, and therefore the irregularities of such foliations must be understood. Suppose that $\varphi: M \to N$ is not a submersion, but that it generates a foliation in the same way that submersions do, through the inverse images of points, so that the leaves are defined by $\varphi^{-1}(n), n \in N$. There will then be points $m \in M$ where $T_m \varphi$ is not surjective (*singular points* for φ), for which $n = \varphi(m)$ is not a regular value (*singular values* for φ). Sard's theorem, according to which R_φ is dense in N (see § 10.1), then implies that singular values will be nowhere dense subsets of N, e.g. points. It can be shown, however, that if n is a regular value and its inverse image $\varphi^{-1}(n)$ is compact, there is a neighborhood U containing n, such that the inverse image of every point in U is diffeomorphic to $\varphi^{-1}(n)$, and that if M is orientable, so is $\varphi^{-1}(n)$ and so is the inverse image of every point in a neighborhood of n (Palais, 1957, p. 16; Milnor, 1963, p. 12).

Foliations generated by mappings which are not submersions can be of various kinds. In what follows, n' will stand consistently for a singular value for φ.

(a) Even if φ is not a submersion, the foliation it generates may be regular, with leaves all diffeomorphic to each other or not. For example, consider

$$\varphi \equiv f: \mathbb{R}^2 \to \mathbb{R}: (x, y) \mapsto x^3.$$

This mapping generates the same foliation as does

$$g: \mathbb{R}^2 \to \mathbb{R}: (x, y) \mapsto x,$$

which is clearly regular (in fact the Cartesian coordinates x, y provide a distinguished chart for g). The foliation is thus regular even though $x = 0$ is a singular value for φ. A similar pathology can be introduced into the example of Eq. (1) by replacing the function f by f^3: what is generated is the same (regular) foliation with one leaf that is not diffeomorphic to the others.

Although a regular foliation, as has just been shown, can be generated by a mapping which is not a submersion, it is nevertheless true that a foliation with singularities cannot be generated by a submersion. If a foliation is regular, however, a submersion can always be found which will generate it (Brickell and Clark, 1970, p. 205). It should be borne in mind, in any case, that what is important from the present point of view is not the irregularity of the mapping which generates it, but the irregularity of the foliation.

(b) The foliation generated by φ may not be regular. For example, let M be the thickened Möbius strip, an open submanifold of $\mathbb{R}^3$ obtained by identifying opposite faces of a cube with a twist, as shown in Fig. 18.3. Let $\varphi: M \to \mathbb{R}: m \mapsto \delta_m$, where δ_m is the square of the distance of m from the center section of M, the one indicated by the dotted line in the figure. Then $\delta_m = 0$ is an irregular value for f, and $f^{-1}(0)$ is a Möbius strip (which is not orientable and hence not diffeomorphic to $f^{-1}(\delta), \delta \neq 0$). This foliation is not regular: every neighborhood which includes

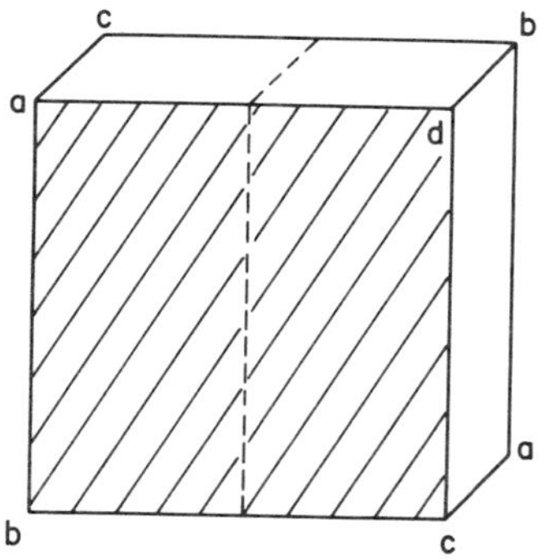

Fig. 18.3. The thickened Möbius strip. The shaded face is identified with the one on the opposite side of the cube, and corners with the same label are also identified

a point $m \in f^{-1}(0)$ intersects some leaves $f^{-1}(\delta)$, $\delta \neq 0$, twice, once on each side of $f^{-1}(0)$. Note that $df = 0$ on $f^{-1}(0)$.

(c) The foliation generated by φ may be singular, as $\varphi^{-1}(n')$ may be of a different dimension from $\varphi^{-1}(n)$, $n \in \mathbb{R}_\varphi$. A simple example is that of Eq. (2), for which $df = 0$ at the origin.

(d) The 'leaves' of the foliation may not even be manifolds; in particular, $\varphi^{-1}(n')$ may not be a manifold. An example is given by $\varphi \equiv f : \mathbb{R}^2 \to \mathbb{R}$, where (note the difference in sign between this and Eq. (1))

$$f(x, y) = [(x-1)^2 + y^2]^{-1/2} + [(x+1)^2 + y^2]^{-1/2}. \tag{3}$$

As is seen in Fig. 18.4, $f^{-1}(2)$ is the figure-eight discussed in Chapter 8, which is not a submanifold of $\mathbb{R}^2$. There is no distinguished chart at the origin.

The example of Eq. (3) has an additional pathology: $f^{-1}(a)$ is not connected for $a > 2$ and is therefore not a leaf in our definition. (Some authors do not require leaves to be connected, so for them these would be leaves.) When one removes the origin of $\mathbb{R}^2$, the first pathology, that of the cross-over point, is eliminated, but this second pathology remains. It is interesting that a similar two-fold pathology occurs when the energy function is used to foliate TQ for the plane pendulum: both the figure-eight and disconnected leaves occur in that case as well (see the discussion in Chapter 7).

4. These examples illustrate the existence of interesting and useful foliations generated (in the sense described at the beginning of § 18.3) by mappings that are not submersions. On the other hand, nonsingular foliations, whose leaves all have the same dimension (hence foliations for which a codimension can be defined), can be represented at least by local submersions. That is, at each $m \in M$ there exists a neighborhood U such that in U the foliation is generated by a submersion $\varphi : U \to N$. Let Φ be a foliation of codimension κ, so that the leaves have dimen-

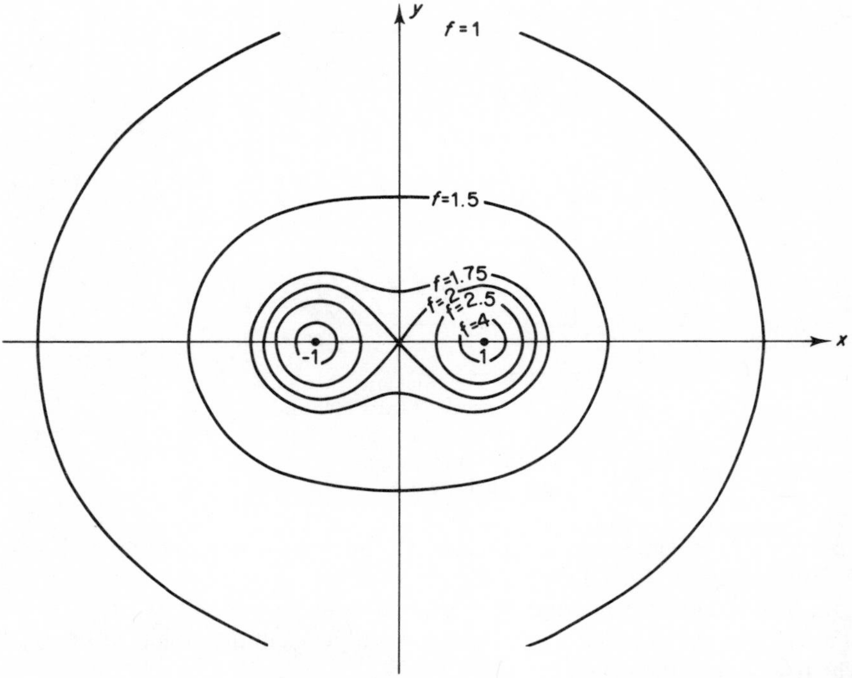

Fig. 18.4. The foliation of Eq. (3)

sion $\mu - \kappa$, and let (U, φ) be a distinguished chart at some point $m \in M$. Then

$$\varphi : U \to \mathbb{R}^{\mu-\kappa} \times \mathbb{R}^{\kappa} : m \mapsto (x^1, \ldots, x^{\mu-\kappa}, a^1, \ldots, a^{\kappa}),$$

where the a^j are constant on each leaf. More specifically, let π_κ be the projection defined by $\pi_\kappa : \mathbb{R}^{\mu-\kappa} \times \mathbb{R}^{\kappa} \to \mathbb{R}^{\kappa}$; then $(a^{-1} \circ \pi_\kappa^{-1})(a^1, \ldots, a^{\kappa})$ for fixed $(a^1, \ldots, a^{\kappa})$ is a connected component of $\ell_\alpha \cap U$ for some α. It follows that the projections

$$\chi_U : U \to \mathbb{R}^{\kappa} : m \mapsto (a^1, \ldots, a^{\kappa}) \tag{4}$$

are (local) submersions, and $\chi_U^{-1}(a^1, \ldots, a^{\kappa})$ is a local leaf, i.e. a connected component of $\ell_\alpha \cap U$. Moreover, since the distinguished charts form an atlas, local submersions of this kind form a family with the following property: if $m \in U \cap U'$, then there exists a local diffeomorphism $\psi_{UU'}$ such that

$$\chi_{U'}(m) = \psi_{UU'} \circ \chi_U(m). \tag{5}$$

To sum up, then, to every nonsingular foliation corresponds at least one family of local submersions satisfying Eqs. (4) and (5). Moreover, every family of local submersions satisfying Eqs. (4) and (5) can be used to generate a foliation, namely the one constructed out of the local leaves.

5. When a foliation is used in analyzing a dynamical system, its leaves ℓ_α are not in general invariant submanifolds, as has already been seen in Chapter 17. It then

becomes important to understand how the system moves from one leaf to another (or, as we shall say, *across* the leaves), i.e. how the index α of the ℓ_α changes with time. This, in turn, requires an understanding of the *quotient* of M with respect to the foliation.

Let $\Phi \equiv \{\ell_\alpha\}$ be a foliation of M. Then Φ defines an equivalence relationship r_Φ on M according to

$$(m_1, m_2) \in r_\Phi \text{ iff } m_1, m_2 \in \ell_\alpha \text{ for some } \alpha, \tag{6}$$

and hence Φ also defines the quotient set M/r_Φ with respect to this equivalence. We shall usually write M/Φ for M/r_Φ and call it the quotient with respect to Φ. Let $\pi: M \to M/\Phi: m \mapsto [m]$ be the canonical projection, where $[m]$ is the equivalence class to which m belongs. The standard quotient-set topology can be induced on M/Φ, and if a differentiable structure can also be found such that π is a submersion, then $M/\Phi \equiv A$ becomes a *quotient manifold* of M. Such a differentiable structure will be called the *quotient manifold structure*.

Remark. Here we are relaxing the requirement that a manifold be Hausdorff as well as second countable. We shall call M/Φ a quotient manifold whenever a differentiable structure can be found on it, independent of the nature of the topology induced.

What this means is that a manifold structure can be found for the index set $\{\alpha\}$, and in most of the examples so far presented this structure has been obvious (the index set has usually been $\mathbb{R}$). But it is not always possible to find such a quotient manifold structure, although when it is, the structure is unique (Brickell and Clark, 1970, p. 93) and A is a well-defined manifold of dimension κ. It is unfortunate that quotient manifolds of Hausdorff manifolds are not necessarily Hausdorff, though those of second countable ones are second countable.

Thus there are three possibilities for A. First, $A \equiv M/\Phi$ can be given a quotient manifold structure such that the induced topology is Hausdorff. Second, A can be given a quotient manifold structure, but the induced topology is not Hausdorff. Third, A cannot be given a quotient manifold structure (or any differentiable structure at all) for other reasons, perhaps more profound, and no motion across the leaves can be defined, as in the case of the irrational winding lines on the torus.

As long as one deals with regular foliations, however, the worst that can happen is that A may not be Hausdorff. If Φ is regular, it is representable by a global submersion $\varphi: M \to N$, and this submersion can be used to define the equivalence relationship r_Φ by

$$(m_1, m_2) \in r_\Phi \text{ iff } \varphi(m_1) = \varphi(m_2). \tag{7}$$

Of course this is the same equivalence relationship as the r_Φ of Eq. (6), namely the one obtained by first going from φ to the foliation Φ it generates and then to the equivalence relationship defined by Φ. But now a bijection η has been established between the quotient set and N:

$$\eta: M/r_\Phi \equiv M/\Phi \to N: [m] \mapsto \varphi(m). \tag{8}$$

This can be used now to induce a differentiable structure on $M/\Phi = A$, and hence as long as A is Hausdorff (important exceptions occur, as will be seen below) it is possible to discuss the motion across the leaves in the usual terms of vector fields on differentiable manifolds.

Two instances of index sets A with quotient manifold structure are provided by the example which opens Chapter 17. First, the function f defined above Eq. (3) of Chapter 17 provides a foliation whose leaves are called the N_c, and $A \equiv \{c\}$ is identified with $\mathbb{R}^+$. Second, it is mentioned that it is convenient to think of $\mathbb{R}^2$ with coordinates q^1, p_1 as the quotient of $\mathbb{R}^4$ with respect to certain other planes which are the leaves of a foliation. In this instance the index set $A = \{q^1, p_1\}$ has the obvious structure.

An instance of the Hausdorff type of difficulty is provided by the example of Eq. (3), as has already been mentioned, or by the energy-function foliation of the phase manifold TQ of the simple pendulum (Chapter 7).

18.2 DISTRIBUTIONS

The vector field of a dynamical system can be thought of as defining, through its trajectories, a foliation of the carrier manifold M. Each trajectory (more accurately, the image of each trajectory) is a submanifold of M, usually of dimension one, though sometimes the dimension is zero, and each such submanifold is a leaf of the foliation. Assume for the purpose of this description that all the trajectories are of dimension one. The leaves then define a one-dimensional tangent space at each point $m \in M$, the tangent space spanned by the tangent vector at m. This set of tangent spaces, called a one-dimensional *distribution*, characterizes the foliation: the foliation, though not the vector field, can be reconstructed if the distribution is given. More generally, a distribution of dimension higher than one is a set of vector spaces, one at each $m \in M$, and under certain conditions these can lead to foliations.

1. Let Φ be a foliation of dimension ν. Then a leaf ℓ_α passes through each $m \in M$, and the tangent space $T_m\ell_\alpha$ is of dimension ν and is spanned by a basis of ν vectors in T_mM. This is true at every point $m \in M$, and at least locally it is possible to construct a set of ν vector fields $X_1, \ldots, X_\nu \in \mathfrak{X}(U)$ such that $X_1(m), \ldots, X_\nu(m)$ span $T_m\ell_\alpha$ at all points $m \in U \subset M$. A *distribution* $\mathscr{D}$ is such a set of vector spaces, a subspace of T_mM at each $m \in M$, all of the same dimension ν, spanned in each neighborhood U by a set of ν independent local vector fields $X_1, \ldots, X_\nu \in \mathfrak{X}(U)$. One can thus think of $\mathscr{D}$ as the set of linear spaces formed by the X_j with coefficients in $\mathscr{F}(U)$, and we shall thus often speak of a local vector field X_U as *belonging to* or *lying in* $\mathscr{D}$: $X_U \in \mathscr{D}$. The X_j are called a *basis* for $\mathscr{D}(m)$.

In addition we shall often speak of global vector fields as belonging to a distribution. The transition from local to global vector fields in each basis goes as follows. Suppose that $X_U \in \mathfrak{X}(U)$ is in $\mathscr{D}$, that is, that $X_U(m) \in \mathscr{D}(m)$, $\forall m \in U$. Any local vector field such as X_U can be extended almost trivially to a global one $X \in \mathfrak{X}(M)$ by multiplying it by a suitable bump function which makes it vanish smoothly outside of U (see AM, pp. 81, 123). Usually, this kind of extension is of little interest, however, for the vector field so obtained is highly singular, vanishing everywhere outside of U. If U' is some neighborhood whose intersection with U is empty, for instance, then $X_{U'} \equiv X_U | U'$ is the null vector field and is only trivially in $\mathscr{D}(m')$, $m' \in U'$. It can certainly not form part of any basis for the distribution at m'. One is therefore ordinarily interested in other extensions, vector fields which vanish at as few points as possible. In any case, we shall say that a vector field $X \in \mathfrak{X}(M)$ belongs to $\mathscr{D}$ iff $X(m) \in \mathscr{D}(m)\ \forall m \in M$. Hence the distribution $\mathscr{D}_\Phi$ associated with the foliation Φ has the property that for each $m \in M$, with ℓ_α passing through m,

$$T_m \ell_\alpha = \mathscr{D}_\Phi(m). \tag{9}$$

An example of such a distribution, the one already mentioned, is obtained from a vector field $X \in \mathfrak{X}(M)$ with no zeros. Then the distribution $\mathscr{D}^X$ obtained from X is defined by $\mathscr{D}^X(m) = \text{span } X(m) \equiv \{aX(m) | a \in \mathbb{R}\}$. This distribution is of dimension one; i.e. $\nu = 1$ for $\mathscr{D}^X$. It is not true, however, that every one-dimensional distribution can be obtained in this way from a global vector field (see Brickell and Clark (1970, p. 193) for an example on the torus).

2. It is clear from this introduction to the concept of a distribution that every foliation Φ leads to a distribution $\mathscr{D}_\Phi$ in accordance with (9). The converse, however, is not true. Not every distribution can be integrated to form a foliation. When it can, i.e. when for a given distribution $\mathscr{D}$ (not initially a $\mathscr{D}_\Phi$ obtained from a foliation) Eq. (9) can be solved for the ℓ_α, then ℓ_α is called an *integral manifold* of $\mathscr{D}$, and $\mathscr{D}$ is said to be *integrable*.

Consider a distinguished chart (U, φ) for a foliation Φ of codimension κ. The vector fields that lie in $T(U \cap \ell_\alpha)$ must obviously be of the form $X = f^j \partial/\partial x^j$, where $f^j \in \mathscr{F}(U)$ and the x^j are distinguished coordinates in $U, j \in \{1, \ldots, \nu = \mu - \kappa\}$, and thus locally all of the vector fields in $\mathscr{D}_\Phi$ are of this form. Moreover, any vector field in $\mathfrak{X}(U)$ which is of this form is also in $\mathscr{D}_\Phi$, and it follows immediately that $X, Y \in \mathscr{D}_\Phi$ implies that $[X, Y] \in \mathscr{D}_\Phi$. Thus a distribution associated with a foliation is closed under commutation, or is *involutive*.

Remark. Involutivity is not a Lie-algebra property in the strict sense of finite-dimensional Lie algebras, for the coefficients are not constants, but functions in $\mathscr{F}(U)$.

What has just been demonstrated is that integrable distributions are involutive. The converse is also true, and it is known as the Frobenius theorem, which will not be proved here (Brickell and Clark, 1970, p. 94; AM, p. 93): every involutive distribution is also integrable.

But not all distributions are involutive, and thus not all can be used to define foliations. For example, $X = \partial/\partial x$ and $Y = \partial/\partial y + x\partial/\partial z$ define a distribution globally on $\mathbb{R}^3$: at each point $(x, y, z) = m \in \mathbb{R}^3$ the subspace of $T_m\mathbb{R}^3$ which is in the distribution is spanned by the two vectors whose components are $(1, 0, 0)$ and

$(0, 1, x)$. This distribution is not involutive, for the commutator $[X, Y] = \partial/\partial z$ does not belong to the distribution. It is easily seen as well that locally this distribution is not integrable, for suppose that there existed a surface $f(x, y, z) = k$ such that X and Y were tangent to it. Then it would follow that $L_X f \equiv \partial f/\partial x = 0$ and $L_Y f \equiv \partial f/\partial y + x\partial f/\partial z = 0$. The only solution to these equations is $f =$ const., so no such surface exists.

Let us pursue this example a little more geometrically. Consider the integral curve of one of the two vector fields, say of X, which passes through the origin of $\mathbb{R}^3$. As shown in Fig. 18.5, this curve lies along the x axis, and through each of its points passes an integral curve of Y. Taken together, all of these integral curves of Y trace out a surface passing through the x axis, the surface labelled A in the figure. Now consider one of the integral curves of Y, say the one passing through the origin. As is shown in the figure, this curve lies along the y axis, and through each of its points passes an integral curve of X. Taken together all of these integral curves of X trace out a surface passing, like A, through both the x axis and the y axis. This surface is labelled B in the figure. Clearly A and B are different surfaces, and this is what is meant when it is said that the distribution is nonintegrable: it does not define a unique integral manifold (or surface in this example) at each point of M (at the origin of $\mathbb{R}^3$ in this example).

Now recall the discussion of the Lie derivative of a vector field in Chapter 5, according to which the commutator of two vector fields describes, roughly speaking, the difference between two paths (see Fig. 5.2). The first of these paths

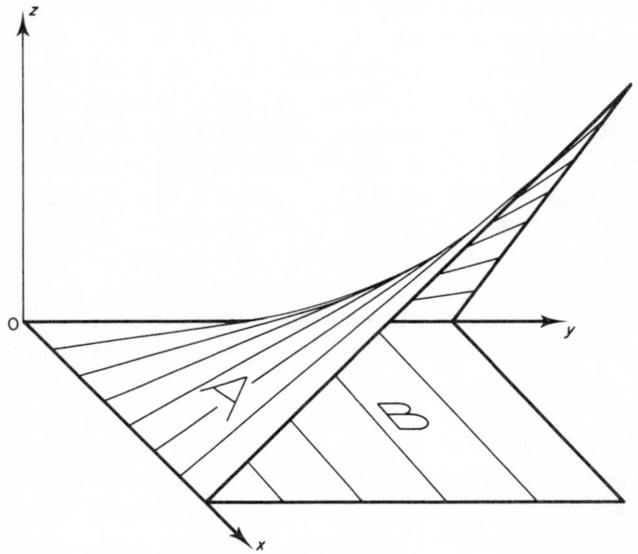

Fig. 18.5. Two surfaces obtained from a nonintegrable (non-involutive) distribution. The lines on surface A, including the y axis, are integral curves of Y. The lines on surface B, including the x axis, are integral curves of X

goes a short distance along an integral curve of one vector field and then a short distance along an integral curve of the other. The second path starts at the same point and travels along integral curves of the vector fields in the opposite order. In the example of Fig. 18.5 one of these paths lies on surface A and the other lies on surface B, so that the commutator lies outside the two surfaces. If the distribution had been integrable, there would have been a single surface passing through the origin, and the difference between the two paths would have lain on that surface and so could have been written as a combination of the two vector fields which define the surface, namely of X and of Y: the commutator would have been in the distribution. This demonstrates the connection between integrability and involutivity.

A distribution can fail to be involutive also for topological reasons. For instance it may be impossible to foliate a certain manifold nonsingularly (that is, with a nonsingular foliation) with codimension $\nu = \mu - \kappa$, and then it can support no distribution of dimension κ. An example of this occurs on the sphere S^2, which cannot be foliated nonsingularly with codimension one, as is well known. Therefore it can support no involutive distribution of dimension one. (Since S^2 is of dimension two, the only nontrivial distributions it can support must have dimension one.) Although it may occasionally become necessary to take such matters into account and to proceed with caution, we shall not go into further detail at this time.

3. The distribution $\mathscr{D}_\Phi$ associated with a foliation Φ can be understood also in terms of the projections χ_U of Eq. (4), which are local submersions. This means that $T\chi_U \neq 0$, and thus that at each point $m \in M$ it is possible to construct the vector space $\ker(T\chi_U)(m) \subset T_m M$ which consists of vectors tangent to the leaf which passes through m. Clearly these vector spaces form a $(\mu - \kappa)$-dimensional distribution which coincides with $\mathscr{D}_\Phi$. For a regular foliation this argument becomes global, since a global submersion can be found, and then $\mathscr{D}_\Phi = \ker(T\chi_U)$ defines a distribution at each $m \in M$. A distribution which is the kernel of a global submersion, obviously integrable and hence involutive, is called a *regular distribution* and leads to a regular foliation.

4. The foliations that are used in the integration of dynamical systems are often obtained from differential forms via distributions, as was demonstrated in Chapter 17. Since differential forms of any degree can be written locally in terms of one-forms, it is helpful to start their discussion with one-forms.

The kernel of a one-form, evaluated at each $m \in M$, is a $(\mu - 1)$-dimensional subspace of $T_m M$, and the set of such subspaces is then a $(\mu - 1)$-dimensional distribution. Given κ independent one-forms, the intersection of the κ subspaces of dimension $(\mu - 1)$ at each $m \in M$ is a $(\mu - \kappa)$-dimensional subspace of $T_m M$, and the set of these subspaces is then a $(\mu - \kappa)$-dimensional distribution $\mathscr{D}$ on M. In fact, locally any $(\mu - \kappa)$-dimensional distribution can be specified in this way, by finding κ independent one-forms whose kernels intersect to yield $\mathscr{D}$.

It has been seen that in the more or less dual situation, in which $\mathscr{D}$ is given locally by $\mu - \kappa$ vector fields rather than by κ one-forms, the Frobenius theorem requires that the vector fields be in involution if the distribution is to be integrable. There is a similar restrictive condition which the one-forms must satisfy for integrability. Two simple examples, one leading to an integrable distribution and the other to a nonintegrable one, are the following.

First, consider the two one-forms $\theta_1 = ydx - xdy$ and $\theta_2 = xdz - zdx$ on $M = \mathbb{R}^3 - \{0\}$. The vector field $X = x\partial/\partial x + y\partial/\partial y + z\partial/\partial z$ lies in the kernels of both one-forms ($Z = \partial/\partial z$ is also in the kernel of θ_1, and $Y = \partial/\partial y$ is also in the kernel of θ_2) and thus spans the distribution they define. Being one-dimensional, the distribution is necessarily involutive and integrable: the foliation is by radial lines in M. Second, consider the one-form $\theta = xdy - dz$, also on $M = \mathbb{R}^3 - \{0\}$. Its two-dimensional kernel is spanned by the vector fields $X = \partial/\partial x$ and $Y = \partial/\partial y + x\partial/\partial z$. These are the two vector fields of the example of Fig. 18.5, and it has already been shown that the distribution they define is not involutive and cannot lead to a foliation of M.

The condition represented by the Frobenius theorem can be stated for one-forms in four equivalent ways. We do not prove them (as we did not prove the Frobenius theorem itself). Let $\theta^1, \ldots, \theta^\kappa$ be a set of linearly independent one-forms, and let $\omega = \theta^1 \wedge \ldots \wedge \theta^\kappa$. Then the intersection of the kernels of the θ^j is an involutive distribution if one of the following equivalent conditions holds

(a) $\theta^j \wedge \omega = 0 \; \forall \; \theta^j$.
(b) There exists a one-form α such that $d\omega = \alpha \wedge \omega$.
(c) There exist local one-forms α^i_j, $i, j \in \{1, \ldots, \kappa)$, such that $d\theta^i = \alpha^i_j \wedge \theta^j$.
(d) There exist functions f^i and g^i_j, $i, j \in \{1, \ldots, \kappa)$, such that $\theta^i = g^i_j df^j$.

For proofs of these consequences of the Frobenius theorem, see Kamber and Tondeur (1975) or Flanders (1969, p. 96). That these conditions apply to the examples just given in easily verified. For instance, in the first example

$$\omega = -x(zdx \wedge dy + xdy \wedge dz + ydz \wedge dx)$$

and $d\omega = -4xdx \wedge dy \wedge dz$, so that the α of condition (b) is $-(4/x)dx$. In the second example $\omega = \theta$, and $d\omega = d\theta = dx \wedge dy$. It is easily established that no α exists such that $d\omega = \alpha \wedge \omega$.

It is seen that the local κ-form ω plays an important role in these considerations. In fact ω could have been used to define the $(\mu - \kappa)$-dimensional distribution which forms both its kernel and the intersection of the kernels of the θ^j. In fact, instead of starting with κ local one-forms, one can start with a global κ-form of constant rank (also equal to κ). Its kernel is then a distribution of dimension $\mu - \kappa$, but it may not be integrable. There is a theorem, however, which makes it possible to check the integrability: locally every κ-form can be written as the exterior product of κ local one-forms, and then the integrability of the distribution can be checked by using any of the four conditions already listed.

Another sufficiency test for integrability is the following. Let ω be a κ-form on M, and let $X, Y \in \mathfrak{X}(M)$ be in ker ω. Then

$$i_{[X,Y]}\omega = (L_X i_Y - i_Y L_X)\omega = i_X i_Y d\omega.$$

Hence $i_X i_Y d\omega = 0$ implies that $[X, Y] \in \ker \omega$. In other words, $\ker \omega$ is an involutive distribution if $i_X i_Y d\omega = 0$ for all $X, Y \in \ker \omega$. For example, $\ker \omega$ is involutive if (these are sufficient, but not necessary conditions) there exists a one-form α such that $d\omega = \alpha \wedge \omega$, or if $d\omega = 0$. This should not be confused, incidentally, with condition (b) above. In condition (b) ω was assumed to be *decomposable* (or a *mononomial*).

When foliations are applied to the reduction of dynamical systems, it will be seen (as may already be somewhat evident from Chapter 17) that defining them in a two-step process through differential forms and distributions is in fact useful.

18.3 FOLIATIONS AND ADDITIONAL STRUCTURE

1. Suppose that (M, ω) is a symplectic manifold. Because the symplectic structure ω is by definition regular, its kernel is empty, and hence the methods of §18.2(3) cannot be applied to ω in an attempt to foliate M. There is, nevertheless, a way in which ω can be used to construct a foliation of M, as will now be shown. The foliation so obtained is called a *Lagragian foliation*, and it will be defined in terms of *Lagrangian manifolds* and *Lagrangian distributions.*

For a preliminary illustration of some of the ideas involved, consider the cotangent bundle (T^*Q, ω_0) of a ν-dimensional manifold Q. Recall that Q can be thought of as a ν-dimensional submanifold of T^*Q, namely the zero section, with the natural injection $i: Q \to T^*Q$. Let $\omega_Q = i^*\omega_0$ be the pullback of ω_0 onto Q. Because $dp_k = 0$ on Q for all k, it follows that $\omega_Q = 0$. A submanifold like Q of a symplectic manifold, one on which the symplectic form ω vanishes in this way, is called a Lagrangian submanifold. More or less complementary to Q is a fiber T_q^*Q over $q \in Q$, which can also be seen as a ν-dimensional submanifold of T^*Q, together with its natural injection $i_q: T_q^*Q \to T^*Q$. Again, the pullback $\omega_q = i_q^*\omega_0$ can be defined, and again $\omega_q = 0$, this time because $dq^k = 0$ for all k on T_q^*Q, which is therefore also a Lagrangian submanifold of (T^*Q, ω_0). Moreover, the set of manifolds $\{T_q^*Q | q \in Q\}$ provides a foliation of the cotangent bundle, all of whose leaves are Lagrangian submanifolds. This is an example of a Lagrangian foliation.

These ideas will be generalized to arbitrary symplectic manifolds by starting with symplectic vector spaces, i.e. symplectic manifolds which are also linear spaces. Let (E, ω) be a symplectic vector space of dimension $\mu = 2\nu$. Let $F \subset E$ be a subspace of E; its *orthogonal complement* (*orthocomplement*) $F^\perp$ in E is defined by

$$F^\perp = \{e \in E \vdash \omega(e, e') = 0 \ \forall \ e' \in F\}. \tag{10}$$

Because ω is antisymmetric, $F \cap F^\perp$ is not in general empty. A subspace F is said to be *isotropic* if $F \subset F^\perp$, i.e. if $w(e, e') = 0 \forall e, e' \in F$; it is said to be *Lagrangian* if $F = F^{\perp}$. Then F is Lagrangian iff it is isotropic and $\dim F = \dim F^{\perp} = \nu$.

It can be shown (AM, p. 403) that if F is a Lagrangian subspace of E, then there exists an isotropic complement F' of F in E, i.e. such that $E = F \times F'$ (or, in vector-space notation, $E = F \oplus F'$). This direct-product relationship between F and F' is similar to the relationship between Q and T_q^*Q.

Now consider a symplectic manifold (M, ω) of dimension $\mu = 2\nu$. Let $N \subset M$ be a submanifold and $i: N \rightarrow M$ be the natural injection, and consider the pullback $\omega_N = i^*\omega$ of ω onto N. Then N is called *isotropic* iff $T_n N$ is isotropic for all $n \in N$, i.e. iff $\omega_N = 0$. A submanifold $N \subset M$ is called *Lagrangian* iff $T_n N$ is Lagrangian for all $n \in N$, i.e. iff $\dim N = \nu$. The direct-product property of vector spaces now becomes the following (AM, p. 409). If L is a Lagrangian submanifold of a symplectic manifold (M, ω), then there exists an isotropic sub-bundle P of $TM|L$ such that $TM|L = TL \times P$.

This way of defining subspaces as Lagrangian if their tangent spaces are Lagrangian can be extended also to distributions. A distribution $\mathscr{D}$ on (M, ω) is called Lagrangian if $\mathscr{D}(m) \subset T_m M$ is Lagrangian at each $m \in M$. If $\mathscr{D}$ is Lagrangian and $X, Y \in \mathscr{D}$, then $i_X i_Y \omega = 0$. If $\mathscr{D}$ is involutive as well as Lagrangian, it defines a *Lagrangian foliation*, one whose leaves are all Lagrangian submanifolds. This means that if ℓ_α is a leaf of the foliation and $i_\alpha: \ell_\alpha \rightarrow M$ is the natural injection, then $\omega_\alpha \equiv i_\alpha^* \omega = 0$. Not all Lagrangian distributions are involutive, however. For example, the distribution on $\mathbb{R}^4 = \{x, y, z, w\}$ which is the one of Fig. 18.5 (previously this was defined as a distribution on $\mathbb{R}^3$), already shown not to be involutive, is Lagrangian with respect to the symplectic form $\omega = dx \wedge dy \wedge dz \wedge dw$.

If $\mathscr{D}$ is both Lagrangian and *symplectic* (defined in chapter 20, but see below), then it is involutive. This may be seen as follows. If $X, Y, Z \in \mathscr{D}$, then it is easily shown that the Lagrangian character of $\mathscr{D}$ implies that

$$i_{[X,Y]} i_Z \omega + i_Y i_X L_Z \omega = 0. \tag{11}$$

In a symplectic distribution every vector field need not be locally Hamiltonian, but it can be written as a sum of locally Hamiltonian vector fields with coefficients in $\mathscr{F}(M)$. For simplicity, let $Z = fW$, where $W \in \mathfrak{X}(M)$ is locally Hamiltonian. Then

$$L_Z \omega = f L_W \omega + df \wedge i_W \omega = df \wedge i_W \omega$$

and

$$i_Y i_X L_Z \omega = (L_X f) i_Y i_W \omega - (L_Y f) i_X i_W \omega = 0$$

because W, as well as X and Y, is in $\mathscr{D}$. Inserting this into (11) yields $i_{[X,Y]} i_Z \omega = 0$ or $[X, Y] \in \mathscr{D}$. Thus $\mathscr{D}$ is involutive.

It has already been seen that the zero section Q of a cotangent bundle T^*Q is a Lagrangian submanifold, and that the set $\{T_q^*Q \; \forall \; q \in Q\}$ is a Lagrangian foliation of T^*Q. Another way to obtain a Lagrangian submanifold of T^*Q is the following. Let $\alpha \in \Omega^1(Q)$ be a closed one-form on Q. Then α can be thought of as a section of T^*Q, that is, $\alpha: Q \rightarrow T^*Q$, and one can therefore define $\omega_\alpha \equiv \alpha^*\omega_0 = -\alpha^* d\theta_0 = -d\alpha^*\theta_0$. Now, it can be shown (see AM, p. 179) that $\alpha^*\theta_0 = \alpha\ \forall \alpha \in \Omega^1(Q)$, so that $\omega_\alpha = -d\alpha = 0$, for α is assumed closed. It is therefore clear that the image $\alpha(Q) \subset T^*Q$ is a Lagrangian submanifold of T^*Q.

Lagrangian submanifolds play a role in Hamilton–Jacobi theory, in action-angle variables, and in quantization. They are discussed in detail in AM, Chapter 5. We shall not go into them here in any greater detail here, though we shall make use of them later.

2. A foliation can be constructed on a manifold M also from a one-form and a distribution. Let $\theta \in \Omega^1(M)$ be a one-form and $\mathscr{D}^X$ be a distribution defined by a finite number of global vector fields $\{X_1, \ldots, X_\kappa\}$, so that $\dim \mathscr{D}^X = \kappa$. The one-form and the distribution together then define a set of functions $f_k \in \mathscr{F}(M)$, $k \in \{1, \ldots, \kappa\}$, by

$$f_j = i_{X_j}\theta. \tag{12}$$

Let $\{Y\}$ be the set of all vector fields $Y \in \mathfrak{X}(M)$ such that

$$L_Y f_k \equiv i_Y df_k = 0\ \forall\ k. \tag{13}$$

This can also be written in the form $Y = \ker TF$, where F is defined by $F \equiv \{f_1, \ldots, f_\kappa\}: M \to \mathbb{R}^\kappa$. If $\{Y\}$, or rather the span of the vector fields in $\{Y\}$, is of the same dimension at all $m \in M$, it forms a distribution. Assume that the dimension is in fact constant, and let the resulting distribution be called $\mathscr{D}^f$. Then $\mathscr{D}^f$ is involutive and thus defines a foliation Φ^f. Indeed, let $Y, Z \in \mathscr{D}^f$. Then

$$i_{[Y,Z]} df_k = L_Y i_Z df_k - i_Z L_Y df_k = -i_Z d L_Y f_k = 0,$$

so that $[X, Y] \in \mathscr{D}^f$. Thus a one-form θ and a distribution $\mathscr{D}^X$ have been used to define a foliation Φ^f on M.

Note that $\mathscr{D}^X$ was not assumed involutive, so that it is possible to define a foliation from a distribution with the aid of a one-form even if the distribution is not involutive. If $\mathscr{D}^X$ is involutive, so that there exist functions $g^i_{jk} \in \mathscr{F}(M)$ such that $[X_j, X_k] = g^i_{jk} X_i$ (if the X_i are dependent, the g^i_{jk} are not uniquely defined), the functions associated by means of (12) with $[X_j, X_k] \in \mathscr{D}^X$ can be expressed in terms of the g^i_{jk} and the f_k:

$$\begin{aligned} i_{[X_j, Y_k]}\theta = g^i_{jk} f_i &= L_{X_j} i_{X_k}\theta - i_{X_k} L_{X_j}\theta \\ &= L_{X_j} f_k - i_{X_k} L_{X_j}\theta. \end{aligned}$$

Finally, if θ is *invariant* under $\mathscr{D}^X$ (that is, if $L_X\theta = 0\ \forall\ X \in \mathscr{D}^X$), then

$$L_{X_j} f_k = g^i_{jk} f_i.$$

We mention this result because it will play a role in the discussion of *function groups* in Chapter 20.

Remarks. 1. If the g^i_{jk} in the above equations are not only functions on M but constants, then $\mathscr{D}^X$ is a (finite-dimensional) Lie algebra. If $\mathscr{D}^X$ is such a Lie algebra and if θ is invariant under $\mathscr{D}^X$, it is easily shown that $[X_j, Y_k] \in \mathscr{D}^X$ for

all X_j in $\mathscr{D}^X$ and all Y_k in $\mathscr{D}^f$, which is a sufficient condition for *projectability* c
the X_k with respect to Φ^f.

2. The mapping $\bar{J}_\theta:\mathscr{D}^X \to \mathscr{F}(M):X_j \mapsto f_j$ generates another mappin
$J_\theta:M \to \mathscr{D}^{X*}$, where $\mathscr{D}^{X*}$ is the dual of $\mathscr{D}^X$, an $\mathscr{F}$-module whose elements ma
$\mathscr{D}^X$ into $\mathbb{R}$. The map J_θ is defined by $J_\theta:m \mapsto \alpha$, where

$$(\alpha(m)\cdot X_j) = f_j(m).$$

This map is independent of the basis $\{X_1,\ldots,X_\kappa\}$ chosen for $\mathscr{D}^X$ and is a wea
form of what will be called the *momentum map* in Chapter 24.

This completes our digression on foliations. The next chapter begins th
applications of foliations to the reduction of dynamical systems by discussing th
general setting for such reductions.

BIBLIOGRAPHY TO CHAPTER 18

As a general introduction to foliations we would suggest Warner's (1971) Chapter 1. Mor
complete and technical information can be found in Bourbaki (1971), Bott (1972), an
Lawson (1974), (1977).

As a general reminder of results related to differentiable submanifolds, immersion
submersions, and the like, see also Bourbaki (1967).

19

Reduction of dynamical systems through regular foliations

This chapter will extend the results of Chapter 17 to more general dynamical systems on more general manifolds. Although more detailed, the treatment here will be analogous to that of linear systems. The difference is essentially that the role played by certain subspaces in the treatment of linear systems will now be played by certain submanifolds (leaves of foliations) and that the quotients with respect to the foliations will not in general be identifiable with submanifolds. Recall, however, that in the example of §17.1, although the system itself was linear, its reduction was not linear in just the way that is described here: the leaves of the foliations involved were not linear subspaces of the initial carrier manifold $\mathbb{R}^4$. Thus the generalization we are about to enter into is not entirely new.

19.1 PROJECTABILITY

1. Here we restate somewhat more generally what was said about projectability of vector fields in §10.2(2). Much, although not all, of what is stated here can be obtained by replacing the vector field X of §10.2(2) by a foliation Φ, or rather by an involutive distribution $\mathscr{D}$ from which that foliation can be obtained.

What will eventually be of interest to us, when we apply the general considerations to dynamical vector fields, is to reduce the complexity of a problem, that is the complexity of integrating the dynamical system, by replacing it by a related system on a manifold of lower dimension, and this will be done by projecting the dynamical system with respect to a suitably chosen foliation. Therefore, let Φ be a foliation of a differentiable manifold M and assume that $M/\Phi \equiv N$ can be (and in fact has been) provided with a manifold structure. Let $\pi: M \to N$ be the projection with respect to the foliation. Then a vector fields $Y \in \mathfrak{X}(M)$ is said to be *projectable with respect to* π (or *with respect to* Φ) *onto the vector field* $\tilde{Y} \in \mathfrak{X}(N)$ iff a vector field $\tilde{Y} \in \mathfrak{X}(N)$ exists such that the following diagram commutes:

$$\begin{array}{ccc} TM & \xrightarrow{T\pi} & TN \\ {\scriptstyle Y}\uparrow & & \uparrow{\scriptstyle \tilde{Y}} \\ M & \xrightarrow[\pi]{} & N \end{array} \tag{1}$$

When $\tilde{Y}$ exists, it is called the projection of Y.

Now let $g \in \mathscr{F}(M)$ be any function constant on the leaves of Φ. Then there exists a unique $\tilde{g} \in \mathscr{F}(N)$ such that

$$g = \tilde{g} \circ \pi, \tag{2}$$

and for every $\tilde{g} \in \mathscr{F}(N)$ the function g defined by (2) is unique and constant on the leaves of Φ. In terms of such functions an equivalent definition of projectability is the following. The vector field $Y \in \mathfrak{X}(M)$ is projectable with respect to Φ iff for every $\tilde{g} \in \mathscr{F}(N)$ there exists a $\tilde{g}' \in \mathscr{F}(N)$ such that

$$g' \equiv L_Y g \equiv L_Y(\tilde{g} \circ \pi) = \tilde{g}' \circ \pi; \tag{3}$$

in other words, the Lie derivative with respect to Y of every function constant on leaves is itself a function constant on leaves. The equivalence can be demonstrated as at Eq. (25) of Chapter 10. That demonstration shows also that the equation

$$L_{\tilde{Y}} \tilde{g} = \tilde{g}' \tag{4}$$

defines the same vector field as does the diagram (1). It is interesting that this equivalent definition can also be expressed by the requirement that a certain diagram commute, namely

$$\begin{array}{ccc} \mathscr{F}(M) & \xrightarrow{L_Y} & \mathscr{F}(M) \\ {\scriptstyle \pi^*}\uparrow & & \uparrow{\scriptstyle \pi^*} \\ \mathscr{F}(N) & \xrightarrow[L_{\tilde{Y}}]{} & \mathscr{F}(N) \end{array}$$

Of particular interest are those vector fields in $\mathfrak{X}(M)$ which project onto the null field in $\mathfrak{X}(N)$. Consider a vector field $Y \in \mathfrak{X}(M)$ which is tangent to the leaves of Φ. Then if $g \in \mathscr{F}(M)$ is constant on the leaves of Φ, it follows that $L_Y g = 0$, so that $\tilde{g}' = 0$ for all $\tilde{g} \in \mathscr{F}(N)$, and then according to (4) that $\tilde{Y} = 0$. Conversely, let $\tilde{Y} \in \mathfrak{X}(N)$ be the projection of some $Y \in \mathfrak{X}(M)$, and assume that $\tilde{Y} = 0$. Then clearly Y is tangent to the leaves of Φ. Thus if two vector fields $X, Y \in \mathfrak{X}(M)$ project down to the same vector field in $\mathfrak{X}(N)$, their difference $X - Y$ is tangent to the leaves of Φ.

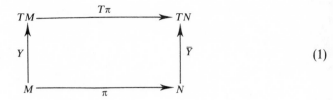

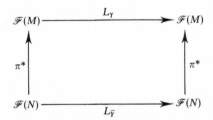

Suppose that $Y \in \mathfrak{X}(M)$ is projectable with respect to Φ. Then fY need not be, $f \in \mathscr{F}(M)$. This is because even though $g' = L_Y g$ is constant on leaves of Φ if g is, $L_{fY}g = fL_Y g = fg'$ need not be. In fact fg' will be constant on leaves iff f is, so that fY will be projectable (assuming that Y is) iff f is constant on leaves of Φ.

Let $c: I \subset \mathbb{R} \to M$ be an integral curve of $Y \in \mathfrak{X}(M)$, and assume that Y is projectable with respect to Φ. Then $\tilde{c} = \pi \circ c: I \to N$ is an integral curve of the projected vector field $\tilde{Y} \in \mathfrak{X}(N)$. This has obvious application to dynamical systems, whose integration consists of finding their integral curves. It can be stated by saying that integration and projection commute: the projections of the integral curves are the integral curves of the projection. The proof requires showing that $\tilde{Y}(\tilde{c}(t)) = T\tilde{c}(t, 1)$. Indeed,

$$\begin{aligned} \tilde{Y}(\tilde{c}(t)) &= \tilde{Y}(\pi \circ c(t)) = T\pi(Yc(t)) = T\pi(Tc(t, 1)) \\ &= T(\pi \circ c)(t, 1) = T\tilde{c}(t, 1). \end{aligned} \tag{5}$$

Let $m, m' \in M$ be on the same leaf of the foliation Φ, so that $\pi m = \pi m' \equiv n \in N$, and let $Y \in \mathfrak{X}(M)$ project down with respect to Φ onto $\tilde{Y} \in \mathfrak{X}(N)$. Then the integral curves c_m and $c_{m'}$ of Y, which are such that $c_m(0) = m$ and $c_{m'}(0) = m'$, project down onto the same integral curve $\tilde{c}_n$ of Y, for both project onto integral curves which pass through n at $t = 0$, and there is only one such integral curve. But then at another value of t both $c_m(t)$ and $c'_{m'}(t)$ lie on a common leaf of Φ, and this is true for all t. In this sense a projectable vector field Y carries leaves of the foliation Φ into leaves: Φ is invariant under Y. More precisely, the one-parameter group φ^Y of diffeomorphism generated by Y (assuming that Y is complete) is a symmetry for Φ.

The converse is also true. Let $Y \in \mathfrak{X}(M)$ be a complete vector field which leaves a foliation invariant in the above sense: the one-parameter group φ^Y of diffeomorphisms generated by Y maps leaves of Φ into leaves. Then φ^Y induces a one-parameter group of diffeomorphisms $\tilde{\varphi}$ on $N = M/\Phi$ (assuming, as always, that N can be given a manifold structure) defined by $\tilde{\varphi} \circ \pi = \pi \varphi^Y$. Let the infinitesimal generator of $\tilde{\varphi}$ be called $Y' \in \mathfrak{X}(N)$, so that one may write $\tilde{\varphi} = \tilde{\varphi}^{Y'}$. The orbits of φ^Y then project in an obvious way onto those of $\tilde{\varphi}^{Y'}$, and therefore the integral curves of Y onto those of Y'. In this way it can be seen that Y' satisfies the diagram of (1) and therefore that $Y' = \tilde{Y}$; this sketches out a proof of the converse. Just as the vector field is called projectable, so is the one-parameter group of diffeomorphisms. Like integration, such diffeomorphisms commute with projection. Another diagram can be used to express this fact:

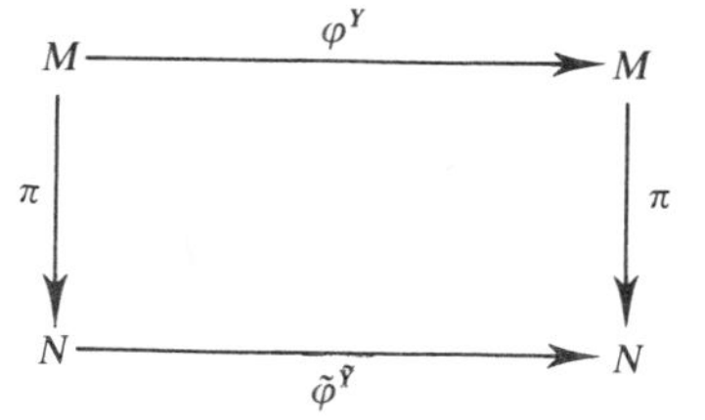

We state without proof (but see Brickell and Clark (1970, p. 126) and recall a similar result with the same reference in §10.2) that every vector $\tilde{Y} \in \mathfrak{X}(N)$ can be obtained by projection from at least one vector field $Y \in \mathfrak{X}(M)$.

2. Now assume that the regular foliation Φ is generated by a submersion $\varphi: M \to N$. Then essentially everything that was said in the preceding §19.1(1) remains valid, except that π must be replaced everywhere by φ. For any such submersion and any chart (ψ, U) on N which sends $n \in N$ into $(x_1, \ldots, x_\kappa)$, the map $\varphi': \varphi^{-1}(U) \to U$ is a local submersion, as is $F: \varphi^{-1}(U) \to \mathbb{R}^\kappa$, $F = (f_1, \ldots, f_\kappa)$, where $f_j = x_j \circ \varphi'$. This function F gives a local characterization of the foliation Φ.

This situation turns out to be particularly useful when N is itself $\mathbb{R}^\kappa$ or part of it and all of the f_j functions are independent, i.e. the df_j are linearly independent at each $m \in M$. The leaves of the foliation are now actually defined by the equation $F = \text{const.}$, and any function $g \in \mathscr{F}(M)$ which is constant on the leaves is then a function of the f_j. What this means is that the exterior derivative of any such function g is of the form

$$dg = g_k df_k, \quad g_k \in \mathscr{F}(M), \quad k \in \{1, \ldots, \kappa\}. \tag{6}$$

Moreover, the independence of the f_j implies that the g_k functions are themselves functions of the f_j. Indeed, it follows from (6) that $dg_k \wedge df_k = 0$; let $A = \{Y \in \mathfrak{X}(M) \vdash i_Y df_k = 0 \,\forall\, k\}$. Then

$$0 = i_Y(dg_k \wedge df_k) = (i_Y dg_k) df_k,$$

and the independence of the df_j, implies that $i_Y dg_k = 0$ for all k. Since this must be true for every $Y \in A$, it follows that $g_k = g_k(F)$, as asserted.

It has been established that a vector field $Y \in \mathscr{F}(M)$ is projectable with respect to Φ iff $L_Y g$ is constant on leaves for every g constant on leaves. From (6) one then obtains

$$L_Y g = i_Y dg = g_k i_Y df_k = g_k L_Y f_k, \tag{7}$$

so that Y is projectable iff $L_Y f_k$ is constant on leaves (or is a function of the f_j) for each k. A particularly simple example of a vector field which is projectable with respect to a foliation generated by a submersion F occurs when $L_Y f_k = 0$ for each k. Then it is immediately evident that $Y = 0$, for this is just the case in which Y is tangent to the leaves of Φ.

Another example is one involving the foliation of $M = \mathbb{R}^3 - \{0\}$ by straight lines passing (almost) through the origin. The quotient of M with respect to this foliation can be identified with the sphere S^2 (visualized, in particular, as the sphere of radius one about the origin in $\mathbb{R}^3$: each line is projected onto its point of intersection with the sphere). Thus $\varphi: \mathbb{R}^3 - \{0\} \to S^2$. Consider the vector field

$$Y = (ax + by)\partial/\partial x - (bx - ay)\partial/\partial y + az\partial/\partial z,$$

where x, y, z are Cartesian coordinates on M and a, b are real constants. Then Y is projectable with respect to this foliation. Indeed, let $g(x, y, z) = G(x/z, y/z)$ be any function constant on those lines for which it is defined (it is a simple exercise to extend this treatment to the rest of M). Then straightforward calculation yields

$$L_Y g(x, y, z) = b(G_1 y/z - G_2 x/z),$$

where G_1 and G_2 are the partial derivatives of G with respect to the first and second variables, respectively. The right-hand side is also a function constant on those lines for which it is defined. If $\xi = x/z$ and $\eta = y/z$ are used as the coordinates of a local chart on S^2, then the projected vector field is given by

$$Y = b(\eta \partial/\partial\xi - \xi \partial/\partial\eta).$$

Note that if $b = 0$, the vector field Y becomes tangent to the leaves of Φ (tangent to the lines), and then $Y = 0$, as expected.

3. When the foliation Φ is associated with, or obtained from an involutive distribution $\mathscr{D}$, the condition for projectability of a vector field can be stated directly in terms of $\mathscr{D}$. Since every vector field $X \in \mathscr{D}$ is tangent to the leaves of the foliation, any $g \in \mathscr{F}(M)$ which is constant on the leaves has the property that $L_X g = 0 \; \forall X \in \mathscr{D}$, and, moreover, every function g with this property is constant on the leaves. Therefore, instead of characterizing functions constant on the leaves by $g = \tilde{g} \circ \pi$, which was done to establish Eq. (3) as a criterion for projectability, one may characterize them by $L_X g = 0 \; \forall X \in \mathscr{D}$. It follows that $Y \in \mathfrak{X}(M)$ is projectable with respect to the regular foliation Φ obtained from $\mathscr{D}$ iff

$$L_X g = 0 \; \forall X \in \mathscr{D} \Rightarrow L_X L_Y g = 0 \; \forall X \in \mathscr{D}. \tag{8}$$

Moreover, a necessary and sufficient condition for (8) is that $[X, Y] \in \mathscr{D}$ for all $X \in \mathscr{D}$. The sufficiency can be seen as follows. Let $L_X g = 0 \; \forall \; X \in \mathscr{D}$. Then

$$L_X L_Y g = L_X L_Y g - L_Y L_X g = L_{[X, Y]} g = 0,$$

so that (8) is satisfied. The condition is also necessary, for in each neighborhood $U \subset M$ it is possible to find a function $g \in \mathscr{F}(U)$ which satisfies $L_X g = 0$ iff $X \in \mathscr{D}$. Assume that $[X, Y] \notin \mathscr{D}$. Then for such $g \in \mathscr{F}(U)$

$$L_X L_Y g = L_{[X, Y]} g \neq 0.$$

A distribution is said to be *invariant* under Y if $[Y, X] \in \mathscr{D}$ for all $X \in \mathscr{D}$. This is some times written in the form

$$[Y, \mathscr{D}] \equiv L_Y \mathscr{D} \subset \mathscr{D}. \tag{9}$$

Thus Y is projectable wit respect to a foliation obtained from an involutive distribution $\mathscr{D}$ if and only if $\mathscr{D}$ is invariant under Y. We shall sometimes say that Y is projectable with respect to $\mathscr{D}$ itself. Note that according to (9) every $X \in \mathscr{D}$ is projectable with respect to $\mathscr{D}$.

For an example of a vector field projectable with respect to an involutive distribution, we turn to the last example of the preceding §19.1(2), that of

$M = \mathbb{R}^3 - \{0\}$ foliated by lines radiating from the origin. This foliation can be obtained from the one-dimensional distribution $\mathscr{D}_X$ spanned by the vector field

$$X = x\partial/\partial x + y\partial/\partial y + z\partial/\partial z.$$

This distribution (being one-dimensional it is necessarily involutive) is invariant under the vector field

$$Z = x \ln |x| \partial/\partial x + y \ln |y| \partial/\partial y + z \ln |z| \partial/\partial z.$$

In fact

$$L_Z X \equiv [Z, X] = -X \in \mathscr{D}_X.$$

Thus Z is projectable with respect to $\mathscr{D}_X$, which is verified by calculating $L_Z g$, where $g(x, y, z) = G(x/z, y/z)$. It is found that

$$L_Z g(x, y, z) = (xG_1/z) \ln |x/z| + (yG_2/z) \ln |y/z|.$$

4. A foliation can be given by a distribution, and a distribution in turn can arise from the kernel of a differential form α on M. Suppose, then, that Φ comes from a distribution $\mathscr{D}_\alpha$ which is spanned by the vector fields in the kernel of some κ-form $\alpha \in \Omega^\kappa(M)$ (we will write $\mathscr{D}_\alpha = \ker \alpha$ in this case). Then the condition for projectability of a vector field with respect to Φ can be stated directly in terms of α itself. If $\mathscr{D}_\alpha = \ker \alpha$, then $X \in \mathscr{D}_\alpha$ iff $i_X \alpha = 0$. In that case invariance of $\mathscr{D}_\alpha$ under a vector field $Y \in \mathfrak{X}(M)$ becomes

$$0 = i_{[X,Y]}\alpha = L_Y i_X \alpha - i_X L_Y \alpha = -i_X L_Y \alpha \;\forall X \in \mathscr{D}_\alpha. \tag{10}$$

In other words, $\ker \alpha \subseteq \ker (L_Y \alpha)$.

A sufficient condition for (10) to be satisfied, a condition which will turn out to be useful later, is that α be *conformally invariant* under Y. This is defined to mean that there exists a function $f \in \mathscr{F}(M)$ such that

$$L_Y \alpha = f\alpha. \tag{11}$$

Indeed, then $i_X L_Y \alpha = f i_X \alpha = 0$, and condition (10) is automatically satisfied. Recall also that a sufficient condition for $\ker \alpha$ to form an involutive distribution, and thereby to yield a foliation, is that α be closed. Thus if $\alpha \in \Omega^\kappa(M)$ is closed and conformally invariant under $Y \in \mathfrak{X}(M)$, then Y is projectable with respect to $\mathscr{D}_\alpha = \ker \alpha$.

A particular case of this is the one in which α is an exact one-form $\alpha = df$, $f \in \mathscr{F}(M)$, and such that $L_Y f = 0$. This is essentially the case of a foliation given by a submersion, already discussed in §19.1(1). From $L_Y f = 0$ it follows that $L_Y df = 0$, so this is a very special case of conformal invariance. Also, as we mentioned in §19.1(1), it is sufficient that $L_Y f = g$ be a function of f, for then $L_Y df = dL_Y f = dg = hdf, h \in \mathscr{F}(M)$, which is conformal invariance more generally.

We now give an example of a foliation obtained from an exact one-form and of a vector field projectable with respect to it. The manifold of this example will be the set of ν-by-ν matrices, namely $\mathbb{R}^{\nu^2}$. Let A be a ν-by-ν matrix with matrix elements a_{jk}, and consider the function det (the determinant) defined by

$$\det A = \nu^{-1}\sum_{jk} A_{jk}a_{jk},$$

where A_{jk} is the minor of a_{jk} in A. Then it is easily shown that

$$d(\det A) = \sum A_{jk}da_{jk},$$

and that the exact one-form $d(\det)$ is therefore zero on the set of points B on which all of the A_{jk} vanish. This set B is closed, for it is the intersection of the sets B_{jk} on which each of the A_{jk} vanishes for fixed $j,k\in\{1,\ldots,\nu\}$, and each B_{jk} is closed. On the submanifold $\mathbb{R}^{\nu^2} - B$ the one-form $d(\det)$ is therefore nonzero.

Remark. B is a subset of the set on which the function det vanishes. Thus if we had been using the function rather than the one-form to define the subsequent foliation, we would have been dealing with a different submanifold.

The exact and hence closed one-form $d(\det)$ can now be used to define, through its kernel, a (ν^2-1)-dimensional involutive distribution, and through this distribution a foliation Φ. The vector field

$$Y = \sum a_{jk}\frac{\partial}{\partial a_{jk}}$$

is projectable with respect to Φ, for, as is easily shown,

$$L_Y d(\det) = \nu d(\det).$$

The foliation can be found by noting that the vector fields that are tangent to it are in the kernel of $d(\det)$; that is, if X is tangent to Φ (i.e. if X is in the distribution from which Φ is obtained), it is in the kernel of $d(\det)$, or

$$i_X d(\det) \equiv L_X(\det) = 0.$$

Thus the leaves of Φ are the submanifolds of constant determinant.

19.2 REDUCTION THROUGH FOLIATION

1. So far in this chapter no mention has been made of dynamical systems in relation to projectability. Now they will be added to what has already been discussed, and the relevance of all that to dynamical systems will be demonstrated and analyzed.

Let $\Delta\in\mathfrak{X}(M)$ be a dynamical vector field, and let Φ be a foliation of M with respect to which Δ is projectable. As before, we continue to assume that $M/\Phi = N$ can be given a differentiable manifold structure, although not that it can be identified with a submanifold of M. If N has dimension ν (the foliation has condimension ν), then Δ projects onto a dynamics $\tilde{\Delta}\in\mathfrak{X}(N)$ of dimension ν. As was pointed out in §19.1(1), the integral curves of Δ project down onto the integral curves of $\tilde{\Delta}$, and therefore by integrating the *reduced* dynamics $\tilde{\Delta}$ one obtains information about the integration of Δ. Because $\nu < \mu$, i.e. because the dimension of N is lower than that of M, it is presumably easier to integrate $\tilde{\Delta}$ than to integrate Δ.

When a dynamical system has been integrated and its dynamical curves c have been found, one says that if the initial point $m = c_m(0) \in M$ is known, the system point $c_m(t) \in M$ is known for all other times t. When only its projection $\tilde{\Delta}$ has been integrated, the same kind of statement can be made only for N: if $\tilde{c}_n(0) \in N$ is known, then $\tilde{c}_n(t) \in N$ is known for all other times t. This statement can be made in terms of M, however: if the initial leaf of Φ is known at time $t = 0$, then the leaf on which the system lies is known for all other times t. Thus what the reduced dynamics tells us is how the system moves from leaf to leaf, but not where it will lie on any of the leaves, and therefore when the reduced system is integrated, the problem is only partially solved. What remains will be discussed in §19.3.

2. A particular case of foliation occurs when it is generated by a submersion

$$F : M \to \mathbb{R}^\nu : m \mapsto \{f_1(m), \ldots, f_\nu(m)\},$$

where $f_k \in \mathscr{F}(M)$, such that F is a constant of the motion, i.e.

$$L_\Delta f_k = 0 \ \forall \ k \in \{1, \ldots, \nu\}.$$

This is the case mentioned immediately after Eq. (7), in which the projected field $\tilde{\Delta} = 0$. In terms used just above, this means that the dynamical system remains on a fixed leaf of the foliation, or that the leaves are invariant submanifolds. The reduced dynamics is trivial, and all that remains is what we have called above the motion on the leaves. Note that in this case F plays the role of π, and $N = F(M) \subset \mathbb{R}^\nu$.

In this case of foliation by a submersion F which is a constant of the motion, each point $n \in N$ determines an invariant submanifold $F^{-1}(n)$, and in this way one obtains a ν-parameter family of dynamical systems, each of dimension $\mu - \nu$ and lying on one of the invariant submanifolds. One might guess that the dynamical systems in such a family are all diffeomorphic to each other, but in fact this is not in general the case, as is shown by this example.

Let $M = I \times T^2$, where $I = (-a, a) \subset \mathbb{R}$ and T^2 is the two-dimensional torus. Let F consist of a single function (so that $\nu = 1$)

$$F : M \to \mathbb{R} : (r, \tau) \mapsto r,$$

where $r \in I$ and $\tau \in T^2$. Let the dynamical vector field be

$$\Delta = \partial/\partial\theta_1 + r\partial/\partial\theta_2,$$

where θ_1 and θ_2 are the usual two coordinates on the torus. Then clearly F is a constant of the motion for Δ, and the leaves of the foliation it generates are diffeomorphic tori. Nevertheless, the dynamics on each torus varies wildly as one goes from torus to torus by varying r, for the orbits change with r from rational to irrational winding lines. In spite of this wild variation, however, the entire dynamical system can be integrated easily by integrating it on each invariant torus separately. On each such torus the integral curves are given by

$$\theta_1(t) = \theta_{10} + t,$$
$$\theta_2(t) = \theta_{20} + rt,$$

where r designates the invariant torus. The entire solution is obtained by adding to these equations the equation that states the invariance of the tori: $r = r_0$.

3. An example illustrating some of what has been said in this chapter, again on $M = \mathbb{R}^3 - \{0\}$, follows. Let the dynamical vector field be (as usual x, y, z are Cartesian coordinates)

$$\Delta = (x - y + r^2 z)\frac{\partial}{\partial x} + (x + y - rz)\frac{\partial}{\partial y} + (-r^2 x + ry + z)\frac{\partial}{\partial z}, \qquad (12)$$

where $r^2 = x^2 + y^2 + z^2$. The one-form

$$\alpha = d(r^2/2) \equiv xdx + ydy + zdz$$

is conformally invariant under Δ, for a straightforward calculation yields $L_\Delta \alpha = 2\alpha$. Clearly α is closed (even exact), so that its kernel spans an involutive distribution with respect to which Δ is projectable. Let an arbitrary vector field $X \in \mathfrak{X}(M)$ be written in the form

$$X = (X_1, X_2, X_3) = X_1 \partial/\partial x + X_2 \partial/\partial y + X_3 \partial/\partial z.$$

The kernel of α is then given by

$$i_X \alpha = xX_1 + yX_2 + zX_3 = 0.$$

Although the kernel of α is two-dimensional, it takes three globally defined vector fields to span it. These may be chosen as $J_1, J_2, J_3 \in \mathfrak{X}(M)$ defined by

$$\begin{aligned} J_1 &= (0, z, -y), \\ J_2 &= (-z, 0, x), \\ J_3 &= (y, -x, 0), \end{aligned} \qquad (13)$$

and then the relevant involutive distribution $\mathscr{D}$ is spanned by the J_k. Involutivity may be checked by taking the commutators of just these three vector fields; one obtains $[J_1, J_2] = J_3$ and cyclic permutations. The projectability of Δ with respect to $\mathscr{D}$ may be checked by taking commutators of Δ with the J_k; this is left to the reader.

The vector fields of (13) may be recognized as the infinitesimal generators of the rotations about the x, y, z axes, respectively, and this gives a hint as to the foliation Φ generated by $\mathscr{D}$: its leaves are the spheres about the origin. In fact $i_X \alpha = 0$, the equation which yields $\mathscr{D}$, can also be written in the form $L_X(r^2) = 0$, and then it is understood that the solutions (13) of this equation must be the vector fields tangent to the spheres. Thus Φ can also be generated by the submersion

$$F: M \to \mathbb{R}^+ : (x, y, z) \mapsto r^2.$$

Now, the radius r can be used to parametrize $M/\Phi = N$, which is of dimension one, and therefore every vector field in $\mathfrak{X}(N)$ can be written in the form $f(r)\partial/\partial r$. To project Δ one may study how it acts on functions which are constant on the leaves of Φ, that is, how it acts on functions of r. A simple calculation yields $L_\Delta g(r) = rg'(r)$. If one writes $g(r) = \tilde{g}(r) \circ \pi, g \in \mathscr{F}(M), \tilde{g} \in \mathscr{F}(N)$, then g and $\tilde{g}$ are clearly the same function of r. Thus the projected vector field $\tilde{\Delta} \in \mathfrak{X}(N)$ is given by

$$\tilde{\Delta} = r\frac{\partial}{\partial r}. \tag{14}$$

Finally, one can integrate the vector field of (14) to find the integral curves $\tilde{c}: \mathbb{R} \to N$. They are given by

$$\tilde{c}_{r_0}(t) = r_0 e^t, \tag{15}$$

where r_0 labels the initial value of r.

Before going on to discuss the integral curves of $\Delta \in \mathfrak{X}(M)$, of which (15) gives only the projection onto N, we should turn more generally to the difficult relation between the total dynamical system Δ and its projection $\tilde{\Delta}$.

19.3 REMARKS ON INTEGRATING THE DYNAMICS ON M

1. As has been mentioned, what $\tilde{\Delta}$ tells one is how the system Δ moves from leaf to leaf of Φ, but not where the trajectory intersects each leaf. When the reduced dynamics has been integrated, the curves $\tilde{c}: \mathbb{R} \to N$ have been found, and hence when the initial point $\tilde{c}(0)$ in N (the initial leaf of Φ) is given, $\tilde{c}(t)$ is known for all subsequent t (we shall assume for the purposes of this discussion that $\tilde{\Delta}$ is complete). But $\tilde{c}(0)$ does not contain all of the initial data. The rest, specifically at which point on the initial leaf the trajectory starts out, are contained in $c(0)$, where $c: \mathbb{R} \to M$ is an integral curve of the unreduced full dynamics Δ. These additional data are needed to solve the full problem, but even they will not suffice. What is needed is some knowledge of how the system moves *along* each leaf, although this is not an accurate way of putting it, as will become evident immediately.

If there existed some natural diffeomorphism connecting all the leaves of Φ it would make sense to ask how the system moves along the leaves, for then this diffeomorphism could be used to map all of the leaves into a representative one and the motion could be studied on this representative leaf and unfolded appropriately to obtain the total motion. However, no such diffeomorphism exists in general, for two reasons. The first is that the leaves may not all be diffeomorphic to each other. The second is that even if they were, there is no particularly natural diffeomorphism. Moreover, even if a convenient diffeomorphism were chosen, the motion on the representative leaf would presumably depend not only on where on the leaf it starts, but also on which of the diffeomorphic leaves is the initial one. Thus trajectories on the representative leaf would intersect and behave in other generally disagreeable ways.

2. The problem is that there are many vector fields in $\mathfrak{X}(M)$ that project down to the reduced dynamics $\tilde{\Delta}\in\mathfrak{X}(N)$, among them the full dynamics Δ, and a knowledge of $\tilde{\Delta}$ does not allow one to choose among them. In addition, each one of these vector fields, in particular Δ, has many integral curves which project down to each integral curve of $\tilde{\Delta}$. We shall now concentrate on the relation between a vector field in $\mathfrak{X}(N)$ and those vector fields in $\mathfrak{X}(M)$ which project down to it.

Let $Y\in\mathfrak{X}(N)$, and write Y^+ for a vector field in $\mathfrak{X}(M)$ which projects down to Y; in the notation we have been using, then, $\tilde{Y}^+ = Y$. We shall say that Y^+ is a *lift* of Y, or that Y^+ is obtained from Y by *lifting*. Given $Y\in\mathfrak{X}(N)$, there are many vector fields $Y^+\in\mathfrak{X}(M)$ that can be obtained from it by lifting, all those which project down to Y. The difference between any two of them, however, obviously projects down to the null vector field on N.

In order to discuss a way to specify a particular lift $Y^+\in\mathfrak{X}(M)$ of a given $Y\in\mathfrak{X}(N)$, consider first a point $m\in M$ and a leaf of Φ which passes through m. The restriction $\mathscr{D}(m)$ to m of the distribution associated with Φ is a vector space of dimension $\kappa=\mu-\nu$, where $\mu=\dim M$ and $\nu=\dim N$. Clearly $\mathscr{D}(m)\subset T_mM$. Let $B(m)\subset T^*_mM$ be the orthodual of $\mathscr{D}(m)$; that is, $B(m)=\{\beta\in T^*_m(M)\vdash (i_X\beta)(m)=0\;\forall X\in\mathscr{D}\}$. Now complete T^*_mM with another vector space $A(m)$, of dimension κ, so that (in vector-space notation) $A(m)\oplus B(m)=T^*_mM$; note that $A(m)$ is not unique. Then for each $X\in\mathscr{D}$ there is at least one $\alpha\in A(m)$ such that $(i_X\alpha)(m)\neq 0$. If $\{\alpha_1,\ldots,\alpha_\kappa\}$ is a basis for $A(m)$, a vector $X(m)\in\mathscr{D}(m)$ is uniquely specified by its 'components' $i_{X(m)}\alpha_j$.

Now consider the restriction to m of all vector fields that can be obtained by lifting a given $Y\in\mathfrak{X}(N)$. Any two vectors so obtained, which may be called lifts of $Y(m)$, will differ by a vector in $\mathscr{D}(m)$ (for recall that the difference between two lifts of Y projects down to the null vector field, or is tangent to the leaves of Φ), and therefore $Y^+(m)$ is completely specified by giving its projection onto $\mathscr{D}(m)$, or its components $i_{Y^+(m)}\alpha$. In particular, of all possible lifts of $Y(m)$ one may choose the one for which (here we drop the argument m of Y^+)

$$i_{Y^+}\alpha_j=0, \tag{16}$$

which will be called the *α-related lift of* $Y(m)$.

These considerations are easily extended from a point to an open neighborhood U of a point, and then Eq. (16) will select one of the vector fields on U which project down to the restriction of Y to the projection of U. It is important to remember that the α-related lift of Y obtained in U by this procedure depends on the choice of $A(m)$ at each point $m\in U$ and more particularly on the basis of one-forms $\{\alpha_1,\ldots,\alpha_\kappa\}$. If such one-forms can be chosen globally, Eq. (16) yields a uniquely defined α-related lift of Y.

Remarks. 1. To assume that such one-forms exist globally is to assume that the leaves of the foliation are parallelizable (see Chapter 25).

2. The α_j can be shown to define a *connection* in a generalized sense. However, this is beyond the scope of this book.

The lift $Y^+ \in \mathfrak{X}(M)$ which is α-related to $Y \in \mathfrak{X}(N)$ by the global version of (16) can also be understood in terms of the foliation associated in the usual way (i.e. through the kernel) with the set of one-forms α_j. Assume that the α_j exist globally and that the intersection of their kernels yields an involutive distribution. They then define a foliation in the way described in §18.2(4). Let this foliation be called Φ_α (associated with $A(m)$ and with the α_j), and the one previously called Φ now be called Φ_β (associated with $B(m)$ and with the one-forms β). Similarly, let the projections with respect to these foliations be called, respectively, $\pi_\alpha : M \to K$ and $\pi_\beta : M \to N$ (this defines K; π_β was previously called simply π), and the associated foliations $\mathscr{D}_\alpha$ and $\mathscr{D}_\beta$. Assume also that K can be given a manifold structure. Since by definition any (local) vector field in $\mathscr{D}_\beta$ is not in the intersection of the kernels of the α_j, it follows that $\mathscr{D}_\alpha \cap \mathscr{D}_\beta = 0$ and hence that the leaves of Φ_α and Φ_β intersect at points. We make another simplifying assumption, namely that the leaves of Φ_α can be parametrized by their points of intersection with the leaves of Φ_β. Then it can be shown that the leaves of Φ_α are diffeomorphic to N. If ℓ is a leaf of Φ_α, let $\chi_\alpha : N \to \ell$ be the diffeomorphism defined by

$$\chi_\alpha^{-1} = \pi_\beta | \ell.$$

Then the α-related lift Y^+ is given by

$$Y^+ = T\chi_\alpha(Y). \tag{17}$$

For an example of this lifting process, let $M = \mathbb{R}^2 - \{0\}$ and $N = \mathbb{R}^+$, and let the projection π_β be defined by

$$\pi_\beta : M \to N : (x, y) \mapsto r = (x^2 + y^2)^{1/2}.$$

The vector field to be lifted is $Y = r\partial/\partial r \in \mathfrak{X}(N)$. So far the situation is very similar to the example of §19.2(4): the leaves of the foliation Φ_β are the circles about the origin $\ell_c = r^{-1}(c)$, $c > 0$. The distribution $\mathscr{D}_\beta$ is spanned by the vector field

$$X = x\partial/\partial y - y\partial/\partial x \in \mathfrak{X}(M).$$

Then B (which in this example can be defined globally from the start, rather than at each point $m \in M$) consists of all one-forms β such that $i_X \beta = 0$ and is spanned by

$$\beta \equiv x dx + y dy.$$

The A of this example, also definable globally, is of dimension one, so that $\alpha \in A$ can be any one-form independent of β and can hence be obtained from the condition $\alpha \wedge \beta \neq 0$. It is easily found that

$$\alpha = (y + gx)dx + (gy - x)dy, \quad g \in \mathscr{F}(M)$$

is the general solution which satisfies this condition. Recall that A is not unique. Each choice of g gives a different α and hence a different A (α spans A).

To obtain the α-related lift of Y, one may first construct the general lift.

Let $Y_0^+ = x\partial/\partial x + y\partial/\partial y \in \mathfrak{X}(M)$. It is easily checked that this vector field projects down to Y (apply it to any function of r alone), and then the general $Y^+ \in \mathfrak{X}(M)$ differs from Y_0^+ by a vector field in $\mathscr{D}_\beta$, namely by a vector field of the form $hX, h\in\mathscr{F}(M)$. From this it is a simple task to show that the α-related (or, as we shall say, the g-related lift) of Y is

$$Y_g^+ = (x - gy)\partial/\partial x + (y + gx)\partial/\partial y, \tag{18}$$

Consider now the integral curves of Y_g^+. Clearly they depend on g, for the vector field is different for each g. (In this two-dimensional case, incidentally, finding the integral curves of Y_g^+ is equivalent to finding the leaves of the foliation Φ_α. In higher dimensions, the integral curves lie on the leaves, but do not completely define them.) For instance, if $g = 0$, the integral curves are those of Y_0^+, just the rays from the origin. If $g = 1$, on the other hand, the integral curves of the lifted vector field Y_1^+ are given by

$$\begin{aligned} x &= e^\tau \cos(\tau + \delta), \\ y &= e^\tau \sin(\tau + \delta), \end{aligned} \tag{19}$$

where τ is a real parameter and δ a real constant (τ_0 and δ can be chosen to have the integral curve start at any point in M). The integral curves are spirals, crossing each of the leaves of Φ_β (circles about the origin) just once. As mentioned, they are also the leaves of Φ_α, and the two foliations cross at points (hardly anything else is possible, of course, in two dimensions). As a third instance, one might choose $g = x$. The lifted vector field for this case is then $Y_x^+ = x(1 - y)\partial/\partial x + (y + x^2)\partial/\partial y$, whose integral curves are much harder to find. In general, finding the foliation Φ_α, involving integration, as it does, is not trivial. Finally, we remark that the integral curves of (19) project down to the integral curves $r = r_0 e^\tau$ on N (compare Eq. (15); r_0 is determined by τ_0).

The lift of a vector field can be defined even if N cannot be immersed as a submanifold into M. Consider, for instance, the example of §14.2(2), where the constant-energy manifolds S^3 of the two-dimensional harmonic oscillator were foliated (this term was not used) by the trajectories of Δ. It was shown that the quotient of S^3 with respect to this foliation is S^2. Suppose that one wants to lift a vector field $Y \in S^2$ onto S^3 with respect to this foliation. Since $\mathscr{D}_\alpha$ (which could now be called $\mathscr{D}_\Delta$, where Δ is the dynamical vector field of the harmonic oscillator) is of dimension one, B is of dimension two. From the known constants of the motion for the harmonic oscillator it is easy to find two independent one-forms that will span B, one-forms such that $i_Y\beta = 0$: we choose $\beta_1 = d(p_1p_2 + q^1q^2)$ and $\beta_2 = (p_1q^2 - q^1p_2)$. Then A, of dimension one in this example, is entirely defined by choosing any α of the form

$$\alpha = p_j dq^j - q^j dp_j + f^\kappa \beta_\kappa, \quad f^1, f^2 \in \mathscr{F}(S^3).$$

The lift defined in this way will be uniquely determined by the choice of the f^κ, and will be different in general for different f^κ. All of this despite the fact that S^2 cannot be immersed into S^3 with a section for π_Δ, the projection with respect to Δ.

3. We now turn to the question of how integration of the reduced system $\tilde{\Delta}\in\mathfrak{X}(N)$ can aid in integrating the original dynamics $\Delta\in\mathfrak{X}(M)$. In Chapter 17 it was suggested that Δ could be written as a sum $\Delta = \tilde{\Delta} + \Delta_f$, but in general this cannot be done, because $\tilde{\Delta}$ is not in $\mathfrak{X}(M)$, but in $\mathfrak{X}(N)$, and there is no way in which it can be interpreted as a vector field on M. On the other hand, $\tilde{\Delta}$ can now be lifted to M, and then Δ could be decomposed in the form $\Delta = \tilde{\Delta}^+ + \Delta_f$, or

$$\Delta_f = \Delta - \tilde{\Delta}^+. \tag{20}$$

Clearly what 'remains' after $\tilde{\Delta}$ has been lifted, namely Δ_f, depends on the particular lifting $\tilde{\Delta}^+$, but since both Δ and $\tilde{\Delta}^+$ project down to $\tilde{\Delta}$, the vector field Δ_f is tangent to the leaves of the foliation and for this reason may be called the dynamics *along* the leaves. In any case, none of the integral curves, whether of Δ, of $\tilde{\Delta}^+$, or of Δ_f, can be obtained without further integration.

We turn again to one of the previous examples. Suppose that the vector field $Y\in\mathbb{R}^+$ is the reduced dynamical vector field $\tilde{\Delta}$ of Eq. (14), obtained now from some vector field $\Delta\in\mathfrak{X}(M)$, where $M = \mathbb{R}^2 - \{0\}$ as before. The integral curves of the reduced dynamical system are then given by (15). The full dynamical vector field Δ might be any one of the lifts of $\tilde{\Delta}$, and we shall present two possibilities.

Suppose first that Δ is the vector field we have called Y_1^+ and suppose also that α has been chosen fortuitously so that the lifted vector field $\tilde{\Delta}^+ = Y_1^+ \equiv \Delta$ and therefore that $\Delta_f = 0$. The integral manifolds of $\mathscr{D}_\alpha$ have already been found, so that the integral curves of Δ can be obtained from the diffeomorphism χ_α of (17). Applied to any point $r\in\mathbb{R}^+$, this diffeomorphism maps it into the point where the circle of radius r crosses the desired leaf of the foliation. A simple calculation for $\chi_\alpha(r)$ then yields

$$x = r\cos(\delta + \ln r),$$
$$y = r\sin(\delta + \ln r).$$

For the integral curves this becomes

$$x = r_0 e^t \cos(t + \delta + \ln r_0),$$
$$y = r_0 e^t \sin(t + \delta + \ln r_0).$$

It may be remarked that these integral curves of Δ could have been found without using χ_α, simply from the facts that $x^2 + y^2 = r^2 = r_0^2 e^{2t}$ and $dx/dt = x - y$ (or the other similar expression for dy/dt obtained from the explicit expression for Δ). In any case, it is seen that obtaining the integral curves of Δ involves at least one integration, either explicitly of the differential equations for the integral curves of Δ or, as demonstrated here, of the distribution $\mathscr{D}_\alpha$.

Consider again Eq. (19). That equation gives a one-chart atlas for M in which τ and δ are the coordinate functions. The differential equations for Δ could also be written in this chart and then they become exceedingly simple, for in this chart Δ is given just by

$$\Delta = \partial/\partial\tau.$$

Integration is immediate: $\tau(t) = \tau_0 + \tau$, $\delta(t) = \delta_0$. This is a hint of the advantage gained from reduction followed by lifting.

Suppose now that Δ is not Y_1^+, but the vector field we have called Y_x^+, yet that the lift of $\tilde{\Delta}$ is chosen as before to be Y_1^+. Of course the lift could have been chosen also to be Y_x^+, but as was pointed out already, the integral manifolds of $\mathscr{D}_\alpha$, which in this case would also have been the integral curves of Δ, would then not be easy to find, at least in the Cartesian chart in which the treatment is begun. However, the $\{\tau, \delta\}$ chart obtained from the lift simplifies the differential equations for the integral curves considerably, for in this second chart the expression for Δ becomes

$$\Delta = \partial/\partial\tau + F(\tau, \delta)\partial/\partial\delta, \tag{21}$$

where $F(\tau, \delta) = -1 + (1/2)e^\tau \cos(\tau + \delta)\cos 2(\tau + \delta)$. It is seen again that the $\tau(t)$ part of the integral curve is immediately obtained. The rest remains nontrivial, however. It can be shown, in fact, that any lift of $\tilde{\Delta}$ is of the form of (21) with some different $F(\tau, \delta)$.

In this example Δ_f does not commute with $\tilde{\Delta}^+$. It can be shown also that if the lift were such that $[\Delta_f, \tilde{\Delta}^+] = 0$, then F would be a function of δ alone, and the dynamics would then have been directly reduced into two independent dynamical systems, one on the manifold $\mathbb{R} = \{\tau\}$ and the other on $S = \{\delta\}$.

It is seen that the lifting process is one that provides a local chart in which the expression for the dynamical vector field becomes particularly simple in the sense that integration is made easier. One should not be led to believe, however, that it is always as simple as in the above examples. In these examples there are two equivalent ways of finding the integral curves of Δ, the first by integrating the differential equations for the curves themselves, and the second by finding the integral manifolds of $\mathscr{D}_\alpha$. This equivalence is the result of the two-dimensionality of the examples, or rather of the fact that the leaves of Φ_α are one-dimensional curves. Another consequence of the two-dimensionality is that the α_j can easily be chosen to given any desired lift of $\tilde{\Delta}$. In fact this is not true even in the two-dimensional case (for instance the null vector field cannot be lifted in this way). Moreover, in the examples the α_j were always globally definable and always yielded an involutive distribution. That the situation is generally considerably more complicated can be seen even with the linear example of Chapter 17.

We briefly list some of the results for that example. Recall the second method of reduction, in which the projection was defined by

$$\pi: M = \mathbb{R}^4 - \{0\} \to N = \mathbb{R}^2 - \{0\} : (q^1, p_1, q^2, p_2) \mapsto (q^2, p_2)$$

and the projected vector field was

$$\tilde{\Delta} = p_2 \partial/\partial q^2 - q^2 \partial/\partial p_2.$$

The general lift of $\tilde{\Delta}$ is found to be

$$\tilde{\Delta}^+ = p_2 \partial/\partial q^2 - q^2 \partial/\partial p_2 + \tau \partial/\partial q^1 + \chi \partial/\partial p_1,$$

where τ and χ are in $\mathscr{F}(M)$. It is then found that the general α_j (there are two of them, for A is of dimension two) are given by

$$\alpha_1 = dq^1 + \mu\theta_1 + \nu\theta_2,$$
$$\alpha_2 = dp_1 + \sigma\theta_1 + \rho\theta_2,$$

where $\mu, \nu, \sigma, \rho \in \mathscr{F}(M)$, $\theta_1 = q^2 dq^2 + p_2 dp_2$ and $\theta_2 = q^2 dp_2 - p_2 dq^2$. The conditions $i_{\tilde{\Delta}^+}\alpha_j = 0$ then give two connections between the two functions in $\tilde{\Delta}^+$ and the four in the α_j, so that the α_j are specified only up to two arbitrary functions even if the lift $\tilde{\Delta}^+$ is specified. One then must insert the requirements of the Frobenius theorem. If this is done in the form $\alpha_j \wedge d\omega = 0$, where $\omega = \alpha_1 \wedge \alpha_2$, the result is

$$\alpha_1 = dq^1 + f\theta_1,$$
$$\alpha_2 = dp_1 + g\theta_1,$$

where f and g are functions of q^1, p_1, and $(q^2)^2 + (p^2)^2$. We do not pursue this example further. The point has been to show that the two-dimensional examples simplify the situation exceedingly.

In conclusion we point out again that the problem of integrating the dynamics is far from solved when the reduced dynamical system has been integrated. The two foliations Φ_α and Φ_β, intersecting at points, provide a coordinate chart for M in which the integration would seem to be simplified, but to find these foliations explicitly is in general no simple task.

20

Foliation of symplectic manifolds and reduction of Hamiltonian systems

This chapter brings together several ideas that have already been discussed: foliation of manifolds with additional structure will be applied to the case in which that structure is symplectic, and the results uncovered will be used to discuss the reduction of Hamiltonian dynamics through foliations generated by submersions. In the process, the 'double value' of constants of the motion, touched on in Chapter 10, will again become manifest. Some of the results of this chapter will play a role also in later developments, in particular in those involving the actions of Lie algebras on symplectic manifolds.

20.1 DISTRIBUTIONS ON SYMPLECTIC MANIFOLDS, FUNCTION GROUPS

1. Consider a symplectic manifold (M, ω) and let $\mathscr{D}^X$ be a δ-dimensional distribution on M. Suppose further that $\mathscr{D}^X$ can be generated by a finite number of globally defined vector fields (though it should be noted that distributions exist that cannot be generated by a finite number of global vector fields), and let κ be the smallest number that will do it. Then let $X_1, \ldots, X_\kappa$ be a set of such globally defined generators. Although in general $\kappa \geqslant \delta$, about each $m \in M$ there is a neighborhood U in which the distribution will be generated by just δ of the X_κ. (Eq. (13) of Chapter 19 is an example of such a distribution. In that distribution $\kappa = 3$ and $\delta = 2$.) We do not assume that $\mathscr{D}^X$ is involutive, but it may be noted that if it is, then in each such neighborhood U the set of δ independent vector fields is also involutive.

Under certain conditions $\mathscr{D}^X$ can be used to define a second distribution on M by combining the X_k with ω to yield a set of one-forms. In fact let $\alpha_1, \ldots, \alpha_\kappa \in \Omega^1(M)$ be one-forms defined by

$$i_{X_k}\omega = \alpha_k. \tag{1}$$

If $\dim(\ker \alpha_1 \cap \ldots \cap \ker \alpha_\kappa)$ is constant, this intersection of kernels of the α_k defines a second distribution $\mathscr{D}^\alpha$:

$$\mathscr{D}^\alpha = \{Y \in \mathfrak{X}(M) \vdash i_Y \alpha_k = 0 \ \forall \ k\}. \tag{2}$$

We shall assume that the intersection has in fact constant dimension and therefore that $\mathscr{D}^\alpha$ exists. Note that the dimension of $\mathscr{D}^\alpha$ is just the dimension of the intersection, and its constancy is guaranteed if $\kappa = \delta$, for then all of the X_k, and hence the α_k, are independent. Thus $\kappa = \delta$ is a sufficient, though certainly not a necessary condition for the existence of the distribution $\mathscr{D}^\alpha$.

Remark. The assumptions we are making around here may seem artificial at this point, but later they may appear less so. For instance, in the discussion of Lie-algebra actions they will translate into assumptions about the nature of the Lie algebras actions involved.

We now assume in addition that $\mathscr{D}^X$ is a *symplectic*, or *locally Hamiltonian* distribution, which means that the κ generators can be (and in fact have been) chosen so that each is a locally Hamiltonian vector field. This does not imply, incidentally, that every $X \in \mathscr{D}^X$ is locally Hamiltonian. Then it follows that $L_{X_k}\omega = 0$, and hence that

$$d\alpha_k = 0. \tag{3}$$

According to §18.2(4) this is a sufficient condition for $\mathscr{D}^\alpha$ to be involutive and to generate a foliation Φ^α of M.

Now another assumption: assume that $\mathscr{D}^X$ is involutive, giving rise to a foliation Φ^X of M. Then it turns out that each X_k is projectable with respect to Φ^α (or with respect to $\mathscr{D}^\alpha$). Indeed, let $Y \in \mathscr{D}^\alpha$, so that $i_Y \alpha_k = 0 \ \forall \ k$. Then

$$\begin{aligned} i_{[X_j, Y]}\alpha_k &= L_{X_j}(i_Y \alpha_k) - i_Y L_{X_j}\alpha_k = -i_Y L_{X_j} i_{X_k}\omega \\ &= -i_Y i_{[X_j, X_k]}\omega - i_Y i_{X_k} L_{X_j}\omega \\ &= -i_Y g^n_{jk} i_{X_n}\omega = -g^n_{jk} i_Y \alpha_n = 0, \end{aligned}$$

where $g^n_{jk} \in \mathscr{F}(M)$ is given by $[X_j, X_k] = g^n_{jk} X_n$, although not uniquely unless $\kappa = \delta$. Thus each X_k is projectable with respect to Φ^α. Note, however, that not every $X \in \mathscr{D}^X$ is so projectable; what has been shown is that each element of any set of locally Hamiltonian generators of $\mathscr{D}^X$ is projectable.

To summarize: the generators X_k of an involutive symplectic distribution $\mathscr{D}^X$ on a symplectic manifold uniquely define a set of one-forms α_k. With some further assumptions, the α_k define a new involutive distribution $\mathscr{D}^\alpha$, and each of the X_k is projectable with respect to this new distribution.

2. The involutivity of $\mathscr{D}^X$ is expressed in terms of its κ generators by the equation

$$[X_j, X_k] = g^n_{jk} X_n. \tag{4}$$

We shall now assume that $\kappa = \delta$, so that the g^n_{jk} are defined by (4). It follows (and was established in the proof of projectability of the X_k) that

$$L_{X_j}\alpha_k = g^n_{jk}\alpha_n. \tag{5}$$

We make one further assumption, namely that the X_k are not only locally, but globally Hamiltonian, i.e. that there exist functions $f_k \in \mathscr{F}(M)$ such that $\alpha_k = df_k$. Then from (5) and the definition of the Poisson bracket at Eq. (8) of Chapter 10 it follows that

$$d\{f_j, f_k\} = -g^n_{jk} df_n. \tag{6}$$

Remark. Later, when Lie algebras are under discussion, the g^n_{jk} will be constants, and then Eq. (6) will be immediately integrable.

It follows from this, in accordance with Eq. (6) of Chapter 19, that $\{f_k, f_j\}$ is a function of the f_k, and moreover that the g^n_{jk} are themselves functions of the f_k (the independence of the f_k follows from the independence of the α_k). A set of functions satisfying (6) is said to form a *function group* (Forsyth, 1959, vol. 5, pp. 344ff.). Function groups become important, for example, when the f_k are constants of the motion for a dynamical system. For such constants of the motion Poisson's theorem states that each $\{f_k, f_j\}$ is also a constant of the motion, and thus a given set of constants of the motion can be extended, by taking Poisson brackets, up to some maximum number σ of independent functions obtained from the original set. Such a maximal set, called *complete*, is a function group. Dimensional considerations show that locally the number σ cannot exceed the dimension of M.

Suppose that $F \in \mathscr{F}(M)$ is a function of the f_k of (6), and consider the vector field $X \in \mathfrak{X}(M)$ defined by $i_X\omega = dF$. Because F is a function of the f_k, there exist functions $h_k \in \mathscr{F}(M)$ such that $dF = h_k df_k$, and it then follows almost immediately that $X = h_k X_k$ and therefore that $X \in \mathscr{D}^X$.

To summarize, a *Hamiltonian* involutive distribution $\mathscr{D}_X$, i.e. one whose generators can be chosen to be Hamiltonian vector fields, yields a function group. The converse is also true: if the f_k form a function group, then a chain of reasoning similar to that used to prove the projectability of the X_k can be used to prove that the Hamiltonian vector fields X_k defined by

$$i_{X_k}\omega = df_k$$

form an involutive distribution. Note that the more general condition $\{f_j, f_k\} = -g^n_{jk} f_n$, $g^n_{jk} \in \mathscr{F}(M)$ (which is, after all, extremely general), is not sufficient to obtain involutivity of the vector fields.

20.2 FOLIATIONS GENERATED BY SUBMERSIONS

1. Foliations generated by submersions are very useful for reducing dynamical systems on symplectic manifolds, and are therefore of more immediate interest than what we have so far been discussing in this chapter. Let $\varphi\ M \to N$ be a submersion, dim $M = \mu$, dim $N = \nu$. For any chart V on N one can define the local submersion

$$F: \varphi^{-1}(V) \to \mathbb{R}^\kappa : m \mapsto (x_1 \circ \varphi, \ldots, x_\kappa \circ \varphi),$$

where the x_k are local coordinates in V. Henceforth it will be assumed that F can be defined globally: we write

$$F: M \to \mathbb{R}^{\kappa}: m \mapsto \{f_1(m), \ldots, f_{\kappa}(m)\},$$

with $f_k \in \mathscr{F}(M)$ and $\kappa \leqslant \mu$. From F one obtains the foliation Φ^F whose leaves are the $N_c = F^{-1}(c)$, with $c \in F(M) \subset \mathbb{R}^{\kappa}$.

The f_k define a set of Hamiltonian vector fields $X_k \in \mathfrak{X}(M)$ through the equations

$$i_{X_k}\omega = df_k. \tag{7}$$

If κ were equal to one, the one vector field so defined would be tangent to the leaves of Φ^F, for from (7) it follows that $L_{X_k} f_k = 0$ for each k. But since there is more than one such vector field, and it is not true in general that $L_{X_j} f_k = 0$ for $j \neq k$, the X_k need not all be tangent to the leaves of the foliation. This is the fundamental difference between the reduction this discussion will lead to, of Hamiltonian systems by means of several constants of the motion, and the reduction already discussed in Chapter 10 with just one constant of the motion. As in Chapter 10, the reduction will now take place in two steps; first onto the N_c, and then from the N_c to manifolds of lower dimension. Therefore the first reduced manifolds of interest are the N_c.

Now, every vector field on N_c can be thought of as coming from at least one vector field on M: if Z is a vector field on M such that $L_Z f_k | N_c = 0$, then at any point $m \in N_c \subset M$ the vector $Z(m)$ lies in $T_m N_c$, and therefore Z is tangent to N_c. Since, in fact, every vector field in $\mathfrak{X}(N_c)$ can be obtained in this way from at least one in $\mathfrak{X}(M)$ which is tangent to N_c, the distinction between those in $\mathfrak{X}(N_c)$ and the corresponding ones in $\mathfrak{X}(M)$ which are tangent to the N_c may be allowed to blur on occasion. (No such blurring may be allowed, on the other hand, for differential forms.)

As has been mentioned, the X_k are not in general tangent to the N_c, but they form a distribution $\mathscr{D}^X$ in M, so that at each point $m \in N_c \subset M$ there are vectors which lie both in $\mathscr{D}^X$ and $T_m N_c$. Let $\mathscr{Z}(m)$ denote the subspace of $T_m M$ which is composed of these vectors; thus $\mathscr{Z}(m) = \mathscr{D}^X(m) \cap T_m N_c$. Let $\mathscr{Z}$ be the collection of the $\mathscr{Z}(m)$. This is illustrated in Fig. 20.1.

Remark. If the f_k form a function group, $\mathscr{D}^X$ is involutive. If they do not, they can be extended to a function group as described in §20.1.

The reason $\mathscr{Z}$ is introduced is that it is related in an important way to a two-form $\omega_c \in \Omega^2(N_c)$ which is induced on N_c by the symplectic structure ω on M. This two-form is defined by

$$\omega_c = i^*\omega, \tag{8}$$

where $i: N_c \to M$ is the (regular) imbedding of N_c in M (see Chapter 8). This pullback ω_c of ω is not in general regular and therefore not a symplectic form on N_c. In fact N_c may be of odd dimension and thus may not be capable of carrying a symplectic structure. The kernel of ω_c is an involutive distribution

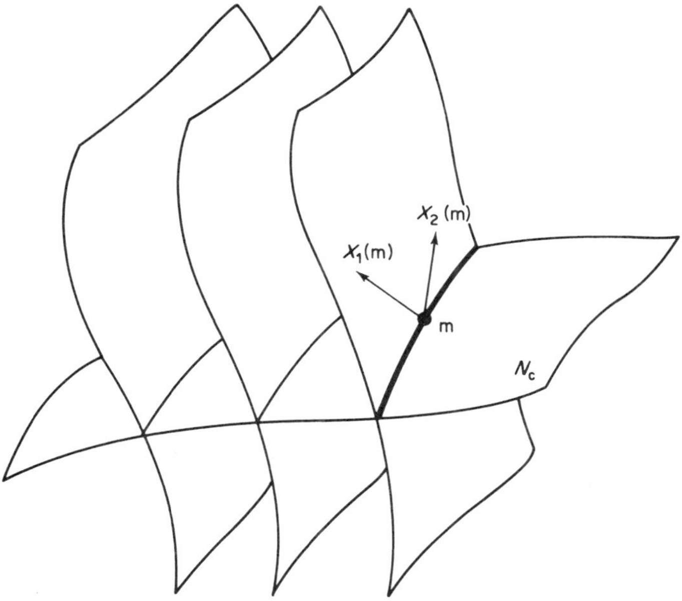

Fig. 20.1. An attempt to illustrate $\mathscr{X}(m)$. The figure shows N_c intersected by three leaves of a foliation of M generated by the X_k. (The distribution $\mathscr{D}^X$ which is generated by the X_k is not always involutive, so the foliation may not exist. It is assumed to exist, however, for the purposes of the figure.) At a point $m \in N_c$ the vectors $X_1(m)$ and $X_2(m)$ are tangent to the leaf, but not to N_c. Vectors in $\mathscr{X}(m)$ are tangent to the heavy curve which is the intersection of N_c and the leaf.

on N_c. More even than that: let $\mathfrak{X}_{\mathscr{L}\mathscr{H}}(N_c)$ be the Lie algebra of locally Hamiltonian vector fields on N_c. (Here we are taking liberties with terminology. Since ω_c, though closed, is not a symplectic form, the fields are not locally Hamiltonian in the strict sense, and the 'Lie algebra' they form is infinite-dimensional, so is not strictly a Lie algebra. What we mean is that $\mathfrak{X}_{\mathscr{L}\mathscr{H}}(N_c)$ is the set of vector fields in $\mathfrak{X}(N_c)$ such that $Y \in \mathfrak{X}_{\mathscr{L}\mathscr{H}}(N_c)$ implies that $L_Y\omega_c = 0$.) Then $\ker \omega_c \in \mathfrak{X}_{\mathscr{L}\mathscr{H}}(N_c)$ and forms an ideal (see §21.2(2)) in it. Indeed, let $X \in \ker \omega_c$ and $Y \in \mathfrak{X}_{\mathscr{L}\mathscr{H}}(N_c)$. Then

$$i_{[X,Y]}\omega_c = L_Y i_X \omega_c - i_X L_Y \omega_c = 0,$$

so that $[X, Y] \in \ker \omega_c$.

2. It follows in particular from this result that $\ker \omega_c$ is an involutive distribution and can thus generate a second foliation Φ_c, this one of N_c. This shows, incidentally, that the symplectic structure on M lends additional foliating power to the submersion F (which is essentially the generalization of what was called the double value of a constant of the motion for a Hamiltonian dynamical system). This second foliation Φ_c and projections with respect to it will be of importance in the reduction procedure of the next section, and it will therefore

be assumed that the quotient $N_c/\Phi_c \equiv M_c$ (discussed in §18.1(5)) can be given a manifold structure. Henceforth M_c will be considered a manifold.

The situation here is complicated by the fact that if this second foliation Φ_c is to exist, the kernel of $\omega_c(m)$ (that is, the subspace of T_mN_c which is annihilated by the antisymmetric matrix $\omega_c(m)$) must have the same dimension at each point $m \in N_c$, or the rank ρ_c of $\omega_c(m)$ must be constant. It will now be shown that $\ker \omega_c(m) = \mathscr{Z}(m)$, which means that the variation of ρ_c from point to point of N_c can be understood if it is understood how $\dim \mathscr{Z}(m)$ varies. First we show that $\mathscr{Z}(m)$ is indeed $\ker \omega_c(m)$ for fixed m. Although not always explicitly stated, the following calculations are for fixed m (e.g. ω_c stands for $\omega_c(m)$). Let $Y \in T_mN_c$; then

$$i_Y i_{X_k} \omega = i_Y df_k = 0 \tag{9}$$

Now let $X = g_k X_k$ be any vector in $\text{span}\{X_k(m)\}$. Then from (9) it follows immediately that $i_X i_Y \omega = 0 \; \forall \, Y \in T_mN_c$, or that the subspace of T_mM spanned by the $X_k(m)$ is in the orthocomplement of T_mN_c with respect to ω. Moreover, because ω is nondegenerate, the dimension of the orthocomplement of T_mN_c is just

$$\dim M - \dim T_mN_c = \mu - (\mu - \kappa) = \kappa,$$

which is the dimension of $\text{span}\{X_k(m)\}$. Thus $\text{span}\{X_k(m)\}$ is in fact the entire orthocomplement of T_mN_c. Since $\text{span}\{X_k(m)\}$ is just $\mathscr{D}^X(m)$, what has been established is that $\mathscr{D}^X(m)$ is the entire orthocomplement of T_mN_c. Finally, a vector Z is in the kernel of ω_c iff it is both in T_mN_c and in its orthocomplement, which is just the definition of $\mathscr{Z}(m)$. This shows that, at each $m \in N_c$,

$$\mathscr{Z}(m) = \ker[\omega_c(m)]. \tag{10}$$

The next step is to study the dimension of $\mathscr{Z}(m)$. Again let $X = g_k X_k(m)$ be any vector in $\text{span}\{X_k(m)\}$. If X is in $\mathscr{Z}(m)$, then it must be in T_mN_c, or

$$i_X df_j = 0 \; \forall \, j.$$

It then follows that

$$g_k i_{X_k} df_j = -g_k\{f_k, f_j\} = 0. \tag{11}$$

The number of independent solutions of this equation for the numbers g_k then gives the dimension of $\mathscr{Z}(m)$. Clearly this number depends on the rank φ of the matrix $F_{jk} = \{f_j, f_k\}$ (recall that the calculation is being performed at a point $m \in N_c$). Then $\dim \mathscr{Z}(m) = \kappa - \varphi$, and this is also $\dim(\ker \omega_c)$. Thus

$$\begin{aligned} \rho_c \equiv \text{rank}\, \omega_c &= \dim(T_mN_c) - \dim \mathscr{Z}(m) \\ &= (\mu - \kappa) - (\kappa - \varphi) \\ &= \mu - 2\kappa - \varphi. \end{aligned} \tag{12}$$

In this expression all that varies from point to point of N_c is φ. Thus the rank of ω_c is constant, and hence the foliation Φ_c exists, iff the rank of the matrix F_{jk} is constant on N_c. If ρ_c is not constant, then N_c', the set of points in N_c on which ρ_c is a maximum, is an open subset of N_c, as follows from the continuity of ω_c, and the results can be applied to N'_c. How this works will eventually be demonstrated in one of the examples.

Remark. There are many situations in which it is obvious that ρ_c is constant. For example, if F is a function group, then the Poisson brackets F_{jk} are all functions of the f_j and are thus constants on N_c and therefore form a constant matrix, necessarily of constant rank.

3. We now give an example of the procedure as described so far. In this example M will be $\mathbb{R}^6$, which will be written as the Kronecker product $\mathbb{R}^3 \times \mathbb{R}^3 = \{\mathbf{q}, \mathbf{p}\}$, where $\mathbf{q}$ and $\mathbf{p}$ represent the usual three-vector notation for the coordinates $\mathbf{q} = \{q^1, q^2, q^3\}$, $\mathbf{p} = \{p_1, p_2, p_3\}$, with $\mathbf{q}$ in the first factor $\mathbb{R}^3$ of the Kronecker product and $\mathbf{p}$ in the second. The submersion $F = \{f_1, f_2, f_3\}$ will be given by

$$\begin{aligned} f_1 &= \mathbf{q}^2/2 \equiv (1/2)\mathbf{q}\cdot\mathbf{q}, \\ f_2 &= \mathbf{q}\cdot\mathbf{p}, \\ f_3 &= \mathbf{p}^2/2 \equiv (1/2)\mathbf{p}\cdot\mathbf{p}. \end{aligned} \tag{13}$$

The symplectic form on M will be defined by

$$\omega = dq^k \wedge dp_k. \tag{14}$$

In this example $\mu = 6$ and $\kappa = 3$, so that $\dim N_c = 6 - 3 = 3$. The distribution $\mathscr{D}^X$ obtained from F is given by the three vector fields which are uniquely defined by the equations $i_{X_k}\omega = df_k$. These X_k are easily found; they are

$$\begin{aligned} X_1 &= (\mathbf{q}, \mathbf{p}; 0, -\mathbf{q}), \\ X_2 &= (\mathbf{q}, \mathbf{p}; \mathbf{q}, -\mathbf{p}), \\ X_3 &= (\mathbf{q}, \mathbf{p}; \mathbf{p}, 0). \end{aligned} \tag{15}$$

The tangent space at each point $(\mathbf{q}, \mathbf{p})$ of $\mathbb{R}^6$ can be identified in an obvious way with $\mathbb{R}^6$ itself. Henceforth in writing vector fields in this example we will suppress the notation for the points; that is, for example, X_1 will be denoted by $(0, -\mathbf{q})$.

The vector bundle TN_c is spanned by the vector fields $Y \in \mathfrak{X}(M)$ which satisfy $i_Y df_k = 0$; vector fields satisfying these equations are tangent to N_c and can therefore be thought of as vector fields on N_c. If $Y \in \mathfrak{X}(M)$ is written (still suppressing the notation for the point) $Y = (\mathbf{y}_q, \mathbf{y}_p)$, the equations become

$$\begin{aligned} i_Y df_1 &= 0 \Rightarrow \mathbf{y}_q\cdot\mathbf{q} = 0, \\ i_Y df_2 &= 0 \Rightarrow \mathbf{y}_q\cdot\mathbf{p} + \mathbf{y}_p\cdot\mathbf{q} = 0, \\ i_Y df_3 &= 0 \Rightarrow \mathbf{y}_p\cdot\mathbf{p} = 0. \end{aligned}$$

Three independent solutions of these equations are

$$\begin{aligned} Y_1 &= (\mathbf{q} \times \mathbf{p}, 0), \\ Y_2 &= (0, \mathbf{q} \times \mathbf{p}), \\ Y_3 &= (\mathbf{q}(\mathbf{q}\cdot\mathbf{p}) - \mathbf{q}^2\mathbf{p}, -\mathbf{p}(\mathbf{q}\cdot\mathbf{p}) + \mathbf{p}^2\mathbf{q}). \end{aligned} \tag{16}$$

Since we have not established a chart on N_c and do not intend to, there is no way to write down ω_c. However, the Y_k span TN_c, and then by contracting each one of them with ω one can calculate $\omega_c(Z, W)$ for any pair of vector fields $Z, W \in \mathfrak{X}(N_c)$. One proceeds as follows. In general $i_Y\omega = \mathbf{y}_q \cdot d\mathbf{p} - \mathbf{y}_p \cdot d\mathbf{q}$ for any $Y \in \mathfrak{X}(M)$ (here, for instance, $d\mathbf{p}$ is the three-vector whose components are dp_1, dp_2, dp_3). Thus

$$\begin{aligned} i_{Y_1}\omega &= (\mathbf{q} \times \mathbf{p}) \cdot d\mathbf{p}, \\ i_{Y_2}\omega &= -(\mathbf{q} \times \mathbf{p}) \cdot d\mathbf{q}, \\ i_{Y_3}\omega &= f_2 \mathbf{q} \cdot d\mathbf{p} - 2f_1 \mathbf{p} \cdot d\mathbf{p} + f_2 \mathbf{p} \cdot d\mathbf{q} - 2f_3 \mathbf{q} \cdot d\mathbf{q} \\ &= f_2 df_2 - 2(f_1 df_3 + f_3 df_1). \end{aligned} \tag{17}$$

Then, for example, $\omega(Y_1, Y_2) = |(\mathbf{q} \times \mathbf{p})|^2$, and $\omega_c(Y_1, Y_2)$ is then obtained by restriction to N_c. (It is easily shown that $\omega_c(Y_1, Y_2) = 4f_1 f_3 - (f_2)^2$.) The last of Eqs. (17) shows that Y_3 is in the kernel of ω_c, for restriction to N_c forces all the df_k to vanish.

The next step is to show that at each point $m \in N_c$ the vector $Y_3(m)$, which we shall write simply Y_3, spans $\ker(\omega_c(m))$, which we shall write simply as $\ker \omega_c$. This will be done by calculating $\dim(\ker \omega_c) = \dim N_c - \rho_c$ (see Eqs. (10) and (12) and recall that $\dim N_c = \dim(T_m N_c)$). Now, ρ_c is determined by φ, the rank of the matrix whose elements are $F_{jk} = \{f_j, f_k\}$. The necessary Poisson brackets are easily calculated, and the matrix at the point whose coordinates are $(\mathbf{q}, \mathbf{p})$ is found to be

$$\begin{vmatrix} 0 & \mathbf{q}^2 & \mathbf{q} \cdot \mathbf{p} \\ -\mathbf{q}^2 & 0 & \mathbf{p}^2 \\ -\mathbf{q} \cdot \mathbf{p} & -\mathbf{p}^2 & 0 \end{vmatrix} \tag{18}$$

whose rank $\varphi = 2$ so long as $\mathbf{q} \neq 0$, $\mathbf{p} \neq 0$. Thus, except at the origin

$$\rho_c = \mu - 2\kappa + \varphi = 6 - 6 + 2 = 2,$$

so that $\dim(\ker \omega_c) = 3 - 2 = 1$ and $\ker \omega_c$ is therefore spanned by Y_3 everywhere except at the origin; thus each N_c with $c \neq 0$ is foliated by the one-dimensional distribution $\ker \omega_c$.

In this example the value of c at the origin is zero, which is not a regular value for F. The origin lies on none of the N_c manifolds with $c \neq 0$, so that nothing has to be removed from any such manifold to obtain the submanifold N'_c on which ρ_c is a maximum. The origin could have been removed at the very beginning of this example, which would then have been not on the M defined at the start, but on $\mathbb{R}^6 - \{0\}$; it was left in up to this point to illustrate the difference between finding N'_c and removing singular points (see § 10.1(3)) for F.

The g_k of Eq. (11) can be found with the aid of the matrix of (18). All solutions of (11) can be written in the form

$$g_1 = -2gf_3, \quad g_2 = -gf_2, \quad g_3 = 2gf_1, \tag{19}$$

where $g \in \mathfrak{X}(M)$ is arbitrary. Then at each point $m \in N_c$ the vector $X = g_k X_k$ lies in $\mathscr{X}(m)$: it is both in span$\{X_k(m)\}$ and in T_mN_c and is thus in ker ω_c, which means that X is a multiple of $Y_3(m)$. It is easily checked that at any $m\in N_c$ one obtains $g_kX_k = -gY_3$.

We do not go further with this example.

4. Returning now to the original discussion, we show that the manifold $M_c = N_c/\Phi_c$ inherits a natural symplectic form from ω_c and hence from ω.

At the end of §19.1(1) it was mentioned that every vector field on the quotient manifold N can be obtained by projecting at least one vector field from the original manifold M. In the present application the original manifold M is N_c, the quotient manifold N is M_c, and the projection is with respect to the foliation Φ_c. Let $\pi_c: N_c \to M_c$ be this projection, and let $\tilde{X}, \tilde{Y} \in \mathfrak{X}(M_c)$ be obtained by projection from $X, Y \in \mathfrak{X}(N_c)$, respectively. Then a two-from Ω_c on M_c will be defined by

$$\pi_*[\Omega_c(\tilde{X}, \tilde{Y})] = \omega_c(X, Y). \tag{20}$$

In order for this definition to be acceptable it must be shown that it is consistent, for X and Y are not uniquely defined by $\tilde{X}$ and $\tilde{Y}$. In addition to demonstrating consistency, we will show that Ω_c is a symplectic form on M_c, i.e. that it is regular and closed.

Recall that the foliation Φ_c is defined by the distribution ker ω_c: a vector field is tangent to the leaves of Φ_c iff it is in ker ω_c. Let $Z\in\ker\omega_c$; then $L_Z g = 0$ for every $g\in\mathscr{F}(N_c)$ which is constant along leaves. This means that the vector fields in ker ω_c are all projectable and that they constitute, moreover, all of the fields which project to the null field on M_c. Thus, given $\tilde{X}, \tilde{Y}\in\mathfrak{X}(M_c)$, the vector fields from which they are projected, namely $X, Y\in\mathfrak{X}(N_c)$, are determined up to the addition of vector fields in ker ω_c, which do not contribute on the right-hand side of (20). Eq. (20) is therefore a consistent definition of Ω_c. That Ω_c is closed follows immediately from the closure of ω_c (and hence from the closure of ω). Finally, if $\Omega_c(\tilde{X}, \tilde{Y}) = 0 \forall \tilde{Y}\in\mathfrak{X}(M_c)$, it must be because the vector field $X\in\mathfrak{X}(N_c)$ from which $\tilde{X}$ is obtained by projection is in ker ω_c. But then $\tilde{X} = 0$, and therefore Ω_c is regular. This also specifies dim M_c, for regularity of Ω_c implies that dim M_c = rank Ω_c, and rank Ω_c is, in turn, equal to rank $\omega_c = \rho_c$, so that dim $M_c = \rho_c$.

The discussion so far has been restricted for simplicity to global submersions of the type $F: M\to\mathbb{R}^K$. It is almost obvious that it applies as well to local submersions and through them to more general submersions $\varphi: M\to N$. The results that can be obtained in this way for the more general submersions are independent of choice of atlas on N.

20.3 REDUCTION OF HAMILTONIAN SYSTEMS

1. Now the above considerations will be applied to the reduction of Hamiltonian dynamical systems through constants of the motion. In essence this will

generalize the discussion of §10.2; there the reduction was achieved by means of one constant of the motion, and here it will be by means of several.

Let $\Delta \in \mathfrak{X}(M)$ be a Hamiltonian dynamical system, and let the submersion F of §20.2 be a κ-fold constant of the motion for Δ. That is, let each f_k be a constant of the motion:

$$L_{\Delta} f_k = 0 \, \forall k \in \{1, \ldots, \kappa\}. \tag{21}$$

It follows from this that Δ is tangent to the N_c and can thus be viewed as a vector field in $\mathfrak{X}(N_c)$. When viewed this way, Δ is what has been called, in an abuse of the terminology (see the discussion following Eq. (8)), Hamiltonian. Indeed, let $h \in \mathscr{F}(M)$ be the Hamiltonian function of Δ, so that $i_{\Delta}\omega = dh$. Then when Δ is viewed as a vector field on N_c this condition becomes

$$i_{\Delta}\omega_c = d(h|N_c),$$

which verifies the assertion.

The fact that Δ is Hamiltonian means that it is projectable with respect to the second foliation Φ_c discussed in §20.2. This is because all locally Hamiltonian vector fields in $\mathfrak{X}(N_c)$ are projectable with respect to Φ_c, as was shown immediately after Eq. (8) when it was established that $\ker \omega_c$ is an ideal in the Lie algebra of locally Hamiltonian vector fields. This kernel is the distribution which defines the foliation, and then because it is an ideal, every locally Hamiltonian vector field satisfies the criterion for projectability of Eq. (9) of Chapter 19.

Let $\tilde{\Delta}$ be the vector field in $\mathfrak{X}(M_c)$ which is obtained from Δ by projection. Then the reduced dynamical system $\tilde{\Delta}$ is Hamiltonian on M_c, and its Hamiltonian function $\tilde{h}_c \in \mathscr{F}(M_c)$ is obtained by restriction and projection:

$$\tilde{h}_c \circ \pi = h|N_c, \tag{22}$$

where π is the projection with respect to Φ_c. This can be seen by calculating the pullback of $i_{\tilde{\Delta}}\Omega_c$:

$$\pi_* i_{\tilde{\Delta}}\Omega_c = i_{\Delta}\pi_*\Omega_c = i_{\Delta}\omega_c = d(h|N_c). \tag{23}$$

This establishes that $\tilde{\Delta}$ is Hamiltonian. If $i_{\tilde{\Delta}}\Omega_c = d\tilde{h}_c$, then from (23) it follows that $\pi_* d\tilde{h}_c = d(h|N_c)$, and then (22) follows immediately.

To summarize, then, if Δ is a Hamiltonian vector field and $F = \{f_2, \ldots, f_{\kappa}\}$ is a constant of the motion for Δ, then (M_c, Ω_c), which is obtained from (M, ω) in the two-step foliation process of §20.2, is the carrier manifold of a Hamiltonian projected vector field $\tilde{\Delta}$. Recall that the dimension of M_c is ρ_c, the rank of ω_c. Eq. (12) shows that this dimension is not $\mu - 2\kappa$, which is what it would have been if §10.2 had generalized in the simplest possible way, but larger by φ, the rank of the Poisson bracket matrix F_{jk}. Thus each single constant of a composite constant of the motion does not have exactly double value unless all of the single constants commute among themselves under the Poisson bracket.

2. A particularly simple example of this reduction procedure is provided by the

usual transition to center-of-mass coordinates for a two-particle system. In that case the configuration space Q is $\mathbb{R}^6$, for which the coordinates of a point in the usual chart are labeled $(\mathbf{q}_1, \mathbf{q}_2)$, the position vectors of the two particles in Cartesian coordinates. Then phase space is $T^*\mathbb{R}^6$, and the coordinates of a representative point may be written $(\mathbf{q}_1, \mathbf{q}_2; \mathbf{p}_1, \mathbf{p}_2)$. Consider a Hamiltonian function $H \in \mathfrak{X}(T^*Q)$ of the form

$$H(\mathbf{q}_1, \mathbf{q}_2; \mathbf{p}_1, \mathbf{p}_2) = (\mathbf{p}_1)^2/(2m_1) + (\mathbf{p}_2)^2/(2m_2) + V(\mathbf{r}), \tag{24}$$

where $(\mathbf{p}_k)^2 \equiv \mathbf{p}_k \cdot \mathbf{p}_k$ and $\mathbf{r} \equiv \mathbf{q}_1 - \mathbf{q}_2$. Then with the usual symplectic structure on $T^*\mathbb{R}^6$, the dynamical vector field becomes

$$\Delta = (\mathbf{p}_1/m_1) \cdot \partial_1 + (\mathbf{p}_2/m_2) \cdot \partial_2 - \partial_r V \cdot \partial_{p_1} + \partial_r V \cdot \partial_{p_2},$$

where ∂_k is the gradient operator with respect to $\mathbf{q}_k$, ∂_r with respect to $\mathbf{r}$, and similarly for $\mathbf{p}_k$. It is well known for this case that the three functions defined by $\mathbf{P} = \mathbf{p}_1 + \mathbf{p}_2$ are constants of the motion for Δ; they can now be used to reduce the dynamical system. This reduction procedure is best discussed in a different chart on $T^*\mathbb{R}^6$, whose coordinates $(\mathbf{R}, \mathbf{r}; \mathbf{P}, \mathbf{p})$ are given by

$$\begin{aligned} \mathbf{R} &= (m_1\mathbf{q}_1 + m_2\mathbf{q}_2)/(m_1 + m_2), \\ \mathbf{r} &= \mathbf{q}_1 - \mathbf{q}_2, \\ \mathbf{P} &= \mathbf{p}_1 + \mathbf{p}_2, \\ \mathbf{p} &= (m_2\mathbf{p}_1 - m_1\mathbf{p}_2)/(m_1 + m_2). \end{aligned}$$

In the new chart H can be written in the form

$$H(\mathbf{R}, \mathbf{r}; \mathbf{P}, \mathbf{p}) = \mathbf{P}^2/(2M) + \mathbf{p}^2/(2m) + V(\mathbf{r}),$$

where $M = m_1 + m_2$ and $m = m_1 m_2/M$ is the *reduced mass*, and the dynamical vector field becomes

$$\Delta = (\mathbf{P}/M) \cdot \partial_R + (\mathbf{p}/m) \cdot \partial_r - \partial_r V \cdot \partial_p. \tag{25}$$

The composite constant of the motion $\mathbf{P}$ foliates $T^*\mathbb{R}^6$. The leaves N_P of this foliation are of dimension 9. The three vector fields X_j obtained from the three components P_j of $\mathbf{P}$ by

$$i_{X_j}\omega = dP_j$$

are given simply in this chart by $X_j = \partial/\partial R_j$. The X_j all commute among themselves (the P_j commute under the Poisson bracket) and all of them are tangent to the N_P submanifolds. Thus each constant of the motion has its full double value. We shall not go further into the details of this well-known reduction, but merely state that after the double reduction has been performed, the reduced manifold M_c can be identified with $\mathbb{R}^6$ in the chart $(\mathbf{r}, \mathbf{p})$, and that the reduced dynamical vector field can then be written in the form

$$\tilde{\Delta} = (\mathbf{p}/m) \cdot \partial_r - \partial_r V \cdot \partial_p, \tag{26}$$

which is the sum of the last two terms of (25).

This example actually represents the case of direct reduction into two independent dynamical systems, for what was called Δ_f in Chapter 19 can here be identified with the first term of (25). This is an independent dynamical system, whose solution is trivial, for it is the free particle in three degrees of freedom. If the lifted vector field $\tilde{\Delta}^+$ is taken to be the original dynamical vector field Δ, one has $[\Delta_f, \tilde{\Delta}^+] = 0$, which was mentioned as a possibility in Chapter 19.

One of the most common examples of a somewhat less trivial reduction is the one that starts from (26) in the case in which V depends only on the magnitude of $\mathbf{r}$, so that $V(\mathbf{r})$ can be replaced by $V(\mathbf{r}^2)$. This is then called the central-force problem. Let us start over again, forgetting the genesis of (26), and write $M = T^*Q$ with $Q = \mathbb{R}^3$ and $\omega = dr^k \wedge dp_k$. Let the dynamical system then be generated by the Hamiltonian function (for simplicity we set $m = 1$)

$$h = \mathbf{p}^2/2 + V(\mathbf{r}^2). \tag{27}$$

As is well known, the three components of the angular momentum $\mathbf{F} = \mathbf{r} \times \mathbf{P}$ are constants of the motion (we write $\mathbf{F}$ for the angular momentum in order to maintain consistency with the general discussion). Then $\mathbf{F} = \{f_1, f_2, f_3\}$ corresponds to

$$f_k = \varepsilon_{kjm}\delta^{mn} r^j p_n \equiv \varepsilon^n_{kj} r^j p_n. \tag{28}$$

Thus in this example $\mu = 6$ and $\kappa = 3$. The three-dimensional N_c submanifolds are given by regular values of $\mathbf{F}$. The three vector fields X_k are easily found from $i_{X_k}\omega = df_k$; they turn out to be

$$X_k = (\mathbf{e}_k \times \mathbf{r}, \mathbf{e}_k \times \mathbf{p}),$$

where $\mathbf{e}_k$ is the vector in three-space whose components are given by $(\mathbf{e}_k)_j = \delta_{jk}$.

To find the dimension of the reduced M_c manifold, one must form the F_{jk} matrix and obtain its rank φ. The commutation rules for the components of the angular momentum are well known, and from the resulting F_{jk} matrix it is found that $\varphi = 2$ and thus that $\dim M_c = 2$. The subspace $\mathscr{L}(m)$ at each $m \in N_c$ can be obtained immediately from $i_X df_j = 0$ (see just above Eq. (11));

$$0 = i_X df_j = i_X i_{X_j}\omega = g_k i_{X_k} i_{X_j}\omega = -g_k\{f_k, f_j\} = -g_k \varepsilon^n_{kj} f_n,$$

so that $g_k = g f_k$, $g \in \mathscr{F}(M)$. Thus the vectors in $\mathscr{L}(m)$ are of the form

$$X = g f_k X_k. \tag{29}$$

Note the familiar result: the $\kappa = 3$ components of the angular momentum reduce the $\mu = 6$ dimensions of the problem not to $\mu - 2\kappa = 0$, but to $\mu - 2\kappa + \varphi = 2$ dimensions. In the classical treatment these two dimensions are usually labeled (r, p_r).

To carry this example further, let $V(\mathbf{r}^2) = \mathbf{r}^2/2$ in the Hamiltonian of (27). This is then the isotropic oscillator in three degrees of freedom (with circular frequency equal to one). Instead of the three f_k, now consider another triple constant of the motion $B = \{b_1, b_2, b_3\}$, where b_k is the energy in the kth degree of freedom: $b_k = [(p_k)^2 + (r^k)^2]/2$. The b_k all commute with each other, so that in this case $\varphi = 0$, and dim $M_c = 0$: the reduced dynamics is trivial. All three of the X_k are tangent to N_c. Note the similarity with reducing out the center-off-mass motion.

On the other hand the same triple constant of the motion can be treated in a different way, that is with a different symplectic form, so that the second foliation will be less trivial. If the symplectic form is taken to be

$$\omega' = dp_2 \wedge dp_1 + dr^2 \wedge dr^1 + dr^3 \wedge dp_3, \tag{30}$$

Δ is ω'-Hamiltonian with Hamiltonian function

$$h' = r^1 p_2 - r^2 p_1 + [(r^3)^2 + (p_3)^2]/2. \tag{31}$$

The three X_k vectors obtained from B are

$$X_1 = (0, r^1, 0; 0, p_1, 0),$$
$$X_2 = (-r^2, 0, 0; -p_2, 0, 0),$$
$$X_3 = (0, 0, p_3; 0, 0, -r^3).$$

These vector fields do not commute with each other and are thus not all tangent to the N_c. To find φ and thereby dim M_c, one calculates the F_{jk} matrix. It is found to be

$$\begin{vmatrix} 0 & f & 0 \\ -f & 0 & 0 \\ 0 & 0 & 0 \end{vmatrix}, \quad f = p_1 p_2 + r^1 r^2, \tag{32}$$

The rank of this matrix is $\varphi = 2$, so that M_c is of dimension two, as when the constant of the motion used was the angular momentum vector and the symplectic form used was the conventional one.

Let us study more carefully the exact form of the foliations in this example. The N_c defined by fixed values of B are three-tori T^3. The triple constant of the motion B also gives three vector fields X_k, and those vectors which are both in the distribution spanned by the X_k and are tangent to one of these tori are of the form $X = g_k X_k$, where the $g_k \in \mathscr{F}(M)$ are solutions of Eq. (11). These vector fields form ker ω_c and will then yield the second foliation, that of each of the N_c. By solving (11) one finds that $X = gX_3$, $g \in \mathscr{F}(M)$ at those points where $f \neq 0$. Where $f = 0$ the rank of the F_{jk} matrix vanishes and ρ_c changes from two to zero: this is an example of what may happen when ρ_c is not the same at all points of N_c.

The second foliation Φ_c and the change in rank of ω_c are perhaps best understood in a different chart on M. Let such a chart, with coordinates $\{R_k, \theta_k\}$, be defined by

$$\begin{aligned} r^k &= R_k \cos\theta_k, \\ p_k &= R_k \sin\theta_k. \end{aligned} \tag{33}$$

In these coordinates the N_c are given by $R_k = \text{const}$. Then the foliation Φ_c obtained from ker ω_c has leaves which are the circles $(r^3)^2 + (p_3)^2 = (R_3)^2 =$ const., and the quotient of N_c with respect to this foliation is the two-torus T^2. To obtain M_c one must subtract out the points at which $f=0$. These are given by $\cos(\theta_1-\theta_2)=0$, two (rational) winding lines on T^2. Then a simple calculation from (30) and (31), using the coordinates of (33), yields

$$\Omega_c = R_1 R_2 \cos(\theta_1-\theta_2) d\theta_2 \wedge d\theta_1$$

and

$$\tilde{h}'_c = R_1 R_2 \sin(\theta_2-\theta_1).$$

The canonical equations are $d\theta_1/dt = d\theta_2/dt = 0$, and it is seen that in this example the singular lines on which $\rho_c = 0$ can be included in the final analysis. The two-form Ω_c is not, however, symplectic on the M_c obtained when the singular lines are included, for it is degenerate, vanishing on those lines.

> *Remark.* It is often the case, as in this example, that the distribution used to foliate the N'_c (see the discussion after Eq. (12)) can also be used to foliate N_c. But then the two-form on $M_c = N_c/\Phi_c$ will be degenerate, also as in this example.

This example shows that a given constant of the motion (in this case B) can lead to two different foliations and thereby to two different reductions if the Hamiltonian system under consideration has two different, though equivalent, canonical descriptions.

3. Finally, we reiterate what was said in Chapter 19. Although the reduction process leads to a dynamical system on a manifold of lower dimension, and therefore presumably to a system that is easier to integrate, it is not entirely clear to what extent this aids in integrating the original dynamical system. The main difference between what has been discussed in this chapter and what was discussed in Chapter 19 is that here the reduced dynamics can be handled by the techniques developed for Hamiltonian systems. There is no obvious advantage to be gained from this in trying to use the integrated reduced dynamics to integrate the full original one.

Section B
Algebra and Group Actions

21

A digression on Lie algebras and their actions on manifolds

It has been seen that vector fields and diffeomorphisms which are symmetries of a dynamical system are useful and important tools in its analysis. Their usefulness increases, becomes deeper and richer, when they can be collected into algebras and groups. It is the enrichment resulting from such additional structure that will be discussed in the last section of this book.

Lie algebras have been mentioned several times in the last few chapters. In fact an involutive distribution is a sort of infinite-dimensional Lie algebra of vector fields, and things become simpler, as we have noted in several places, when the dimension becomes finite and the distribution can be treated as a true Lie algebra. Such a distribution can be thought of as a *realization* or *action* on a manifold of an abstract Lie algebra, what would be called, more familiarly perhaps, a *representation* if the manifold were a linear space. Thus it is natural for us to start the treatment of Lie algebra actions with a brief discussion of abstract Lie algebras, with some special cases as examples. It is only in later chapters that we shall make the transition from Lie algebras to Lie groups and to a brief discussion of the relevant parts of group theory.

The order in which we treat these matters is certainly not the traditional one for discussing the applications of Lie groups and algebras to physics. One usually starts with symmetry groups and then goes on to their infinitesimal aspects, namely to their Lie algebras. Here, because we start from the connection between distributions and Lie algebras, we do it in the opposite order. Group actions and algebra actions on manifolds are related in more or less the same way that diffeomorphisms are to vector fields: the former in some sense represent the integrals of the latter, or the latter are the infinitesimal aspects or generators of the former. This will be discussed more fully in Chapter 23.

21.1 INTRODUCTION: DISTRIBUTIONS AND LIE ALGEBRAS

Although an involutive distribution $\mathscr{D}$ is in general a sort of infinite-dimensional Lie algebra of vector fields on a manifold M, it is sometimes possible to find a finite number of vector fields X_k which generate $\mathscr{D}$ and form a true (that is, finite-dimensional) Lie algebra. It is important that it is this special set of generating vector fields X_k which forms the Lie algebra, not $\mathscr{D}$ (despite some of the words we may have used on occasion), and a different set of generating vector fields for $\mathscr{D}$ will in general not form a Lie algebra, though they may form a different one.

We start, as we do often, with an example. Let $M = \mathbb{R}^3 - \{0\}$, and let x, y, z be the usual Cartesian coordinates on M. Let $f \in \mathscr{F}(M)$ be defined by $f(x, y, z) = (x^2 + y^2 + z^2)/2$, and let $\mathscr{D}$ be the distribution $\ker(df)$. Then although $\mathscr{D}$ is of dimension two, it requires three globally defined vector fields to generate it (this is essentially the same distribution as the one at Eq. (13) of Chapter 19). For example

$$\begin{aligned} X_1 &= y\partial/\partial z - z\partial/\partial y, \\ X_2 &= z\partial/\partial x - x\partial/\partial z, \\ X_3 &= x\partial/\partial y - y\partial/\partial x \end{aligned} \tag{1}$$

span $\mathscr{D}$, and it is well known that they form a realization of the Lie algebra $su(2, C)$ (which is also the Lie algebra of the rotation group in three dimensions and is therefore also denoted $so(3)$). To invert the order here, one could say that starting with $su(2, C)$, one obtains its realization by the vector fields of (1), which then leads to an involutive distribution $\mathscr{D}$ and hence to a foliation of M, in this case by spheres about the origin.

Remark. As was mentioned at the beginning of Chapter 19, this is an example of a distribution whose dimension δ is smaller than the number κ of globally defined vector fields required to span it. In any neighborhood, of course, a subset of just δ of them will do.

It should be noted that almost any other set of generating vector fields for $\mathscr{D}$ will not form a Lie algebra. For example, even the three vector fields $Y_k = fX_k$, where f is the function defined above, fail to do so, for $[Y_1, Y_2] = fY_3$, etc., and the factor f destroys the Lie-algebra property.

More generally, suppose that a distribution $\mathscr{D}$ on a manifold M is spanned by a set of vector fields X_k which *close under the bracket* to form a Lie algebra:

$$[X_j, X_k] = c^l_{jk} X_l \tag{2}$$

(the c^l_{jk} are real numbers, called the *structure constants* of the Lie algebra; see §21.2). Let $Y_k = fX_k, f \neq 0 \in \mathscr{F}(M)$. Then straightforward calculation yields

$$[Y_k, Y_j] = fc^l_{jk} Y_l + \frac{1}{f}\{(L_{Y_k} f) Y_j - (L_{Y_j} f) Y_k\}, \tag{3}$$

which is no longer a Lie algebra.

Later it will be particularly important to find realizations of Lie algebras by Hamiltonian vector fields, especially when dealing with Hamiltonian dynamical systems. If the X_k of Eq. (2) are all Hamiltonian, then $\mathscr{D}$ is what we have called a

Hamiltonian distribution: there exist functions $h_k \in \mathscr{F}(M)$ such that $i_{X_k}\omega = dh_k$, where ω is a symplectic form on M. On the other hand, the Y_k of Eq. (3) will then not in general be Hamiltonian, although

$$i_{Y_k}\Omega = dh_k,$$

where $\Omega = \omega/f$. Since, however, Ω is not in general closed, it is not a symplectic form, and the Y_k cannot strictly be called even Ω-Hamiltonian. It is interesting that in two dimensions (i.e. if $\dim M = 2$), on the other hand, every two-form is closed, so that the Y_k are in fact Hamiltonian.

Because of the importance of Hamiltonian realizations, the reverse procedure is of some interest. Suppose, that is, that a set of Hamiltonian generators Y_k has been found for $\mathscr{D}$ which do not close to form a Lie algebra. Is it then possible to find another set which is both Hamiltonian and closes to form a Lie algebra? In general it is not possible, although sometimes it may become so. An example in which it is possible, in the special case of two dimensions, is the following. Let $M = \mathbb{R}^2 - \{0\}$, with a canonical chart $\{q, p\}$ and the usual symplectic form $\omega = dq \wedge dp$. Let the Y_k be given by

$$\begin{aligned}
Y_1 &= \tfrac{1}{2}(q\partial/\partial p - p\partial/\partial q),\\
Y_2 &= \tfrac{1}{2}(q\partial/\partial p + p\partial/\partial q),\\
Y_3 &= \tfrac{1}{2}(q\partial/\partial q - p\partial/\partial p).
\end{aligned}$$

The distribution $\mathscr{D}$ generated by these three vector fields is of dimension two; it is an improper distribution in the sense that the subspace of T_mM which lies in $\mathscr{D}$ at each $m \in M$ is all of T_mM itself. Note that $\mathscr{D}$ can be generated by sets containing only two vector fields (e.g. $\partial/\partial q$ and $\partial/\partial p$), but that no two of the Y_k will generate $\mathscr{D}$ on all of M. The Y_k both are Hamiltonian and close to form a Lie algebra. The Hamiltonian functions are

$$\begin{aligned}
h_1 &= -(q^2 + p^2)/4,\\
h_2 &= (p^2 - q^2)/4,\\
h_3 &= \tfrac{1}{2}qp,
\end{aligned}$$

and the Lie algebra is given by

$$[Y_1, Y_2] = Y_3, \quad [Y_2, Y_3] = -Y_1, \quad [Y_3, Y_1] = Y_2.$$

This algebra, by the way, is the same one as that of Eq. (1) if it is considered an algebra over the complex numbers (that is, if the abstract algebra is defined over a complex vector space), but it is a different one, called $so(2, 1)$ if it is over the reals.

Now consider the vector fields $X_k = h_1 Y_k$, $k = 1, 2, 3$. The X_k no longer form a Lie algebra. In fact because the Lie derivatives of h_1 with respect to Y_2 and Y_3 (and hence with respect to X_2 and X_3) involve h_3 and h_2, all three of the h_k enter into the expressions obtained in this case from Eq. (3). On the other hand, the X_k are Hamiltonian with respect to the symplectic form $\Omega = \omega/h^1$: one obtains $i_{X_k}\Omega = dh_k$. Thus it is seen that in the unlikely event that one had been given the symplectic form Ω and the X_k as generators of $\mathscr{D}$, one could have found another set of generators, the Y_k, which are Hamiltonian, but now with respect to the different symplectic form ω, and which have the further property of forming a Lie algebra.

We now proceed to a more general and formal discussion of Lie algebras and of their realizations, or actions, on manifolds.

21.2 LIE ALGEBRAS

1. This section is devoted to some definitions. A γ-dimensional *algebra A* over $\mathbb{R}$ is a γ-dimensional vector space V over $\mathbb{R}$ on which a bilinear *composition map* $V \times V \to V$ is defined, denoted here by a center dot $(\cdot)$, such that for all $a, b, c \in V$ and for all $r \in \mathbb{R}$

$$(a+b)\cdot c = a\cdot c + b\cdot c,$$
$$a\cdot(b+c) = a\cdot b + a\cdot c,$$
$$r(a\cdot b) = (ra)\cdot b = a\cdot(rb). \tag{4}$$

An algebra is *associative* if $a\cdot(b\cdot c) = (a\cdot b)\cdot c$, but we will be concerned mostly with nonassociative algebras.

A *Lie algebra* $\mathfrak{G}$ over $\mathbb{R}$ is an algebra over $\mathbb{R}$, in general not associative, in which the composition map, traditionally denoted by square brackets (that is, $a\cdot b$ is written $[a, b]$) and called the Lie bracket, is antisymmetric

$$[a, b] = -[b, a] \tag{5}$$

and satisfies the *Jacobi identity*

$$[a,[b,c]] + [b,[c,a]] + [c,[a,b]] = 0 \tag{6}$$

for all $a, b, c \in \mathfrak{G}$.

Remark. Equation (5) is equivalent to $[a, a] = 0 \forall a \in \mathfrak{G}$.

Let $\{e_1, \ldots, e_\gamma\}$ be a basis for $\mathfrak{G}$ (i.e. a basis for the vector space V; henceforth we shall generally use $\mathfrak{G}$ to denote both the vector space and the algebra). Then there exist constants $c^h_{jk} \in \mathbb{R}$ such that

$$[e_j, e_k] = c^h_{jk} e_h. \tag{7}$$

The c^h_{jk} are called the *structure constants* of the Lie algebra. It follows clearly from linearity that the structure constants uniquely determine the Lie algebra $\mathfrak{G}$ (that is, they uniquely determine the composition map). But $\mathfrak{G}$ determines the structure constants only up to transformations induced on them by changes of basis. Eqs. (5) and (6) impose the following identities on the structure constants:

$$c^h_{jk} = -c^h_{kj},$$
$$c^r_{hj} c^s_{rk} + c^r_{jk} c^s_{rh} + c^r_{kh} c^s_{rj} = 0. \tag{8}$$

Three common examples of Lie algebras are the following.

(a) Let $V = \mathbb{R}^3$ and the composition map be the usual vector cross product. The Lie algebra so obtained is $su(2, C)$ (or $so(3)$), already seen in §21.1. In the usual Cartesian coordinates the structure constants are the ε_{hjk}, which form the Levi–Civita tensor density.

(b) Let A be any associative algebra over $\mathbb{R}$. The *Lie algebra* A_L *of* A is obtained when the Lie bracket is taken to be the commutator bracket:

$$[a,b] \equiv a \cdot b - b \cdot a \;\forall\; a,b \in A. \tag{9}$$

It is easily verified that Eqs. (5) and (6) are satisfied and that A_L so defined is indeed a Lie algebra. It is interesting that this is the kind of thing that is done in defining the commutator bracket for vector fields, which is what gives $\mathfrak{X}(M)$ a sort of infinite-dimensional Lie algebra structure. This will become clearer in the discussion of realizations of Lie algebras on manifolds.

(c) Let A be any algebra over $\mathbb{R}$, not necessarily associative, and consider the set $\mathscr{D}(A)$ of *derivations* on A. A derivation is a linear map $D: A \to A$ such that

$$D(a \cdot b) = D(a) \cdot b + a \cdot D(b).$$

In an obvious way $\mathscr{D}(A)$ has a vector-space structure; for instance $(D_1 + D_2)(a) = D_1(a) + D_2(a)$. Although the composition $D_1 \circ D_2$ of two derivations is not in general in $\mathscr{D}(A)$, the commutator

$$[D_1, D_2] \equiv D_1 \circ D_2 - D_2 \circ D_1 \tag{10}$$

is. This endows $\mathscr{D}(A)$ with a Lie algebra structure, and the Lie algebra so obtained is called the *Lie algebra of derivations in* A, or simply the *derivation algebra* of A. This leads to another view of what we have called the infinite-dimensional Lie-algebra structure of $\mathfrak{X}(M)$ in (b) above. Take $A = \mathscr{F}(M)$, an infinite-dimensional algebra over $\mathbb{R}$. As has been said on occasion, every local derivation on $\mathscr{F}(M)$ can be represented by the Lie derivative with respect to a vector field in $\mathfrak{X}(M)$, and thus $\mathscr{D}(\mathscr{F}(M))$ can be identified with the set of Lie derivatives, so that each element of $\mathscr{D}(\mathscr{F}(M))$ corresponds to a vector field in $\mathfrak{X}(M)$.

Two Lie algebras $\mathfrak{G}_1$ and $\mathfrak{G}_2$ are *isomorphic* (we write $\mathfrak{G}_1 \approx \mathfrak{G}_2$) if there exists a bijection $\varphi: \mathfrak{G}_1 \to \mathfrak{G}_2$ such that $\varphi([a,b]) = [\varphi(a), \varphi(b)] \;\forall\; a,b \in \mathfrak{G}_1$. We shall often fail to distinguish between isomorphic Lie algebras; for our purposes two algebras are the same if they are isomorphic. The two algebras are said to be *homomorphic* if φ is surjective (onto), but not injective (one-to-one).

2. A subspace $\mathfrak{H}$ of $\mathfrak{G}$ is a *subalgebra* of $\mathfrak{G}$ if it is closed under the Lie bracket, i.e. iff $[a, b] \in \mathfrak{H} \;\forall\; a, b \in \mathfrak{H}$, or, as we shall write, iff $[\mathfrak{H}, \mathfrak{H}] \subset \mathfrak{H}$. If $\mathfrak{H}$ is a subalgebra, its *normalizer* is the set defined by $N_{\mathfrak{H}} = \{a \in \mathfrak{G} \vdash [a, \mathfrak{H}] \subset \mathfrak{H}\}$, so that $[N_{\mathfrak{H}}, \mathfrak{H}] \subset \mathfrak{H}$. The normalizer of a subalgebra $\mathfrak{H}$ is therefore the largest subalgebra of $\mathfrak{G}$ which contains $\mathfrak{H}$ as an ideal. (An *ideal* is a subalgebra $\mathfrak{H}$ such that $[\mathfrak{H}, \mathfrak{G}] \subset \mathfrak{H}$.) For example, if $\mathfrak{H}$ is an ideal of $\mathfrak{G}$, then $N_{\mathfrak{H}} = \mathfrak{G}$. A subalgebra which is its own normalizer is called a *Cartan subalgebra*.

If $\mathfrak{H}$ is an ideal of $\mathfrak{G}$, the set of *cosets* of $\mathfrak{H}$ is the set $\{(a + \mathfrak{H}) | a \in \mathfrak{G}\}$. On this set it is easy to define the standard operations for a linear space over $\mathbb{R}$. One writes

$$(a+\mathfrak{H})+(b+\mathfrak{H}) \equiv ((a+b)+\mathfrak{H}),$$
$$r(a+\mathfrak{H}) \equiv (ra+\mathfrak{H}).$$

Thus the set of cosets is a vector space. Because $\mathfrak{H}$ is an ideal, so that $[a,\mathfrak{H}] \subset \mathfrak{H}$, a Lie-algebra structure can be defined on this space as well:

$$[(a+\mathfrak{H}),(b+\mathfrak{H})] \equiv ([a,b]+\mathfrak{H}).$$

This Lie algebra is called the *quotient algebra* $\mathfrak{G}/\mathfrak{H}$ (also the *factor algebra*).

Two examples of ideals are the *center* and the *derived algebra* of $\mathfrak{G}$. The center $\mathfrak{C}$ is the set of elements such that $[\mathfrak{C},\mathfrak{C}]=0$. The derived algebra $\mathfrak{G}'$ of $\mathfrak{G}$ is defined by $\mathfrak{G}'=[\mathfrak{G},\mathfrak{G}]$. The derived algebra is also the first term of the *lower central series*

$$\mathfrak{G} \supset \mathfrak{G}^2 \equiv \mathfrak{G}' = [\mathfrak{G},\mathfrak{G}] \supset \mathfrak{G}^3 = [\mathfrak{G}^2,\mathfrak{G}] \supset \cdots \supset \mathfrak{G}^k = [\mathfrak{G}^{k-1},\mathfrak{G}] \supset \cdots,$$

all of whose terms are ideals of $\mathfrak{G}$. A Lie algebra is said to be *nilpotent* if $\mathfrak{G}^k=0$ for some positive integer k. A familiar nilpotent Lie algebra is the *Heisenberg algebra*, with elements $\{q^j, p_j, \mathbf{1}\}$, whose bracket is defined by

$$[q^j, p_k] = \delta^j_k \mathbf{1}, [q^j, \mathbf{1}] = [p_j, \mathbf{1}] = 0.$$

Clearly $\mathbf{1}$ is in the center of the algebra, and the third term of the lower central series vanishes.

The direct sum of two Lie algebras is defined as for vector spaces, but with the additional Lie-algebra structure treated in the following way. Let $\mathfrak{G}_1$ and $\mathfrak{G}_2$ be two Lie algebras. Their direct sum $\mathfrak{G}=\mathfrak{G}_1 \oplus \mathfrak{G}_2$ is the set of ordered pairs $a=(a_1,a_2)$, $a_k \in \mathfrak{G}_k$, with the bracket given by

$$[a,b] = ([a_1,b_1],[a_2,b_2]).$$

It is easily seen that the set of elements of the type $(a_1,0)$ forms an ideal in $\mathfrak{G}$ and that this ideal is isomorphic to $\mathfrak{G}_1$; a similar statement can be made about elements of the type $(0,a_2)$ and $\mathfrak{G}_2$. Thus $\mathfrak{G}$ is the direct sum of its ideals $\mathfrak{G}_1$ and $\mathfrak{G}_2$ (remember that we are not distinguishing between isomorphic Lie algebras). More generally, if a Lie algebra is the direct sum as a vector space of its ideals, then it is also their direct sum as a Lie algebra.

An algebra $\mathfrak{G}$ can also be the direct sum as a vector space of two subalgebras only one of which is an ideal. It is then called a *semidirect sum* of the two subalgebras. Let the ideal be $\mathfrak{H}$ and the other subalgebra be $\mathfrak{U}$; then every element of $\mathfrak{G}$ can be written in the form $g=a+h, a\in\mathfrak{U}, h\in\mathfrak{H}$. In that case

$$[g_1,g_2] = [a_1,a_2] + ([a_1,h_2]+[h_1,a_2]+[h_1,h_2]),$$

where the first term on the right-hand side is in $\mathfrak{U}$ and the second (the one in parentheses) is in $\mathfrak{H}$. An example of such a semidirect product is the Lie algebra $e(2)$ of the group of motions of the plane (the Euclidean group in two dimensions), a three-dimensional algebra whose bracket may be given by

$$[a,h_1] = -h_2, \quad [a,h_2] = h_1, \quad [h_1,h_2] = 0. \tag{11}$$

The ideal $\mathfrak{H}$ is Abelian and is spanned by h_1 and h_2, and the other subalgebra $\mathfrak{U}$ is spanned by the element a (being one-dimensional, $\mathfrak{U}$ is also Abelian).

Here we have used the term *Abelian* without defining it. An algebra $\mathfrak{G}$ is Abelian if $\mathfrak{G}'=0$, or if the bracket of any two elements vanishes. It is called *simple* if it is not Abelian and has no nontrivial ideals (that is, other than the null vector and $\mathfrak{G}$ itself). It is called *semisimple* if it has no nontrivial Abelian ideals. Clearly simple algebras are semisimple, and Abelian ones are neither. Any ideal of a semisimple Lie algebra is also semisimple.

There are several reasons why simple and semisimple Lie algebras are of particular importance and interest. The Lie algebras of the classical transformation groups are mostly simple, and these groups play an important role in the analysis of symmetries and the reduction of dynamical systems. In semisimple algebras, and hence also in simple ones, the image under the bracket is the algebra itself; that is, $\mathfrak{G}'=\mathfrak{G}$. To see how this may be important, consider a symplectic manifold (M,ω) and on it a *Lie algebra* $\mathfrak{G}$ *of vector fields.* A Lie algebra of vector fields is a linear space of fields in $\mathfrak{X}(M)$ (not an $\mathscr{F}$-module; the coefficients must be in $\mathbb{R}$) for which the Lie bracket is taken to be the commutator bracket, and such that the space is closed under this bracket. That is, the vector fields must satisfy an equation like (7) with the e_j representing vector fields and the bracket being the commutator. Note that even an involutive distribution need not be a Lie algebra, for the c^h_{jk} of (7) must be constants.

Let this $\mathfrak{G}$ have a basis $\{X_1,\ldots,X_\gamma\}$ of locally Hamiltonian vector fields. Then by linearity every element of $\mathfrak{G}$ is locally Hamiltonian. Every element of $\mathfrak{G}'$, however, is Hamiltonian, not just locally. Indeed, every element of $\mathfrak{G}'$ is of the form $[Y,Z]$ for some $Y,Z\in\mathfrak{G}$; then

$$i_{[Y,Z]}\omega = L_Y i_Z\omega - i_Z L_Y\omega = L_Y i_Z\omega = L_Y\alpha = d(i_Y\alpha),$$

where we have used the fact that Y and Z are locally Hamiltonian: $L_Y\omega=0$, and if $i_Z\omega=\alpha$, then $d\alpha=0$. It follows that if $\mathfrak{G}$ is semisimple and has a locally Hamiltonian basis, then every element of $\mathfrak{G}$ is Hamiltonian.

There is a well-developed theory of simple and semisimple Lie algebras, some of whose results are the following. There exists a definite criterion for semisimplicity (the *Cartan criterion*) involving something called the *Killing form*, to be discussed later in the context of the adjoint representation. Every semisimple Lie algebra $\mathfrak{G}$ can be written uniquely as the direct sum of simple ones which are its ideals. Every simple Lie algebra $\mathfrak{G}$ has a particularly useful basis which contains a maximum number of commuting elements; these commuting elements span the Cartan subalgebra (defined above), whose dimension is called the *rank* of $\mathfrak{G}$. For these and other details, the reader is referred to the bibliography at the end of this chapter. See also Chapter 22.

21.3 ACTIONS OF LIE ALGEBRAS ON MANIFOLDS

1. Lie algebras of vector fields have already been introduced and have been called realizations or actions of Lie algebras on manifolds. More formally, an *action*

of a Lie algebra $\mathfrak{G}$ on a manifold is a smooth map $\varphi:\mathfrak{G} \times M \to TM$ which maps each pair (a, m) into a vector $X_a(m)$ in $T_m M$ in such a way that the *associated mapping*

$$\tilde{\varphi}:\mathfrak{G} \to \mathfrak{X}(M): a \mapsto X_a \tag{12}$$

is a Lie algebra homomorphism, meaning that it is linear and that

$$X_{[a,b]} = [X_a, X_b]. \tag{13}$$

For the sake of clarity, we collect here four mappings (the last two of which are hereby defined) to which we shall often refer:

$$\begin{aligned} \varphi&:\mathfrak{G} \times M \to TM:(a, m) \mapsto X_a(m), \\ \tilde{\varphi}&:\mathfrak{G} \to \mathfrak{X}(M): a \mapsto X_a, \\ \varphi_a&: M \to TM: m \mapsto X_a(m), \\ \varphi_m&: \mathfrak{G} \to T_m M: a \mapsto X_a(m). \end{aligned} \tag{14}$$

An example of the action of a Lie algebra on a manifold is provided by the three vector fields of Eq. (1) taken together with Example (a) which follows Eq. (8). In $\mathfrak{G}$ the bracket of two elements is given by $[\mathbf{a}, \mathbf{b}] = \mathbf{a} \times \mathbf{b}$ (the usual vector cross-product). The three vector fields may be collected and written in the form $\mathbf{X} = (X_1, X_2, X_3)$, and then $X_1(\mathbf{m})$, for instance, is given by

$$X_1(\mathbf{m}) = m_y \partial/\partial z - m_z \partial/\partial y,$$

$\mathbf{m} \in M = \mathbb{R}^3 - \{0\}$. In this notation, then,

$$\varphi:\mathfrak{G} \times M \to TM:(\mathbf{a}, \mathbf{m}) \mapsto \mathbf{a} \cdot \mathbf{X}(\mathbf{m}) \tag{15}$$

and

$$\tilde{\varphi}:\mathfrak{G} \to \mathfrak{X}(M): \mathbf{a} \mapsto \mathbf{a} \cdot \mathbf{X}.$$

Note that the action of a Lie algebra on a manifold M is not, as the words might seem to imply, a map from M into M itself, but into TM. (The terminology has to do with the actions of groups on manifolds, which are in fact maps from M into M.) Note also that a Lie algebra may have more than one action on a given manifold, and therefore, for example, an algebra as such cannot be a symmetry for a dynamical system, but a particular action of that algebra can be.

Suppose that M and N are manifolds and that $\chi: M \to N$ is a diffeomorphism. (The discussion which follows is easily applied to smooth maps other than diffeomorphisms.) Then χ is *equivariant* with respect to the actions φ of $\mathfrak{G}$ on M and ψ of the same algebra $\mathfrak{G}$ on N if the following diagram commutes for all $a \in \mathfrak{G}$:

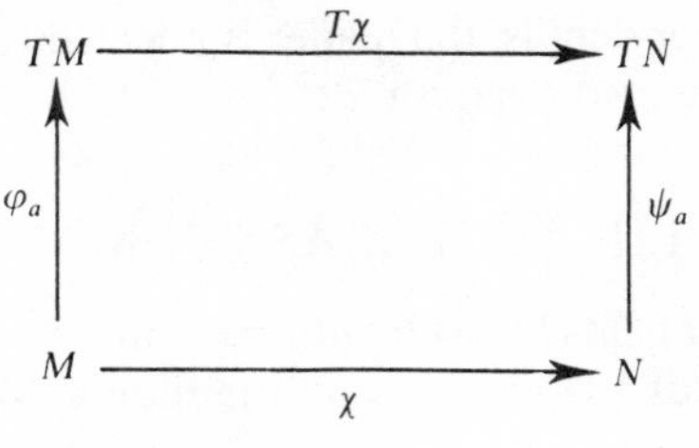

that is, if $\psi_a \circ \chi(m) = T\chi \circ \varphi_a(m)$. Then if $\tilde{\varphi}(a) = X_a \in \mathfrak{X}(M)$ and $\tilde{\psi}(a) = Y_a \in \mathfrak{X}(N)$, it follows from the definition of the push-forward of a vector field (at the end of §5.3) that

$$Y_a = \chi_* X_a. \tag{16}$$

In other words, χ is equivariant if the vector fields in $\tilde{\psi}(\mathfrak{G})$ are the push-forwards of the vector fields in $\tilde{\varphi}(\mathfrak{G})$.

Let $\chi: M \to N$ be any C^∞ mapping and $X \in \mathfrak{X}(M)$, $Y \in \mathfrak{X}(N)$ be vector fields such that

$$\chi^*(L_Y f) = L_X(\chi^* f) \;\forall f \in \mathscr{F}(N).$$

Then X is said to be *χ-related* to Y. (If χ is a surjective submersion, then this is equivalent to saying that X projects onto Y.) The terminology extends to actions of Lie algebras as well. An action φ of $\mathfrak{G}$ on M is *χ-related* to an action ψ of the same algebra $\mathfrak{G}$ on N iff there exists a C^∞ map $\chi: M \to N$ which is equivariant with respect to those actions. This is important because we shall be dealing with mappings which are not surjective. If $M = N$ and χ is a diffeomorphism, the two actions φ and ψ of $\mathfrak{G}$ on M may be called *equivalent*. We shall be interested eventually in actions which are equivalent and preserve additional structure (e.g. symplectic structure), so shall not carry this discussion further at this point.

2. As a Lie algebra action defines a set of vectors in $T_m M$ for each $m \in M$, it might be expected to give rise to a distribution. Unfortunately this is not always the case, for the dimension of the subspace of $T_m M$ which is spanned by these vectors may change from point to point of M, and a distribution in our definition must have the same dimension at all points. Partly to take account of this possibility one defines the *orbits* of an action φ of a Lie algebra $\mathfrak{G}$ on a manifold M in the following way. The action φ gives a subspace span$\{X_j(m)\} \equiv S(m) \subset T_m M$ at each $m \in M$. Any connected submanifold $S^\alpha \subset M$ containing m and such that $T_n S^\alpha = S(n) \;\forall\; n \in S^\alpha$ is a *local orbit* of m under φ. If S^α and S^β are two local orbits of m, so is their union $S^\gamma = S^\alpha \cup S^\beta$. The *orbit* of m under φ is then the maximal local orbit of m, indicated by $\mathfrak{G} \cdot m$. A nontrivial theorem resulting from 'Lie's second fundamental theorem' (Palais, 1957) says that there is an orbit passing through each point $m \in M$.

A *foliating action* is one all of whose orbits have the same dimension. This means that the action provides a distribution, and the fact that the action is a Lie algebra implies that the distribution it generates is involutive and hence leads to a foliation of the manifold M on which it is defined. Much of the analysis in this chapter will be made on the assumption that the actions involved are foliating. Without this assumption, an action will not necessarily yield even a distribution. (Similarly, not every distribution corresponds to a Lie algebra action—for instance, a noninvolutive one or one that is generated by an infinite number of vector fields). One could restrict one's considerations to the subset M' of M through whose points pass orbits of the maximum dimension

$\delta \leqslant \gamma$. It is rather obvious that M' is an open submanifold of M, and since the dimension of the orbits is by definition constant on M', the action restricted to M' is foliating.

A simple example of an action and its orbits is obtained by taking M to be $\mathbb{R}$ and $\mathfrak{G}$ to be the one-dimensional Lie algebra (whose vector space is also $\mathbb{R}$ and whose bracket vanishes). Let the action then be given by $\varphi{:}\mathbb{R}\times\mathbb{R}\rightarrow T\mathbb{R}{:}(a,m)\mapsto(m,am)$, or $\varphi_a(m)=X_a(m)=am\partial/\partial m$, where $m\in\mathbb{R}$ is taken as the coordinate on M. Then there are three orbits under this action: the two one-dimensional ones for $m<0$ and $m>0$, and the single zero-dimensional one $m=0$.

For another example of orbits we return to Eq. (15). Let $r^2=m_x^2+m_y^2+m_z^2$. Then the X_j are clearly tangent to the level sets of this function $r^2\in\mathscr{F}(M)$, which thereby define local orbits under the action defined by (15). The orbits are therefore all two-dimensional, the spheres about the origin, and they foliate M. Note that if the origin had been included in M it would have represented an additional orbit, of dimension zero.

Let φ be the action of $\mathfrak{G}$ on M. The *isotropy algebra* at $m\in M$ is the subset $\mathfrak{G}_m$ of $\mathfrak{G}$ defined by

$$\mathfrak{G}_m=\{a\in\mathfrak{G}\vdash\varphi_a(m)=0\}. \tag{17}$$

It is easily seen that $\mathfrak{G}_m$ is a subalgebra of $\mathfrak{G}$. A *fixed point* $m'\in M$ of the action φ is a point such that

$$\varphi_a(m')=0\ \forall\ a\in\mathfrak{G}. \tag{18}$$

It is easily seen that if m is a fixed point, then $\mathfrak{G}_m=\mathfrak{G}$.

Two important classes of Lie algebra actions are those that are effective and those that are free. An action φ is *effective* if $\tilde{\varphi}{:}\mathfrak{G}\rightarrow\tilde{\varphi}(\mathfrak{G})\subset\mathfrak{X}(M)$ is an isomorphism (i.e. if $\tilde{\varphi}$ is injective). Although this means that there is a one-to-one correspondence between the elements of $\mathfrak{G}$ and the vector fields which represent them in $\mathfrak{X}(M)$, two linearly independent elements of $\mathfrak{G}$ can be represented by vector fields which are not independent. If $a,b\in\mathfrak{G}$ are independent, then there is no $k\in\mathbb{R}$ such that $\tilde{\varphi}(a)\equiv X_a=kX_b\equiv k\tilde{\varphi}(b)$, but there may be an $f\in\mathscr{F}(M)$ such that $X_a=fX_b$. An effective action may have fixed points, but no element of $\mathfrak{G}$ (other than zero) can be in the isotropy algebra of all $m\in M$. Another way of saying this is that an action is effective if $X_a=0$ implies that $a=0$, and this shows that the actions of semisimple Lie algebras are necessarily effective. Indeed, in an action which is not effective, the subset K of elements in $\mathfrak{G}$ which are mapped under $\tilde{\varphi}$ into the null vector field forms an Abelian ideal of $\mathfrak{G}$, and this excludes semisimple algebras.

An action φ is *free* if $\varphi_m{:}\mathfrak{G}\rightarrow T_mM$ is injective for all $m\in M$. This means that there is a one-to-one correspondence between the elements of $\mathfrak{G}$ and the points that represent them in T_mM for all $m\in M$. An action which is free has no fixed points and the isotropy algebra of every point $m\in M$ consists of the null element of $\mathfrak{G}$. There is no function $f\in\mathscr{F}(M)$ such that $X_a=fX_b$ for two linearly independent elements $a,b\in\mathfrak{G}$, for if there were such a function, the nonnull element $f(m)b-a=c\in\mathfrak{G}$ would satisfy $\varphi_m(c)\equiv X_c(m)=0$. Thus free actions are realized by γ independent vector fields, their orbits are γ-dimensional, and they provide a γ-dimensional foliation of M. Another way of putting it is to say that an action is free if $X_a(m)=0$ for any m implies that $a=0$.

Every free action is effective. Indeed, assume that an action φ is not effective. Then elements a, b are contained in $\mathfrak{G}$ such that $\varphi_a(m)=\varphi_b(m)$ for all $m \in M$. But then for $c=a-b \neq 0$ it follows that $\varphi_c(m)=0$ for all $m \in M$, or that c is in the isotropy algebra of all m. Therefore the action is not free. From the discussion of free and effective actions it should also be clear that not every effective action is free, for a function $f \in \mathscr{F}(M)$ could exist such that $X_b = fX_a$ for an effective action, but not for a free one.

Suppose that $\varphi: \mathfrak{G} \times M \to TM$ is an action which is not effective. Then $\ker \tilde{\varphi} \neq 0$ and is an ideal of $\mathfrak{G}$. As at Eq. (11), therefore, the quotient algebra $\mathfrak{G}/\ker \tilde{\varphi}$ can be constructed and a new action $\varphi': \mathfrak{G}/\ker \tilde{\varphi} \times M \to TM$ defined by

$$\tilde{\varphi}':(a+\ker \tilde{\varphi}) \mapsto X_a \forall a \in \mathfrak{G}.$$

It is clear now that φ' is effective, as $\tilde{\varphi}'\ (a+\ker \tilde{\varphi})=0$ implies that $a=0$.

An action is called *transitive* if it has only one orbit, which must therefore pass through all $m \in M$. Then M is called *homogeneous* under the action. This can occur only if $\gamma \geqslant \mu = \dim M$. An action is called *regular* if all of its orbits are regular submanifolds.

In the example of Eq. (15) the action is effective (only the null element of $\mathfrak{G}$ is represented by the null vector field), but not free. Indeed, the three fields are not independent, for $\mathbf{m} \cdot \mathbf{X}=0$. Alternatively, define $\mathbf{e}_k \in \mathfrak{G}$ by $\varphi(\mathbf{e}_k)=X_k$; then the isotropy algebra of $(0,0,c) \in M$ is easily seen to be $\{\mathbf{e}_3\}$.

This action, as has already been pointed out, is foliating, and the leaves are the spheres about the origin. When the action is restricted to one of these spheres, the orbits become not only effective, but transitive, for each sphere is an orbit for the action. In a common local chart for the sphere (spherical polar coordinates with φ the azimuth and θ the colatitude) the action is defined by

$$X_1 = \sin \varphi \partial/\partial\theta + \cot\theta \cos \varphi \partial/\partial\varphi,$$
$$X_2 = \cos \varphi \partial/\partial\theta - \cot\theta \sin \varphi \partial/\partial\varphi,$$
$$X_3 = \partial/\partial\varphi.$$

The isotropy algebra of $(\frac{1}{2}\pi, 0)$, for instance, is $\{\mathbf{e}_1\}$. It is left to the reader to show that the three vector fields are not independent.

3. When the manifold M on which the action takes place is a linear space V and the action φ_a is linear for each $a \in \mathfrak{G}$, the action φ is called a *representation* of $\mathfrak{G}$ on V. Since V is linear, the tangent space at each point can be identified in the usual way with V itself, and the associated mapping $\tilde{\varphi}$ can be thought of as a map from $\mathfrak{G}$ into the linear operators on V in the following way.

In this linear case the vector field X_a which is obtained by applying $\tilde{\varphi}$ to $a \in \mathfrak{G}$ can be written in a chart on V in the form

$$X_a = A^h_j x^j \partial/\partial x^h, \tag{19}$$

where the constant matrix of the A^h_j is said to represent a in this chart or basis. Thus vector fields can be identified with matrices in a given basis, and through them, as is easily seen, with the linear operators on V whose representatives in that basis are the matrices. In this way, when dealing with linear representations, $\mathfrak{X}(M)$ can be replaced by $\mathrm{Lin}(V, V)$.

It can be confusing and destructive to think of the linear case as being in some way too similar to the general nonlinear one. In general the vector field X_a can be defined in a *local* chart on M by

$$X_a(x) = f^h_a(x)\partial/\partial x^h.$$

Now, one may attempt to make this look like the linear case by writing $f^h_a(x)$ locally in the form $A^h_j(x)x^j$, but this is misleading, for the f^h_a need not vanish for $x^j = 0$.

We do not go in any generality into the linear case, for what will interest us mostly is the possibility that V is $\mathfrak{G}$ itself. We add without proof, however, that in a linear representation the Lie bracket $[a, b]$ of two elements $a, b \in \mathfrak{G}$ is represented by the commutator bracket $AB - BA$ of the operators which represent them (compare Eq. (9)).

Consider $\mathfrak{G}$ in both of its aspects, as a Lie algebra and as a vector space, and for the purpose of what follows, identify $\mathfrak{G}$ with its tangent space at each point. Then the *adjoint representation ad* of $\mathfrak{G}$ is defined by

$$ad{:}\mathfrak{G} \times \mathfrak{G} \to \mathfrak{G}{:}(a, b) \mapsto [a, b]. \tag{20}$$

For each $a \in \mathfrak{G}$ the map ad_a sends $\mathfrak{G}$ linearly into $\mathfrak{G}$ according to

$$ad_a{:}\mathfrak{G} \to \mathfrak{G}{:}b \mapsto [a, b]. \tag{21}$$

One says that in the adjoint representation the element a of $\mathfrak{G}$ is represented by the linear operator (acting on the vector space $\mathfrak{G}$) which is defined by (21), or that a is represented by $[a,\ \ ]$. In a given basis on $\mathfrak{G}$, with the structure constants given by (7), $ad_a(b)$ has components $a^k b^j c^h_{kj}$ (here the a^k and b^k are the components of a and b). Therefore in this basis the matrix of ad_a has elements $a^k c^h_{kj}$, corresponding to A^h_j of Eq. (19), from which the vector field is immediately obtained.

What was said above about the general linear case then implies that

$$ad_{[a,b]} = ad_a \cdot ad_b - ad_b \cdot ad_a \equiv [ad_a, ad_b]_{\mathrm{L}}, \tag{22}$$

where $[\ ,\ ]_{\mathrm{L}}$ is the bracket in the Lie algebra obtained from the associative algebra of linear operators on $\mathfrak{G}$ as described at Eq. (9).

Another representation based on the vector-space nature of $\mathfrak{G}$ is constructed not on $\mathfrak{G}$ itself, but on its dual space $\mathfrak{G}^*$. This is called the *coadjoint representation*, and is defined by

$$ad^*{:}\mathfrak{G} \times \mathfrak{G}^* \to \mathfrak{G}^*{:}(a, \alpha) \mapsto \beta, \tag{23}$$

where β is defined by

$$\langle \beta, b \rangle = \langle \alpha, [a, b] \rangle \ \forall\ b \in \mathfrak{G}, \tag{24}$$

and $\langle\ ,\ \rangle$ is the inner product of elements of $\mathfrak{G}^*$ with elements of $\mathfrak{G}$. Again, as at Eq. (21), one can define the map ad_a^* for each $a \in \mathfrak{G}$ by

$$ad_a^*:\mathfrak{G}^* \to \mathfrak{G}^*:\alpha \mapsto \beta, \tag{25}$$

with β defined by (24). It is easily verified (use the Jacobi identity) that

$$ad_{[a,b]}^* = ad_b^* \cdot ad_a^* - ad_a^* \cdot ad_b^* \equiv -[ad_a^*, ad_b^*]_L. \tag{26}$$

Note the minus on the right-hand side of (26); this is sometimes called an *antihomomorphism.*

For the matrix representation corresponding to (19), let the chart on $\mathfrak{G}^*$ have coordinates ξ_k. Then a vector field on $\mathfrak{G}^*$ may be written in the form

$$X_a(\xi) = A_j^{*h}\xi_h \partial/\partial\xi_j, \tag{27}$$

where the A_j^{*h} form the constant matrix which represents $a \in \mathfrak{G}^*$. If the chart chosen on $\mathfrak{G}^*$ is dual (see the Remark following Eq. (28)) to the chart on $\mathfrak{G}$ whose structure constants are the particular ones we have been using, the matrix of the A_j^{*h} in the coadjoint representation turns out to consist of the same elements A_j^h as those in the adjoint representation, but now with the sum on the other index (so that it is actually the transpose matrix that is used). In other words, if a^k and b^k are the components of $a, b \in \mathfrak{G}$, and β_k are the components of $\beta \in \mathfrak{G}^*$ in the dual chart, then

$$\begin{aligned} [ad_a(b)]^h &= (a^k c_{kj}^h) b^j, \\ [ad_a^*(\beta)]_j &= (a^k c_{kj}^h)\beta_h. \end{aligned} \tag{28}$$

Remark. The basis $\{\varepsilon^j\}$ in $\mathfrak{G}^*$ dual to the basis $\{e_j\}$ in $\mathfrak{G}$ is defined by $\langle \varepsilon^j, e_k \rangle = \delta_k^j$. Dual charts are those whose coordinates form dual bases.

As will be seen in some of the examples, the adjoint and coadjoint representations are quite different. In particular, the orbits can have different dimensions.

4. The Killing form, mentioned already, but not yet described, can now be defined in terms of the adjoint representation. It is the bilinear form $K(a,b)$ on $\mathfrak{G}$ (i.e. a bilinear map from $\mathfrak{G} \times \mathfrak{G}$ into $\mathbb{R}$) defined by

$$K(a, b) = \mathrm{tr}(ad_a \cdot ad_b). \tag{29}$$

It can be shown that

$$K(a, [b, c]) = K(b, [c, a]) = K(c, [a, b]),$$

and from this it follows (use (22)) that

$$K(ad_c(a), b) = -K(a, ad_c(b)). \tag{30}$$

In spite of the negative sign here, ad_c is then said to be Hermitian with respect to the Killing form. It can be shown, but will not be shown here, that the Killing form is nondegenerate on a semisimple Lie algebra, so that it provides a test of semisimplicity.

For fixed $a \in \mathfrak{G}$ the Killing form $K(a,\ )$ is an element, say α, of $\mathfrak{G}^*$, and one can write $K(a, b) = \langle \alpha, b \rangle$. In this sense K is a map from $\mathfrak{G}$ into $\mathfrak{G}^*$. Let this map be called

$$\chi : \mathfrak{G} \to \mathfrak{G}^* : a \mapsto K(a,\ ).$$

Then χ turns out to be *antiequivariant* with respect to ad_a and ad_a^*, for some calculation will show that

$$ad_a^* \circ \chi = -\chi \circ ad_a \ \forall\ a \in \mathfrak{G}.$$

In terms of diagrams this can be represented by

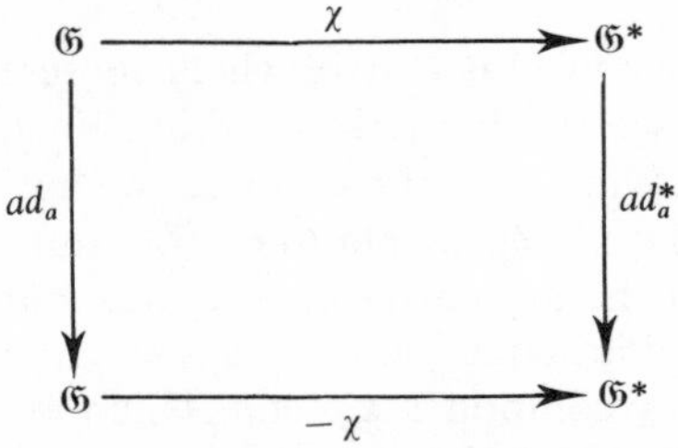

and it may thus be seen that the negative sign arises from the familiar requirement of 'reversing the arrows' on going from a space to its dual; as the arrow in the diagram is in the original direction, the sign must be changed. Antiequivariance, like equivariance, allows one to transfer the bracket from $\mathfrak{G}$ to $\mathfrak{G}^*$, only the minus sign must be included:

$$\chi[a, b] = -[\chi(a), \chi(b)] = -[\alpha, \beta].$$

When the Killing form is nondegenerate, χ becomes an isomorphism. As a consequence, for semisimple Lie algebras the orbits of the adjoint and coadjoint representations are isomorphic and have therefore the same dimension. Indeed, the isomorphism χ can be used to transfer the orbit of $b \in \mathfrak{G}$ under ad_a to the corresponding orbit of $\chi(b) \in \mathfrak{G}^*$ under ad_a^*.

A Lie algebra is called *compact* if its Killing form is negative definite, which then of course implies semisimplicity. The terminology comes from the fact that the Lie group associated with such a Lie algebra is compact in the usual topological sense when the group is seen as a differential manifold.

5. For an example of some of these ideas, we turn again to $e(2)$, the Lie algebra defined at Eq. (11). We rewrite the bracket of $e(2)$ in the form

$$[e_1, e_2] = -e_3, \quad [e_2, e_3] = 0, \quad [e_3, e_1] = -e_2. \tag{31}$$

Note that $\{e_2, e_3\}$ is an Abelian ideal of $e(2)$, which is therefore not semisimple. The structure constants in this basis are

$$c_{12}^3 = c_{31}^2 = -1,$$

and all others, except those obtained by antisymmetry, vanish. Then from (28), by setting $a = e_k$, $k = 1, 2, 3$ (so that $a^j = \delta^j_k$), one obtains

$$ad_{e_1} = \begin{vmatrix} 0 & 0 & 0 \\ 0 & 0 & 1 \\ 0 & -1 & 0 \end{vmatrix}, \quad ad_{e_2} = \begin{vmatrix} 0 & 0 & 0 \\ 0 & 0 & 0 \\ 1 & 0 & 0 \end{vmatrix}, \quad ad_{e_3} = \begin{vmatrix} 0 & 0 & 0 \\ -1 & 0 & 0 \\ 0 & 0 & 0 \end{vmatrix}. \tag{32}$$

It can be shown from this that (in the customary three-vector notation for the cross-product) in general

$$ad_a(b) = (0, [a \times b]^2, [a \times b]^3).$$

The last equation shows that the orbit of b depends on b. For example, the orbit of $0 \in \mathfrak{G}$ is zero-dimensional, the origin itself (0 is a fixed point). To obtain other orbits, consider first b of the form $b = (0, b^2, b^3)$. Then $ad_a(b) = (0, -a^1b^3, a^1b^2)$. Since the orbits are obtained by considering all $a \in \mathfrak{G}$, it is sufficient to treat orbit points of the form $(0, -b^3, b^2)$, that is to set $a^1 = 1$. Now recall that in this linear representation the entire vector space is identified with the tangent space at each point. Thus $(0, -b^3, b^2)$ represents a vector field which must be integrated to find its integral curves. This is easily done, and the integral curves are then found to be circles in the (e_2, e_3) plane. These orbits are therefore of dimension one. Now consider general b. In that case the vectors $ad_a(b)$ can point in any direction in the plane passing through b and lying parallel to the (e_2, e_3) plane. Integration now yields not curves but planes, all of them parallel to the (e_2, e_3) plane. These orbits are therefore of dimension two.

In the coadjoint representation the matrices are obtained simply by transposing those of the adjoint representation. Then it is found that

$$ad^*_a(\beta) = ([a \times \beta]_1, -a^1\beta_3, a^1\beta_2). \tag{33}$$

In the coadjoint representation, therefore, elements $\beta \in \mathfrak{G}^*$ of the form $\beta = (\beta_1, 0, 0)$, that is all points on the one-axis in $\mathfrak{G}^*$, are fixed points of the representation. All other orbits are easily seen to be cylinders whose axis is this one-axis. The orbits are thus of two types, zero-dimensional and two-dimensional. What is particularly interesting, however, is the difference between the orbits of the adjoint and coadjoint representations, a difference that can occur only when the algebra is not semisimple.

21.4 THE CONNECTION BETWEEN LIE ALGEBRA ACTIONS AND DYNAMICS

The aim of this section is to relate reduction of dynamical systems, as described in Chapters 19 and 20, to the actions of Lie algebras. The usual way to apply 'group theoretical methods in mechanics' is to find a symmetry group or algebra (actually a symmetric action of a group or algebra) and use it as a tool in integration, usually by reduction of some kind. As was shown in Chapters 19 and 20, symmetries represent a special case for reduction, and what we intend to demonstrate now is how the actions of Lie algebras are involved in more general reduction through foliations. The arguments will be developed briefly here, leaving most of the elaboration to later chapters, in particular to the next one, in which symplectic actions will be discussed at length.

1. Reduction, it will be recalled, involves projecting the dynamics Δ with respect to a foliation Φ. Such foliations are obtained in various ways, but the only way which is of relevance at this point is the one involving involutive distributions directly.

An action φ of a Lie algebra $\mathfrak{G}$ on a manifold M provides a distribution $\mathscr{D}^{\varphi}$ if the action is foliating. So long as it is foliating (i.e. so long as its orbits are of the same dimension at all points of M), as is implied by the terminology itself, $\mathscr{D}^{\varphi}$ is involutive and gives rise to a foliation Φ^{φ}. If φ is not foliating, on the other hand, its domain can be restricted on M so that on the restricted domain φ becomes foliating. Let us assume, therefore, that the action of φ is foliating.

Even if φ is foliating, however, $\mathscr{D}^{\varphi}$ may not be regular, and then Φ^{φ} is not a regular foliation and may fail to provide a quotient manifold $N = M/\Phi^{\varphi}$ (or at any rate one that is Hausdorff; it will be seen eventually that for actions of Lie groups criteria can be given to overcome this kind of difficulty). But if N can be defined as a Hausdorff manifold, the check for projectability is simplified, for it can be performed on the finite number $X_1, \ldots, X_\gamma$ of generators of $\mathscr{D}^{\varphi}$ which make up the Lie algebra. What need be checked is that $L_\Delta X_j \in \mathscr{D}^{\varphi}$ for $j \in \{1, \ldots, \gamma\}$.

It is seen, then, that there are some advantages to starting with an action of a Lie algebra rather than from a general involutive distribution, but that these advantages are not overwhelming.

2. Suppose further that the Lie algebra action φ under consideration is a symmetry for Δ, that is, that $L_\Delta X_j = 0 \forall j$. Then Δ is necessarily projectable with respect to this action of $\mathfrak{G}$ (note: it is the action, not $\mathfrak{G}$ itself that enters), provided as before that N can be defined as a Hausdorff manifold. The projected vector field $\tilde{\Delta}$ is not zero, for $X \in \mathscr{D}^{\varphi}$ does not imply that $L_\Delta X = 0$. It is seen that to be a symmetry for Δ is a sufficient condition for projectability, but is not necessary.

It is interesting, on the other hand, that projectability implies that Δ (or rather its flow) carries leaves of Φ^{φ} into leaves, or that the foliation is invariant under Δ; as we have seen already, the condition for projectability can be written in the form $L_\Delta \Phi^{\varphi} = 0$. In this sense the action is a symmetry for Δ.

If Δ is projectable with respect to Φ^{φ}, it is possible that the $L_\Delta X_j \equiv [\Delta, X_j]$ lie not only in $\mathscr{D}^{\varphi}$, but in $\tilde{\varphi}(\mathfrak{G})$, or are linear combinations of the X_j. Then the set $\{X_1, \ldots, X_\gamma, \Delta\}$ itself forms a second Lie algebra $\mathfrak{G}_2$, thereby yielding an action of $\mathfrak{G}_2$. If this action is foliating, Δ is automatically projectable with respect to the new foliation.

3. Consider an action φ of $\mathfrak{G}$ on M with respect to which the dynamics Δ is

projectable, and let ψ be another action of $\mathfrak{G}$ on M such that there exists a diffeomorphism $\chi: M \to M$ which is equivariant with respect to φ and ψ. Then Δ is not in general also projectable with respect to Φ^{ψ}. Indeed, let $Y_j = \chi_* X_j$, where X_j is, as above, one of the generators of $\tilde{\varphi}(\mathfrak{G})$ (and hence Y_j is one of the generators of $\tilde{\psi}(\mathfrak{G})$). Then

$$L_\Delta Y_j \equiv [\Delta, Y_j] = [\Delta, \chi_* X_j] = \chi_*[\chi_*^{-1}\Delta, X_j]. \tag{34}$$

Since the last bracket is not necessarily in $\mathscr{D}^{\varphi}$, this is not necessarily in $\mathscr{D}^{\psi}$. If, however, χ is a symmetry for Δ or even if $\chi_*\Delta = k\Delta$, the situation changes and Δ is clearly projectable also with respect to Φ^{ψ}.

In this context two actions of $\mathfrak{G}$ could be defined as equivalent if they are related by an equivariant diffeomorphism which is also a symmetry for Δ.

BIBLIOGRAPHY TO CHAPTER 21

There are many books written as introductions to abstract Lie algebras and their actions, some with emphasis on their representations on vector spaces. We mention, in particular, Jacobson (1962) and Serre (1965). Other treatments which emphasize actions as opposed to abstract algebras can be found in Chapters 2 and 3 of Helgason (1962); Chapter 2 of Bishop and Crittenden (1964); Hausner and Schwarts (1968); Chapters 12 and 13 of Brickell and Clark (1970); and Chapter 4 of Boothby (1975).

In general these references approach Lie algebras from Lie groups. We will mention some of these books again in connection with Lie groups at the end of Chapter 23.

22

Actions of Lie algebras on symplectic manifolds

We have often distinguished from all others the locally Hamiltonian vector fields, those which preserve the symplectic form ω of the manifold on which they live, provided, of course, that the manifold is symplectic. We now do roughly the same for the actions of Lie algebras. Chapter 21 is related to the present chapter in more or less the same way that Chapter 19, dealing with reduction in general, is related to Chapter 20, which specialized to reduction on symplectic manifolds. The important advantage obtained on symplectic manifolds is what has been called the 'double value' of constants of the motion. We should now speak of this more accurately as the double foliation obtained from such constants, the first by the level sets N_c of the function and the second, a foliation of these level sets, by vector fields generated by these functions and ω.

It will be recalled that in Chapter 20 these vector fields did not foliate M (see, for instance, the caption of Fig. 20.1). Now, because the vector fields will come from the actions of Lie algebras (and because of some additional assumptions) they will. This is because if a Lie algebra action generates a distribution (i.e. if the dimension of the orbits is independent of the point in M), the distribution is necessarily involutive and gives rise to a foliation.

Throughout this chapter, then, we shall concentrate on actions φ which have these two properties: they are foliating and symplectic. That they are foliating means that they yield a distribution, necessarily of some finite dimension (the *order* of the action) $\delta \leqslant \gamma$, and hence by involutivity a foliation, also of dimension δ. Even if φ is not foliating, as has been said already, the treatment can be restricted to the open submanifold $M' \subset M$ on which δ is a maximum, or at least to a connected component of M'. In fact we shall automatically assume wherever necessary throughout this chapters that such a restriction has already been made, and that all orbits have the same dimension δ. That an action is symplectic means that $X \in \tilde{\varphi}(\mathfrak{G})$ implies $L_X\omega = 0$.

22.1 SYMPLECTIC, HAMILTONIAN, AND OTHER ACTIONS

1. In keeping with what is said above, let $\varphi:\mathfrak{G}\times M\rightarrow TM$ be an action of the Lie algebra $\mathfrak{G}$ on the symplectic manifold (M,ω), and assume that it gives rise to a distribution $\mathscr{D}^X$ and to a nonsingular foliation Φ^X, both of dimension δ. The distribution and hence the foliation are defined in the usual way through the vector fields of $\tilde{\varphi}(\mathfrak{G})$ and have nothing to do with the fact that M is symplectic.

That M is symplectic becomes useful, however, because it allows the definition of another involutive distribution (this may be called the 'double value' of the action) if the one-forms

$$\alpha_j = i_{X_j}\omega, \quad j=1,\ldots,\gamma, \tag{1}$$

are closed for all $X_j=\tilde{\varphi}(e_j)$, where the e_j form a basis in $\mathfrak{G}$. Then the distribution $\mathscr{D}^\alpha$ defined through the α_j by Eq. (2) of chapter 20 is of dimension $\mu-\delta=2\nu-\delta$ and is involutive, inducing a $(2\nu-\delta)$-dimensional distribution Φ^α on M. Now, a necessary and sufficient condition for the α_j to be closed is that the X_j be locally Hamiltonian. An action made up of such vector fields is called *symplectic*: φ is a symplectic action of $\mathfrak{G}$ on (M,ω) iff $\tilde{\varphi}(\mathfrak{G})\in\mathfrak{X}_{\mathscr{L}\mathscr{H}}(M)$.

More generally, an action can be called *Hamiltonian* iff $\tilde{\varphi}(\mathfrak{G})\in\mathfrak{X}_{\mathscr{H}}(M)$; it can be called *Liouville* iff $\tilde{\varphi}(\mathfrak{G})\in\mathfrak{X}_{\mathrm{Liouv}}(M)$, etc. Clearly every Hamiltonian action is symplectic, but the converse is not true. Because of their 'double value', essentially only symplectic actions will be considered in the sequel, although Hamiltonian ones will play a particularly important role among them. This is because, as may be recalled, the derived algebra $\mathfrak{G}'$ of $\mathfrak{G}$ is represented in a symplectic action by Hamiltonian vector fields. Therefore if $\mathfrak{G}'=\mathfrak{G}$, in particular if $\mathfrak{G}$ is simple, every symplectic action of $\mathfrak{G}$ is Hamiltonian.

Consider a locally Hamiltonian action $\varphi:\mathfrak{G}\times M\rightarrow TM$, and let the local functions f_j be defined by $i_{X_j}\omega=df_j$. Of course these conditions define the f_j only up to additive constants, but once these constants have been chosen the action φ defines the map $\varphi_{\mathscr{F}}:\mathfrak{G}\times M\rightarrow\mathbb{R}:(a,m)\mapsto f_a(m)$. This definition works, it should be noted, because in a symplectic (or Hamiltonian) action all elements of $\mathfrak{G}$, not just the basis elements, are represented by locally Hamiltonian (or Hamiltonian) vector fields. Eqs. (4) and (6) of Chapter 20 and the Remark following them show that

$$d\{f_j,f_k\} = -c^h_{jk}df_h, \tag{2}$$

where the c^h_{jk} are structure constants of $\mathfrak{G}$. Eq. (2) implies that

$$\{f_j,f_k\} = -c^h_{jk}f_h - d_{jk}, \tag{3}$$

where the d_{jk} are constants determined by the choice of the arbitrary constants in the definition of the f_j. It therefore follows that the associated mapping $\tilde{\varphi}_{\mathscr{F}}$ of $\varphi_{\mathscr{F}}$, defined by

$$\tilde{\varphi}_{\mathscr{F}}:\mathfrak{G}\rightarrow\mathscr{F}(M):a\mapsto f_a \tag{4}$$

is not, as might have been guessed, a Lie-algebra homomorphism in which the Poisson bracket is the Lie bracket in $\mathscr{F}(M)$; it all depends on the choice of constants in the definition of the f_j. If those constants can be chosen so as to make the d_{jk} vanish, $\tilde{\varphi}_{\mathscr{F}}$ can be made a homomorphism. Otherwise $\tilde{\varphi}_{\mathscr{F}}$ will be called an *affine* homomorphism.

Given one set of the f_j, every other set is obtained by adding arbitrary constants: $f'_j = f_j + c_j$, $c_j \in \mathbb{R}$. The f'_j still satisfy (2), but the constants d_{jk} of (3) are transformed to

$$d'_{jk} = d_{jk} - c^h_{jk} c_h. \tag{5}$$

Now one may try to choose the c_j so that the d'_{jk} all vanish and thus to make $\tilde{\varphi}_{\mathscr{F}}$ a true Lie-algebra homomorphism. When this is possible, the Hamiltonian action φ is called *strongly Hamiltonian*, but in general this is not possible. That is, there exist *weakly Hamiltonian* actions for which no choice of the f_j will eliminate the d_{jk}. We return to this after some examples.

The first example is on the torus T. Let the usual coordinates θ, φ define a local chart on T and let $\omega = d\theta \wedge d\varphi$. Note that although θ and φ are defined only locally, the one-forms $d\theta$ and $d\varphi$ are globally defined, and therefore so is ω. Let $\mathfrak{G}$ be the two-dimensional Abelian Lie algebra, and let $\tilde{\varphi}(e_1) = \partial/\partial\theta$ and $\tilde{\varphi}(e_2) = \partial/\partial\varphi$. Then the action is only locally Hamiltonian, because the functions $f_1 \equiv \varphi$ and $f_2 \equiv -\theta$ are only locally defined.

The second example is similar, but now on $M = \mathbb{R}^2$ with Cartesian coordinates q, p. Let $\omega = dq \wedge dp$ and $\mathfrak{G}$ be the Abelian Lie algebra in two dimensions as in the previous example. Similarly, choose the action given by $\varphi(e_1) = \partial/\partial q$, $\varphi(e_2) = \partial/\partial p$, with $f_1 = p$ and $f_2 = -q$. Then $\{f_1, f_2\} = 1$, so that $d_{12} = -1$, and since all of the structure constants vanish it follows from (5) that d_{12} cannot be changed by any other choice of the f_j functions. Thus this action is only weakly Hamiltonian. The relation of this example to the Heisenberg algebra of §21.2(2) is suggestive.

In the third example, let $M = T^*\mathbb{R}^3 = \mathbb{R}^3 \times \mathbb{R}^3$ with coordinates $\mathbf{q}$ and $\mathbf{p}$ and with $\omega = dq^k \wedge dp_k$. The Lie algebra will be $so(3)$, and the action will be given by

$$\varphi(e_k) \equiv X_k = \varepsilon^h_{kj}(q^j \partial/\partial q^h - p_h \partial/\partial p_j), \tag{6}$$

where $\varepsilon^h_{kj} = \delta^{hn}\varepsilon_{kjn}$. It is well known that the X_j satisfy $[X_j, X_k] = \varepsilon^h_{jk} X_h$ and thus provide an action of $so(3)$ on M. It is also well known that the three components of the angular momentum

$$J_k \equiv f_k = \varepsilon^h_{kj} q^j p_h \tag{7}$$

are Hamiltonian functions for the X_k (i.e. $i_{X_k}\omega = dJ_k$) and that

$$\{J_j, J_k\} = \varepsilon^h_{jk} J_h. \tag{8}$$

Thus the action is strongly Hamiltonian, and it is foliating on the submanifold M' of M where the X_k are linearly independent. The linear dependence of the X_k can be studied by considering the 3×6 matrix of their components, namely

$$\begin{vmatrix} 0 & -q^3 & q^2 & 0 & -p_3 & p_2 \\ q^3 & 0 & -q^1 & p_3 & 0 & -p_1 \\ -q^2 & q^1 & 0 & -p_2 & p_1 & 0 \end{vmatrix}.$$

The rank of the matrix becomes less than three only when $\mathbf{q}$ and $\mathbf{p}$ are linearly dependent, which occurs only when the total angular momentum vanishes. Thus M' is the submanifold of M obtained by removing the points at which

$$\mathbf{J}^2 = J_1^2 + J_2^2 + J_3^2 = 0.$$

In order to discuss the manifold M' on which this action is foliating, let $F_1 = \theta$, the angle between $\mathbf{q}$ and $\mathbf{p}$, $F_2 = \mathbf{q}^2$ and $F_3 = \mathbf{p}^2$. Then $F \equiv (F_1, F_2, F_3)$ defines a submersion of M' in $\mathbb{R}^3$ by $F:(\mathbf{q},\mathbf{p}) \mapsto (F_1, F_2, F_3) \in \mathbb{R}^3$, and the leaves of $\mathscr{D}^X$ are the submanifolds $F^{-1}(r)$, $r \in \mathbb{R}^3$; this last follows from the fact that $L_{X_k} F_j = 0$ for all k and j and that the F^{-1} submanifolds are three-dimensional. Thus $M' = F^{-1}(r) \times \mathbb{R}^3$. Finally, $F^{-1}(r)$ is diffeomorphic to the group $SO(3)$ of orthogonal matrices in three dimensions with determinant one. To see this, consider some arbitrary fixed $(\mathbf{q}_0, \mathbf{p}_0) \in F^{-1}(r)$, and let A be a matrix of $SO(3)$. Define

$$\psi_A : F^{-1}(r) \to F^{-1}(r)$$

by

$$\psi_A(\mathbf{q}, \mathbf{p}) = (A\mathbf{q}, A\mathbf{p}).$$

Then it is clear that $\psi : SO(3) \to F^{-1}(r) : A \mapsto \psi_A(\mathbf{q}_0, \mathbf{p}_0)$ is a C^∞ bijection, and hence that $F^{-1}(r)$ is diffeomorphic to $SO(3)$. Thus M' is diffeomorphic to $SO(3) \times \mathbb{R}^3$.

2. We now turn to the general problem of the d_{jk} of Eq. (3), a problem which has been discussed extensively in the literature. The first question is whether, given a Hamiltonian action φ of $\mathfrak{G}$ on (M, ω), it is possible to eliminate the d_{jk} by a suitable choice of the f_j, in other words, to tell whether φ is strongly Hamiltonian. To eliminate the d_{jk} one must solve (5) for the c_h with $d'_{jk} = 0$; that is, one must solve the $\frac{1}{2}\gamma(\gamma - 1)$ equations

$$c_{jk}^h c_h \equiv (C_j)_k^h c_h = d_{jk} \tag{9}$$

for the γ unknown c_h. These equations always have solutions when the C_j matrices are invertible, but they may also have solutions more generally. In part this is true because the d_{jk} are not all independent: they are antisymmetric in their indices and the Jacobi identity imposes

$$c_{js}^r d_{rk} + c_{kj}^r d_{rs} + c_{sk}^r d_{rj} = 0. \tag{10}$$

We shall not enter into this problem in any detail, but report some of the results. A special case is the one in which $\mathfrak{G}$ is semisimple, for then the nonvanishing of the Killing form can be shown to guarantee the invertibility of the C_j. Therefore every Hamiltonian action of a semisimple Lie algebra is strongly Hamiltonian. Recall that symplectic actions of semisimple algebras are automatically Hamiltonian, so that every symplectic action of such an algebra is strongly Hamiltonian.

More generally the solution to the problem has to do with the cohomology theory of Lie algebras. Some of the results obtained are the following. Let two sets of d_{jk} be called equivalent if they can be transformed into each other by

Eq. (5). The number of equivalence sets of d_{jk} is then equal to the number of elements in the second cohomology group H^2 of $\mathfrak{G}$. For example, for simple Lie algebras $H^2 = 0$ and thus contains only one element, so that all of the d_{jk} of such an algebra are equivalent to zero (quite generally, there is always one equivalence set which contains $d_{jk} = 0$). For the Abelian algebra in two dimensions, on the other hand, H^2 has two elements, so that there must be two equivalence classes of d_{jk}. One of them was exhibited in the first two examples at the end of §(1): the actions were only weakly Hamiltonian. Another action of the same algebra, this time on $\mathbb{R}^4 = \{q^1, p_1, q^2, p_2\}$ with $\omega = dp_k \wedge dq^k$, is given by $\tilde{\varphi}(e_k) = \partial/\partial q^k$. With $f_k = p_k$ one obtains $\{f_1, f_2\} = 0$, so that this action is strongly Hamiltonian.

The second question is what to do about weakly Hamiltonian actions, or even how to deal with strongly Hamiltonian ones when there is some reason to prefer nonzero values for the d_{jk} (for instance if certain Hamiltonian functions f_j are preferable to others). What can be done then is to go to what is called a *central extension* of $\mathfrak{G}$. In general a Lie algebra $\hat{\mathfrak{G}}$ is an *extension* of another $\mathfrak{G}$ by a third $\mathfrak{G}_1$ if $\hat{\mathfrak{G}}$ contains an ideal isomorphic to $\mathfrak{G}_1$ (which we shall call $\mathfrak{G}_1$ in keeping with our policy not to distinguish between isomorphic Lie algebras), such that $\hat{\mathfrak{G}}/\mathfrak{G}_1$ is isomorphic to $\mathfrak{G}$. If $\mathfrak{G}_1$ is in the center of $\hat{\mathfrak{G}}$, the extension is called *central.*

Suppose now that the d_{jk} do not all vanish. Then a central extension $\hat{\mathfrak{G}}$ of $\mathfrak{G}$, one dimension higher, can be provided by defining the structure constants $\hat{c}^h_{jk}$ of $\hat{\mathfrak{G}}$ by

$$\hat{c}^h_{jk} = c^h_{jk}, \quad h, j, k \leqslant \gamma; \quad \hat{c}^h_{\gamma+1,k} = 0, \quad \hat{c}^{\gamma+1}_{jk} = d_{jk}. \tag{11}$$

The antisymmetry of the d_{jk} and Eq. (10) then guarantee that the $\hat{c}^h_{jk}$ so defined are antisymmetric and satisfy the Jacobi identity, and hence that $\hat{\mathfrak{G}}$ so defined is indeed a Lie algebra. It is left to the reader to verify that it is actually a central extension of $\mathfrak{G}$; it is easily seen that the new basis element $e_{\gamma+1}$ commutes with all the others.

Given a weakly Hamiltonian action of $\mathfrak{G}$ on a manifold M, it can be considered a strongly Hamiltonian action of $\hat{\mathfrak{G}}$ on M, with $\tilde{\varphi}(e_{\gamma+1}) \equiv X_{\gamma+1} = 0$ and $f_{\gamma+1} = 1$. A glance at Eq. (3) will verify this statement. The resulting action of $\hat{\mathfrak{G}}$ is certainly not effective, for $X_{\gamma+1} = 0$.

A simple example of a central extension is from the Abelian algebra in two dimensions to the Heisenberg algebra. In the second example at the end of § (1) we had $d_{12} = -1$. In the central extension, then,

$$\hat{c}^1_{12} = \hat{c}^2_{12} = 0; \quad \hat{c}^n_{3k} = 0; \quad \hat{c}^3_{12} = -1,$$

from which

$$[e_1, e_2] = -e_3, \quad [e_3, e_k] = 0.$$

It is interesting to see how the action φ in that example is interpreted as a strongly Hamiltonian action of the Heisenberg algebra on $\mathbb{R}^2$.

In sum, three ways have been discussed for dealing with Hamiltonian actions in which the d_{jk} do not all vanish. The first, possible only if the action is strongly

Hamiltonian, is to eliminate the d_{jk} by a new choice of the f_j. The second is to leave the d_{jk} as they are, allowing $\tilde{\varphi}_{\mathscr{F}}$ to be an affine homomorphism. The third is to pass from the action of $\mathfrak{G}$ to the action of its central extension $\hat{\mathfrak{G}}$ defined by (11).

3. Interest in symplectic actions stems, as has been pointed out, from the possibility of obtaining a second foliation Φ^α from the action. In later chapters the interest in Hamiltonian actions will also be clarified; it has to do largely with the momentum map, soon to be discussed. Nevertheless, although they may be of interest in principle, symplectic and Hamiltonian actions would be of little practical interest unless there were some assurance that they occur sufficiently often in practice. It turns out, as will now be shown, that they are found in the cotangent bundles T^*Q of manifolds Q.

Let Q be a manifold of dimension ν and let φ be the action of a Lie algebra $\mathfrak{G}$ on Q. As before, the vector fields corresponding to the basis elements are $\tilde{\varphi}(e_j) = X_j \in \mathfrak{X}(Q), j = 1, \ldots, \gamma$. Recall the definition in §11.2 of the lift $X^* \in \mathfrak{X}(T^*Q)$ of a vector field $X \in \mathfrak{X}(Q)$. Such a lifted vector field is always Hamiltonian. Indeed, it is easily shown (for instance from the local form of Eq. (21) of Chapter 11) that $L_{X^*}\theta_0 = 0$, where $\omega = -d\theta_0$ (we shall write simply ω rather than ω_0, for this is the only symplectic form that will be considered on T^*Q), from which

$$i_{X^*}\omega = d(i_{X^*}\theta_0) \equiv df.$$

It follows that the *lifted action* φ^* of $\mathfrak{G}$ on T^*Q defined by

$$\varphi^*(e_j) = X_j^* \in \mathfrak{X}(T^*Q), \tag{12}$$

is Hamiltonian, for the vector fields which make it up are all Hamiltonian.

Remark. We have not shown explicitly that φ^* so defined is in fact an action of $\mathfrak{G}$ on T^*Q, i.e. that $[X_j^*, X_k^*]$ is the lift of $[X_j, X_k]$. The reader may wish to check this.

It turns out, in fact, that φ^* is strongly Hamiltonian. Indeed, with $f_j = i_{X_j^*}\theta_0$ as above, one obtains

$$\begin{aligned} i_{[X_j^*, X_k^*]}\theta_0 &= L_{X_j^*} i_{X_k^*}\theta_0 - i_{X_k^*} L_{X_j^*}\theta_0 = L_{X_j^*} f_k \\ &= i_{c_{jk}^h X_h^*}\theta_0 = c_{jk}^h i_{X_h^*}\theta_0 = c_{jk}^h f_h. \end{aligned}$$

In sum, an action φ^* on T^*Q lifted from an action φ on Q, both of the same Lie algebra $\mathfrak{G}$, is necessarily strongly Hamiltonian with respect to ω.

There are many symplectic manifolds on which symplectic actions are necessarily Hamiltonian. For instance Poincaré's lemma (AM, p. 118) states that a connected manifold is simply connected iff every closed one-form is exact. The carrier manifolds of dynamical systems are chosen throughout to be connected (or connected components are considered separately) and many are even simply connected, such as $\mathbb{R}^k$ and its tangent and cotangent bundles, or S^k, $k > 1$, and its tangent and cotangent bundles. On these manifolds, therefore, the α_j of Eq. (1) are automatically exact so long as they are closed, and thus on them every symplectic action is also Hamiltonian.

22.2 INTRODUCTION TO THE MOMENTUM MAP

In §20.2 a set of independent functions $F = \{f_1, \ldots, f_\kappa\}$ on a symplectic manifold (M, ω) is used to generate a submersion $F: M \to \mathbb{R}^\kappa$ and from it a foliation Φ^F, whose leaves may be called the N_a, $a \in \mathbb{R}^\kappa$. Then the Hamiltonian vector fields X_j defined by $i_{X_j}\omega = df_j$ are used to foliate the N_a, but not M itself, for the X_j are not necessarily in involution.

Now the situation is somewhat different. The action φ of a γ-dimensional Lie algebra $\mathfrak{G}$ on (M, ω), Hamiltonian and foliating, of order $\delta \leqslant \gamma$, yields γ Hamiltonian vector fields $X_j = \tilde{\varphi}(e_j)$ in involution. But in each neighborhood U only δ of these vector fields are independent, so the foliation Φ^X of M which they generate is of dimension δ. Related in the usual way to the X_j, namely by the equation $i_{X_j}\omega = df_j$, are their Hamiltonian functions f_j. But in each neighborhood U only δ of these functions are independent (i.e. only δ of the df_j are linearly independent), so the foliation Φ^F which they generate by local submersions is of codimension δ (of dimension $2v - \delta$). This will be discussed in more detail later, in the next section, but it is already evident that these two foliations are a manifestation of the 'double value', or rather the double foliation, cited so often in connection with symplectic manifolds.

1. Let φ now be any Hamiltonian action of a Lie algebra $\mathfrak{G}$ on a symplectic manifold (M, ω), and for each $a \in \mathfrak{G}$ let $\tilde{\varphi}(a) = X_a$. Let the function $f_a: M \to \mathbb{R}$ be defined by $i_{X_a}\omega = df_a$. A word of caution: this equation does not define a unique f_a. Specifically, recall that for a basis element e_j of $\mathfrak{G}$ the function f_j is defined only up to an additive constant, the c_j of Eq. (5). Suppose that the c_j have already been chosen and that $a = a^j e_j$. Then $f_a \equiv a^j f_j$, and the f_a so defined will be different for different choices of the c_j.

By choosing a basis $\varepsilon^1, \ldots, \varepsilon^\gamma$ in $\mathfrak{G}^*$ one can construct something like the opposite for the dual space, namely the map

$$f_j \varepsilon^j : M \to \mathfrak{G}^*,$$

where the f_j are obtained as above from the elements of the basis in $\mathfrak{G}$ which is dual to the $\{\varepsilon^j\}$. It is easily seen that the map so defined is independent of the particular choice of basis in $\mathfrak{G}^*$. This thought suggests the following definition.

The *momentum map* μ from M to $\mathfrak{G}^*$ is defined by

$$\mu: M \to \mathfrak{G}^*: m \mapsto \alpha_m, \quad \text{where} \quad \langle \alpha_m, a \rangle = f_a(m) \forall a \in \mathfrak{G}. \tag{13}$$

For each given $m \in M$ this map is linear, for

$$\langle \alpha_m, a + b \rangle = f_a(m) + f_b(m).$$

If it were true in general that μ were a submersion, it could be used almost immediately to foliate M, leading to a foliation which might be called Φ^μ and which would depend, through the definition of μ, on the f_j. But in general μ is not a submersion. Nevertheless, if it turns out that φ is μ-related (in general μ is not surjective; if it were, we would have written 'φ is projectable') to an action of $\mathfrak{G}$ on $\mathfrak{G}^*$, the resulting action can be studied as an alternative, or at least in addition, to φ itself (it will turn out that φ is in fact μ-related to the coadjoint action). Both of these aspects of the momentum map will now be taken up.

2. Recall the definition of a submersion. The momentum map $\mu: M \to \mathfrak{G}^*$ is a submersion if $T_m\mu: T_m M \to T_{\mu(m)}\mathfrak{G}^*$ is surjective for all $m \in M$. When μ is a submersion, the level sets $N_\alpha \equiv \mu^{-1}(\alpha)$, $\alpha \in \mu(M)$, are $(2\nu - \gamma)$-dimensional submanifolds of M and therefore provide a regular foliation Φ^μ of M (recall that $\dim M = 2\nu$). But since μ is not in general a submersion, Φ^μ does not in general exist.

The problems of foliation that arise when a map, in particular μ, fails to be a submersion have been enumerated in Chapter 18. (Even when μ fails to be a submersion it may give rise to a regular foliation and thus to a manageable and useful Φ^μ, but the definition of μ makes it undesirable to remove pathologies in some of the ways described in Chapter 18. For instance multiplication by functions would destroy the algebraic structure required to preserve the Lie algebra action.) Nevertheless the X_a vector fields in $\tilde{\varphi}(\mathfrak{G})$ on M are always μ-related to vector fields on $\mathfrak{G}^*$, whether or not μ is a submersion of constant rank. In any case, it is helpful to understand the singularities of μ in order to find regions of $\mathfrak{G}^*$ which can define regular leaves on parts of M.

It is difficult to say anything very definite about the singularities of μ in general. The problem is dealt with in books on singularities of maps from one manifold to another, the main ideas can be summarized in the following way. The map μ is called *locally trivial* at $\alpha \in \mu(M)$ if there is a neighborhood U containing α such that $\mu^{-1}(\beta) \equiv N_\beta$ is a regular submanifold of M for all $\beta \in U$ and there is a smooth map $\psi: \mu^{-1}(U) \to N_\alpha$ such that $\mu \times \psi: \mu^{-1}(U) \to U \times N_\alpha$ is a diffeomorphism. This makes it possible to use values in U to label the set of manifolds N_β, $\beta \in U$. A point $\alpha \in \mu(M)$ is said to be in the *bifurcation set* Σ_μ if μ is not locally trivial at α. This bifurcation set can be shown (AM, p. 340) to contain the set S_μ of singular values of μ, the complement in $\mathfrak{G}^*$ of the set of regular values R_μ. It also contains those values $\alpha \in \mu(M)$ at which the topological type of $\mu^{-1}(M)$ changes, like the value zero in the example of Eq. (1) of Chapter 10. Thus the bifurcation set reflects the singularities of μ, but it does not describe them perfectly. It is not true, for instance, that if $\alpha \in \Sigma_\mu$ the topological type of N_β will change as β passes through α. Nor is it possible to predict in general the nature of Σ_μ or of S_μ or how these sets are arranged in $\mathfrak{G}^*$. More will be possible when the action is integrable to a group action, but that comes later. It is interesting, however, that both of these sets remain unaltered when the c_j of Eq. (5) are changed.

We now give an example of a momentum map, which will be seen not to be a submersion. Let $M = T^*\mathbb{R}^2$, with $\omega = dq^1 \wedge dp_1 + dq^2 \wedge dp_2$. The Lie algebra $\mathfrak{G}$

is $so(2, 1)$ of the last example in §21.1, its action given by

$$\begin{aligned}\tilde{\varphi}(e_1) &\equiv X_1 = p_1\partial/\partial q^1 + p_2\partial/\partial q^2,\\ \tilde{\varphi}(e_2) &\equiv X_2 = -q^1\partial/\partial p_1 - q^2\partial/\partial p_2,\\ \tilde{\varphi}(e_3) &\equiv X_3 = q^1\partial/\partial q^1 - p_1\partial/\partial p_1 + q^2\partial/\partial q^2 - p_2\partial/\partial p_2.\end{aligned} \tag{14}$$

In this example the basis in $so(2, 1)$ has been chosen so that

$$[e_1, e_2] = e_3, \quad [e_2, e_3] = -2e_2, \quad [e_3, e_1] = -2e_1.$$

It is left to the reader to show that this is indeed $so(2, 1)$ as previously defined. The Hamiltonian functions for the vector fields of (14) are now chosen to be

$$f_1 = \tfrac{1}{2}\mathbf{p}^2, \quad f_2 = \tfrac{1}{2}\mathbf{q}^2, \quad f_3 = \mathbf{q}\cdot\mathbf{p},$$

where we have used the usual vector notation $\mathbf{p} \equiv (p_1, p_2)$, etc. From the definition of the momentum map at Eq. (13) it follows that

$$\langle \mu(\mathbf{q}, \mathbf{p}), e_j \rangle = f_j(\mathbf{q}, \mathbf{p}),$$

or that if $\{\varepsilon^1, \varepsilon^2, \varepsilon^3\}$ is the basis in $\mathfrak{G}^*$ which is dual to the e_j-basis in $\mathfrak{G}$, then

$$\mu(\mathbf{q}, \mathbf{p}) = f_j(\mathbf{q}, \mathbf{p})\varepsilon^j. \tag{15}$$

To show that μ is not a submersion we must show that there are points in M where $T_m M$ is not surjective, that is where the df_j are linearly dependent or the f_j dependent. This can be done by calculating $df_1 \wedge df_2 \wedge df_3$, which is found to contain a functional factor of the form $q^2 p_1 - p_2 q^1$ (or $\mathbf{q} \times \mathbf{p}$ in vector notation). This means that when $\mathbf{q} \times \mathbf{p} = 0$, or when $\mathbf{q}$ and $\mathbf{p}$ are collinear, the df_j are linearly dependent. Another way to see this is directly from the f_j functions: when $\mathbf{q}$ and $\mathbf{p}$ are collinear, $f_3 = 2(f_1 f_2)^{1/2}$ and the functions are not independent. This last equation also identifies S_μ: it is a hyperbolic paraboloid in $\mathfrak{G}^*$, the boundary of $\mu(M)$. Note, incidentally, that $\mu(M)$ is clearly not all of $\mathfrak{G}^*$. For example, only elements of $\mathfrak{G}^*$ with positive first and second components are found in $\mu(M)$.

3. It has been mentioned that the action φ of $\mathfrak{G}$ on (M, ω) is μ-related to an action on $\mathfrak{G}^*$. It will now be shown that this is true for every Hamiltonian action φ. What must be shown, according to the discussion following Eq. (16) of Chapter 21, is that for each vector field $X \in \tilde{\varphi}(\mathfrak{G})$ on M there is a vector field $\tilde{X} \in \mathfrak{X}(\mathfrak{G}^*)$ such that for every function $F \in \mathscr{F}(\mathfrak{G}^*)$

$$\mu^*(L_{\tilde{X}}F) = L_X(\mu^* F).$$

Let $\{\varepsilon^j\}$ be the basis on $\mathfrak{G}^*$ dual to the basis $\{e_j\}$ on $\mathfrak{G}$. Then each ε^j defines a (linear) function $\mathcal{J}_j$ on $\mathfrak{G}^*$ according to $\mathcal{J}_j(\beta) = \beta_j$, for $\beta = \beta_j \varepsilon^j$ any element of $\mathfrak{G}^*$; in fact $\mathcal{J}_j \equiv \langle\ , e_j\rangle$. Then $\mu^*\mathcal{J}_j \in \mathscr{F}(\mathfrak{G})$ is the function f_j on the right-hand side of (13) with a replaced by e_j. Indeed, according to (13) and the definition of $\mathcal{J}_j$,

$$(\mu^*\mathcal{J}_j)(m) = \mathcal{J}_j \circ \mu(m) = \langle \mu(m), e_j \rangle = f_j(m).$$

Now let φ be a strongly Hamiltonian action of $\mathfrak{G}$ on M, so that $\{f_j, f_k\} = c^h_{kj} f_h$

(note the order of the indices, by which we bypass the negative sign), and define

$$\tilde{X}_j = c^h_{kj}\tilde{f}_h \partial/\partial\beta_k. \tag{16}$$

Then $\tilde{X}_j$ is μ-related to X_j. To see this, let $F \in \mathscr{F}(\mathfrak{G}^*)$. We calculate

$$(\mu^* L_{\tilde{X}_j} F)(m) = c^h_{kj} f_h(m)[(\partial F/\partial\beta_k)\circ\mu](m)$$

and

$$\begin{aligned}[L_{X_j}(F\circ\mu)](m) &= L_{X_j}F(f_1(m),\ldots,f_\gamma(m)) = \{f_j, f_k\}\partial F/\partial f_k \\ &= c^h_{kj} f_h(m)[(\partial F/\partial\beta_k)\circ\mu](m).\end{aligned}$$

Here we have used $L_{X_j} f_k = \{f_k, f_j\}$, This establishes that the $\tilde{X}_j$ are μ-related to the X_j, and the result can be extended by linearity to the rest of $\mathfrak{G}$. Now, since one does not always deal with strongly Hamiltonian actions, or even when one does, the c_j are not always chosen so that the d_{jk} of Eq. (3) vanish. In that case the vector fields $\tilde{X}_j \in \mathfrak{G}^*$ have to be modified. We state without proof that in general the vector fields which are μ-related to the $X_j \in \tilde{\varphi}(\mathfrak{G})$ are given by

$$\tilde{X}_j = (c^h_{kj}\tilde{f}_h + d_{kj})\partial/\partial\beta_k.$$

Because, as was pointed out above, $\tilde{f}_h(\beta) = \beta_h$, this can be written in the form

$$\tilde{X}_j = (c^h_{kj}\beta_h + d_{kj})\partial/\partial\beta_k. \tag{17}$$

Comparison of (17) and Eq. (28) of Chapter 21 will show that this is essentially the coadjoint representation on $\mathfrak{G}^*$ (an affine coadjoint representation if the d_{kj} do not all vanish).

Although it has now been shown that every Hamiltonian action φ of a Lie algebra $\mathfrak{G}$ on a symplectic manifold (M,ω) is μ-related to the coadjoint representation on $\mathfrak{G}^*$, where μ is the momentum map, in general it is not the entire coadjoint representation that is involved. This is because μ maps M in general not onto $\mathfrak{G}^*$, but onto only part of it, as was seen in one of the examples, and will shortly be discussed in some detail. However, if φ is of order $\delta = \gamma$ (or, equivalently, if φ is free), $\mu(M)$ is all of $\mathfrak{G}^*$ and φ is actually projected onto the coadjoint representation. This is an interesting result, true for every free Hamiltonian action.

The more usual situation is the one in which $\delta < \gamma$, so that $\mu(M) \subset \mathfrak{G}^*$. Then φ is μ-related to the coadjoint representation, but also to any other action of $\mathfrak{G}$ that might exist on $\mathfrak{G}^*$ which, when restricted to $\mu(M)$, is the same as the restriction to $\mu(M)$ of the coadjoint representation. In this case we shall be interested in how $\mu(M)$ lies in $\mathfrak{G}^*$ and in the actual vector fields which are obtained on $\mu(M)$ itself. This will be explained more fully and treated in §22.3.

4. The point of the momentum map is to obtain something like the double foliation used in Chapter 20 to reduce dynamical systems, but starting now from a Hamiltonian action, rather than a submersion. For this purpose assume that μ generates a foliation on at least some region of M (remember that μ is not in general a submersion), and let $N_\alpha = \mu^{-1}(\alpha)$ be one of its leaves corresponding to a

regular value $\alpha \in \mathfrak{G}^*$. One can attempt now to obtain a second foliation, this one of the leaf, in a way which will now be explained.

This foliation makes use of the isotropy algebra of α. Recall that in Chapter 20 the foliation of the N_c required certain vector fields, combinations of the X_j, which were tangent to the N_c. Such vector fields lie in the isotropy algebra. Indeed, let $\mathfrak{G}_\alpha$ be the isotropy algebra of α; that is, if $b \in \mathfrak{G}_\alpha$, then $\tilde{X}_b(\alpha) = 0$. It follows that $L_{X_b} N_\alpha = 0$, or that every field in $\tilde{\varphi}(\mathfrak{G}_\alpha)$ is tangent to N_α. These vector fields therefore lie in TN_α. They are also in the kernel of the pullback ω_α of ω on N_α by the regular imbedding $i: N_\alpha \to M$ (the demonstration of this is essentially the same as in Chapter 20). It is seen, then, that these vector fields have the properties needed to foliate the N_α as in Chapter 20.

To pass on from here to the quotient $N_\alpha/\mathfrak{G}_\alpha$ is not possible without knowing something more about the nature of the orbits of $\mathfrak{G}_\alpha$. There is nothing more that can be said in general at this point, therefore; we will return to this question in the discussion of group actions, when more can be obtained.

22.3 THE MOMENTUM MAP: DETAILS AND EXAMPLES

In general the Hamiltonian functions f_j are not all independent (the df_j, are not linearly independent). If the action φ, of order δ, is Hamiltonian and foliating, the f_j define a submersion of M into a δ-dimensional manifold N not necessarily diffeomorphic to $\mathbb{R}^\delta$. Let this submersion be called F. In this subsection we shall study F, construct it, and show that it enters into the diagram

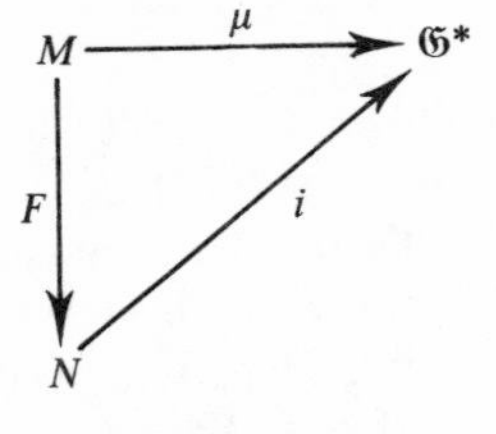

Part of the task will also be to clarify the map i.

1. Let φ be a Hamiltonian action of a Lie algebra $\mathfrak{G}$ on a symplectic manifold (M, ω). Let $F = \{f_1, \ldots, f_\gamma\}$ be a set of Hamiltonian functions for φ, defined in the usual way,

$$i_{X_j}\omega = df_j,$$

where $X_j = \tilde{\varphi}(e_j)$ and $\{e_j\}$ is a basis for $\mathfrak{G}$. Consider the map $F: M \to N \subset \mathbb{R}^\gamma$. If the action is of order δ, as was mentioned above, only δ of the f_j are independent, and N is a δ-dimensional manifold, which can be defined through a set of open neighborhoods in $\mathbb{R}^\delta$ together with the transition functions. For this purpose, let U_m be a neighborhood of some point $m \in M$ in which the independent functions are $f \equiv \{f_{j_1}, \ldots, f_{j_\delta}\}$. Then because the f_{j_k} are independent, $V = f(U_m)$ is an open neighborhood of $\mathbb{R}^\delta$. This neighborhood V constitutes a local chart of the manifold N being defined. One can do the same thing for some other point $m' \in M$, obtaining $V' = f'(U_{m'})$, where $f' \equiv \{f'_{j_1}, \ldots, f'_{j\delta}\}$ is a set of δ of the f_j, these independent in $U_{m'}$. If $U_m \cap U_{m'}$ is not empty, the transition function from V to V' may be defined by $\psi_{VV'} = f' \circ f^{-1}$, which is well defined even if f^{-1} is not, for the set of points in the intersection where $f =$ const. is also the set where $f' =$ const. When it is noted that $\psi_{VV'}$ is C^∞, it is seen that this procedure defines the manifold N, for it gives an atlas on it.

Once N is defined, so is the map $F: M \to N$, for any point $m' \in U_m$ is mapped into $F(m')$ given by $f(m') \in \mathbb{R}^\delta$ in the local chart at m. Now consider $N_n = F^{-1}(n), n \in N$. This is a regular submanifold of M, of dimension $2\nu - \delta$ (recall that dim $M = 2\nu$), a set of points on which all of the $f_j, j = 1, \ldots, \gamma$, are constant. In this way the N_n foliate M.

So far no algebraic properties have been used, but now it will be shown that N can be immersed in a natural way into $\mathfrak{G}^*$. Let $\{\varepsilon^j\}$ be the basis in $\mathfrak{G}^*$ which is dual to $\{e_j\}$, $j = 1, \ldots, \gamma$. Then the immersion of N into $\mathfrak{G}^*$ may be defined by

$$i: N \to \mathfrak{G}^*: n \mapsto f_j(m)\varepsilon^j, \quad m \in F^{-1}(n) \tag{18}$$

But according to (13) (compare also the equation just above (13) and the example of Eq. (15)) this is just the point in $\mathfrak{G}^*$ into which the momentum map sends $m \in M$, and therefore we arrive at

$$\mu = i \circ F, \tag{19}$$

which establishes the validity of the diagram at the beginning of this section. The sum in Eq. (18) is over all values of j from $j = 1$ to $j = \gamma$, although the f_j functions are not independent. At each point $m \in M$ a knowledge of a certain set of δ of them determines the other $\gamma - \delta$, and therefore $i(N)$ is a δ-dimensional submanifold of $\mathfrak{G}^*$. Henceforth, unless otherwise stated, $i(N)$ will be identified with N and μ with F.

2. The further study of N as a submanifold of $\mathfrak{G}^*$ can be carried out in the $\{\varepsilon^j\}$ basis. Let the element $\beta \in \mathfrak{G}^*$ have coordinates β_j in this basis. If β lies on $N \subset \mathfrak{G}^*$, then it is seen from (18) that $\beta_j = f_j(m)$ for some point $m \in M$, and therefore the coordinates β_j of points on N are connected by the relations between the f_j functions: a knowledge of a certain δ of them will determine the other $\gamma - \delta$. Without loss of generality one may choose the first δ to determine the others in some neighborhood $F^{-1}(U_n \cap N) \equiv \mu^{-1}(U_n \cap N)$. The relations between the β_j can then be written in the form

$$\begin{aligned} &\beta_{\delta+1} - g_{\delta+1}(\beta_1, \ldots, \beta_\delta) \equiv G_1 = 0, \\ &\vdots \\ &\beta_\gamma - g_\gamma(\beta_1, \ldots, \beta_\delta) \equiv G_{\gamma-\delta} = 0, \end{aligned} \tag{20}$$

where the functions $g_{\delta+1}, \ldots, g_\gamma$ are of course those which express the last $\gamma - \delta$ of the f_j in terms of the first δ of them.

Let us assume the action to be strongly Hamiltonian and the d_{jk} to vanish. From (17) or Eq. (28) of Chapter 21 one can write the vectors of the coadjoint representation in the form

$$\tilde{X}_j = c^h_{kj}\beta_h \partial/\partial\beta_k, \tag{21}$$

and it can then be shown that the G_b functions, and hence the relation between the coordinates on N, are invariant under the coadjoint representation. Indeed,

$$L_{\tilde{X}_j} G_b = \beta_h c^h_{kj} \delta^j_{\delta+b} - \beta_h c^h_{ks} \partial g_{\delta+b}/\partial\beta_s,$$

where $j = 1, \ldots, \gamma$; $b = 1, \ldots, \gamma - \delta$; $s = 1, \ldots, \delta$. On the other hand, because $g_{\delta+b}$ represents the $f_{\delta+b}$,

$$\{f_j, g_{\delta+b}\} = -c^h_{j,\delta+b} f_h = \{f_j, f_s\}\partial g_{\delta+b}/\partial\beta_s = -c^h_{js} f_h \partial g_{\delta+b}/\partial\beta_s.$$

From the fact that $f_h(m) = \beta_h$ it then follows that $L_{\tilde{X}_j} G_b = 0$. This shows that the vector fields of the coadjoint representation are tangent to N, so that one can indeed speak of their restrictions to N as we have above. Although the $\tilde{X}_j$, defined on all of $\mathfrak{G}^*$, are complete on $\mathfrak{G}^*$, they need not be complete on N, and thus N may not be an orbit of the coadjoint representation. Let S^α be the orbit of $\alpha \in \mathfrak{G}^*$ under the coadjoint representation. Then $S^\alpha \cap N$, if it is not empty, is a submanifold of N of dimension $\rho_\alpha = \dim S^\alpha$.

The action of the momentum map has now been exhibited explicitly: the original manifold M on which φ acts is mapped into N, and the vector fields $X_k \in \tilde{\varphi}(\mathfrak{G})$ are mapped into the restrictions of the $\tilde{X}_k$ of Eq. (21) to N.

3. The end of §22.2(4) mentions the foliation of M provided by the momentum map. This foliation will be designated Φ^μ; its leaves are the $(2\nu - \delta)$-dimensional regular submanifolds $N_\alpha \equiv \mu^{-1}(\alpha)$. In addition, the foliation Φ^X provided by the vector fields in $\tilde{\varphi}(\mathfrak{G})$ is assumed already to exist (φ is assumed foliating). The leaves of Φ^X will be called W_a, where a lies in some as yet arbitrary set which labels the leaves; by definition, $\dim W_a = \delta$.

The situation in §20.2 was similar, except that Φ^X did not necessarily exist, for the vector fields X_k defined a distribution which was not necessarily involutive. Now, however, the X_k do form an involutive distribution (at least on M'), and hence Φ^X exists. It is the intersection of the leaves of this distribution with those of Φ^μ that is of interest, what was called $\mathscr{L}$ in §20.2. Then, because the W_a did not necessarily exist, we were forced to discuss $\mathscr{L}(m)$ at each point separately. Now the situation is richer, and the added richness comes from the Lie algebra action to which the X_k belong.

It is convenient, for the discussion which follows, to refer to Fig. 22.1.

Each of the N_α is mapped by μ into a point $\alpha \in N \subset \mathfrak{G}^*$, as is obvious from the definition of the N_α. Suppose there is some vector field $X_b \in \tilde{\varphi}(\mathfrak{G})$ whose flow carries N_α into N_β. Then in the coadjoint representation there is a vector field $\tilde{X}_b$ whose flow carries α into β, and hence $\beta \in S^\alpha$. In other words, $S^\alpha = S^\beta$. Moreover there are vector fields in $\tilde{\varphi}(\mathfrak{G})$ which carry N_α into N_β if and only if some leaf W_a intersects both N_α and N_β, so each W_a is mapped into an orbit of the coadjoint representation (into a *coorbit*). If W_a and W_b both intersect N_α, the momentum map sends them both into the same coorbit, that of α. In this way, each coorbit

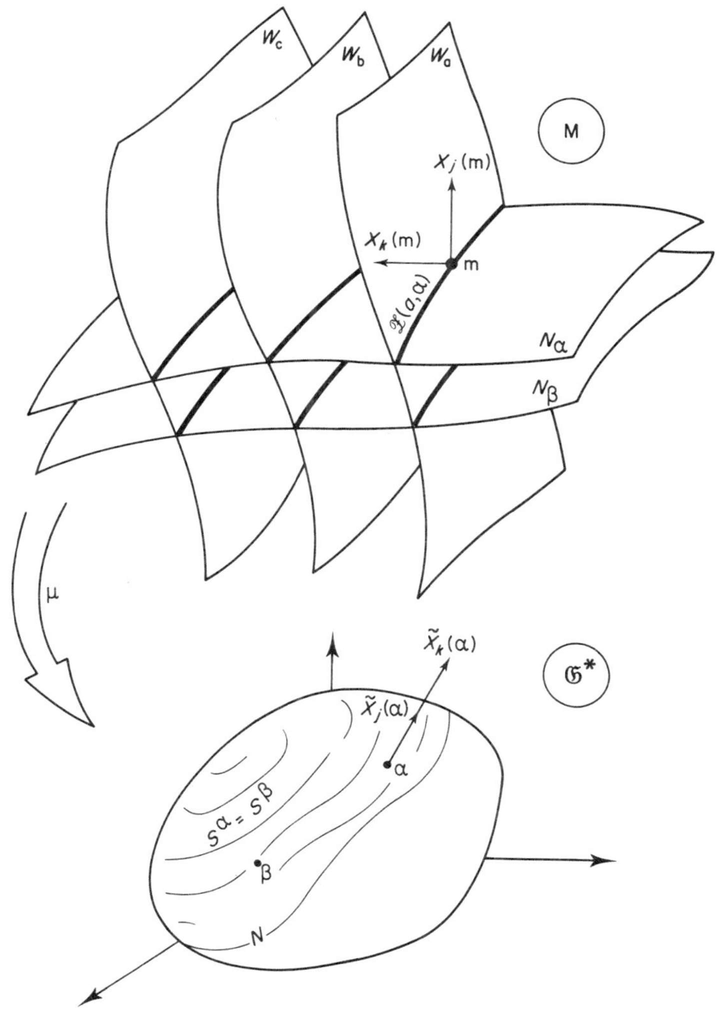

Fig. 22.1. An attempt to illustrate the momentum map. The upper part of the figure represents M and the lower part represent $\mathfrak{G}^*$. In the lower part N represents $\mu(M) \subset \mathfrak{G}^*$. A part of N is the orbit S^α, the same orbit as S^β, for N_α and N_β intersect the same W_a manifolds in M. N_α is mapped into α and N_β into β, and each of the W_a is mapped into S^α, for each intersects N_α. Although the vector fields X_j and X_k may be independent, their projections $\tilde{X}_j$ and $\tilde{X}_k$ need not be. In general, in fact, $\dim S^\alpha = \rho_\alpha < \dim W_a = \delta$, and hence the same inequality holds true for their tangent spaces at each point. We list the dimensions of the various manifolds: $\dim M = 2\nu$, $\dim W_a = \delta$, $\dim N_\alpha = 2\nu - \delta$, $\dim \mathscr{L}(a, \alpha) = \delta - \rho_\alpha$, $\dim \mathfrak{G}^* = \gamma$, $\dim N = \delta$, $\dim S^\alpha = \rho_\alpha$

is associated with a set which is composed both of leaves W_a of Φ^X and of leaves N_α of Φ^μ which they intersect.

Now consider the intersections. First, it is clear that neither the W_a nor the N_α intersect among themselves, for Φ^X and Φ^μ are assumed to be regular foliations. Let the intersection of W_a and N_α be called $\mathscr{L}(a,\alpha)$. Since $\mathscr{L}(a,\alpha)\subset N_\alpha$, the momentum map sends the intersection into the point α. We will now prove the following properties of the $\mathscr{L}(a,\alpha)$. (i) The intersection $\mathscr{L}(a,\alpha)$ is a submanifold whose dimension is $\delta-\rho_\alpha$, where $\rho_\alpha=\dim S^\alpha$. (This implies that if $\rho_\alpha\neq\rho_\beta$, then $\dim\mathscr{L}(a,\alpha)\neq\dim\mathscr{L}(a,\beta)$.) (ii) If ρ_α is maximal, there is a neighborhood of S^α which is the union of orbits all of the same dimension, and the inverse image under μ of this neighborhood is an open neighborhood of M (where the W_a and N_α intersect, of course, always in $\mathscr{L}(a,\alpha)$ of the same dimension).

From §20.2 it is known that $\mathscr{L}(a,\alpha)=\ker\omega_\alpha$; this can be demonstrated now as it was in Chapter 20, except that now one need use only those f_k functions which are independent in each neighborhood. Also as in § 20.2, $\ker\omega_\alpha$ can be studied by means of the Poisson bracket of the f_k. Let U_m be a neighborhood of $m\in N_\alpha\subset M$ and assume the f_k ordered so that the first δ of them are independent in U_m. Then as has been seen,

$$\dim(\ker\omega_\alpha)=\delta-\operatorname{rank}F_{jk}, \tag{22}$$

where $F_{jk}=\{f_j,f_k\}$, $j,k\in\{1,\ldots,\delta\}$. But $\{f_j,f_k\}=-c^h_{jk}f_h$, $h=1,\ldots,\gamma$, and this is a constant matrix on each of the N_α because all of the f_h are constant on the N_α. Thus the rank of F_{jk} is constant, and so is $\dim(\ker\omega_\alpha)=\dim\mathscr{L}(a,\alpha)\equiv\sigma_\alpha$. Now recall that each W_a is mapped into the coorbit S^α and that each N_α consists of all of the points in M which are mapped into the point $\alpha\in S^\alpha$. But then $\mathscr{L}(a,\alpha)$ is the entire set of points in W_a which is mapped into α, and therefore $\dim(S^\alpha)$ satisfies $\dim(W_a)-\dim\mathscr{L}(a,\alpha)=\dim(S^\alpha)$, or

$$\dim\mathscr{L}(a,\alpha)=\sigma_\alpha=\delta-\rho_\alpha. \tag{23}$$

Comparison of (22) and (23) shows that $\rho_\alpha=\operatorname{rank}F_{jk}$, which is enough to establish that the region of maximum ρ_α is an open neighborhood of $\mathfrak{G}^*$. The rest follows immediately. Note that the region of maximum ρ_α is the region in which the maximum number of $\tilde{X}_k$ are independent.

4. We now give some examples.

Example 22.1.

Consider the following action of $e(2)$ (see Eq. 31 of Chapter 21) on $\mathbb{R}^4$, with the usual coordinates (q^k,p_k), $k=1,2$, and the usual symplectic form $\omega=dq^k\wedge dp_k$:

$$\begin{aligned}\tilde{\varphi}(e_1)&\equiv X_1=-p_2\partial/\partial p_1+p_1\partial/\partial p_2-q^2\partial/\partial q^1+q^1\partial/\partial q^2,\\ \tilde{\varphi}(e_2)&\equiv X_2=\partial/\partial q^1,\quad \tilde{\varphi}(e_3)\equiv X_3=\partial/\partial q^2.\end{aligned} \tag{24}$$

This action is strongly Hamiltonian with Hamiltonian functions

$$f_1=q^1p_2-q^2p_1\equiv\mathbf{q}\times\mathbf{p},\quad f_2=p_1,\quad f_3=p_2, \tag{25}$$

and is foliating on $M' = \mathbb{R}^4 - \{q^1, q^2, 0, 0\}$, where the three vector fields are independent. The momentum map μ sends each point of M' with coordinates (q^k, p_k) into the point $\alpha \in \mathfrak{G}^*$ whose coordinates in the $\{\varepsilon^j\}$ basis are (f_1, f_2, f_3) as given by (25). Thus $N = \mathbb{R}^3 - \{\alpha_1, 0, 0\} \subset \mathfrak{G}^*$; N is $\mathfrak{G}^*$ without the 1-axis. The foliation Φ^X is found from (24). Its leaves W_a, of dimension $\delta = 3$, are given by

$$p_1^2 + p_2^2 \equiv \mathbf{p}^2 = a \in \mathbb{R}^+,$$

and it is easily seen that the W_a are diffeomorphic to $S \times \mathbb{R}^2$. The N_α, of dimension one ($2\nu = 4$ in this example) for all $\alpha \in \mathfrak{G}^*$, are given by

$$\mathbf{q} \times \mathbf{p} = \alpha_1, \quad p_1 = \alpha_2, \qquad p_2 = \alpha_3.$$

The intersections $\mathscr{L}(a, \alpha)$ are nonempty only if $\alpha_2^2 + \alpha_3^2 = a$, in which case $\mathscr{L}(a, \alpha)$ is N_α itself. This indicates that the orbit S^α contains all points $\beta \in \mathfrak{G}^*$ for which $\beta_2^2 + \beta_3^2 = \alpha_2^2 + \alpha_3^2$, and is thus a cylinder about the 1-axis in $\mathfrak{G}^*$. This may be compared with the discussion around Eq. (33) of Chapter 21. What is missing here is the set of zero-dimensional orbits; but they are excluded because the 1-axis is not in N. The projected vector fields can be found either by applying the X_k to the f_j or by using (21). It is found that

$$\tilde{X}_1 = \beta_2 \partial/\partial\beta_3 - \beta_3 \partial/\partial\beta_2,$$
$$\tilde{X}_2 = \beta_3 \partial/\partial\beta_1, \quad \tilde{X}_3 = -\beta_2 \partial/\partial\beta_1.$$

Example 22.2.

We return to the example of Eq. (6), an example in which the topology of M' is more interesting and which deals with the angular momentum and is thus less artificial. In this case the momentum map, according to Eq. (7), sends each point $(\mathbf{q}, \mathbf{p}) \in M$ into the point $\alpha \in \mathfrak{G}^*$ whose coordinates in the $\{\varepsilon^k\}$ basis are $\alpha_k = J_k$. (It should be noted that the F of the previous discussion is not the map $F: M \to N$ which is now being identified with μ.) Thus N in this example is all of $\mathfrak{G}^*$. It was shown in the earlier discussion that the leaves W_r (then called $F^{-1}(r)$) of Φ^X are diffeomorphic to $SO(3)$. The leaves N_α of Φ_μ are the submanifolds of M' on which the angular momentum $\mathbf{J}$ is fixed. Their topology can be found by noting that for fixed $\mathbf{J}$, both $\mathbf{q}$ and $\mathbf{p}$ lie in the plane perpendicular to $\mathbf{J}$, and that $\mathbf{q}$ can take on any value in that plane other than zero. For each value of $\mathbf{q}$, the vectors $\mathbf{p}$ which yield the fixed $\mathbf{q} \times \mathbf{p} = \mathbf{J}$ lie on a line, and therefore the desired topology is that of $(\mathbb{R}^2 - \{0\}) \times \mathbb{R}$ or, equivalently, $S \times \mathbb{R}^2$ or $\mathbb{R}^3 - \{x, 0, 0\}$.

To find the intersection $\mathscr{L}(r, \alpha)$ one need only fix $\mathbf{q}^2$ and $\mathbf{p}^2$ on N_α, for $\mathbf{J}^2$ is already fixed. But that means that the angle θ between $\mathbf{q}$ and $\mathbf{p}$ can take on only two values, whose sum is π (unless $\theta = \pi/2$, for which there is only one value). Consider one of these two values. When θ is fixed all pairs $(\mathbf{q}, \mathbf{p})$ that are available can be obtained from each other by rotating them together about $\mathbf{J}$, and then one can identify the resulting circle of values with the circle S in the topology $S \times \mathbb{R}^2$ of N_α. Actually, for $\theta \neq \pi/2$ there are two disconnected circles, but we shall ignore this complication. Therefore, except for this complication, $\mathscr{L}(r, \alpha)$ is a circle in N_α. Recall from the earlier discussion that the r in W_r lies in $\mathbb{R}^3$, and that the first component of r was precisely θ and uniquely determines $\mathbf{J}^2$. Thus each W_r intersects an N_α if and only if the $\mathbf{J}^2$ in W_r is equal to the $\mathbf{J}^2$ in N_α, namely to $\alpha_1^2 + \alpha_2^2 + \alpha_3^2$. From this one deduces that the coorbits are spheres about the origin in $\mathfrak{G}^*$, which is easily verified directly. We do not discuss the projected vector fields; they can be found without difficulty from (21).

Example 22.3.

The properties of the momentum map depend on the specific action of a given Lie algebra, not only on the algebra $\mathfrak{G}$ and on the manifold M. Consider as an illustration another action of $so(3)$ on $\mathbb{R}^6$, which we shall define by specifying the Hamiltonian functions. Let the new functions f_k be given by

$$\begin{aligned} f_1 &= J_1/B, \\ f_2 &= (J_2^2 + J_3^2)^{1/2} B^{-1} \cos[B \arccos J_2 (J_2^2 + J_3^2)^{-1/2}], \\ f_3 &= (J_2^2 + J_3^2)^{1/2} B^{-1} \sin[B \arccos J_2 (J_2^2 + J_3^2)^{-1/2}], \end{aligned} \tag{26}$$

where B is an arbitrary function of $\mathbf{J}^2$ (and then $B = 1$ yields the action of Example 22.2). It can be verified that the f_k satisfy the correct Poisson bracket relations and that they are independent on a certain open region $M' \subset \mathbb{R}^6$ so long as B is chosen judiciously. One finds easily that

$$f_1^2 + f_2^2 + f_3^2 = \mathbf{J}^2/B^2,$$

so that if B is chosen equal to $(\mathbf{J}^2)^{1/2}$, the f_k are no longer independent (the number of independent ones becomes two). It can then be shown that the entire manifold M' is mapped by μ onto the sphere of radius one in $\mathfrak{G}^*$, whereas M' in Example 22.2 was mapped onto all spheres about the origin of $\mathfrak{G}^*$. We do not actually work out the momentum map in this example.

22.4 THE EXISTENCE OF HAMILTONIAN ACTIONS

For reasons having to do with the transition to quantum mechanics and in particular with dynamical symmetries in quantum mechanics it is of interest to determine the limits to the existence of Hamiltonian actions of Lie algebras on manifolds. There are certain advantages to obtaining Poisson-bracket realizations of larger and larger algebras in which the Hamiltonian functions are invariant under the dynamical system.

1. The results of the preceding section give immediate limitations on the sizes of these algebras, limitations which are imposed by the dimensions of the various manifolds involved. With the aid of Fig. 22.1, for instance, it becomes

obvious that

$$\dim W_a \equiv \delta \geqslant \dim S^\alpha \equiv \rho_\alpha \geqslant \rho,$$

where ρ is the minimal coorbital dimension in $\mathfrak{G}^*$. In addition,

$$\dim \mathscr{X}(a, \alpha) = \delta - \rho_\alpha \leqslant \dim N_\alpha \equiv 2\nu - \delta,$$

so that

$$2(\delta - \nu) \leqslant \rho_\alpha \leqslant \delta \leqslant \gamma, \tag{27}$$

where we have used $\dim W_a \equiv \delta \leqslant \dim \mathfrak{G} \equiv \gamma$. For some special cases these inequalities are not surprising. For example, if $\mathfrak{G}$ is Abelian, $\rho_\alpha = 0$ and then $\delta \leqslant \nu$, which says that on a symplectic manifold of dimension 2ν there cannot be more than ν independent Hamiltonian vector fields which commute among themselves.

2. More generally, for reasons that will become evident in the discussion of group actions, one is interested in the action of semisimple or even simple algebras, so that the kind of question we are considering now may be stated as follows: can one find a symplectic action of a given simple Lie algebra $\mathfrak{G}$ on a given (symplectic) manifold (M, ω)? It may be recalled that all (nontrivial) actions of simple Lie algebras are effective, and that all of their symplectic actions are strongly Hamiltonian. This means that if one can find such an action, that action can be made strongly Hamiltonian, so that $\varphi_{\mathscr{F}}: \mathfrak{G} \rightarrow \mathscr{F}(M)$ is strictly a homomorphism, not affine.

Further discussion of this question will be limited to the classical Lie algebras (the Lie algebras of the classical transformation groups) in their real Cartan forms, as will be described below. It can be shown, moreover, that the results obtained in this way will remain valid more generally for all algebras that can be brought to these forms by complex transformations.

Every simple Lie algebra $\mathfrak{G}$ has a basis

$$\{h_1, \ldots, h_r; \quad e_1, \ldots, e_{\gamma - r}\}$$

whose first r elements $\{h_j\}$ span the r-dimensional maximal Abelian subalgebra of $\mathfrak{G}$ (called the *Cartan subalgebra* of $\mathfrak{G}$; r is called the *rank* of $\mathfrak{G}$) and in which the bracket and hence the structure constants as well are given by

$$\begin{aligned} [h_j, h_k] &= 0 \\ [h_j, e_w] &= s_j(w) e_w, \\ [e_t, e_w] &= N^j_{tw} h_j, \end{aligned} \tag{28}$$

where, of course, all of the $s_j(w)$ and N^j_{tw} are real. For all simple Lie algebras is turns out that $\gamma \geqslant 2r$, or $\gamma - r \geqslant r$. In Eqs. (28) the indices j, k take on values from 1 to r, and w, t from 1 to $\gamma - r$. It is always possible to choose r of the e_w so that the $s_j(w)$ belonging to those e_w form r linearly independent vectors $\mathbf{s}(w)$. Without loss of generality these may be the first r, so that w runs from 1 to r for them and we may write $\mathbf{s}(k)$ for these vectors, called the *roots* of the e_k. The root vectors thus span an r-dimensional space (the *root* space); the matrix $s_j(k)$ is nonsingular.

Consider now a symplectic action φ of a simple Lie algebra $\mathfrak{G}$ of rank r on a manifold (M,ω). Let $\tilde{\varphi}(h_j)=X_j^H, \tilde{\varphi}(e_j)=X_j^E$, and let the Hamiltonian functions of the X_j^H and X_j^E be H_j and E_j, respectively. Then from (28) it follows that

$$\begin{aligned}[X_j^H, X_k^H]&=0,\\ [X_j^H, X_k^E]&=s_j(k)X_k^E.\end{aligned} \tag{29}$$

There are other commutation relations, but they are not important for the present discussion. Eq. (29) can be used to show that the orbits S^α into which the W_a are mapped by the momentum map must have dimension higher than $2r$, and therefore that not only γ, but also δ must be greater than $2r$. This is because (29) implies a similar relation for the functions H_j and E_k as for their vector fields: for instance $\{H_j, E_k\} = -s_j(k)E_k$. These relations, in turn, determine the F_{jk} matrix of Eq. (22), and that matrix, or rather its rank, determines the dimension of S^α. The determinant of that matrix can be shown to be the square of the determinant of the $s_j(k)E_k$, and that this is nonzero (wherever the product of all of the E_k is nonzero) follows from the linear independence of the $\mathbf{s}(k)$.

Thus one arrives at a theorem which says that in any symplectic action of a simple Lie algebra of rank r, at least $2r$ of the vector fields are pointwise independent, and therefore so are their Hamiltonian functions. Now on a manifold of dimension 2ν there can be no more than 2ν independent functions, and therefore as a corollary it follows that the rank r of the Lie algebra in such a symplectic action cannot exceed ν.

3. In many situations it is useful to have at one's disposal quantities which are formed out of the elements of the Lie algebra and which commute with them. In the case of the simple algebras certain homogeneous polynomials formed of the elements serve this purpose (they are called *Casimir operators*). They are used, among other things, to characterize representations of the algebras and in certain applications in quantum mechanics. Here their interest lies in their connection with foliations obtained from Hamiltonian actions.

These polynomials are formed not with the composition map (the bracket) of the Lie algebra, but with a different kind of multiplication, yet to be defined, which will combine properly with the bracket. Thus the next step leads out of the realm of the Lie bracket, extending the kind of operations that can be performed on the algebra. This can be done in terms of the adjoint representation, which explains, at least in part, the *operator* terminology. Rather than go into this rigorously and in all detail, we proceed by example.

An example of a Casimir operator for $so(2, 1)$ (see Eq. (14)), now given in the form

$$[e_1, e_2]=e_2, \quad [e_2, e_3]=e_1, \quad [e_3, e_1]=e_3, \tag{30}$$

is

$$C=e_1^2+e_2e_3+e_3e_2. \tag{31}$$

The commutation can be checked by using a sort of Leibnitz rule for the bracket of the Lie algebra applied to a formal product (this, essentially, is the meaning of combining properly), i.e.

$$[a, bc] = b[a, c] + [a, b]c.$$

This is clearly what is wanted in representations, in which the elements of the algebra are represented by linear operators and the bracket by the commutator.

Suppose now that a Casimir operator C is given for a Lie algebra $\mathfrak{G}$, and that φ is a Hamiltonian action of $\mathfrak{G}$ on (M, ω). There are two things that can be done to transform C into an object having to do with M. First, the elements e_k which appear in the expression

$$C = \Sigma a_{jh \ldots p} e_j e_h \ldots e_p \tag{32}$$

can be replaced by their actions $\tilde{\varphi}(e_k) = X_k$, and then C is replaced by what may be called $C(X)$, an operator in which the products can be interpreted as successive application of Lie derivatives. Then $C(X)$ becomes a linear operator of higher order with some properties which relate to the particular action and which come from the fact that C is a Casimir operator on $\mathfrak{G}$. Second, the e_k in (32) can be replaced by the Hamiltonian functions f_k, and then C is replaced by what may be called $C(f)$. Then from the fact that $L_{X_j} f_k = \{f_k, f_j\} = c^h_{jk} f_h$ it follows that

$$L_{X_j} C(f) = 0 \; \forall \, j, \tag{33}$$

or that $C(f)$ is constant on the leaves W_a of Φ^X. From the fact that $C(f)$ is a function of the f_j it follows that it is constant also on the N_α. Thus it is constant on that part of M which is sent by μ into a coorbit S^α, or on $\mu^{-1}(S^\alpha)$. This can be useful, clearly, in analyzing the foliation of M.

A Casimir operator can be used similarly to analyze the foliation of $\mathfrak{G}^*$, for if the e_k in (32) are replaced by the β_k, so that C is replaced by what may be called $C(\beta)$, it follows from (21) similarly that

$$L_{\mathfrak{X}_j} C(\beta) = 0 \; \forall \, j.$$

Defined in terms of abstract Lie algebras, Casimir operators are restricted to polynomials, for otherwise it would be difficult to define their commutators with elements of the algebra. More generally, a *Casimir function* of a Hamiltonian action of a Lie algebra $\mathfrak{G}$ on a manifold (M, ω) can be defined (see Helgason (1962, p. 451) for an equivalent definition) as a function $f \in \mathscr{F}(M)$ such that

$$df = g^j df_j, \quad g^j \in \mathscr{F}(M)$$
$$L_{X_j} f = \{f, f_j\} = 0 \; \forall \, j, \tag{34}$$

where the f_j are the Hamiltonian functions of the action. Such a function is sometimes called an *indicial function* in the theory of function groups.

A relevant theorem, stated without proof: Let φ be a Hamiltonian action of a Lie algebra $\mathfrak{G}$ on a symplectic manifold (M, ω). If the order of φ is either $2r$ (minimal) or $2v$ (maximal), then all Casimir functions of the action, in particular $C(f)$ if C is a Casimir operator, are multiples of one, identically constant. The maximal case is almost trivial from the geometric point of view, for the action must then be transitive on M (i.e. $W_a = M$) and hence any Casimir function must be a constant.

22.5 SYMPLECTIC FORMS ON $\mathfrak{G}^*$ AND POISSON BRACKETS

It has been shown that the submersion $\mu: M \to \mathfrak{G}^*$ maps the action φ into the coadjoint representation of $\mathfrak{G}$ on $\mathfrak{G}^*$. The next step is to see how μ maps the symplectic form ω, defined on M, onto a form on $\mathfrak{G}^*$. This step is complicated by the fact that μ maps M only onto $N \subset \mathfrak{G}^*$. We shall therefore proceed indirectly, by defining a Poisson bracket on N through ω and the pullbacks of functions in $\mathscr{F}(N)$ to functions in $\mathscr{F}(M)$, and then using this Poisson bracket to define a symplectic form on each coorbit of $\mathfrak{G}^*$. It turns out that the symplectic form obtained in this way coincides with the one defined on $\mathfrak{G}^*$ by Kirillov (1962), Kostant (1965), and Souriau (1966).

1. Let $\{F, G\}_M$ be the Poisson bracket of two functions $F, G \in \mathscr{F}(M)$. The Poisson bracket $\{f, g\}_N$ of two functions $f, g \in \mathscr{F}(N)$ will be defined through its pullback and the pullbacks of f and g by

$$\mu^*\{f, g\}_N = \{\mu^* f, \mu^* g\}_M = -\frac{\partial f}{\partial f_j} c^h_{jk} f_h \frac{\partial g}{\partial f_k}. \tag{35}$$

The bracket so defined is antisymmetric and satisfies the Jacobi identity (Eq. (6) of Chapter 21), but it may be degenerate. The condition for nondegeneracy is that for every open set $U \subset N$ and $f, g \in \mathscr{F}(N)$,

$$\{f, g\}_N = 0 \;\forall f \Rightarrow g = \text{const.} \tag{36}$$

There is a general theorem (AM, p. 416) which says that if this condition is satisfied, there exists a unique symplectic form on N such that (35) is the Poisson bracket obtained from it. It will be seen eventually that a condition for nondegeneracy is satisfied on coorbits.

A Poisson bracket can be defined similarly on all of $\mathscr{F}(\mathfrak{G}^*)$. If $\psi, \chi \in \mathscr{F}(\mathfrak{G}^*)$, their Poisson bracket may be defined by

$$\{\psi, \chi\}_{\mathfrak{G}^*} = -\frac{\partial \psi}{\partial \beta_j} c^h_{jk} \beta_h \frac{\partial \chi}{\partial \beta_k}. \tag{37}$$

This bracket is also antisymmetric and satisfies the Jacobi identity, but like (35) it may also be degenerate. In fact, the brackets of (35) and (37) are related: the restriction of (37) to N is (35). To show this, one must first show that (37) can in fact be restricted to N, i.e. that if ψ, χ and ψ', χ' are two pairs of functions in $\mathscr{F}(\mathfrak{G}^*)$ which have the same values on N (on N, that is, $\psi = \psi'$ and $\chi = \chi'$), then their Poisson brackets as defined by (37) have the same values on N. This may be seen by noting that in general

$$\{\xi, \eta\}_{\mathfrak{G}^*} = -\frac{\partial \xi}{\partial \beta_j} L_{\tilde{X}_j} \eta = \frac{\partial \eta}{\partial \beta_j} L_{\tilde{X}_j} \xi,$$

where $\xi, \eta \in \mathscr{F}(\mathfrak{G}^*)$. From this one arrives at

$$\{\psi, \chi\}_{\mathfrak{G}^*} - \{\psi', \chi'\}_{\mathfrak{G}^*} = L_{\tilde{X}_j}(\chi - \chi')\frac{\partial \psi}{\partial \beta_j} - L_{\tilde{X}_j}(\psi - \psi')\frac{\partial \chi'}{\partial \beta_j}.$$

This is zero because the $\tilde{X}_j$ are tangent to N and because on N both $\psi = \psi'$ and $\chi = \chi'$. Note the Poisson bracket can be restricted in the same way not only to N, but to any submanifold of $\mathfrak{G}^*$ to which the $\tilde{X}_j$ are tangent, for instance to coorbits. To establish the relation of (35) to (37) now, let $f, g \in \mathscr{F}(N)$ be such that $f = \psi | N$ and $g = \chi | N$. Then it is clear that

$$\{f, g\}_N = (\{\psi, \chi\}_{\mathfrak{G}^*}) | N. \tag{38}$$

2. These results can now be used to define a Poisson bracket on coorbits in the same way, but this one will be nondegenerate and hence will lead to a symplectic form on each coorbit. Let $f, g \in \mathscr{F}(S^\alpha)$ be given by $f = \psi | S^\alpha$ and $g = \chi | S^\alpha$. Then the Poisson bracket on the coorbit S^α of α is defined by

$$\{f, g\}_\alpha = (\{\psi, \chi\}_{\mathfrak{G}^*}) | S^\alpha = -\left(\frac{\partial \psi}{\partial \beta_j} L_{\tilde{X}_j}\chi\right) | S^\alpha. \tag{39}$$

Now suppose that $\{f, g\}_\alpha = 0$ for all $f \in \mathscr{F}(S^\alpha)$. Choose, in particular, ψ to be β_k, and then repeat this for all of the β_k in sequence. Then it follows that

$$L_{\tilde{X}_j}\chi | S^\alpha = 0 \ \forall j,$$

and since the $\tilde{X}_j$ span the tangent space of S^α at each of its points, $g \equiv \chi | S^\alpha = \text{const}$. Thus the condition for nondegeneracy is fulfilled.

The coorbits are therefore symplectic manifolds. The symplectic form Ω_α defined in this way on the coorbit can be shown to coincide with the one defined by Kirillov, Kostant, and Souriau: $\Omega_\alpha(\tilde{X}_j, \tilde{X}_k) = P([e_j, e_k])$. We call the point of the coorbit α; they call it P.

As might be expected from the way Ω_α is defined in terms of the Poisson bracket on S^α, which itself is defined by a chain of reasoning in terms of the symplectic form ω on M, there is a connection between ω and Ω_α. The connection is similar to the one in Chapter 20 between ω and the symplectic form Ω_c on M_c. This connection can be traced out in the following way. Let Ω_α be pulled back to $M_\alpha = \mu^{-1}(S^\alpha)$, and let the pullback be called ω_α:

$$\omega_\alpha = \mu^* \Omega_\alpha.$$

Then ω and ω_α coincide when restricted to a leaf of Φ^X, i.e. to any one of the W_a. We do not prove this assertion or pursue the connection any further.

22.6 A DETAILED EXAMPLE: THE KEPLER PROBLEM

So far this chapter has not dealt with dynamical vector fields. The example we now present concerns a strongly Hamiltonian action of a Lie algebra on phase space, an action constructed with the aid of constants of the motion for the Kepler problem. Let H be the Hamiltonian function for the Kepler problem: $i_\Delta \omega = dH$, where Δ is the dynamical vector field. The algebra $\mathfrak{G}$ will be realized by vector fields X_j for which H is an invariant and whose Hamiltonian functions f_j are in turn invariants for the dynamics: $L_{X_j}H = 0$ and $L_\Delta f_j = 0$. Hence Δ will be tangent to the N_α submanifolds, and the distribution $\mathscr{D}^X$ generated by the X_k is tangent to the $H = \text{const.}$ manifolds. It is then found that $\mathscr{D}^X$ foliates the $H = \text{const.}$ manifolds and that Δ projects with respect to $\mathscr{D}^X$ onto the quotient of these manifolds with respect to the foliation.

The manifold of this example is $M = \mathbb{R}^6$ with coordinates $(\mathbf{q}, \mathbf{p})$ and the usual symplectic form $\omega = dq^k \wedge dp_k$. The Hamiltonian function is defined by

$$H(\mathbf{q}, \mathbf{p}) = \tfrac{1}{2}\mathbf{p}^2 - 1/q, \tag{40}$$

where $q = (\mathbf{q}^2)^{1/2}$. As before, we write $\mathbf{J} = \mathbf{q} \times \mathbf{p}$, and $\mathbf{J}$ is now a vector constant of the motion. Another vector constant of the motion is the Runge–Lenz vector $\mathbf{A} = \mathbf{p} \times \mathbf{J} - \mathbf{q}/q$, and on occasion we shall refer to a third one, $\mathbf{B} = \mathbf{A} \times \mathbf{J}$.

Simoni *et al.* (1974) have enumerated the possible actions of $so(4)$ and $e(3)$ on M with vector fields that are obtained from $\mathbf{J}$, $\mathbf{A}$, and $\mathbf{B}$. Of their enumeration we shall analyze only one set of the $so(4)$ actions. The $so(4)$ algebra can be described in a basis $\{e_1, e_2, e_3, s_1, s_2, s_3\}$ in which the e_j span an $so(3)$ subalgebra with the usual commutation relations; the bracket for the entire algebra is given by

$$[e_j, e_k] = \varepsilon^h_{jk} e_h, \quad [s_j, e_k] = \varepsilon^h_{jk} s_h, \quad [s_j, s_k] = \varepsilon^h_{jk} e_h. \tag{41}$$

A strongly Hamiltonian action φ of $so(4)$ on $\mathbb{R}^6$ can then be characterized by giving its Hamiltonian functions. Let f_k be the Hamiltonian function of $X_k \equiv \tilde{\varphi}(e_k)$ and g_k be the Hamiltonian function of $Y_k \equiv \tilde{\varphi}(s_k)$. Then φ is defined by

$$f_k = J_k, \quad g_k = A^c_k, \tag{42}$$

where

$$\mathbf{A}^c = \left[\frac{1}{2H}\left\{-1 + \frac{C(H)}{\varepsilon^2}\right\}\right]^{1/2} \mathbf{A}, \tag{43}$$

where ε is the eccentricity, defined by the usual equation $1 + 2H\mathbf{J}^2 = \varepsilon^2$. In order that $\mathbf{A}^c$ be real, the function $C(H)$ must satisfy the conditions

$$\begin{aligned} &\text{if } H < 0 \quad \text{then} \quad C(H) \leqslant \varepsilon^2; \\ &\text{if } H > 0 \quad \text{then} \quad C(H) \geqslant \varepsilon^2. \end{aligned} \tag{44}$$

Otherwise $C(H)$ is an arbitrary real function.

The first task is to clean up the phase space M by requiring that the functions involved all be well-defined and that the number of independent vector fields remain fixed, so that the action is foliating. What this will mean, as it always does, is that the phase space will be somewhat reduced.

First, H is not well-defined for $\mathbf{q} = 0$. Thus M is reduced from $\mathbb{R}^6$ to $\mathbb{R}^6 - \{0, \mathbf{p}\} \equiv M_1$.

Second, the reality condition (44) cannot be satisfied simultaneously for $H < 0$ and $H > 0$. Only the case $H < 0$ will be considered, forcing reduction to $M_2 = M_1 \cap \{\mathbf{q}, \mathbf{p} \vdash H < 0\}$. Then C will be further restricted to $C \leqslant 0$.

Third, $\mathbf{J}$ and $\mathbf{A}^c$, in terms of which the momentum map will be given, must be well-defined functions. Then if $C(H) \neq 0$ a further reduction is needed, to $M_3 = M_2 - \{\mathbf{q}, \mathbf{p} \vdash \varepsilon = 0\}$. When $C = 0$, then $M_3 = M_2$.

Finally, M_3 has to be reduced to its submanifold on which the action has the maximum number of independent vector fields. To discuss this it helps to exhibit the Casimir operators. They are

$$\begin{aligned} C_1 &= \delta^{jk}(e_j e_k + s_j s_k), \\ C_2 &= \delta^{jk} e_j s_k. \end{aligned} \tag{45}$$

Replacing the elements of $\mathfrak{G} = so(4)$ by the Hamiltonian functions, as at Eq. (33), one obtains

$$\begin{aligned} C_1(f) &= \mathbf{J}^2 + (\mathbf{A}^c)^2 \equiv \tfrac{1}{2}[C(H) - 1]/H, \\ C_2(f) &= \mathbf{J} \cdot \mathbf{A} \equiv 0. \end{aligned} \tag{46}$$

When $C = 0$, only the second of these conditions is a relationship between the functions; then just five of them are independent, and M_3 must be further reduced to $M_4 = M_3 - P$, where P is the closed set without interior on which the number of independent vector fields is less than five. This set is not empty, for it contains at least those points on which $\mathbf{J} = 0$, as discussed in the example of Eq. (6). Thus M_3 actually is reduced in this case. The dynamics Δ is complete on M_4 because of the removal of those points where $\mathbf{J} = 0$. (The problem would have been that if the $\mathbf{J} = 0$ points had not been removed, Δ would pass through the origin of M, but the origin has already been removed.)

When $C(H) \neq 0$, there is a special case in which $C(H) = (1 + 2H)v$, $v \in \mathbb{R}^+$. In this case, in addition to $C_2(f) = 0$, Eq. (46) implies that $C_1(f) = v$, and then only four of the functions are independent. Then M_3 is reduced to $M_4 = M_3 \cap B - P'$, where B takes account of the reality condition and is given by $B = \{\mathbf{q}, \mathbf{p} \vdash H < -\tfrac{1}{2}\}$, and P' is the set on which the number of independent vector fields is less than four. From now on we shall omit the subscript 4 on M_4, writing simply M.

The structure of the $so(4)$ coorbits is easy to find, partly because the algebra is semisimple and hence the adjoint and coadjoint representations have the same (isomorphic) orbits. For this reason one may speak of $\mathfrak{G}$ as though it were $\mathfrak{G}^*$, and we shall do so. It is helpful to transform to a basis in $so(4)$ which exhibits the way the algebra is the direct product of $so(3)$ algebras. Such a basis is $\{a_k, r_k\}$, $k = 1, 2, 3$, defined by $a_k = \tfrac{1}{2}(e_k + s_k)$ and $r_k = \tfrac{1}{2}(e_k - s_k)$, for which the bracket in the Lie algebra becomes

$$[a_j, a_k] = \varepsilon^h_{jk} a_h, \quad [r_j, r_k] = \varepsilon^h_{jk} r_k, \quad [a_j, r_k] = 0. \tag{47}$$

The two Casimir operators can now be written in the form $\mathbf{a}^2 \equiv \delta^{jk} a_j a_k$ and $\mathbf{r}^2 \equiv \delta^{jk} r_j r_k$. Let the coordinates of a point α in $\mathfrak{G}$ be α^k and ρ^k, so that $\alpha = \alpha^k a_k + \rho^k r_k$. Then the coorbits (remember that we are speaking of $\mathfrak{G}$ as though it were $\mathfrak{G}^*$) are given by $\boldsymbol{\alpha}^2 = \lambda_1 \geqslant 0$ and $\boldsymbol{\rho}^2 = \lambda_2 \geqslant 0$ (see the discussion below Eq. (33)), where $\boldsymbol{\alpha}^2 = \delta_{jk} \alpha^j \alpha^k$, etc. These coorbits are therefore of the form $S^2 \times S^2$, of dimension four, so long as neither of the λ_j vanishes. They are of the form

S^2, of dimension two, if one of them vanishes, and points (of dimension zero) if both vanish.

Consider the momentum map μ for the case in which $C(H) = 0$. Let the coordinates of $\alpha \in \mathfrak{G}$ in the $\{e_k, s_k\}$ basis be ε^k and σ^k. The identity $\mathbf{J} \cdot \mathbf{A} = 0$ yields $\delta_{jk}\varepsilon^j\sigma^k = 0$ on N, from which it follows that $\boldsymbol{\alpha}^2 = \boldsymbol{\rho}^2$, or $\lambda_1 = \lambda_2$. In other words, N contains only those coorbits of type $S^2 \times S^2$ in $\mathfrak{G}^*$ for which the two spheres have equal radius $r = \lambda_1 = \lambda_2$. This single restriction means that N is of dimension five. But not all point which satisfy this restriction are on N. For instance, the points for which $\mathbf{J} = 0$ are not in M, and these correspond to points for which $\alpha^k = -\rho^k$; such points must be removed from $S^2 \times S^2$.

In terms of the manifolds described and represented in Fig. 22.1, the momentum map for this case has the following properties. The leaves of Φ^X, conveniently labeled with the value of the Hamiltonian and hence called W_H, the surfaces of constant energy, are of dimension five. Since dim $M = 6$ in this example, it follows that dim $N_\alpha = 6 - 5 = 1$. The N_α contain the points in M with a given angular momentum $\mathbf{J}$ and a given value of the Runge–Lenz vector $\mathbf{A}$. The condition $\mathbf{J} \cdot \mathbf{A} = 0$ connects these vectors and causes N_α to be one-dimensional; if $\mathbf{J}$ and $\mathbf{A}$ were independent, N_α would consist of only one point. In this example $\mathscr{L}(H, \alpha)$ is N_α itself. This also indicates that the dimension of the coorbits S^α must be four, as undeed it is, for as has already been seen the coorbits are of the form $S^2 \times S^2$, consisting of two spheres of the same radius, one in the subspace of $\mathfrak{G}^*$ spanned by the a_k, the other in the subspace spanned by the r_k. The dynamical vector field is tangent to the N_α, for its angular momentum, Runge–Lenz vector, and energy are fixed.

The picture changes when $C(H) = (1 + 2H)v$. In this case $\lambda_1 + \lambda_2 = \frac{1}{2}$, so that the two spheres are fixed. All of M is mapped into a single orbit. The dimension of W_H is now four, for only four of the functions are now independent, and the dimension of N_α is now $6 - 4 = 2$. The coorbit is still of dimension four, so that the dimension of $\mathscr{L}(H, \alpha)$ is now $4 - 4 = 0$: the N_α intersect the W_H at points. The W_H (H is perhaps no longer a reasonable designation) foliate M as before, but they now also foliate the $H^{-1}(E)$ surfaces, and Δ, which is tangent to these surfaces, is projectable with respect to this foliation.

We leave this example at this point. The next chapter begins the extension of some of the results of the last two chapter to results that can be obtained when the Lie algebras are integrable to groups.

23

A digression on Lie groups and their actions on manifolds

Next in our analysis will be to study what more is obtained when an action of a Lie algebra can be *integrated* to yield an *action of a Lie group* on the same manifold M. The connection between Lie groups and algebras is well known: the algebras exhibit in a sense the infinitesimal structure of the groups, and the groups, conversely, represent the integrals of the algebras. This will be described briefly in this chapter. Our real interest, however, is not so much in the groups and algebras themselves, as in their actions on manifolds, so the pure group-theoretical point of view will not suffice. It turns out that not every Lie algebra action can be integrated to a Lie group action, which is unfortunate, for group actions provide further useful tools in the reduction of dynamical systems.

It is known that every Lie group G has its Lie algebra $\mathfrak{G}$ and that every $\mathfrak{G}$ is the Lie algebra of at least one Lie group. Thus it is natural to look for a Lie group action (to be precisely defined later) which can be associated with a given Lie algebra action. Such group actions play an important, even an essential part in the quantum-mechanical analysis of symmetry and degeneracy of energy levels, as well in programs of geometric quantization. In classical mechanics, the action of a group gives a way to carry local results in neighborhoods $U \subset M$ to the entire orbit of the group. It also adds strength and meaning to the momentum map. Nevertheless, we repeat, there is not always a group action corresponding to an algebra action. This implies, among other things, that *universal symmetries* which have been proposed for large families of dynamical systems involve Lie algebra actions and can yield only local transformation groups (Fradkin, 1967; Stehle and Han, 1967).

Often, especially when the carrier manifold has the structure of a vector space (like $\mathbb{R}^{\nu}$, for instance), one actually starts from the action of a classical transformation group. Examples are well known: the orthogonal group, the Lorentz group, etc. In such cases one goes easily back and forth between the actions of the Lie group and the Lie algebra. Our interest in such examples is in seeing what the difference is between dealing with the group action and the algebra action.

Sometimes M itself, the carrier manifold of the dynamics, has the structure of a Lie group (again, a simple example is $\mathbb{R}^\nu$; a somewhat less simple one is $SO(3)$, the carrier manifold in rigid body motion, which will be discussed in Chapter 25). This will then lead to an interesting study of the action of Lie groups on Lie groups and of dynamical systems on Lie groups.

23.1 LIE GROUPS

1. We treat the subject of abstract Lie groups quite briefly, relying heavily on AM, Chapter 4. Our presentation will emphasize the manifold structure of Lie groups. This subsection is devoted mostly to some definitions.

A *γ-dimensional Lie group G* is a γ-dimensional manifold with a group operation defined on it, called multiplication and with the usual group properties (closure, associativity, existence of the identity and of inverses) and such that both

$$G \times G \to G:(g_1, g_2) \mapsto g_1 g_2,$$

and

$$G \to G: g \mapsto g^{-1}$$

are C^∞ maps for all $g_1, g_2, g \in G$ (it turns out that the second of these maps is automatically C^∞ if the first is). We shall often speak of the group and its manifold interchangeably, more or less as we do of a Lie algebra and the vector space on which it is defined.

It can be shown that the manifold of a Lie group is necessarily Hausdorff, but it is not necessarily second countable. Thus when we speak of Lie groups as manifolds, we are relaxing the condition of second countability. On the other hand it can be shown that connected Lie groups are always second countable. The connected component of the identity in a Lie group is itself a Lie group and therefore, according to the above, second countable. Such connected components play an important role both in the theory and in the applications.

A *Lie subgroup H* of a Lie group G is a subgroup of G which is an immersed submanifold (see §8.1(3)) of G. The topology induced by the differentiable structure on H need not coincide with the subset topology. It can be shown that if H is a *closed* subgroup of G, it is a submanifold (see §8.1(2)). This will be discussed more fully later.

An *invariant subgroup H* of a Lie group G is a subgroup such that $g^{-1} Hg \subset H \; \forall \; g \in G$.

If G and H are Lie groups, the map $\varphi: G \to H$ is a *Lie group homomorphism* if it is C^∞ and if $\varphi(g_1 g_2) = \varphi(g_1)\varphi(g_2)$. If in addition φ is a bijection, it is a *Lie group isomorphism.* As in the case of Lie algebras, we shall often fail to distinguish between isomorphic Lie groups.

We do not give any examples yet (see §(3)), but mention some physically important Lie groups: the general linear group $GL(n, \mathbb{R})$ of invertible n-by-n matrices, whose group operation is matrix multiplication; the group of translations of $\mathbb{R}^\nu$, whose operation is successive translation (this group is Abelian); the rotation group $SO(n)$ in n dimensions, whose group operation is also matrix multiplication (or successive rotation); the Galilei group; the Lorentz group. Most of these are familiar.

2. The group operation on G makes it possible to move points around in ways that are not permitted on manifolds without this additional structure. *Left translation* by or with respect to an element $g \in G$ is defined as the map

$$L_g : G \to G : h \mapsto gh. \tag{1}$$

For each $g \in G$, the left translation L_g is by definition a diffeomorphism in a Lie group, and it can therefore be used also to move vector fields and forms around the group manifold. Then $TL_g : TG \to TG$ maps the tangent space at h into the tangent space at gh. That is,

$$T_h L_g \equiv TL_g | T_h G : T_h G \to T_{gh} G, \tag{2}$$

and then

$$T_e L_g : T_e G \to T_g G, \tag{3}$$

where e is the identity element of G. Thus every element $w \in T_e G$ can be transported to $T_g G$ through the group structure of G. This possibility can now be used to induce a Lie algebra structure on $T_g G$ from the Lie algebra structure which we have already seen is possessed by the vector fields on manifolds, in particular on the manifold G.

Let $w \in T_e G$ as before, and consider the vector field $X_w \in \mathfrak{X}(G)$ defined by

$$X_w(g) = T_e L_g w \in T_g G \;\forall\; g \in G. \tag{4}$$

It is easily seen that X_w is uniquely defined by $w \equiv X_w(e)$, that it is complete, and that it is *left invariant*:

$$(L_g)_* X_w = X_w \forall g \in G. \tag{5}$$

In fact any left-invariant vector field $X \in \mathfrak{X}(G)$ is by definition of the form X_w and thus satisfies (5), for all one need do is write w for $X(e)$. Moreover, the left invariant vector fields form a subalgebra of the 'Lie algebra' of vector fields, for if X and Y are both left invariant, then

$$(L_g)_*[X, Y] = [(L_g)_* X, (L_g)_* Y] = [X, Y].$$

In this way one is led to a Lie algebra $\mathfrak{G}$ defined in terms of $T_e G$ as follows. Let $w, v \in T_e G$; then

$$[w, v] = [X_w, X_v](e). \tag{6}$$

Suppose $\{X_j\}$ is a set of left-invariant vector fields obtained as above by left translation of a basis of vectors in $T_g G$. Then since there exist constants c^h_{jk} such that $[X_j, X_k] = c^h_{jk} X_h$ at the identity and since everything remains the same when it is left translated, the c^h_{jk} constants enter into the same equation at any other point in G. These are structure constants of the Lie algebra $\mathfrak{G}$. When it is necessary to distinguish it from other Lie algebras, we shall write $\mathfrak{G}_G$ for $\mathfrak{G}$. One thus finds that (as usual, up to isomorphisms) to each Lie group belongs its unique Lie algebra.

Remark. There is nothing special about left as opposed to right translation. If we had used right translation in Eq. (1), defined by $R_g h = hg$, the result would have been the same, but for the Lie algebra of right-invariant vector fields.

In addition to left-invariant vector fields, there exist left-invariant differential forms. In particular, the one-forms dual to left-invariant vector fields are themselves left-invariant. If $\{X_j\}$ is, as above, a basis of left-invariant vector fields, then the one-forms $\{\theta^k\}$ satisfying the equations $\theta^k X_j = \delta^k_j$ are a set of γ independent left-invariant one-forms which will play an important role in the calculus on TG and T^*G.

3. We now turn to two examples of Lie groups and their left-invariant vector fields.

(a) The Euclidean group E_2 in two dimensions

This group is the set of all rotations and translations of the plane $\mathbb{R}^2$, with successive application taken as the group operation. Let $g(\theta, \mathbf{a})$ be the rotation about the origin by an angle θ followed by a translation by the vector $\mathbf{a}$. Then a point $\mathbf{x} \in \mathbb{R}^2$ is transformed according to

$$g(\theta, \mathbf{a}): \mathrm{x} \mapsto \mathbf{x}' = R_\theta \mathbf{x} + \mathbf{a},$$

where $R_\theta \mathbf{x}$ is the point with coordinates

$$(R_\theta \mathbf{x})_1 = x_1 \cos\theta + x_2 \sin\theta,$$

$$(R_\theta \mathbf{x})_2 = -x_1 \sin\theta + x_2 \cos\theta;$$

here x_1 and x_2 are the coordinates of $\mathbf{x}$. Obviously the group operates on $\mathbb{R}^2$ in a C^∞ way. The identity element is clearly $e = g(0, 0)$, and the group product is given by

$$g(\theta, \mathbf{a}) g(\varphi, \mathbf{b}) = g(\theta + \varphi, \mathbf{a} + R_\theta \mathbf{b}). \tag{7}$$

The group manifold of E_2 is seen to be of dimension three and diffeomorphic to $S \times \mathbb{R}^2$. Left translation is given by Eq. (7), for $g(\theta, \mathbf{a}) g(\varphi, \mathbf{b})$ is just $L_{\theta,a} g(\varphi, \mathbf{b})$. From this one can construct a basis of left-invariant vector fields. We go through the calculation in some detail as an example of how this can be done. First one names a basis in $T_e E_2$. Since $T_e E_2$, like E_2, is of dimension three, the basis must consist of three vectors. Consider the basis $\{e_k\}$ consisting of the vectors whose components are given by $(e_k)^j = \delta^j_k$ in coordinates the first of which refers to S and the other two to $\mathbb{R}^2$. Now write Eq. (4) in the form

$$X_w(g) = T_e L_g(e, w) = (L_g e, DL_g e \cdot w) = (g, DL_g e \cdot w),$$

and apply this to $g = g(\theta, \mathbf{a})$. From Eq. (7) and the expression for R_θ one obtains

$$DL_{\theta,a}g(\varphi, \mathbf{b}) = \begin{vmatrix} 1 & 0 & 0 \\ 0 & \cos\theta & -\sin\theta \\ 0 & \sin\theta & \cos\theta \end{vmatrix},$$

and then

$$\begin{aligned} X_1(\theta, \mathbf{a}) &= \partial/\partial\theta, \\ X_2(\theta, \mathbf{a}) &= \cos\theta\,\partial/\partial a_1 + \sin\theta\,\partial/\partial a_2, \\ X_3(\theta, \mathbf{a}) &= -\sin\theta\,\partial/\partial a_1 + \cos\theta\,\partial/\partial a_2. \end{aligned} \tag{8}$$

The commutation rules are easily obtained:

$$[X_1, X_2] = X_3, \quad [X_2, X_3] = 0, \quad [X_3, X_1] = X_2.$$

Compare Eq. (11) of Chapter 21.

(b) The unitary unimodular group $SU(2)$ in two complex dimensions

The group $U(2)$ can be defined as the set of linear transformations on the complex plane $\mathbb{C}^2$ which leaves $|z|^2 \equiv z_1 z_2^* + z_1^* z_2$ invariant. The elements of this group are the complex *unitary* matrices σ, that is, two-by-two matrices which satisfy the condition $\sigma^+\sigma = 1$, where σ^+ is the Hermitian conjugate of σ. The condition of unitarity immediately implies that $|\det\sigma| = 1$. The *unimodular* subgroup $SU(2)$ of this group, on which we shall concentrate, is the subset of those elements for which $\det\sigma = 1$. It follows that $\sigma \in SU(2)$ can be written in the form

$$\sigma = \begin{vmatrix} u & -v \\ v^* & u^* \end{vmatrix} \tag{9}$$

with the condition $|u|^2 + |v|^2 = 1$. If one writes $u = \alpha_0 - i\alpha_3$, $v = -\alpha_2 + i\alpha_1$, $\alpha_k \in \mathbb{R}$ (the signs are chosen for convenience), this condition becomes

$$(\alpha_0)^2 + (\alpha_1)^2 + (\alpha_2)^2 + (\alpha_3)^2 \equiv (\alpha_0)^2 + \boldsymbol{\alpha}^2 = 1, \tag{10}$$

where $\boldsymbol{\alpha}$ is the three-vector with components $(\alpha_1, \alpha_2, \alpha_3)$. One thus sees that the manifold of $SU(2)$ is diffeomorphic to the three-sphere S^3, connected and compact.

The four real numbers $(\alpha_0, \boldsymbol{\alpha})$ are called the Cayley–Klein parameters for the group $SU(2)$ (Smirnov, 1961, p. 305). The multiplication law of the abstract group can be worked out in terms of these parameters and is found to be (Saletan and Cromer, 1971)

$$g(\alpha_0, \boldsymbol{\alpha})g(\beta_0, \boldsymbol{\beta}) = g(\alpha_0\beta_0 - \boldsymbol{\alpha}\cdot\boldsymbol{\beta}, \alpha_0\boldsymbol{\beta} + \beta_0\boldsymbol{\alpha} + \boldsymbol{\alpha}\times\boldsymbol{\beta}), \tag{11}$$

where $\boldsymbol{\alpha}\times\boldsymbol{\beta}$ represents the ordinary vector cross-product. It should be borne in mind that the Cayley–Klein parameters are not independent, but are connected by Eq. (10). It could not be otherwise, for the manifold, diffeomorphic to the three-sphere, is of dimension three and can support only three independent parameters in each neighborhood. Eq. (11) can now be used to find a set of left-invariant vector fields in the same way that Eq. (7) was used to obtain the vector fields of (8), but although the manifold is of dimension three, four vector fields will be obtained in this way, one corresponding to each of the parameters. We omit many of the details of the calculation. The matrix which gives the action of DL_g is

$$\begin{vmatrix} \alpha_0 & -\alpha_1 & -\alpha_2 & -\alpha_3 \\ \alpha_1 & \alpha_0 & -\alpha_3 & \alpha_2 \\ \alpha_2 & \alpha_3 & \alpha_0 & -\alpha_1 \\ \alpha_3 & -\alpha_2 & \alpha_1 & \alpha_0 \end{vmatrix}$$

If this matrix is called M, its components $M_{\mu\nu}$ are given by

$$\begin{aligned} M_{k0} &= -M_{0k} = \alpha_k, \\ M_{jk} &= \varepsilon_{jkh}\alpha_h, \quad j \neq k, \\ M_{\mu\mu} &= \alpha_0, \end{aligned} \tag{12}$$

where Greek indices run from 0 to 3 and Latin ones from 1 to 3. The left-invariant vector fields are then given by $X_\mu = M_{\nu\mu}\partial_\nu$, where $\partial_\nu \equiv \partial/\partial\alpha_\nu$. One then finds that

$$\begin{aligned} X_0 &= \alpha_0\partial_0 + \boldsymbol{\alpha}\cdot\boldsymbol{\partial} \equiv 0, \\ X_j &= \alpha_0\partial_j - \alpha_j\partial_0 - [\boldsymbol{\alpha} \times \boldsymbol{\partial}]_j. \end{aligned} \tag{13}$$

That X_0 is identically equal to zero on the group manifold follows from Eq. (10). The commutation relations are

$$[X_j, X_k] = -2\varepsilon_{jkh}X_h. \tag{14}$$

It is easily shown that the Lie algebra of these vector fields is what was called $so(3)$ and $su(2,\mathbb{C})$ in Chapter 21.

There are other parameters, other coordinates, in which to express the elements of $SU(2)$ and the vector fields. The usual ones, having to do with the relation of $SU(2)$ to the rotation group $SO(3)$ in three dimensions, are the so-called Euler angles. We shall not go into all that here, except to mention that there is no one-chart atlas for the manifold.

23.2 FOLIATIONS AND SUBGROUPS OF $GL(n, \mathbb{R})$

Both examples at the end of the preceding subsection were groups of linear transformations. There is a theorem (the Ado–Isawara theorem) which states that every Lie group G can be realized as a group of linear transformations on an $\mathbb{R}^n$ for suitable n. (In this and some of the subsequent sections we shall use the traditional n, rather than ν, for the dimension of the space involved.) Thus a discussion of linear transformation groups is more general than it may at first seem, many of the results obtained in such a discussion being applicable to all Lie groups. We therefore wish to describe in some detail some important groups of linear transformations, their manifold structure and the left-invariant vector fields and forms on them.

1. The largest linear transformation group on $\mathbb{R}^n$ is called the *general linear group* $GL(n, \mathbb{R})$, given by

$$GL(n, \mathbb{R}) = \{A \in \mathrm{Lin}(\mathbb{R}^n, \mathbb{R}^n) \vdash \det A \neq 0\}. \tag{15}$$

If it is viewed as a set of matrices, the group operation is matrix multiplication, and the matrix elements form an open submanifold of $\mathbb{R}^{n^2}$. The dimension of the group is thus n^2.

To find the left-invariant vector fields, choose a basis in T_eG, much as was done in the two examples of §23.1(3). Let such a basis $\{e_{wt}\}$ consist of the matrices all of whose elements are zero except the element in the (w, t) position, which is equal to one, and let $X_{wt}(A) = T_eL_A(e_{wt})$ define the left-invariant vector field obtained in the usual way (according to Eq. (4)) from e_{wt}, $w, t \in \{1, \ldots, n\}$. Then a calculation similar to the one performed on the previous examples yields

$$X_{wt}(A) = a_{kw} \frac{\partial}{\partial a_{kt}}, \tag{16}$$

where the a_{wt} are the matrix elements of A. The bracket in the Lie algebra $\mathfrak{G}$ of $G = GL(n, \mathbb{R})$ is then easily found to be given by

$$[X_{wt}, X_{uv}] = \delta_{ut} X_{wv} - \delta_{wv} X_{ut}. \tag{17}$$

Now one can write down the one-forms θ^{wt} (we are not concerned at this point whether the indices appear as subscripts or superscripts, but we continue to sum on repeated indices) dual to these left-invariant vector fields, which are therefore themselves left-invariant. The duality is expressed by

$$\theta^{uv} X_{wt} = \delta^u_w \delta^v_t.$$

It is immediately obvious then that if A^{-1} is the inverse of A,

$$\theta^{wt} = (A^{-1})_{wk} da_{kt} \equiv (A^{-1} dA)_{wt}. \tag{18}$$

2. We now turn to the relation between various foliations of $GL(n, \mathbb{R})$ and its subgroups. Recall that regular foliations of a manifold M can be generated by submersions $\varphi: M \to N$, and in particular by submersion via a function $f: M \to \mathbb{R}$.

A function which by definition has no zeros on $GL(n, \mathbb{R})$ is

$$\det: GL(n, \mathbb{R}) \to \mathbb{R}: A \mapsto \det A. \tag{19}$$

We have already calculated its exterior derivative near the end of §19.1(4):

$$d(\det A) = A_{wt} da_{wt}, \tag{20}$$

where A_{wt} is the minor of a_{wt} in A. It was mentioned there that $d(\det)$ vanishes on a subset of the set where det vanishes, so it is clear that the exact one-form $d(\det)$ fails to vanish on $GL(n, \mathbb{R})$. This implies that det is indeed a submersion, and it generates a regular foliation of $GL(n, \mathbb{R})$ whose diffeomorphic leaves are of dimension $n^2 - 1$, each leaf corresponding to a value of the function det. These leaves are regular submanifolds of $GL(n, \mathbb{R})$: the topology induced by their differentiable structures is the subset topology, and the group operation therefore remains C^∞ on the leaves. The product of two elements on a given leaf, however, does not in general lie on that leaf, for the determinant of the product is the product of the determinants.

An exception is the leaf belonging to $\det A = 1$, which therefore inherits from $GL(n, \mathbb{R})$ an obvious Lie group structure. The resulting Lie group, a subgroup of $GL(n, \mathbb{R})$ whose dimension is $n^2 - 1$, is called the *special linear group* $SL(n, \mathbb{R})$. It will be seen eventually that the fact that $SL(n, \mathbb{R})$ is a Lie subgroup of $GL(n, \mathbb{R})$ is related to the fact that each leaf of the foliation is closed in the $GL(n, \mathbb{R})$ topology.

The left-invariant vector fields of Eq. (16) are projectable with respect to the foliation generated by det:

$$[L_{X_{rt}}\det](A) \equiv [i_{X_{rt}}d(\det)](A) = a_{wr}A_{wt} = \delta_{rt}\det A,$$

so that det is conformally invariant under each of the X_{wt}. Note that most of the left-invariant vector fields project onto the null vector field on the quotient $\mathbb{R} - \{0\}$, but that the diagonal ones project down to (det) $(\partial/\partial\det)$.

3. Another foliation of $GL(n, \mathbb{R})$ is obtained from the mapping

$$\varphi: GL(n, \mathbb{R}) \to \mathrm{Lin}(\mathbb{R}^n, \mathbb{R}^n): A \mapsto AA^{\mathrm{T}} - \mathbf{1}, \tag{21}$$

where A^{T} is the transpose of A and $\mathbf{1}$ is the n-by-n unit matrix. As $\varphi(A)$ is symmetric (equal to its transpose), there are at most $\frac{1}{2}n(n+1)$ independent functions in this map, functions which may be specified by

$$\varphi_{wt}: GL(n, \mathbb{R}) \to \mathbb{R}: A \mapsto (AA^{\mathrm{T}} - \mathbf{1})_{wt} = A_{wu}A_{tu} - \delta_{wt},$$

$w \leqslant t$. It can be shown that all of these functions are independent; we omit the proof. Then an argument similar to the one used in §(2) for $SL(n, \mathbb{R})$ will show that the φ_{wt} are submersions and that (21) therefore generates a regular foliation of $GL(n, \mathbb{R})$ whose leaves are (regular) submanifolds of dimension $\frac{1}{2}n(n-1)$.

This time it is the leaf corresponding to $\varphi_{wt} = 0$ for all $w \leqslant t$, $w, t \in \{1, \ldots, n\}$, which inherits the group structure, for this is the leaf which contains the unit matrix $\mathbf{1}$. The resulting Lie subgroup of $GL(n, \mathbb{R})$ is called the *orthogonal group* $O(n, \mathbb{R})$. It is easily seen that the orthogonal group is compact and closed in the topology of $GL(n, \mathbb{R})$ (one may start with the observation that if $A \in O(n, \mathbb{R})$, then $\det A = \pm 1$).

4. The subgroups of $GL(n, \mathbb{R})$ we are discussing are called *algebraic* subgroups: they are given by means of algebraic relations. We shall now describe another such algebraic subgroup, this time of $GL(2n, \mathbb{R})$, the *linear symplectic group in* $2n$ *dimensions* $Sp(2n, \mathbb{R})$. Note that the dimension must now be even. Consider the mapping $\psi: GL(2n, \mathbb{R}) \to \mathrm{Lin}(\mathbb{R}^{2n}, \mathbb{R}^{2n})$ defined by

$$\psi: A \mapsto A^{\mathrm{T}}JA - J, \tag{22}$$

where J is the $2n$-by-$2n$ antisymmetric ($J^{\mathrm{T}} = -J$) matrix

$$J = \begin{vmatrix} 0 & -\mathbf{1}_n \\ \mathbf{1}_n & 0 \end{vmatrix}, \tag{23}$$

in which $\mathbf{1}_n$ is the unit matrix in n dimensions. It is easily checked that $\psi(A)$ is antisymmetric and therefore that ψ contains at most $2n^2 - n$ independent functions $\psi_{wt}: GL(2n, \mathbb{R}) \to \mathbb{R}$. Again, as in the example of §(3), it can be shown that these are indeed $2n^2 - n$ independent submersions, so that (22) generates a foliation whose leaves are of dimension $2n^2 + n$. The leaf $\psi^{-1}(0)$ which contains the unit matrix on $\mathbb{R}^{2n}$ is a closed noncompact Lie subgroup of $GL(2n, \mathbb{R})$: this is $Sp(2n, \mathbb{R})$.

Remarks. 1. The symplectic group discussed at p. 167 of AM is a generalization of $Sp(2n, \mathbb{R})$ to nonlinear transformations and to replacing J by a general symplectic form ω.

2. Other subgroups of the general linear group could be defined by generalizing the above constructions to mappings that leave some suitable tensor τ invariant in the sense that, in matrix notation, $A^{\mathrm{T}}\tau A = \tau$.

23.3 LIE GROUPS AND LIE ALGEBRAS

1. The Lie algebra $\mathfrak{G}$ of Eq. (6) is called the Lie algebra *of* the group G from which it was constructed. Thus each Lie group has its unique Lie algebra.

Since we have started with Lie algebras rather than Lie groups, it would be useful for us to invert the construction of §23.1(2) in order to obtain a Lie group from a given Lie algebra. In general there is more than one Lie group for each Lie algebra $\mathfrak{G}$, but every $\mathfrak{G}$ is the Lie algebra of a connected, simply connected Lie group G which is unique up to Lie group isomorphisms (Robin, 1974, App. C, Th. 6). This theorem is a consequence of Ado's theorem (Jacobson, 1962, p. 202) according to which every finite-dimensional Lie algebra has a faithful finite-dimensional representation. Although it is not the whole story, the simple connectedness referred to in the theorem is important. For example, the Abelian groups $\mathbb{R}$ and S have the same Lie algebras, but there is no nontrivial Lie group homomorphism $S \to \mathbb{R}$.

Roughly speaking, the attempt to get to the group from the algebra involves identifying $\mathfrak{G}$ with T_eG and then associating each element of T_eG with a left-invariant vector field in $\mathfrak{X}(G)$. The map that is used to try to reconstruct the group in this way is called the *exponential map*. Let $w \in T_eG$ and consider the integral curve c_w of X_w such that $c_w(0) = e$. It can be shown that $c_w(t)c_w(s) = c_w(t+s)$, which means that c_w is (isomorphic to) the additive group $\mathbb{R}$; it is called a one-parameter subgroup of G and is written $c_w(t) = \exp(tw)$. The exponential map is then defined by

$$\exp: T_eG \to G: w \mapsto \exp(w). \tag{24}$$

It is clear that if one tries to reconstruct the group from the algebra by identifying $\mathfrak{G}$ with T_eG and then using the exponential map, one cannot reach more than the connected component of the identity in G, but there are examples (see, e.g., AM, Example 4.1.9(c), p. 257) which show that even in relatively straightforward cases there are points in the connected component of the identity which cannot be reached in this way. (It turns out that the entire connected component of the identity can be reconstructed in this way if G is compact.)

An important property of the exponential map is that if $\varphi:G_1 \to G_2$ is a Lie group homomorphism, the following diagram commutes:

$$\begin{array}{ccc} T_eG_1 & \xrightarrow{T\varphi} & T_eG_2 \\ \downarrow{\scriptstyle \exp_1} & & \downarrow{\scriptstyle \exp_2} \\ G_1 & \xrightarrow[\varphi]{} & G_2 \end{array}$$

where $\exp_k$ is the exponential map from $\mathfrak{G}_k$ to G_k.

2. From the construction in §23.1(2) of $\mathfrak{G}$ from G it is evident that the left-invariant vector fields provide an action of the abstract Lie algebra $\mathfrak{G}$ on G. (Incidentally, the same can be said of the right-invariant vector fields as for the left-invariant ones, but the action is in general different.) This action has several interesting properties. First, the vector fields are independent (the action is free). Second, the vector fields are complete, which implies that their flows provide γ families of diffeomorphisms of G to G (recall that $\gamma = \dim G$). It will be seen in the next section that this family of flows is a *Lie group action*, in this case of G on itself. We wish to give a brief description of this Lie group action, which turns out to be right translation, and of its relation to the left-invariant vector fields.

Let $c(t)$ be a curve on G. It can be used together with right translation to obtain a family of diffeomorphisms on G according to

$$R_{c(t)}g \equiv gc(t) \equiv g'(t), \quad t\in\mathbb{R}, \quad g\in G.$$

The infinitesimal generator $X\in\mathfrak{X}(G)$ of this family of diffeomorphisms is given by

$$X(g'(t)) = dg'(t)/dt = gdc(t)/dt = gc(t)c^{-1}(t)dc(t)/dt = g'(t)c^{-1}(t)dc(t)/dt. \quad (25)$$

Now apply a left translation by $h\in G$ to this vector field: $(L_h)_*X$ is obtained by applying L_h to all of the group elements in (25). Then $c^{-1}dc/dt$ becomes

$$(hc)^{-1}d(hc)/dt = c^{-1}h^{-1}hdc/dt = c^{-1}dc/dt,$$

so that $(L_h)_*X(g) = X(L_hg)$, or X is left-invariant. Therefore the infinitesimal generators of the right translations are the left-invariant vector fields. The opposite result can be shown similarly: the infinitesimal generators of the left translations are the right-invariant vector fields.

We present an example of this relation between left-invariant vector fields and right translations on $GL(2,\mathbb{R})$. Consider the left-invariant vector field

$$X = a_{11}\frac{\partial}{\partial a_{11}} + a_{21}\frac{\partial}{\partial a_{21}}.$$

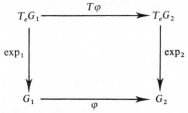

A glance at Eq. (16) shows that this is the left-invariant vector field which is denoted there by X_{11}. Its integral curves are given by

$$A(t) \equiv \begin{vmatrix} a_{11} & a_{12} \\ a_{21} & a_{22} \end{vmatrix} = \begin{vmatrix} c_{11}e^t & c_{12} \\ c_{21}e^t & c_{22} \end{vmatrix},$$

where the c_{jk} are constants. This can also be written in the form

$$A(t) = \begin{vmatrix} c_{11} & c_{12} \\ c_{21} & c_{22} \end{vmatrix} \cdot \begin{vmatrix} e^t & 0 \\ 0 & 1 \end{vmatrix}.$$

In this expression the first matrix is a constant matrix, a fixed element of $GL(n, \mathbb{R})$, and the second is a curve, also in $GL(n, \mathbb{R})$. Clearly this is precisely of the form described.

3. In §23.1(3) it was shown that the Lie algebras of E_2 and $SU(2)$ are $e(2)$ and $su(2)$. Eq. (17) defines the Lie algebra $gl(n,\mathbb{R})$ of $GL(n,\mathbb{R})$. Recall that the X_{wt} are obtained from the e_{wt} matrices defined above Eq. (16). A brief calculation will verify that the e_{wt} satisfy the same Eq. (17) as do the X_{wt}, but with the bracket now standing for the commutator bracket of the matrices. A similar result obtains for the Lie subgroups of $GL(n, \mathbb{R})$ described in §23.2, with the commutator bracket representing the bracket of the Lie algebra and appropriate matrices representing its elements.

In particular, the Lie algebra of $SL(n, \mathbb{R})$ may be characterized by

$$sl(n, \mathbb{R}) = \{A \in \mathrm{Lin}(\mathbb{R}^n, \mathbb{R}^n) \vdash \mathrm{tr}(A) = 0\}.$$

The Lie algebra of $O(n, \mathbb{R})$ may be characterized by

$$o(n, \mathbb{R}) = \{A \in \mathrm{Lin}\,(\mathbb{R}^n, \mathbb{R}^n) \vdash A^{\mathrm{T}} + A = 0\},$$

consisting of the antisymmetric matrices.

The Lie algebra of $Sp(2n, \mathbb{R})$ may be characterized by

$$sp(2n, \mathbb{R}) = \{A \in \mathrm{Lin}\,(\mathbb{R}^{2n}, \mathbb{R}^{2n}) \vdash A^{\mathrm{T}}J + JA = 0\}.$$

4. As a last remark concerning the relation between Lie groups and their algebras, we mention the following theorem (see for instance Hausner and Schwartz (1968, pp. 72ff.)). Let $\mathfrak{G}$ be the Lie algebra of the Lie group G and let $\mathfrak{H}$ be a subalgebra of $\mathfrak{G}$. Then there is a unique connected Lie subgroup H of G whose Lie algebra is $\mathfrak{H}$.

Of more interest to us, however, than the relation between abstract Lie groups and algebras is the relation between their actions on manifolds.

23.4 ACTIONS OF LIE GROUPS ON MANIFOLDS

1. The theory of Lie groups has its origins in what would now be called the theory of representations, for it started with transformations on linear spaces, now often thought of as representations of abstract groups. Actions of abstract groups on

manifolds are a generalization, and they are related to representations in the same way as the actions of Lie algebras are related to their representations.

An *action* Ψ of a Lie group G on a manifold M is a C^∞ map

$$\Psi{:}G\times M\to M{:}(g,m)\mapsto\Psi_g(m),\quad \Psi_g\in\mathrm{Diff}(M), \tag{26}$$

such that

$$\tilde{\Psi}{:}G\to\mathrm{Diff}(M){:}g\mapsto\Psi_g \tag{27}$$

is a group homomorphism in the sense that $\Psi_g\circ\Psi_h=\Psi_{gh}$.

We give two examples of actions of G on itself. First, one that we have all but seen, defined through left translation as

$$L{:}G\times G\to G{:}(g,h)\mapsto gh,$$

$g,h\in G$, so that $L_g=g\cdot\in\mathrm{Diff}(G)$ (the center dot stands here for the group operation). Compare the description of right translation in §23.3(2). This is, of course, the same as the definition of L_g at Eq. (1). Second, an action can be defined through a combination of left and right translations. This action is specified by

$$I{:}G\times G\to G{:}(g,h)\mapsto ghg^{-1}. \tag{28}$$

This action leaves the identity element of G unchanged: $I(g,e)=e\ \forall\ g\in G$. Thus $I_g(e)=e$, so that T_eI_g maps T_eG into T_eG. Because T_eG can be identified with the Lie algebra $\mathfrak{G}$ of G, this defines a map, called the *adjoint mapping* of $\mathfrak{G}$ into $\mathfrak{G}$ by $Ad_g=T_eI_g{:}\mathfrak{G}\to\mathfrak{G}$. In an almost obvious way this gives rise to a new action, this time of G on $\mathfrak{G}$, called the *adjoint action*:

$$Ad{:}G\times\mathfrak{G}\to\mathfrak{G}{:}(g,a)\mapsto Ad_g(a). \tag{29}$$

The adjoint action has several interesting and useful properties. Because it is an action of G and hence a Lie group homomorphism, one obtains

$$Ad_{gh}(a)=Ad_g\circ Ad_h(a)=Ad_g(Ad_h(a)).$$

It is easily checked that it also provides a Lie algebra homomorphism, namely

$$Ad_g([a,b])=[Ad_g(a),Ad_g(b)].$$

The adjoint mapping can be used also to define the *coadjoint mapping* of G on the dual $\mathfrak{G}^*$ of $\mathfrak{G}$. This is specified by

$$Ad_g^*{:}\mathfrak{G}^*\to\mathfrak{G}^*{:}\alpha\mapsto Ad_g^*(\alpha)\ \forall\ \alpha\in\mathfrak{G}^*,$$

where $Ad_g^*(\alpha)$ is defined by

$$\langle Ad_g^*(\alpha),a\rangle=\langle\alpha,Ad_g(a)\rangle. \tag{30}$$

By extension one gets the *coadjoint action of* G on $\mathfrak{G}^*$:

$$Ad^*{:}G\times\mathfrak{G}^*\to\mathfrak{G}^*{:}(g,\alpha)\mapsto Ad_g^*(\alpha),$$

and it can be shown that

$$Ad_g^*([\alpha,\beta])=-[Ad_g^*(\alpha),Ad_g^*(\beta)].$$

Both $\mathfrak{G}$ and $\mathfrak{G}^*$ are vector spaces, and it is easily demonstrated that the adjoint and coadjoint actions of G are linear; they are therefore representations. It is found that there is a change in order in the coadjoint of representation, as there is a negative sign in the last equation (and in Eq. (26) of Chapter 21):

$$Ad^*_{gh} = Ad^*_h \circ Ad^*_g. \tag{31}$$

2. Like Lie algebra actions, Lie group actions have fixed points, orbits, etc.

The *orbit* of $m \in M$ under the action Ψ is denoted $G \cdot m$ and defined by $G \cdot m = \{\Psi_g(m) | g \in G\}$. It is the set of points in M which can be reached from m by the action Ψ of G, and in spite of the notation it depends on Ψ as well as on G. The orbit of a Lie group action is easier to define than that of a Lie algebra action because the map in the group action is into M itself, rather than into TM. The quotient M/G of the action (which also depends on Ψ, though the notation does not reflect it) is the set of equivalence classes for the equivalence relationship defined through the orbits: $m_1, m_2 \in M$ are in the same equivalence class if $m_1, m_2 \in G \cdot m$ for some $m \in M$.

A *fixed point* $m \in M$ under the action is a point whose orbit is m itself, i.e. such that $\Psi_g(m) = m \; \forall \; g \in G$.

The *isotropy group* of Ψ at m, denoted G_m, is defined by $G_m = \{g \in G \vdash \Psi_g(m) = m\}$. It can be verified that G_m is a closed subgroup and therefore a submanifold of G. (Some authors call G_m the *little group* of m.) The isotropy group at a fixed point is all of G itself.

We now turn to describing several types of actions.

An action Ψ is *effective* (also called faithful or essential) if $\Psi_g = \mathbf{1}$ implies that $g = e$. What this means is that $\tilde{\Psi}$, defined in Eq. (27), is a *monomorphism*, an isomorphism on the image. An example is the *natural action* of $O(3, \mathbb{R})$ (or, as we shall often write, of $O(3)$) on $\mathbb{R}^3$ as defined by the matrices applied to the vectors of $\mathbb{R}^3$. There may be fixed points of an effective action (the origin in this example), but there are no elements $g \in G$ other than e which belong to all isotropy groups.

An action is *free* if $G_m = e$ for all $m \in M$. This means that Ψ_g moves evey point m unless $g = e$. Examples are (a) the action of $\mathbb{R}$ on itself defined by $\Psi: \mathbb{R} \times \mathbb{R} \to \mathbb{R}: (x, y) \mapsto (x + y)$, and (b) the rotations of $\mathbb{R}^2 - \{0\}$. A free action is always effective, but an effective action need not be free. For example the rotations of $\mathbb{R}^3 - \{0\}$ are an effective action, but the isotropy group of any point in $\mathbb{R}^3$ consists of all rotations whose axes pass through that point.

An action is *transitive* if $G \cdot m = M \; \forall \; m \in M$ (actually it is sufficient that $G \cdot m$ be M for just one point m, and the rest follows from the group operation). Then there is only one orbit in M, and that orbit is all of M itself, which implies that given any two points $m, m' \in M$, there is at least one $g \in G$ such that $\Psi_g(m) = m'$. In general there is more than one $g \in G$ that will do, but if there is only one for each pair m, m', the action is sometimes called *simply transitive*.

An action is *proper* (Boothby, 1975, p. 81) if

$$\Psi': G \times M \to M \times M: (g, m) \mapsto (m, \Psi_g(m))$$

is such that the inverse image of every compact set is itself compact. This seemingly very technical property plays an important part later, in discussions of orbit spaces and quotient spaces.

Consider as an example irrational flow on the torus T^2, i.e. the flow of a vector field whose integral curves are irrational winding lines. The vector field may be written in the form

$$X = r\partial/\partial\theta_1 + s\partial/\partial\theta_2,$$

$r, s \in \mathbb{R}$, with r/s irrational. The group whose action on the torus will be described is the additive group $\mathbb{R}$ with $\Psi_t(m) = c_m(t)$, $t \in \mathbb{R}$, $m \in T^2$, where c_m is the integral curve of X for which $c_m(0) = m$. Then Ψ', defined above, is given by

$$\Psi' : \mathbb{R} \times T^2 \to T^2 \times T^2 : (t, m) \mapsto (m, \Psi_t(m)).$$

Now let C be a patch on the torus, the closure of something like U in Fig. 8.2, and consider the compact set $K = m \times C$ in $T^2 \times T^2$, where m is an arbitrary point on the torus. The inverse image of K under Ψ' is not compact. Indeed,

$$\Psi'^{-1}(K) = \bigcup_{k=-\infty}^{\infty} I_k \times m,$$

where each $I_k \in \mathbb{R}$ represents an interval of time during which c_m remains in C. This inverse image is not compact, and hence the action is not proper.

A similar case is that of the harmonic oscillator in one degree of freedom. Again one may choose K to be the Kronecker product of a point and a patch on $M = \mathbb{R}^2 - \{0\}$, and for the action obtained in the same way, but from the dynamical vector field of the oscillator instead of the vector field X on T^2, one obtains a similar expression for $\Psi'^{-1}(K)$. Now, however, the intervals I_k are all equal and spaced periodically on $\mathbb{R}$, and the action can be viewed as the action of S, rather than of $\mathbb{R}$, on M. This new action is proper.

3. The difference between the harmonic oscillator and the flow just discussed torus is related to a theorem concerning free and proper actions and the nature of the quotient M/G and of the projection $\pi : M \to M/G$. Before listing some such theorems, we return to a remark made in §23.1(1).

If H is a subgroup of G and is closed in the subset topology, then it is a Lie subgroup of G and is a regular submanifold, i.e. the topology induced by its differentiable structure coincides with the subset topology. That it is a Lie subgroup of G is proved (see, e.g., Hausner and Schwartz (1968)) by showing that its Lie algebra $\mathfrak{H}$ is a subalgebra of $\mathfrak{G}$, the Lie algebra of G. The second part of the assertion states that the natural injection is a regular imbedding. A sketch of the proof may be given as follows. The left-invariant vector fields which make up $\mathfrak{H}$ foliate G into diffeomorphic leaves of dimension equal to $\dim \mathfrak{H} \equiv \dim H$ (diffeomorphic because left translation takes leaves into leaves). The leaf passing through the identity e is of course H itself, and therefore all leaves of this foliation are diffeomorphic to H. Let G/H be the quotient with respect to this foliation (it is not in general a group unless H is a normal subgroup). If H is a regular submanifold of G, then the topology of G/H is the quotient topology:

the projection $G \to G/H$ is smooth and a point in G/H must be closed. This implies that H is closed, and conversely.

There are several theorems concerning the action of Lie groups and their closed subgroups on manifolds. We shall not prove them, but list some here without proof. Their proofs may be found in AM (pp. 261ff.), and in Brickell and Clark (1970, around p. 243). We shall call on them and discuss them if they become necessary in the later exposition.

Let H be a closed subgroup of G (and it is then a Lie subgroup), and define an action of H on G by

$$\Psi: H \times G \to G: (h, g) \mapsto hg,$$

where $h \in H$, $g \in G$. Then G/H is a Hausdorff manifold and $\pi: G \to G/H$ is a submersion.

If Ψ is a proper action of G on M, then $G \cdot m$ is a closed submanifold of M.

If Ψ is a proper and free action of G on M, then M/G is a manifold, not necessarily Hausdorff, and $\pi: M \to M/G$ is a submersion.

Let Ψ be an action of G on M and R be the subset of $M \times M$ defined by $R = \{(m, \Psi(m)) | g \in G,\ m \in M\}$. Then R is a closed submanifold of $M \times M$ iff $\pi: M \to M/G$ is a submersion. (If Ψ is free and proper, then R is closed.)

Remark. A principal G-bundle is a manifold M on which a group G acts freely and such that $\pi: M \to M/G$ is a submersion. Thus the conditions of this last theorem are fulfilled in principal G-bundles. This is of some physical interest because gauge theories involve 'ambient spaces' which are in fact principal G-bundles.

23.5 RELATION BETWEEN THE ACTIONS OF LIE GROUPS AND THE ACTIONS OF LIE ALGEBRAS

1. It has been seen that every Lie group G has its Lie agebra $\mathfrak{G}$. Both Lie groups and Lie algebras have actions on manifolds. Is there some way to transport an action of G on M to an action of $\mathfrak{G}$ on M?

We may start by paraphrasing AM. An action Ψ of G on M may be thought of as a homomorphism of G to Diff(M), and one can look for the induced homomorphism of Lie algebras, that is of $\mathfrak{G}$ to $\mathfrak{X}(M)$. This is really only a qualitative analogy, however, for the group Diff(M) of all diffeomorphisms of M is (Helgason, 1962, p. 87) too big to form a Lie group in any reasonable topology.

The action Ψ of G on M is transported to the *associated action* ψ of $\mathfrak{G}$ on M through the exponential map of Eq. (24). The *infinitesimal generators* of Ψ are defined by

$$X_a(m) = \frac{d}{dt} \Psi_{\exp(ta)}(m)|_{t=0}\ \forall\ a \in \mathfrak{G}, \quad m \in M. \tag{32}$$

The vector field $X_a \in \mathfrak{X}(M)$ so defined is complete, and the map $a \mapsto X_a$ provides the associated action by

$$\psi:\mathfrak{G} \times M \to TM:(a,m) \mapsto X_a(m).$$

If a is e_j, an element of a basis in $\mathfrak{G}$, the vector field so obtained may be called X_j; according to (24) and the discussion above it this is the left-invariant vector field whose integral curve passing through the identity is just $\exp(te_j)$. Then in the identification of $\mathfrak{G}$ with T_eG, the basis element e_j is identified with $X_j(e)$. This demonstrates that ψ is indeed an action of $\mathfrak{G}$ on M. An equivalent definition is given by

$$X_j(m) = T_e\Psi_m(e_j), \tag{33}$$

where e_j is again an element of $\mathfrak{G}$ in the identification of $\mathfrak{G}$ with T_eG. We shall not go into a formal proof that this is indeed a Lie algebra action.

The properties of Ψ and ψ are not always related in a simple way. We mention two theorems (Palais, 1957). The first states that if G is simply connected and M is compact, then ψ is effective iff Ψ is effective. The second is the following. Let M and N be manifolds and let $\chi:M \to N$, be equivariant with respect to the two Lie group actions $\Phi:G \times M \to M$ and $\Psi:G \times N \to N$. Then χ is equivariant also with respect to the associated Lie algebra actions φ and ψ. In terms of diagrams, commutation of

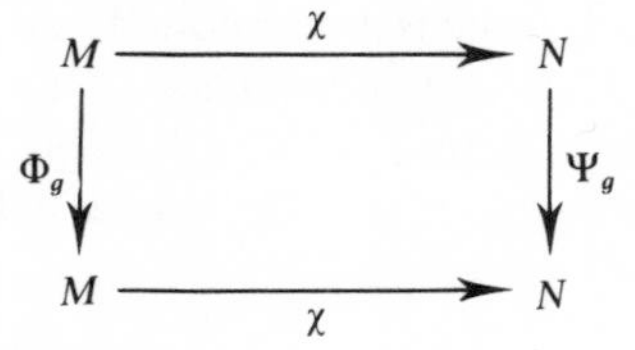

for all $g \in G$ implies commutation of

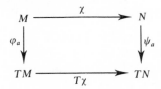

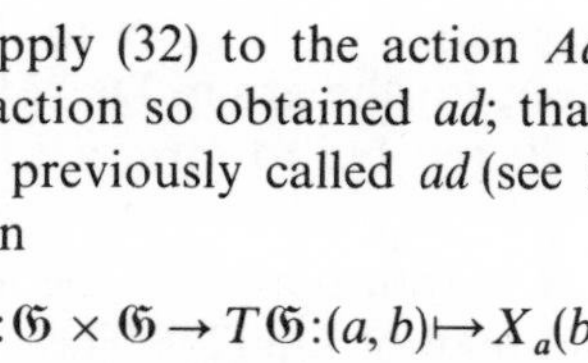

for all $a \in \mathfrak{G}$.

As an example, let us apply (32) to the action Ad defined at Eq. (29). We shall call the Lie algebra action so obtained ad; that it is the same action as the adjoint representation previously called ad (see Eq. (20) of Chapter 21) is demonstrated by AM. Then

$$ad:\mathfrak{G} \times \mathfrak{G} \to T\mathfrak{G}:(a,b) \mapsto X_a(b), \tag{34}$$

where $X_a(b) = (d/dt)Ad_{\exp(ta)}(b)|_{t=0} \in \mathfrak{X}(\mathfrak{G})$. In a similar way, the coadjoint representation of $\mathfrak{G}$ on $\mathfrak{G}^*$ is the Lie algebra action associated with Ad^*, defined around Eq. (30).

Other examples of associated actions will be given in the discussion of some Lie groups which are of particular physical interest.

2. Suppose now that what is given is an action ψ of $\mathfrak{G}$ on M, where $\mathfrak{G}$ is the Lie algebra of a known Lie group G, but that ψ has not been obtained by applying (32) to some action Ψ of G on M. Is it possible to find an action Ψ of G on M such that ψ is its associated action (to *integrate* ψ)? This is a question of particular interest to us, for distributions, foliations, and other geometric objects on the carrier space often lead to Lie algebras of vector fields, which are then actions of Lie algebras. It can be extremely useful to obtain from these the actions of global Lie groups on the manifold. The reader is referred to Palais (1957), especially to Chapters II (Local and Infinitesimal Transformation Groups) and III (Globalizable Infinitesimal Transformation Groups).

For the general question as posed here, with no qualifications concerning the type of action, the topology of G or of M, etc., the answer is negative: in general the action of a Lie algebra cannot be integrated to the action of a global Lie group. For instance, think of an action of $\mathfrak{G}$ which contains incomplete vector fields. This could not be obtained from an action of G by applying (32), for that produces only complete vector fields. (The flow box of the vector fields in such an action would, however, define a *local Lie group*.)

An example (Boothby, 1975, p. 126) will show what kind of difficulties can arise. Let $\mathfrak{G}$ be the one-dimensional Lie algebra, and let ψ be the action of $\mathfrak{G}$ on $M = \mathbb{R}^2 - \{0\}$ for which $\psi(a) = (a\partial/\partial x, 0)\ \forall\ a \in \mathfrak{G}$. Now, $\mathfrak{G}$ is the Lie algebra of the additive Lie group $\mathbb{R}$ and if the origin had not been removed, the Lie group action of $\mathbb{R}$ on $\mathbb{R}^2$ whose associated action is ψ would be given, as is easily verified, by

$$\Psi_r(x, y) = (x + r, y),$$

where $r \in G$ and $(x, y) \in \mathbb{R}^2$. On M, however, with the origin removed from $\mathbb{R}^2$, when Ψ_r acts on $(-r, 0)$ it leads out of M. If not for the missing point at the origin, Ψ_r as defined above would be the Lie group action whose associated Lie algebra action is ψ. It is seen, then, that there are sometimes obstructions to integrating Lie algebra actions.

Two helpful theorems, both based on results by Palais, are the following.

Let ψ be a Hamiltonian action of a Lie algebra $\mathfrak{G}$ on M, and let the vector fields $X_1, \ldots, X_\delta$ which generate it be complete. Then the entire Lie algebra action is composed of complete vector fields. (In general neither commutation nor linear combination conserve completeness.)

Let G be a connected, simply connected Lie group and $\mathfrak{G}$ its Lie algebra. Then every action ψ of $\mathfrak{G}$ on a compact manifold M can be integrated to a unique action Ψ of G on M. The compactness of M is essential to the theorem, for it guarantees that the vector fields of $\mathfrak{G}$ will be complete. It is not essential for integrability, however; this is just a sufficient condition.

For an example of a Hamiltonian Lie algebra action which is not integrable, let M be $\mathbb{R}^{2\nu}$ equipped with the usual symplectic structure, and consider the Hamiltonian vector fields obtained from the two Hamiltonian functions (the submanifold of $\mathbb{R}^{2\nu}$ on which g is not C^∞ must be removed)

$$f = -\Sigma_k p^2 q^k, \quad g = \exp(\Sigma_k 1/p_k).$$

The two vector fields generate a Lie algebra action according to

$$[X_f, X_g] = X_g, \tag{35}$$

but X_f is not complete. This is most simply seen in two dimensions (that is with $\nu = 1$), for which

$$X_f = -2pq\partial/\partial q + p^2\partial/\partial p,$$

whose integral curves are given by

$$q = q_0[1 - (t - t_0)p_0]^2,$$
$$p = p_0[1 - (t - t_0)p_0]^{-1}.$$

It is seen that X_f is not complete and hence that there is no group action whose associated action is that of these two vector fields.

An example of an integrable action on a noncompact manifold is the one of $so(3)$ given in Eq. (1) of Chapter 21, in which all vector fields are complete and Hamiltonian. The action of $so(3)$ restricted to the sphere S^2 with one point removed, as at the very end of §21.3(2), is not integrable, for only X_3 is complete.

What is needed is a generalization of the existence and uniqueness theorem for ordinary differential equations. Palais shows that such a generalization is always possible when M is Hausdorff, in the sense that a Lie algebra action can always be integrated to the action of a *local Lie group*. What this means roughly is that there is a neighborhood U of the $e \in G$ such that the action ψ of $\mathfrak{G}$ on M can be integrated to an action Ψ_U of G so long as one applies it only to elements of U.

23.6 EXAMPLES OF LIE GROUP ACTIONS

We end this chapter with a description of some Lie groups that are of special physical interest. For these groups we shall describe the left-invariant vector fields and make some comments about lifts to their tangent and cotangent bundles.

We will deal with the following groups. (1) The group T_n of translations of n-dimensional Euclidean space. (2) The group $O(n)$ (or $SO(n)$; we shall omit the $\mathbb{R}$ of $O(n, \mathbb{R})$) of orthogonal (or proper orthogonal) real transformations in n dimensions. (3) The Euclidean group E_n, a combination of T_n and $O(n)$. (4) The Galilei group of space–time transformations in Newtonian relativity. (5) The proper homogeneous and inhomogeneous Lorentz groups of space–time transformations in special relativity. In this section we continue to use the traditional n rather than ν to denote the dimensions of the spaces.

1. The translation group T_n. Let A_n be an affine space of n dimensions, and suppose that T_n, the n-dimensional Abelian Lie group, acts on A_n effectively and simply transitively. This action Ψ endows A_n with a vector-space structure in the following way. Single out one point of A_n and call it the origin (let it be denoted by 0). Then for any $m \in M$ there is exactly one element g of T_n such that $\Psi_g(0) = m \equiv m(g)$. The vector-space structure on A_n can be obtained from the definition

$$m(g) + m(h) \equiv m(gh).$$

The details of the vector-space structure of A_n will be left to the reader, and we shall proceed on the assumption that A_n is a vector space.

Clearly T_n can be identified with $\mathbb{R}^n$ and through $\mathbb{R}^n$ it is also endowed with a vector space structure. In fact the action of T_n on A_n can be described entirely in terms of vector addition. Let the group operation in T_n be called addition and denoted $+$. We shall use the same set of letters for the elements of A_n and T_n (e.g. $m(g)$ will be called simply g). Then the action Ψ is given by

$$\Psi: T_n \times A_n \to A_n: (g, h) \mapsto g + h \equiv \Psi_g(h). \tag{36}$$

The fact that the two vector-space structures, of T_n and of A_n, are isomorphic simplifies the task of constructing the associated action of the Lie algebra $\mathfrak{G}_{T_n}$ of T_n. Let $\{x^1, \ldots, x^n\}$ be a coordinate chart for T_n which is such that if $g, h \in T_n$ have coordinates g^k, h^k, respectively, then $m = g + h$ has coordinates $m^k = g^k + h^k$. Then it is easily seen that the tangent space $T_0 T_n$ at the identity (we write 0 for the identity element) is spanned by the vector fields $\partial/\partial x^k$.

Now consider the left-invariant vector field $X_a = a^k \partial/\partial x^k$. It generates the one-parameter subgroup of T_n consisting of the elements ta whose coordinates in this chart are ta^k, $t \in \mathbb{R}$. In the identification of $T_0 T_n$ with $\mathfrak{G}_{T_n}$, the vector field $\partial/\partial x^k$ may be identified with a basis element e_k of the Lie algebra, and the vector field X_a identified with the element $a_{\mathfrak{G}} = a^k e_k$. (The subscript $\mathfrak{G}$ is necessary here to distinguish between the element of the Lie algebra and that of the Lie group, as in the following equation. Some confusion arises because both vector addition in $\mathfrak{G}_{T_n}$ and the group operation in G are denoted by addition.) It then follows that

$$\exp(ta_{\mathfrak{G}}) = ta.$$

It is now a simple task to use (32) and (36) to obtain the action ψ of $\mathfrak{G}_{T_n}$ which is associated with the action Ψ of T_n: it is defined by

$$\tilde{\psi}(a_{\mathfrak{G}}) \equiv X_a = a^k \partial/\partial x^k, \tag{37}$$

where now $\{x^k\}$ is the chart induced on A_n by the given chart in T_n. (We will not in general go into this much detail when deriving the relation between Ψ and ψ.)

The action ψ can be *lifted* to an action either on TA_n or on T^*A_n by lifting all of the vector fields in $\tilde{\psi}(T_n)$ in accordance with either Eq. (36) of Chapter 5 or Eq. (20) of Chapter 11. Let $X_k \in \mathfrak{X}(A_n)$ denote the vector field $\tilde{\psi}(e_k) = \partial/\partial x^k$. Then the lifts to both the tangent bundle and the cotangent bundle have the same expressions:

$$X^T(e_k) \equiv X_k^T = \partial/\partial x^k;$$
$$X^*(e_k) \equiv X_k^* = \partial/\partial x^k.$$

Neither the derivatives with respect to the $\dot{x}^k$ on the tangent bundle nor with respect to the p_k (conjugate to the x^k) on the cotangent bundle enter.

On the cotangent bundle the action is clearly Hamiltonian, for with $\omega = -dp_k \wedge dx^k$ one obtains

$$i_{X_k^*}\omega = dp_k.$$

The two foliations Φ^X and Φ^α now coincide (see the beginning of §22.1(1) for their definitions): they are the level sets of the $\{p_k\}$, hyperplanes parallel to the zero section of T^*A_n.

2. The rotation group SO(n). Let V be an orthogonal vector space of dimension n, and let $\{e_1, e_2, \ldots, e_n\}$ be an orthonormal basis for V, so that $e_j \cdot e_k = \delta_{jk}$ (we use the center dot for the inner product in V). Of course the basis provides a chart (or coordinate system)

$$x: V \to \mathbb{R}^n : v = x^k e_k \mapsto (x^1, \ldots, x^n). \tag{38}$$

If $\{e'_1, \ldots, e'_n\}$ is another orthonormal basis and

$$A_{jk} = e'_j \cdot e_k, \tag{39}$$

then the matrix A of the A_{jk} satisfies the equation

$$A^T A = A A^T = \mathbf{1},$$

so that the A matrices make up the orthogonal group described in §23.2(3). It is easily seen from (39) that

$$e'_j = A_{jk} e_k, \quad e_k = A_{jk} e'_j. \tag{40}$$

The orthogonal matrices thus yield a transformation from one orthogonal basis to another. The subgroup of those matrices A for which $\det A = 1$ is called the *proper* or *special* orthogonal group or simply the *rotation group* $SO(n)$. Two charts $x: V \to \mathbb{R}^n$ and $y: V \to \mathbb{R}^n$ are said to be *compatible* with respect to the rotation group if $x \circ y^{-1}: \mathbb{R}^n \to \mathbb{R}^n$ is a rotation.

One immediately apparent action of $SO(n)$ is its action on V. Let g be an element of the abstract group $SO(n)$, and let A_g be the matrix to which it corresponds. Then the action on V is given by

$$\Psi: SO(n) \times V \to V: (g, v) \mapsto A_g v. \tag{41}$$

(For simplicity we shall sometimes write gv for $A_g v$.) This action is linear, for the A_g are linear operators on V. Another action is left translation, under which (g, h) is mapped into gh, or (A_g, A_h) into $A_{gh} = A_g A_h$. This second action is not linear, for $SO(n)$ is not a linear space (that is, the sum of two rotation matrices is not a rotation matrix).

In most physical applications $n = 3$, and we now specialize to this case.

The one-parameter subgroups of the three-dimensional rotation group are the rotations about an axis, or in the plane orthogonal to the axis. In the chart x of Eq. (38) infinitesimal generators of the action on V of such one-parameter subgroups, spanning the vector space at the identity, are left-invariant vector fields of the form

$$X_k = \varepsilon^h_{kj} x^j \partial/\partial x^h; \tag{42}$$

this one generates rotations about the kth axis. These vector fields then define the action ψ of the Lie algebra $so(3)$ associated with the action Ψ of the Lie group $SO(3)$ which is defined by (41) with $n = 3$.

The action ψ can be lifted to actions on both TV and T^*V. In the resulting actions the vector fields of (42) are lifted canonically to (note the use of the superscript T here: it is the lift of the vector field, not the transpose of a matrix)

$$X^T_k = \varepsilon^h_{kj}(x^j \partial/\partial x^h + \dot{x}^j \partial/\partial \dot{x}^h)$$

and

$$X^*_k = \varepsilon^h_{kj}(x^j \partial/\partial x^h - p_h \partial/\partial p_j),$$

on the tangent and cotangent bundles, respectively, where p_j is conjugate to x^j. With $\omega = -dp_k \wedge dx^k$, the action on T^*V is Hamiltonian, and the Hamiltonian function of X^T_k is

$$J_k = \varepsilon^h_{kj} x^j p_h.$$

Note that X^T_k is composed of the vector field $X_k \in \mathfrak{X}(V)$ of Eq. (42) and the vector field $\bar{X}_k = \varepsilon_{kj}{}^h \dot{x}^j \partial/\partial \dot{x}^h$ which acts on the fibers of TV. One can then generalize the action of $SO(3)$ on V (in fact this can be done more generally, with $n \neq 3$) by allowing different elements of the Lie algebra to act on the base space and on the fiber. That is, one can construct the action of $SO(3) \times SO(3)$ on TV in accordance with

$$(g_1, g_2; v, \dot{v}) \mapsto (g_1 v, g_2 \dot{v}), \tag{43}$$

where $g_k \in SO(3)$ and $(v, \dot{v}) \in TV$. Then the canonical lift of the group action, obtained by integrating the canonical lift of the algebra action, is the diagonal form of (43), given by $g \mapsto (g, g)$, in which it becomes

$$(g, g; v, \dot{v}) \equiv (g; v, \dot{v}) \mapsto (gv, g\dot{v}).$$

In this way one can construct several different actions of $SO(3)$ on TV; two in addition to the diagonal form of (43) are

$$(g; v, \dot{v}) \mapsto (gv, \dot{v}),$$
$$(g; v, \dot{v}) \mapsto (v, g\dot{v}).$$

There are others involving, for instance, the transpose of g (actually of A_g).

3. *The Euclidean group E_n.* We continue to specialize to $n = 3$, write $\mathbb{R}^3$ for V, and consider the joint action of T_3 and $SO(3)$, which is an interesting example of a semidirect product. Some of the results, not so much for the algebra as for the group, are easily extended to $\mathbb{R}^n$ from $\mathbb{R}^3$, which is why we have written E_n instead of E_3.

This time we start with ψ, the action of the algebra. It has been seen that the infinitesimal generators of $SO(3)$ and T_3 are given, respectively, by

$$X_k = \varepsilon^h_{kj} x^j \partial/\partial x^h, \tag{44a}$$

and

$$Y_k = \partial/\partial x^k. \tag{44b}$$

The commutation relations between the X_k and Y_k are obtained by taking derivatives; they are

$$[X_k, Y_j] = -\varepsilon^h_{kj} Y_h. \tag{45}$$

These commutation relations will now be used to define the Lie alebra $e(3)$ of E_3; the group will be obtained by integrating the action which is defined through (45).

The elements of $e(3)$ consist of pairs (r, a), $r \in so(3)$, $a \in t(3)$, the Lie algebra of T_3; the bracket in $e(3)$ is given by

$$[(r, a), (s, b)] = ([r, s], [r, b] + [a, s]). \tag{46}$$

Here on the right-hand side the first bracket is the one in $so(3)$ and the second two are obtained from (45) by writing $a = a^k Y_k$ and $r = r^k X_k$ (compare the equation just before Eq. (11) of Chapter 20; $t(3)$ is Abelian). The r^k and a^k are the coordinates of $r \in so(3)$ and $a \in t(3)$. Then the action ψ is given by

$$\tilde{\psi}: e(3) \to T\mathbb{R}^3 : (r, a) \mapsto r^k X_k + a^k Y_k. \tag{47}$$

This Lie algebra action can be integrated to obtain the action Ψ of the entire Lie group E_3. Let $x = (x^1, x^2, x^3)$ be an element of $\mathbb{R}^3$ and let (g, a) be an element of E_3, with $g \in SO(3)$ and $a \in T_3$ (recall that the elements of T_n and of its Lie algebra can be written in the same way). Then by integration one obtains

$$\Psi: E_3 \times \mathbb{R}^3 \to \mathbb{R}^3 : \{(g, a), x\} \mapsto gx + a. \tag{48}$$

From this one can obtain the group operation in E_3. It is given by

$$(g, a)(h, b) = (gh, gb + a). \tag{49}$$

The action Ψ can be interpreted also as an action of E_3 on T_3, for T_3 can be identified with $\mathbb{R}^3$.

Remark. It can be shown that (49) is a special case of the following general relationship between a Lie group G and its Lie algebra $\mathfrak{G}$. Because of the group structure, as has been seen, TG is diffeomorphic to $G \times \mathfrak{G}$. Let (g, a) be an element of TG, $g \in G$, $a \in \mathfrak{G}$. Then TG can be given a group structure defined by the composition rule

$$(g, a)(h, b) = (gh, a + Ad_g b),$$

where Ad_g is the adjoint action defined at Eq. (29).

We shall not go into the way the action of E_3 on $\mathbb{R}^3$ lifts to actions on $T\mathbb{R}^3$ and $T^*\mathbb{R}^3$; these lifts are combinations of the lifts already discussed of $SO(3)$ and T_3.

4. The Galiei group. The Galilei group is the group of space–time transformations in Newtonian mechanics. It is a ten-parameter group, topologically isomorphic to $SO(3) \times \mathbb{R}^7$, whose composition law is given by

$$(A_1, \mathbf{v}_1, \tau_1, \mathbf{a}_1)(A_2, \mathbf{v}_2, \tau_2, \mathbf{a}_2) = (A_1 A_2, \mathbf{v}_1 + A_1\mathbf{v}_2, \tau_1 + \tau_2, \mathbf{a}_1 + \tau_2\mathbf{v}_1 + A_1\mathbf{a}_2). \quad (50)$$

Here the $\mathbf{v}_k$ and $\mathbf{a}_k$ are in $\mathbb{R}^3$, the τ_k in $\mathbb{R}$, and the A_k in $SO(3)$. This composition law is inherent in the representation of the group by matrices of the form

$$(A, \mathbf{v}, \tau, \mathbf{a}) \mapsto \left[\begin{array}{c|ccc|c} 1 & 0 & 0 & 0 & \tau \\ \hline \mathbf{v} & & A & & \mathbf{a} \\ \hline 0 & 0 & 0 & 0 & 1 \end{array}\right] \quad (51)$$

where $\mathbf{v}$ and $\mathbf{a}$ are interpreted as column vectors.

Eq. (51) may seem to define an action (in particular a representation) of the group on $\mathbb{R}^5$, but in fact the action takes place on $\mathbb{R}^4$: the bottom row of the matrix serves only to square it off and does not enter into the action. We shall denote by Ψ the representation on $\mathbb{R}^4$ which is so defined. Let the chart in $\mathbb{R}^4$ in which the action is given by (51) be written in the form $\{t, \mathbf{x}\}$, and consider the time function $t:(t, \mathbf{x}) \mapsto t$. Then t is equivariant with respect to Ψ and time translation T_τ on $\mathbb{R}$; that is, the following diagram commutes:

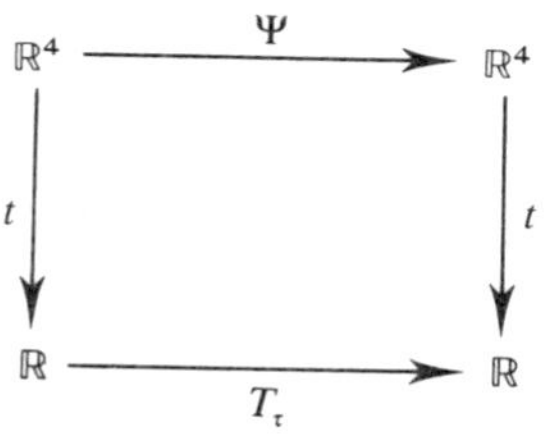

The function t provides a foliation of $\mathbb{R}^4$ each of whose leaves is $\mathbb{R}^3$ composed of space variables. The equivariance expressed by (52) thus means that the Galilei group preserves the notion of simultaneity.

The A and $\mathbf{a}$ parts of the matrix in (51) act on each of the $\mathbb{R}^3$ leaves the way E_3 acts on $\mathbb{R}^3$ in §(3), and this action projects onto the zero time translation in $\mathbb{R}$. The additional part of the action on $\mathbb{R}^3$, generated by the $\mathbf{v}$ part of the matrix, is made up of what are called *boosts*. These correspond to transforming from one frame of reference to another moving with respect to the first at a constant velocity $\mathbf{v}$. For completeness, we list the action of the group on $\mathbb{R}^4$:

$$(t, \mathbf{x}) \mapsto (t + \tau, A\mathbf{x} + \mathbf{v}t + \mathbf{a}). \quad (53)$$

The Lie algebra action ψ associated with Ψ can be characterized by giving the infinitesimal generators of the rotations, the translations, the time translations, and the boosts. For the rotations and translations these generators are exactly those given by Eq. (44). The infinitesimal generator of the time translations is

$$Y_t = \partial/\partial t, \quad (54a)$$

and for the boosts it is

$$Z_k = t\partial/\partial x^k. \tag{54b}$$

The commutation relations can be obtained by taking derivatives.

The lift of ψ to $T\mathbb{R}^4$ is characterized simply. Again, the lifts of the rotations and translations are obtained from previous results; the others can be obtained from (54). The full characterization is

$$\begin{aligned} \text{rotations: } X_k^T &= \varepsilon_{kj}^h(x^j\partial/\partial x^h + \dot{x}^j\partial/\partial \dot{x}^h), \\ \text{translations: } Y_k^T &= \partial/\partial x^k, \\ \text{time translations: } Y_t^T &= \partial/\partial t, \\ \text{boosts: } Z_k^T &= t\partial/\partial x^k + \dot{t}\partial/\partial \dot{x}^k. \end{aligned} \tag{55}$$

The commutation relations can be obtained by taking derivatives.

The finite action of the group on $T\mathbb{R}^4$ can be obtained by integration. It is given by

$$(t, \mathbf{x}; \dot{t}, \dot{\mathbf{x}}) \mapsto (t + \tau, A\mathbf{x} + \mathbf{v}t + \mathbf{a}; \dot{t}, A\dot{\mathbf{x}} + \mathbf{v}). \tag{56}$$

Equation (56) can be used as an aid in finding the action of the group on the cotangent bundle. We shall not actually go into that action, but make a brief comment. On the basis of one-forms $d\mathbf{x}$, $d\dot{\mathbf{x}}$, dt, $d\dot{t}$ the action deduced from (56) is

$$\begin{aligned} \Psi_g^*(d\mathbf{x}) &= Ad\mathbf{x} + \mathbf{v}dt, \\ \Psi_g^*(d\dot{\mathbf{x}}) &= Ad\dot{\mathbf{x}}, \\ \Psi_g^*(dt) &= dt, \\ \Psi_g^*(d\dot{t}) &= d\dot{t}, \end{aligned}$$

where $g = (A, \mathbf{v}, \tau, \mathbf{a})$ is an element of the Galilei group. As $\dot{t} = 1$ by assumption, one is not interested in the entire cotangent bundle, but only in a part of it isomorphic to $T^*\mathbb{R}^3 \times \mathbb{R}$. Assume that a dynamical system is given whose equations of motion are

$$\begin{aligned} d\mathbf{x}/dt &= \dot{\mathbf{x}}, \\ md\dot{\mathbf{x}}/dt &= \mathbf{F}, \end{aligned}$$

and on $T^*\mathbb{R}^3 \times \mathbb{R}$ consider the two-form

$$\omega = (md\dot{x}^k - F^k dt) \wedge (dx^k - \dot{x}^k dt).$$

This is a contact form on this seven-dimensional manifold (see the Appendix on time at the end of Part I). It can be shown that ω is invariant under the action of the Galilei group. These considerations are relevant to the treatment of time-dependent systems, but we shall not go further into this now.

5. *The Poincaré group.* We shall start with the homogeneous Lorentz group, which acts on the same space $\mathbb{R}^4$ as does the Galilei group.

As is well known, every homogeneous Lorentz transformation can be written as A_1BA_2, where A_1 and A_2 are rotations acting on $\mathbb{R}^3$, and B is a Lorentz boost along a fixed direction (which we shall here call x) given by

$$x' = \frac{x - vt}{[1 - v^2/c^2]^{1/2}},$$
$$t' = \frac{t - vx/c^2}{[1 - v^2/c^2]^{1/2}}. \tag{57}$$

The Lorentz group can be characterized as the subgroup of $GL(4, \mathbb{R})$ which preserves the metric $(\mathbf{x}, t)\cdot(\mathbf{x}, t) = \mathbf{x}^2 - c^2t^2$; it is a noncompact group also called $O(3, 1)$. A common way to deal with this group is to introduce the *rapidity* parameter ζ defined by

$$v \equiv c \tanh \zeta, \tag{58}$$

which ranges through all of $\mathbb{R}$ as v ranges through $[-c, c]$. From the definition of ζ one obtains

$$\cosh \zeta = \frac{1}{[1 - v^2/c^2]^{1/2}},$$

$$\sinh \zeta = \frac{v/c}{[1 - v^2/c^2]^{1/2}},$$

which means that (57) can be written in the form

$$\begin{aligned} x' &= x \cosh \zeta - ct \sinh \zeta, \\ ct' &= -x \sinh \zeta + ct \cosh \zeta. \end{aligned} \tag{59}$$

It is now clearly seen that, while transformations among the space coordinates are simply rotations in $SO(3)$, those which mix space and time coordinates are *hyperbolic rotations.*

Lorentz transformations do not, of course, preserve time foliations (or simultaneity) as do the Galilei transformations, so that it is usual to replace the $\{\mathbf{x}, t\}$ notation for $\mathbb{R}^4$ by the *covariant notation* $\{x^\mu\}$, where μ runs from zero to three: $x^0 = ct$ and x^j has the same meaning as before for $j = 1, 2, 3$. A Lorentz transformation can then be written in the form

$$x'^\mu = \Lambda^\mu{}_\nu x^\nu, \tag{60}$$

where the $\Lambda^\mu{}_\nu$ are obtained by combining (59) appropriately with rotations. In these coordinates the preserved metric is given by $g_{\mu\nu}x^\mu x^\nu$, where $g_{\mu\nu}$ is diagonal with elements $(-1, 1, 1, 1)$, and the $\Lambda^\mu{}_\nu$ satisfy $g_{\mu\lambda}\Lambda^\mu{}_\nu\Lambda^\lambda{}_\kappa = g_{\nu\kappa}$.

Two charts x and y on $\mathbb{R}^4$ are *Lorentz compatible* if $x \circ y^{-1}$ is a Lorentz transformation. Suppose two such charts are connected by a rotation and motion at constant velocity $\mathbf{v} = (v^1, v^2, v^3)$. Then with $\zeta = (\zeta^1, \zeta^2, \zeta^3)$ and each ζ^k given in terms of v^k by (58), the $\Lambda^\mu{}_\nu$ of (60) are found to be

$$\begin{aligned}\Lambda^j{}_k &= \delta^{jh} A_{hk} - \zeta^j\zeta^h A_{hk}(1-\cosh\zeta)/\zeta^2,\\ \Lambda^j{}_0 &= -(\zeta^j \sinh\zeta)/\zeta,\\ \Lambda^0{}_j &= -(\zeta^k A_{kj}\sinh\zeta)/\zeta,\\ \Lambda^0{}_0 &= \cosh\zeta,\end{aligned} \tag{61}$$

where $\zeta = (\zeta^2)^{1/2}$. Here the A_{jk} are matrix elements of $A \in SO(3)$. The δ^{jh} in the first equation raises one of the indices of A_{hk}; it arises from the way we have written Eq. (40).

So far we have described the action of the Lorentz group $O(3,1)$ on $\mathbb{R}^4$. Now we describe the action of its Lie algebra $o(3,1)$. The infinitesimal generators of the rotations are the same as before: we shall now call them R_k (as above, Latin indices run from one to three). Then

$$R_k = \varepsilon_{kj}{}^h x^j \partial/\partial x^h. \tag{62}$$

The infinitesimal generators of the boosts in the three directions are given by

$$K_k = t\partial/\partial x^k + x^k \partial/\partial t. \tag{63}$$

The commutation relations can be obtained by taking derivatives. They are the usual ones for the generators of the rotations and

$$\begin{aligned}[][K_j, R_k] &= \varepsilon_{kj}{}^h K_h,\\ [K_j, K_k] &= \varepsilon_{jk}{}^h R_h.\end{aligned} \tag{65}$$

To lift these actions to $T\mathbb{R}^4$ is straightforward. The action of the group lifts in a simple way to actions on the fibers; if one writes $\{\dot{x}^\mu\}$ for the coordinates on the fibers obtained from the $\{x^\mu\}$ on the base spae, the action on the fibers is given by

$$\dot{x}'^\mu = \Lambda^\mu{}_\nu \dot{x}^\nu.$$

This is related to (60) in an obvious way.

We say very little about the lift to $T^*\mathbb{R}^4$. Let $p^0 \equiv -H/c$, where H is the Hamiltonian function of some dynamical system. Then the one-form $\theta_0 = \mathbf{p}\cdot d\mathbf{q} - H dt = p_\mu dx^\mu$ is invariant under the lifted action. The dynamical interest of this one-form is evident from the Appendix on time. See, for instance, Eq. (A11) of that Appendix. Note that H itself is not invariant: in the transformation from one coordinate system to another it picks up terms from $\mathbf{p}$.

The full Poincaré group is obtained by adding translations to the Lorentz group. Let $a \in \mathbb{R}^4$ be a constant *four-vector* and Λ be the matrix of the $\Lambda^\mu{}_\nu$. Then the elements of the Poincaré group are of the form (Λ, a), and the group composition law is

$$(\Lambda', a')(\Lambda, a) = (\Lambda'\Lambda, a' + \Lambda' a). \tag{66}$$

This is again a semidirect product, manifestly similar to the semidirect product for E_3, as expressed by Eq. (49).

Let $T_\mu = \partial/\partial x^\mu$ be the infinitesimal generator of translations in the μ direction.

Then the commutation rules which must be added to those already obtained for the homogeneous Lorentz group are

$$[R_j, T_k] = \varepsilon_{jk}{}^h T_h, \quad [R_j, T_0] = 0,$$
$$[K_j, T_k] = \delta_{jk} T_0, \quad [K_j, T_0] = T_j. \tag{67a}$$

This can be put in covariant notation by writing $R_{\mu\nu}$ for the infinitesimal generator of rotations or boosts in the $(\mu\nu)$ plane. Then (67a) can be written

$$[R_{\mu\nu}, T_\lambda] = g_{\mu\lambda} T_\nu - g_{\nu\lambda} T_\mu. \tag{67b}$$

BIBLIOGRAPHY TO CHAPTER 23

Most of the relevant concepts about Lie groups and their actions will be found in AM, Chapter IV.

Some of the texts already mentioned in connection with Lie algebras serve also as introductions to Lie groups and their actions. Among these, in particular, are Helgason (1962), Bishop and Crittenden (1964), Hausner and Schwartz (1968), Brickell and Clark (1970), and Boothby (1975). See also Palais (1957).

A specific introduction to the theory of Lie groups (mostly representations) in application to the classification of elementary particles is Behrends, Dreitlein, Fronsdal and Lee (1962).

24

Actions of Lie groups on symplectic manifolds

A symplectic structure, as has been seen so often, lends additional richness to a manifold and to the relations among the differential and geometric objects on it. That this is true for Lie algebra actions has been seen in Chapter 22. Now it will be investigated for the actions of Lie groups.

This chapter will concentrate on symplectic manifolds and those actions of Lie groups on them which preserve the symplectic structure ω. All that has been said so far about the action of Lie groups on manifolds remains, of course, true on symplectic manifolds, but we shall discuss now those properties which are particular to actions which preserve ω. Much of what will be covered in this chapter has to do with the actions of Lie algebras, but now for such actions which are integrable to Lie group actions.

24.1 LIE GROUP ACTIONS WHICH PRESERVE DIFFERENTIAL FORMS

1. Let M be a manifold and α some geometric structure on M (e.g. a one-form, a vector field, a symplectic structure). We shall write $\mathrm{Diff}(M,\alpha)$ for the set of diffeomorphisms on M which preserve α. An action $\Psi: G \times M \to M$ of a Lie group G on a symplectic manifold (M,ω) is called a *symplectic action* iff

$$\Psi^* \omega = \omega \ \forall \ g \in G. \tag{1}$$

Then with every symplectic action Ψ is associated the map

$$\tilde{\Psi}: G \to \mathrm{Diff}(M,\omega) \tag{2}$$

in the same way that such a map is associated with any action of a Lie group on a manifold. That is, for each $g \in G$ the map $\tilde{\Psi}(g)$ is a symplectomorphism.

If ω is an exact two-form, i.e. if there exists a $\theta \in \mathfrak{X}^*(M)$ such that $\omega = -d\theta$, the group $\mathrm{Diff}(M,\theta) \subset \mathrm{Diff}(M,\omega)$ obviously also preserves ω. Suppose that M is the cotangent bundle T^*Q of a manifold Q and that a Lie group G acts on Q. The action Ψ can be *lifted* to an action on T^*Q by lifting each diffeomorphism in $\tilde{\Psi}$ in accordance with Eqs. (7–8) of Chapter 5. The resulting lifted action then preserves the natural one-form θ_0 on T^*M (see Remark 2 after Eq. (1) of Chapter 12). This is true, for instance, if Q is G itself and the action on $Q = G$ is taken to be the left or right action.

On TQ, because there is no canonical one-form or symplectic structure and one has therefore to construct them, the situation is more complicated, depending on that construction, in particular on the Lagrangian function.

Let Ψ be a symplectic action of a Lie group G on a symplectic manifold (M, ω), and let $\psi : \mathfrak{G} \times M \to TM$ be the associated action of its Lie algebra. Then it follows from the definition of the Lie derivative that

$$L_X\omega = 0 \;\forall\; X \in \tilde{\psi}(\mathfrak{G}). \tag{3}$$

Similarly if Ψ preserves θ, the vector fields in $\tilde{\Psi}$ satisfy $L_X\theta = 0$. To put it succinctly,

$$\begin{aligned} \tilde{\Psi} &: G \to \mathrm{Diff}(M, \omega) \Rightarrow \tilde{\psi} : \mathfrak{G} \to \mathfrak{X}_{\mathscr{L}\mathscr{H}}(M), \\ \tilde{\Psi} &: G \to \mathrm{Diff}(M, \theta) \Rightarrow \tilde{\psi} : \mathfrak{G} \to \mathfrak{X}_{\mathscr{H}}(M). \end{aligned} \tag{4}$$

The Lie algebra actions on the first line of (4) have been called symplectic, and those on the second line, Hamiltonian (see §22.1(1)).

A symplectic Lie group action thus leads to a symplectic Lie algebra action, and all of the results that have already been found for such algebra actions therefore also hold for these group actions. But there are additional results obtainable for symplectic group actions, having to do with the integrability of the symplectic algebra actions. We will turn to those eventually.

2. We now specialize to $Q = G$ a Lie group. There are several statements that can be made in this case which are not true more generally. For one, the lift to TG of the left or right action on G is always symplectic if the Lagrangian is constructed in a specific way. More even than that, the lift preserves the Lagrangian one-form. We shall describe such a construction for the case of the left action of G on itself. Some of the details (e.g. the existence of a basis of μ left-invariant one-forms) depend on the parallelizability of TG; for such details the reader is referred to Chapter 25.

Left-invariant one forms were mentioned briefly at the end of §23.1(2). A one-form $\theta \in \mathfrak{X}^*(M)$ is left-invariant iff $i_X\theta = \text{const.}$ for every left-invariant vector field $X \in \mathfrak{X}(M)$. Let $\{\theta^j\}, j \in \{1, \ldots, \mu\}$, be a basis of left-invariant one-forms on G; here $\mu = \dim G$. From them one can construct a set of μ functions on TG in the following way. Let $v_g \in T_gG$, and define the $\xi^k \in \mathscr{F}(TG)$ by

$$\xi^k(g \cdot v_g) \equiv \langle \theta^k(g) | v_g \rangle, \tag{5}$$

where $\langle \;|\; \rangle$ is the inner product in T_gG. These ξ^k are more or less generalized coordinates on the fibers of TG: they give the components of v_g along the $\theta^k(g)$. If one thinks of the points on the base space G as the g^k, then one may think of the ξ^k as generalized $\dot{q}^k$. One might even choose to write, say, $\dot{Q}^k$ for them, except that they do not necessarily correspond to any coordinates Q^k on G. Being defined through left-invariant one-forms on G, the ξ^k are functions on TG which are invariant under the lift of the left action of G on itself.

We make two assertions without proof at this time (but see Chapter 25). Let $\theta^{k\prime} \equiv (\tau_G)^* \theta^k$ be the pullback of θ^k to TG. Then the first of these assertions is that the set

$$\{d\xi^k, \theta^{k\prime}\}, \quad k \in \{1, \ldots, \mu\} \tag{6}$$

constitutes a basis of one-forms for TG at every point, and these one-forms are invariant under the lift of the left action. The second assertion is that every second-order vector field $\Delta \in \mathfrak{X}(TG)$ satisfies

$$i_\Delta \theta^{k\prime} = \xi^k.$$

Now consider the regular Lagrangian function

$$\mathscr{L} = \delta_{jk} \xi^j \xi^k, \tag{7}$$

which is clearly invariant under the lift of the left action. To the extent that the ξ^k are generalized coordinates on the fibers, or generalized $\dot{q}^k$, this Lagrangian looks like a kinetic energy. It is obviously smooth and well-defined everywhere on TG. One can now obtain the Lagrangian one-form $\theta_{\mathscr{L}}$ from the $\mathscr{L}$ of (7); it is found to be

$$\theta_{\mathscr{L}} = \delta_{jk} \xi^j \theta^{k\prime}. \tag{8}$$

Its exterior derivative is then a symplectic form on TG. It is given by

$$\omega_{\mathscr{L}} \equiv -d\theta_{\mathscr{L}} = -\delta_{jk} d\xi^j \wedge \theta^{k\prime} - \delta_{jk} \xi^j d\theta^k. \tag{9}$$

It is obvious that $\theta_{\mathscr{L}}$ and $\omega_{\mathscr{L}}$, like $\mathscr{L}$ itself, are invariant under the lift of the left action.

It will now be shown that the $d\theta^{k\prime}$ can be expressed as linear combinations of the two-forms $\theta^{j\prime} \wedge \theta^{h\prime}$, so that $\omega_{\mathscr{L}}$ can be written in terms of exterior products of one-forms in the basis of Eq. (5). To do so, we drop down from TG to G and show that the $d\theta^k$ can be expressed as linear combinations of the $\theta^j \wedge \theta^h$. Let $\{X_k\}$ be the set of (left-invariant) vector fields on G which are dual to the θ^k. Because they are left-invariant they form an action of the Lie algebra $\mathfrak{G}$ of the group G on G itself. Then

$$\begin{aligned} L_{X_j}(i_{X_h} \theta^k) &= i_{X_h}(L_{X_j} \theta^k) + i_{[X_j, X_h]} \theta^k \\ &= i_{X_h} i_{X_j} d\theta^k + i_{c^m_{hj} X_m} \theta^k \\ &= i_{X_h} i_{X_j} d\theta^k + c^k_{hj}, \end{aligned}$$

where the c^k_{hj} are structure constants for $\mathfrak{G}$. Since the θ^k form a basis of one-forms, this shows that

$$d\theta^k + \tfrac{1}{2} c^k_{jh} \theta^j \wedge \theta^h = 0. \tag{10}$$

This is known as the Maurer–Cartan equation. By pulling it back to TG, the same equation is obtained for the $\theta^{k\prime} \in \mathfrak{X}^*(TG)$.

Eq. (10), or rather its analogue for the $\theta^{k'}$, can now be inserted into (9) to yield the desired expression for $\omega_{\mathscr{L}}$:

$$\omega_{\mathscr{L}} = -\delta_{jk} d\xi^j \wedge \theta^{k'} + \tfrac{1}{2}\delta_{jk} c^j_{hm} \xi^k \theta^{h'} \wedge \theta^{m'}. \tag{11}$$

This is a symplectic structure invariant under the lift of the left action of G. If right-invariant one-forms had been used at the start instead of left-invariant ones, the ξ^k defined by (5) would have been different functions on the fibers, and the Lagrangian function defined by (7) would have been a different function on TG. Then the symplectic structure obtained would have been invariant under the right action of G.

3. Other special aspects of a Lie group G as opposed to other manifolds have to do with reduction procedures. There are certain quotient manifolds in Lie groups which carry a symplectic structure, and it will be seen eventually that they are diffeomorphic to orbits of the coadjoint representation of the group.

One way to obtain such symplectic quotient manifolds would be to choose a one-form θ on G and then define a foliation in the usual way through $\ker d\theta$ and a symplectic form through the projection of $d\theta$. This procedure gives nothing really different from foliation by the kernel of any other two-form, and in order to obtain a useful result one would have to assume that the foliation so derived is regular. The situation is different, however, if the one-form θ is left invariant; then no additional assumptions need be made. One finds then that $\ker d\theta$ is a regular distribution and that the quotient with respect to the distribution (with respect to the foliation obtained from it) is a smooth manifold with a smooth projection.

For the proof of this assertion, consider a left-invariant one-form θ. To say that it is left-invariant is to say that $\theta(g)$, $g \in G$, is determined by $\theta(e)$. In fact from $i_X\theta = \text{const.}$, $X \in \mathfrak{X}(G)$ left-invariant, it follows immediately that

$$\theta(g) = (L_{g^{-1}})^* \theta(e) \ \forall\ g \in G. \tag{12}$$

Now, $\omega = -d\theta$ is obviously also left-invariant, for $dL_g^* = L_g^* d$. Through the usual identification of the Lie algebra $\mathfrak{G}$ of G with T_eG, the two-form ω yields an alternating map $\omega(e): \mathfrak{G} \times \mathfrak{G} \to \mathbb{R}$. Consider $\ker \omega(e) \subset \mathfrak{G}$. Because ω is closed (in fact exact) the left-invariant vector fields in $\ker \omega \subset \mathfrak{X}(G)$ close to form a Lie algebra, and hence so do the vectors in $\ker \omega(e)$. This Lie algebra is a subalgebra $\mathfrak{H}_\theta$ of $\mathfrak{G}$ which depends on the original choice of θ. Since left-invariant vector fields are complete, $\mathfrak{H}_\theta$ integrates to a connected Lie subgroup H_θ of G. What must be shown is that H_θ is closed, so that the closed subgroup theorem can be applied to show that the quotient G/H_θ is a smooth manifold (see the first of the theorems listed at the end of §23.4).

To show that H_θ is closed we shall exhibit a subgroup of H which is closed and in which the component of the identity is H_θ. This subgroup H is defined by

$$H = \{h \in G \vdash (R_h)^* \theta = \theta\}. \tag{13}$$

Recall (§23.3(2)) that the left-invariant fields are the infinitesimal generators of the right action. Let $X \in \mathfrak{X}(G)$ be left-invariant: then X lies in the Lie algebra $\mathfrak{H}$ of H, or rather in the subspace of T_eG which is identified with the subalgebra $\mathfrak{H} \subset \mathfrak{G}$ which is the Lie algebra of the subgroup $H \subset G$. Moreover,

$$L_X\theta = 0 = i_X d\theta + di_X\theta = -i_X\omega;$$

$\mathfrak{H}$ and $\mathfrak{H}_\theta$ coincide. It follows that H_θ, obtained by integrating $\mathfrak{H}_\theta$, is the connected component of the identity in H.

We now show that H is closed. Recall that the left-invariant one-forms are dual to the left-invariant vector fields. If $\{X_k\}$ and $\{\theta^k\}$ are dual bases, just as the $X_k(e)$ form a basis for T_eG, so the $\theta^k(e)$ form a basis for T_e^*G, and as T_eG can be identified with $\mathfrak{G}$, so T_e^*G can be identified with $\mathfrak{G}^*$. The right action of the group defines what might be called an action on the left-invariant one-forms, and hence on $\mathfrak{G}^*$, given on the basis by $(R_g)^*\theta^k$ (as in Eq. (13)), and H is then the isotropy group, under this action, at the one-form θ being considered. That H is closed follows from the smoothness of the action on $\mathfrak{G}^*$ and from the definition of H as the inverse image of a point under the map (also smooth) $G \to \mathfrak{G}^*: g \mapsto (R_g)^*\theta$ obtained from that action. Since H is closed, so is H_θ.

In conclusion, it has been shown that any left-invariant one-form $\theta \in \mathfrak{X}^*(G)$ defines, through the kernel of $d\theta$, a quotient space which is a symplectic manifold.

Although preserved by the left action of G, the left-invariant one-forms are not, or course, invariant under the right action. It should be more or less evident that the right action on them is related to an action of the group on the dual of the algebra, for any left-invariant one-form such as θ becomes an element of $\mathfrak{G}^*$ in the identification of $\mathfrak{G}^*$ with T_e^*G. It turns out that the action on the dual to which the right action is related is the coadjoint action, as we shall now try to show. The invariance of the left-invariant one-forms under the left action causes a modification of the conjugating action I (see Eq. (28) of Chapter 23) to act on them in the same way as the right action. This modification I' is defined by

$$I' = (I_g)^{-1}: G \to G: h \mapsto g^{-1}hg. \tag{14}$$

It is then easy to see that for any left-invariant one-form θ

$$(I')^*\theta = (R_g)^*\theta. \tag{15}$$

Now, the coadjoint action Ad^* is defined at Eq. (30) of Chapter 23 in terms of the adjoint action Ad, which is itself defined above Eq. (29) of Chapter 23 in terms of the conjugating action I. In the same way as Ad was defined in terms of the tangent map of I_g at the identity, so Ad^* could have been defined in terms of the cotangent map of I'_g at the identity, and this would yield the left-hand side of (15). Thus the right action on the left-invariant one-forms coincides, through the identification of T_e^*G with $\mathfrak{G}^*$, with the coadjoint action on the dual of the Lie algebra. (As we have indicated on several occasions, a similar statement can be made with right and left interchanged in the actions and in the invariances.)

Now let θ be some left-invariant one-form and let $\alpha_\theta \in \mathfrak{G}^*$ be the vector which corresponds to θ in the identification of $\mathfrak{G}^*$ with T_e^*G. If $G\cdot\alpha$ is the orbit of α_θ under the coadjoint action, the isotropy groups of all elements on $G\cdot\alpha$ are isomorphic; in fact they are conjugate subgroups. If H, as in Eq. (13), is the isotropy group of θ under the right action, or of α_θ under the coadjoint action, then the orbit $G\cdot\alpha$ can be characterized by H; it is also diffeomorphic to G/H. Thus we arrive at: every orbit of G in $\mathfrak{G}^*$ under the coadjoint action is a symplectic homogeneous space. The symplectic structure is obtained by transporting it from G/H, i.e. by viewing $d\theta$ as a two-form on $\mathfrak{G}^*$.

A final point in this connection: it is important that one start from a one-form θ rather than a two-form ω which is left-invariant. Indeed, the following is a counterexample on the additive group of the three-torus T^3. Let $\theta^j, j=1,2,3$, be a basis of left-invariant one-forms, and consider the two-form

$$\omega = \theta^1 \wedge \theta^2 + \sqrt{2}\theta^1 \wedge \theta^3. \tag{16}$$

Clearly ω is left-invariant. That ω is closed follows from the Maurer–Cartan equation (10) and from the fact that T^3 is Abelian, so that all of the $d\theta^j$ vanish. The one-dimensional and therefore involutive distribution defined by $\ker\omega$ is generated by

$$\sqrt{2}X_2 - X_3,$$

where $X_j\theta^k = \delta_j^k$. Thus the integral submanifolds are irrational winding lines, so that the foliation is not regular. This illustrates that it may be important to start with a left-invariant two-form which is obtained from a left-invariant one-form. On the other hand it has been shown (Chu, 1974) that if G is a simply connected Lie group, then every left-invariant closed two-form defines a quotient space which is a symplectic manifold. The torus in the counterexample is connected, but not simply connected.

24.2 REDUCTION WITH SYMPLECTIC ACTIONS: THE MOMENTUM MAP

When a group acts on a manifold it may foliate it into the orbits under the action; for example the rotation group foliates $\mathbb{R}^3$ into two-spheres. The quotient with respect to this foliation is M/G. The theorems at the end of §23.4(3) show that under certain conditions M/G is a smooth manifold with smooth projection $\pi: M \to M/G$. If M is the phase manifold of a dynamical system, M/G will often play the role of a reduced phase manifold, the carrier manifold of a reduced dynamical system. For this reason there is some interest in how structures, and more generally properties, of M carry down to M/G. The further condition that M be symplectic and that the action of G also be symplectic allows one to come to further conclusions.

1. Assume, therefore, that G acts symplectically on a symplectic manifold (M, ω) in such a way that M/G is a smooth manifold and that the projection

$$\pi: M \to M/G \tag{17}$$

is smooth. We show first that the Poisson bracket on M is carried, in a sense projects, onto a Poisson bracket on the quotient. Indeed, let the action of G on M be denoted, as usual, by Ψ, and let $f \in \mathscr{F}(M/G)$. Then from the definition of the action and of M/G it follows that for $g \in G$

$$(\Psi_g)^* \pi^* f \equiv (\pi \circ \Psi_g)^* f = \pi^* f.$$

Now consider f_1, f_2 both in $\mathscr{F}(M/G)$, and the Poisson bracket of their pullbacks $\{\pi^* f_1, \pi^* f_2\}$. Applying the action to this, one obtains

$$(\Psi_g)^* \{\pi^* f_1, \pi^* f_2\} = \{(\Psi_g)^* \pi^* f_1, (\Psi_g)^* \pi^* f_2\} = \{\pi^* f_1, \pi^* f_2\}. \tag{18}$$

But $(\Psi_g)^* F = F$ for $F \in \mathscr{F}(M)$ implies that F is the pullback of some function in $\mathscr{F}(M/G)$. Thus the Poisson bracket of two pullback functions is itself a pullback function: there is an $h \in \mathscr{F}(M/G)$ such that

$$\{\pi^* f_1, \pi^* f_2\} = \pi^* h. \tag{19}$$

This equation makes it possible to define a Poisson bracket on the quotient manifold M/G which is obtained from the one on M:

$$\{f_1, f_2\}_{M/G} = h, \tag{20}$$

where $h \in \mathscr{F}(M/G)$ is defined by (19). This Poisson bracket on M/G is well-defined because π is onto and because we assume, as usual, that the leaves are connected. It is easy to check that it also satisfies the Jacobi identity.

Remark. A manifold like M/G with a Poisson bracket, called a *Poisson manifold*, need not be symplectic, though it is obvious that all symplectic manifolds are also Poisson. An example of a Poisson manifold is the dual $\mathfrak{G}^*$ of a Lie algebra $\mathfrak{G}$. The Poisson bracket on $\mathfrak{G}^*$ is generated by the Lie bracket on $\mathfrak{G}$, for the elements of $\mathfrak{G}$ are linear functions on $\mathfrak{G}^*$ (it can then be extended to nonlinear functions by the Leibnitz rule). Clearly $\mathfrak{G}^*$ cannot always be symplectic, for it may be of odd dimension.

2. We now return to the momentum map, which was first introduced by Souriau (1966) as a generalization of Noether's theorem precisely for symplectic Lie group actions. From our point of view it will be seen that the momentum map helps transport the symplectic structure from M to the reduced phase manifold.

A symplectic action of a Lie group G on a symplectic manifold (M, ω) induces a symplectic Lie algebra action $\mathfrak{G} \to \mathfrak{X}_{\mathscr{L}\mathscr{H}}(M)$, as described at Eq. (4). If the Lie algebra action is moreover Hamiltonian, i.e. if $\mathfrak{G} \to \mathfrak{X}_{\mathscr{H}}(M)$, then one says that G admits a momentum map. The momentum map, it may be recalled, was constructed in Chapter 22 by choosing a basis $\{e_j\}$ in $\mathfrak{G}$, finding the set of vector fields $\{X_j\}$ which correspond to it in the Hamiltonian Lie algebra action ψ of $\mathfrak{G}$ on M, finding the Hamiltonian functions f_j of the X_j, and defining the momentum map μ by

$$\mu : M \to \mathfrak{G}^* : m \mapsto \varepsilon^j f_j(m), \tag{21a}$$

where $\{\varepsilon^j\}$ is the $\mathfrak{G}^*$ basis which is dual to $\{e_j\}$. The definition may be given in a basis-independent way as well: let $x \in \mathfrak{G}$ and $\tilde{\psi}(x)$ be $X \in \mathfrak{X}(M)$. Then

$$i_X\omega(m) = d\langle \mu(m), x \rangle, \tag{21b}$$

for according to (21a) the Hamiltonian function of X is $\langle \mu, x \rangle$. The difference between the use made of this definition now and in Chapter 22 is that now the action ψ of $\mathfrak{G}$ on M is obtained from the Lie group action Ψ of G on M through the vector fields which are left-invariant (or equivalently right-invariant) under Ψ. Note again that the Hamiltonian functions are not unique, so that the momentum map is not unique, but if μ and μ' are two momentum maps obtained from the same action, then $\mu' = \mu + \mu_0$, where μ_0 is some fixed element of $\mathfrak{G}^*$.

The treatment from here onward is essentially the same as the treatment in Chapter 22, where only the Lie algebra action was available. One of the results that follows from the integrability of that action to a Lie group action is that the group action Ψ of G on M is μ-related to the coadjoint action Ad^* of G on $\mathfrak{G}^*$, or to the action obtained by integrating the affine coadjoint representation of $\mathfrak{G}$ on $\mathfrak{G}^*$ (see the discussion around Eq. (17) of Chapter 22, but recall that this qualification is not necessary if G is semisimple). What this means is that if one writes Ad^* for the coadjoint action or for its affine variant, the following diagram commutes:

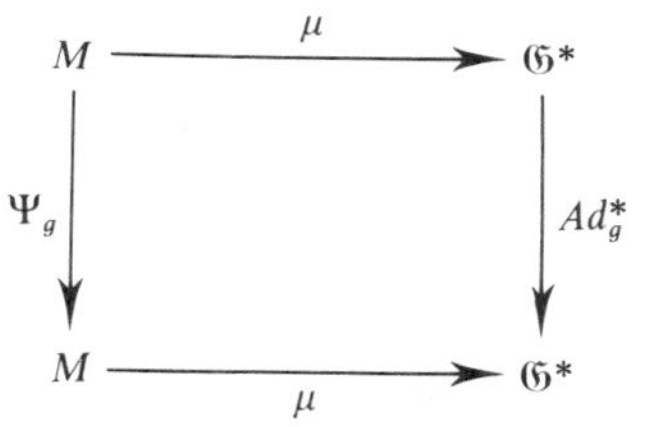

Another result obtained from the integrability of the Lie algebra action ψ, or the existence of the group action Ψ, is the following. Let $N_\alpha = \mu^{-1}(\alpha)$, $\alpha \in \mathfrak{G}^*$, as in Chapter 22. Then if $\alpha, \beta \in \mathfrak{G}^*$ are on the same orbit, i.e. if $\beta \in G\cdot\alpha$, then N_α and N_β are diffeomorphic. We outline a proof of this assertion under the simplifying assumption that α and β lie on a one-parameter subgroup under the coadjoint action. Let $\tilde{X} \in \mathfrak{X}(\mathfrak{G}^*)$ be the infinitesimal generator of this one-parameter subgroup, and let $X \in \mathfrak{X}_{\mathscr{H}}(M)$ be a vector field which is carried into $\tilde{X}$ by the momentum map μ. Take $N_\alpha \subset M$ as a set of initial conditions for the flow of X. Then because $\tilde{X}$ carries α into β, equivariance implies that X carries N_α into N_β; this map of N_α into N_β is necessarily a diffeomorphism. This result implies, incidentally, that the restriction of the momentum map to the inverse image of an orbit, i.e. $\mu|\mu^{-1}(G\cdot\alpha)$, is of constant rank.

To say it differently, the integrability of the action ψ of $\mathfrak{G}$, that is the existence of the action Ψ of G, makes available the complete vector fields on M which correspond to the generators of orbits in $\mathfrak{G}^*$. Their completeness leads to diffeomorphisms which move leaves around in M in a way that corresponds through equivariance with the way Ad^* moves points around in $\mathfrak{G}^*$.

3. What we have called the double value of functions in reducing symplectic manifolds appears again.

Let Ψ be, as before, a symplectic action of a Lie group G on a symplectic manifold (M, ω) and assume that this action admits a momentum map μ which is equivariant with respect to Ψ and Ad^*. Let G_α be the isotropy group of $\alpha \in \mathfrak{G}^*$ under Ad^*; then G_α is closed in G, so it is a Lie subgroup of G. Consider the action of G_α on $N_\alpha \equiv \mu^{-1}(\alpha)$. If this action is free and proper (§23.4(2)), the set $M_\alpha = N_\alpha/G_\alpha$ is a quotient manifold; in any case, we make the assumption that it is. Denote by π_α the projection map $N_\alpha \to M_\alpha$. Then M_α is a symplectic manifold with symplectic structure Ω_α given by

$$(i_\alpha)^*\omega = (\pi_\alpha)^*\Omega_\alpha, \tag{22}$$

where $i_\alpha : N_\alpha \to M$ is the natural injection. The proof is essentially the same as the proof in §20.2 (see in particular §20.2(4)) for reduction of symplectic manifolds by constants of the motion. For the present proof the assumption that M_α is a quotient manifold can be replaced by the assumption that the action of G_α on N_α is free and proper, as mentioned above.

Further, one may introduce the submanifolds $P_\alpha \equiv \mu^{-1}(G \cdot \alpha)$. The submanifold N_α is closed in P_α, and the natural injection of P_α into M is an immersion, though it may not be an imbedding. Let an equivalence relation be defined on P_α by: $x \approx y$, $x, y \in P_\alpha$, iff $\mu(x) = \mu(y)$ and iff there exists $g \in G$ such that $\Psi_g(x) = y$; let Q_α denote the set of equivalence classes and $\tilde{\pi}$ the projection of P_α into Q_α. Then the restriction of M to P_α in an obvious way defines a map $\tilde{\mu}$ from Q_α to $G \cdot \alpha$. It can then be shown (Marle, 1975) that Q_α has a smooth manifold structure such that $\tilde{\pi}$ and $\tilde{\mu}$ are smooth submersions and that G acts smoothly on Q_α. Moreover, there exists a symplectic form ω_{Q_α} on Q_α such that $\pi^*\omega_{Q_\alpha}$ is the two-form induced on P_α by ω, and G acts on the new symplectic manifold $(Q_\alpha, \omega_{Q_\alpha})$ with the new momentum map $\tilde{\mu}: Q_\alpha \to \mathfrak{G}^*$. In addition, Q_α has a local direct product structure; i.e. for each $\beta \in G \cdot \alpha$ there exists an open neighborhood V containing β in $G \cdot \alpha$ and a diffeomorphism of the neighborhood $\tilde{\mu}^{-1}(V) \subset Q_\alpha$ onto $M_\alpha \times V$. Restricted to $\tilde{\mu}^{-1}(V)$, the symplectic form ω_{Q_α} is the product of Ω_α with the symplectic form on $G \cdot \alpha$.

24.3 REDUCTION ON T^*Q. DYNAMICS

In this section we discuss two aspects of reduction with symplectic Lie group actions, the first having to do with so-called point transformations on T^*Q and the second with invariance of dynamical systems.

1. The so-called *point transformations* on T^*Q, obtained from transformations on the configuration manifold Q, are the most familiar to physicists; the Q-transformations are extended (*lifted*, in our terminology) to transformations on the carrier manifold of the dynamical system, which could be TQ, but in the present discussion will be taken to be T^*Q. For example, this is the usual setting for the Noether theorem. We remind the reader again that the lift of a

diffeomorphism on Q to another on T^*Q is defined at Eqs. (7–8) of Chapter 5 together with the diagram that follows them and that near the beginning of Chapter 12 it is asserted that a diffeomorphism on T^*Q preserves the natural one-form θ_0 iff it is the lift of a diffeomorphism on Q.

Now consider an action Ψ of a Lie group G on T^*Q, and assume that Ψ is what may be called a *Q-action*, that is an action obtained by lifting an action Ψ_Q of G on Q. Then the associated action ψ of the Lie algebra $\mathfrak{G}$ of G on T^*Q is necessarily Hamiltonian. This means that if $X \in \mathfrak{X}(T^*Q)$ is in ψ, there exists a function $f_X \in \mathscr{F}(T^*Q)$ such that

$$i_X\omega \equiv -i_X d\theta_0 = df_X \tag{23}$$

(we leave the subscript 0 off ω). Moreover, ψ is also a Q-action, obtained by lifting the action ψ_Q of $\mathfrak{G}$ on Q which is associated with Ψ_Q. It is interesting to obtain an expression in local coordinates for f_X in terms of the components of X. For this purpose, suppose that X is obtained by lifting a vector field $X_Q \in \mathfrak{X}(Q)$ in ψ_Q which is given in local coordinates by

$$X_Q = A^j \partial/\partial q^j,$$

$A^j \in \mathscr{F}(Q)$. Then in local coordinates

$$X = A^j \partial/\partial q^j - p_j(\partial A^j/\partial q^k)\partial/\partial p_k, \tag{24}$$

and from $\theta_0 = p_j dq^j$ and Eq. (23) it follows that

$$f_X = A^j p_j. \tag{25}$$

We now go on to investigate the reduced phase manifold for this case of Q-action on T^*Q. We shall do this for a single vector field X and one function f_X, that is for a one-parameter subgroup, making a comment at the end about the situation for more general Lie groups (we shall drop the subscript X on f, since this is the only function involved). For the one-parameter group the momentum map is, by definition,

$$\mu : m \mapsto f(m) \in \mathfrak{G}^*,$$

where $m \in T^*Q$, and we shall use the symbols f and μ for this map interchangeably. From the expression

$$df = A^j dp_j + p_j dA^j, \tag{26}$$

it is clear that all values are regular except for fixed points of the action generated by X_Q and Q, i.e. points where the A^j vanish. We proceed on the assumption that there are no fixed points of this action, so that all of the values are in fact regular.

The linear p_j dependence of f means that the submanifold $N_\alpha \equiv f^{-1}(\alpha) \subset T^*Q$, $\alpha \in \mathfrak{G}^*$, is transversal to the fibers, i.e. that $T_m(N_\alpha)$ and $\ker T\tau_Q^*(m)$ together span $T_m(T^*Q)$ at each point $m \in N_\alpha$. This, in turn, implies that

$$\tau_Q^* | N_\alpha : N_\alpha \to Q$$

is onto Q. Let G now denote the one-parameter group whose action on Q is generated by X_Q (or whose action on T^*Q is generated by X). Then the reduced phase space N_α/G is projected by τ_Q^* onto the quotient Q/G, because by construction X is projectable under τ_Q^* and projects onto X_Q. Finally, N_α/G is in fact diffeomorphic to $T^*(Q/G)$. This is because the linear dependence of f on the p_j implies that a new coordinate system can be defined on T^*Q in which f replaces one of the p_j.

We shall not go into further detail. For Lie groups in general, of dimension higher than one, the considerations have to be restricted to the open submanifold consisting of orbits of the maximum dimension. With that proviso, the result is essentially the same: the reduced phase space associated with the momentum map arising from a Q-action of G is diffeomorphic to $T^*(Q/G)$.

Remark. It is clear from the above that it is necessary that Q/G be a quotient manifold. In general the treatment must be restricted to the open submanifold of Q, if it exists, where the action of G induces a smooth quotient.

2. Suppose now that Ψ is a symplectic action of a Lie group G on a symplectic manifold (M, ω), and that Ψ admits a momentum map μ. Suppose further that M has on it a Hamiltonian dynamical system $\Delta \in \mathfrak{X}(M)$ with Hamiltonian function $H \in \mathscr{F}(M)$, i.e. that

$$i_\Delta \omega = dH,$$

and that Δ is invariant under the action of G, i.e. that

$$(\Psi_g)^*\Delta = \Delta \, \forall g \in G. \tag{27}$$

From this equation and the fact that Ψ is a symplectic action it follows that $d(\Psi_g)^*H = dH$, which implies that

$$(\Psi_g)^*H = H + C_g, \tag{28}$$

where C_g is some constant depending on g (it is possible, of course, that $C_g = 0$ for all $g \in G$). In other words, except for possible additive constants, H is invariant under Ψ.

We shall now be interested in the way Δ projects into the different symplectic quotients associated with Ψ.

In general one finds that Δ is projectable with respect to

$$\pi: M \to M/G,$$

as follows from

$$(\Psi_g)^*L_\Delta \pi^* f = L_\Delta (\Psi_g)^* \pi^* f = L_\Delta \pi^* f.$$

This means that the Lie derivative with respect to Δ of a function which is invariant under Ψ is itself a function which is invariant under Ψ, and hence Δ is projectable. If f is the pullback by π of a function in $\mathscr{F}(M/G)$, then

$$(\Psi_g)^*\{H, f\} = \{(\Psi_g)^*H, (\Psi_g)^*f\}, \tag{29}$$

which implies that the projected dynamics is described by a Poisson bracket on M/G. If M/G is even-dimensional, the projected dynamics is Hamiltonian.

The dynamical vector field Δ is projectable also with respect to the momentum map $\mu: M \to \mathfrak{G}^*$. This may be seen by considering first any element $a \in \mathfrak{G}$ and the vector field $X_a = \tilde{\psi}(a)$. The invariance of Δ under Ψ implies that $[\Delta, X_a] = 0$, and then if $f_a \in \mathscr{F}(M)$ is the Hamiltonian function of X_a, i.e. if $i_{X_a}\omega = df_a$, then

$$L_\Delta df_a = 0. \tag{30}$$

Since the Hamiltonian functions of the vector fields in ψ actually define the momentum map μ, this implies that Δ is projectable with respect to μ. Eq. (30) also implies that $L_\Delta f_a$ is a constant. If a is now extended to a set of vectors which make up a basis in $\mathfrak{G}$, the resulting Hamiltonian functions corresponding to f_a make up a set of coordinate functions on $\mathfrak{G}^*$. The constancy of $L_\Delta f_a$ then implies that Δ projects down to a vector field in $\mathfrak{X}(\mathfrak{G}^*)$ whose components are constant and which is therefore complete.

This result is easily applied to the Poisson bracket and yields an explicit expression for the projected vector field. Let $\{x^j\}$ be such a set of coordinate functions on $\mathfrak{G}^*$, and let $\{X_j\}$ be the set of Hamiltonian vector fields whose Hamiltonian functions are the $\mu^* x^j$. Then

$$\{\mu^* x^k, H\} = L_\Delta \mu^* x^k = c^k,$$

and hence for any function $f \in \mathscr{F}(\mathfrak{G}^*)$

$$\{\mu^* f, H\} = c^k \mu^*(\partial f / \partial x^k).$$

This shows that the projected vector field is given in these coordinates on $\mathfrak{G}^*$ by

$$\tilde{\Delta} = c^k \partial/\partial x^k \equiv \{\mu^* x^k, H\} \partial/\partial x^k. \tag{31}$$

Because $L_\Delta x^k = -L_{X_k} H$ and the X_k are the infinitesimal generators of the group action, the c^k are related through exponentiation to the C_g of Eq. (28). We do not go into that, but remark that Δ is the null vector field if all of the c^k vanish; this corresponds to reduction by constants of the motion. The case in which the c^k are not all zero has been studied in connection with so-called noninvariance groups (Vitale, 1966).

It is interesting to consider a different kind of Hamiltonian function, of the type

$$H = H_1 + H_2 \circ \mu, \tag{32}$$

where H_1 is invariant under the action Ψ of G and $H_2 \in \mathscr{F}(\mathfrak{G}^*)$. The Hamiltonian vector field obtained from this H will clearly preserve $P_\alpha = \mu^{-1}(G \cdot \alpha)$. A particular type of such an H is one in which H_2 is linear:

$$(H_2 \circ \mu)(m) = \langle \mu(m), a \rangle, \quad a \in \mathfrak{G}. \tag{33}$$

It turns out then that H_1 can be thought of as the Hamiltonian of a given dynamical system in a fixed frame, and that the total Hamiltonian H of (32) can be thought of as the Hamiltonian of the same dynamical system, but in a moving frame. The relative motion of the two frames is given by the one-parameter group of diffeomorphisms of M obtained from $\exp(ta)$, $t \in \mathbb{R}$. This is a method that has been used (e.g. Robbin, 1973) to study 'relative equilibrium' of dynamical systems with symmetries.

25

Parallelizable manifolds. Dynamics on Lie groups

The definition of a *parallelizable* manifold was given in the Remark above Eq. (21) of Chapter 5: a manifold M is parallelizable if there exists on it a set of μ independent vector fields, where $\mu = \dim M$. It was also pointed out that if TM is a *trivial* vector bundle (that is, if TM is diffeomorphic to $M \times \mathbb{R}^\mu$), then M is parallelizable. The study of parallelizable manifolds and of calculus on parallelizable manifolds is useful in dealing with Lie groups, for Lie groups are parallelizable. Indeed, the μ left-invariant (or equivalently, right-invariant) vector fields are independent.

25.1 CALCULUS ON PARALLELIZABLE MANIFOLDS

1. Let M be a parallelizable manifold, and let $\{X_j\}$ be a set of independent vector fields on M. The X_j may be said to provide a *parallelization* of M. Since there is more than one set of independent vector fields on M, there is more than one parallelization. If M contains an atlas with only one chart (e.g. if $M = \mathbb{R}^\mu$), it is obviously parallelizable, for the set can be chosen to be $\{\partial/\partial x^j\}$; the elements of a set chosen in this way clearly commute. But there exist parallelizable manifolds that cannot be described by just one chart (e.g. the circle S), and then it may be impossible to choose the elements of $\{X_j\}$ so that they commute. Since they are always independent, however, it is always possible to write

$$[X_j, X_k] = g^h_{jk} X_h, \quad g^h_{jk} = -g^h_{kj} \in \mathscr{F}(M); \tag{1}$$

this equation, in particular the g^h_{jk}, will contain important information for the calculus on parallelizable manifolds.

Let $w_m \in T_m M$ and $w_p \in T_p M$, $m, p \in M$. Then w_m and w_p are vectors *parallel* with respect to the parallelization $\{X_j\}$ iff $w_m = c^j X_j(m)$, $c^j \in \mathbb{R}$, implies that $w_p = c^j X_j(p)$. A vector field $X \in \mathfrak{X}(M)$ is a *parallel vector field* (again, with respect to a given parallelization) iff $X = c^j X_j$, $c^j \in \mathbb{R}$, that is iff it consists entirely of parallel vectors.

Remarks. 1. It has been pointed out that if TM is a trivial vector bundle, then M is parallelizable. The converse is also true (Brickell and Clark, 1970, p. 119, Problem 7.3.1).

2. Let $X = \xi^j X_j$, where $\{X_j\}$ is a parallelization of M and $\xi^j \in \mathscr{F}(M)$, and let $Y \in \mathfrak{X}(M)$. Then

$$[Y, X] = [Y, \xi^j X_j] = (L_Y \xi^j) X_j + \xi^j [Y, X_j] \equiv \nabla_Y X + \xi^j [Y, X_j],$$

where ∇_Y is called the *linear connection* associated with the parallelization (see Brickell and Clark, 1970, p. 154).

2. We now list some standard formulas which apply locally, within each chart (or globally in a one-chart atlas if such an atlas exists) of a manifold which is not parallelizable, but which apply globally in a parallelizable manifold.

As has already been seen, every $X \in \mathfrak{X}(M)$ can be written in the form

$$X = \xi^j X_j, \quad \xi^j \in \mathscr{F}(M). \tag{2}$$

Eq. (1) follows from this observation. In local coordinates the X_j can be chosen to be the $\partial/\partial x^j$, and then if this choice is inserted into (1) all of the g^h_{jk} vanish.

Every ρ-form α on M can be written in the form

$$\alpha = a_{j_1 \cdots j_\rho} \theta^{j_1} \wedge \cdots \wedge \theta^{j_\rho}, \tag{3}$$

where the $a_{j_1 \cdots j_\rho}$ are in $\mathscr{F}(M)$ and the θ^j are μ independent one-forms obtained by duality from the X_j:

$$i_{X_j} \theta^k = \delta^k_j. \tag{4}$$

Eqs. (1)–(4) make it possible to express the entire apparatus of the calculus in terms of the sets $\{X_j\}$, $\{\theta^j\}$, and $\{g^h_{jk}\}$. For exmple, the external derivative of any $h \in \mathscr{F}(M)$ can be written $dh = h_j \theta^j$ with $h_j \in \mathscr{F}(M)$, and then

$$L_{X_j} h = dh(X_j) = h_k \theta^k (X_j) = h_j. \tag{5}$$

Let $X = f^j X_j$, as above. Then from (5) one obtains

$$L_X h = h_j f^j. \tag{6}$$

In local coordinates many of these formulas look more familiar. Let $X_j = \partial/\partial x^j$ locally, and write $X = X^j \partial/\partial x^j$. Then θ^j becomes dx^j, and $dh = (\partial h/\partial x^j) dx^j$. Eq. (6) then becomes $L_X h = (\partial h/\partial x^j) X^j$.

Two particularly important formulas, to be used often, are

$$i_{X_k} d\theta^j = -g^j_{kh} \theta^h \tag{7}$$

and

$$d\theta^j = -\tfrac{1}{2} g^j_{kh} \theta^k \wedge \theta^h. \tag{8}$$

These can be obtained by first using (4) to show that

$$i_{X_h} i_{X_k} d\theta^j = g^j_{hk}.$$

One can then show that if α and β are ρ-forms such that $L_{X_j} \alpha = L_{X_j} \beta$ for all

j, then $\alpha = \beta$. From this it is a simple matter to obtain (7), and then in the same way from (7) to obtain (8). The local form of (7) and (8) is trivial: it says only that $d(dx^j) = 0$. From (7) and (4) it follows almost immediately that

$$L_{X_j}\theta^k = g^k_{hj}\theta^h,$$

and from this one can obtain a general expression for $L_X\alpha$, where $X \in \mathfrak{X}(M)$ and $\alpha \in \mathfrak{X}^*(M)$ are arbitrary. Write $\alpha = a_j\theta^j$ and $X = \xi^j X_j$, $a_j, \xi^j \in \mathscr{F}(M)$, as before. Then

$$L_X\alpha = a_j d\xi^j + \xi^k[L_{X_k}a_j - a_n g^n_{kj}]\theta^j. \tag{9}$$

An example of a parallelizable manifold is the torus T^2. It has been pointed out several times that the two angle coordinates on the torus, which may be called φ^1 and φ^2, provide only a local chart, but that the vector fields $X_1 = \partial/\partial\varphi^1$ and $X_2 = \partial/\partial\varphi^2$ are globally defined, as are their dual one-forms $\theta^1 = d\varphi^1$ and $\theta^2 = d\varphi^2$. These vector fields provide a parallelization of the torus such that the g^h_{jk} functions all vanish. On the other hand if X_1 is chosen to be $X_1 = f\partial/\partial\varphi^1$, where f is a nonzero function only of φ^2, and X_2 remains as before, then X_1 and X_2 again provide a parallelization, but now the g^h_{jk} functions do not all vanish ($g^1_{21} = f'/f$). Then θ^1 becomes $(1/f)d\varphi^1$, and it is a simple matter to check any of the above formulas, for example (8).

25.2 EXPLICIT CONSTRUCTION OF PARALLELIZATIONS

1. Let M be parallelizable: there exist μ independent vector fields $\{X_j\}$ in $\mathfrak{X}(M)$. In order to obtain a parallelization of TM (or of T^*M) one must find 2μ independent vector fields $\{Y_k\}$ in $\mathfrak{X}(TM)$ (or in $\mathfrak{X}(T^*M)$. We shall place two algebraic requirements on the parallelization $\{Y_j\}$:

(i) Among the Y_k there should be μ vector fields which have the same commutation rules as do the X_k. That is, if the X_k satisfy (1), then there exist Y_k such that

$$[Y_j, Y_k] = (\tau^*_M g^h_{jk})Y_h. \tag{10}$$

(ii) The commutation relations of these Y_k with the others and the commutators of the others among themselves, in short all of the commutation relations, should be expressible in terms of the original parallelization of M, i.e. in terms of the g^h_{jk}.

We shall describe a method for parallelizing TM and a similar one for T^*M which will satisfy requirements (i) and (ii). For a group G, the g^h_{jk} functions become the structure constants, and this allows a different type of parallelization of TG and T^*G. This parallelization will be used in the discussion of the rigid rotator.

2. We start by considering TM. Let $\{X_j\}$ be a parallelization of M and consider

a vector $v_m \in T_m M$. There is a unique set of μ numbers $\xi^j(m)$ such that $v_m = \xi^j(m) X_j(m)$. Each $\xi^j(m)$ for fixed j is a function on $T_m M$, for it maps $T_m M$ into $\mathbb{R}$: it gives the *component along* $X_j(m)$ of any vector v_m in $T_m M$. These μ functions can be globalized from $T_m M$ to TM, for the X_j in terms of which they are defined are global vector fields.

From the duality of the X_j and the θ^j it follows that

$$\xi^j(m) = \theta^j(m) \cdot v_m.$$

Expressions for the ξ^j in local coordinates help to clarify the meaning of this equation. In general, locally one may write $X_j = A_j{}^k \partial/\partial x^k$ and $\theta^j = B_k{}^j dx^k$, where the $B_j{}^k$ and $A_j{}^k$ are local functions. Duality then implies that $A_j{}^h B_h{}^k = \delta^k_j$, or that A and B are inverse matrices. Then $\partial/\partial x^k = B_k{}^j X_j$ and $dx^k = A_j{}^k \theta^j$. Then if one writes v_m in the form $v_m = \dot{x}^k \partial/\partial x^k$, so that the components of v_m along the $\partial/\partial x^k$ are $\dot{x}^k$, one arrives at

$$\xi^j = B_k{}^j \dot{x}^k.$$

Now consider the 2μ one-forms $\{\theta^{k\prime} \equiv (\tau_M)^* \theta^k, d\xi^k\}, k = 1, \ldots, \mu$. These 2μ one-forms are the duals of the vector fields which make up the desired parallelization of TM. Specifically, the set $\{Y^j\}$ is made up of vector fields $X_k^+, \dot{X}_k$, $j, k \in \{1, \ldots, \mu\}$ defined by

$$\begin{aligned} i_{X_j^+} \theta^{k\prime} &= \delta^k_j, \quad i_{X_j^+} d\xi^k = 0; \\ i_{\dot{X}_j} \theta^{k\prime} &= 0, \quad\;\; i_{\dot{X}_j} d\xi^k = \delta^k_j. \end{aligned} \tag{11}$$

The commutators of these vector fields can be calculated with the aid of Eq. (8) if one keeps in mind that the expansion of a vector field in the Y_j is unique. It is found that

$$[X_j^+, X_k^+] = g^{h\prime}_{jk} X_h^+, \quad [X_j^+, \dot{X}_k] = [\dot{X}_j, \dot{X}_k] = 0, \tag{12}$$

where $g^{h\prime}_{jk} \equiv (\tau_M)^* g^h_{jk} \in \mathscr{F}(TM)$.

The X_j^+ and X_j can be written out explicitly in local coordinates; it is a matter of finding the coefficients of the expansion of these vector fields in the $\partial/\partial x^k$ and $\partial/\partial \dot{x}^k$. Straightforward calculation leads to

$$\begin{aligned} X_j^+ &= A_j{}^k \frac{\partial}{\partial x^k} + A_j{}^h \dot{x}^n \frac{\partial A_r{}^k}{\partial x^h} B^r_n \frac{\partial}{\partial \dot{x}^k}, \\ \dot{X}_j &= A_j{}^k \frac{\partial}{\partial \dot{x}^k}. \end{aligned} \tag{13}$$

The second of these equations can also be written in the form

$$\dot{X}_j = \partial/\partial \xi^j \tag{14}$$

if $\{x^j, \xi^j\}$ are chosen as local independent coordinates on TM. Since, moreover, both the $\dot{X}_j$ and the ξ^j are globally defined, this notation is globally valid.

It is to be noted that X_j^+ is not the canonical lift X_j^T of X_j to TM. Indeed, the local form of X_j^T is

$$X_j^T = A_j^{\ k}\frac{\partial}{\partial x^k} + \dot{x}^h\frac{\partial A_j^{\ k}}{\partial x^h}\frac{\partial}{\partial \dot{x}^k}.$$

The reason the X_j^+ rather than the X_j^T are used to parallelize TM is that unless the g_{jk}^h are constants, the commutators of the canonically lifted vector fields X_j^T differ from those of the X_j vector fields on the base manifold; in fact one finds (this is probably easiest to do in local coordinates) that

$$[X_j^T, X_k^T] = g_{jk}^h X_h^T + (L_{X_r^T} g_{jk}^h)\xi^r \frac{\partial}{\partial \xi^h}. \tag{15}$$

When the vector fields form a Lie algebra and the g_{jk}^h are constants, the lifted fields yield a suitable parallelization.

3. Second, we turn to T^*M. The procedure here is almost exactly the same as for TM. Let $\alpha_m \in T_m^*M$ be written in the form $\alpha_m = P_j(m)\theta^j$. The P_j can be globalized; they play the same role on T^*M as the ξ^j did on TM, providing μ maps from T^*M to $\mathbb{R}$. For fixed m the $P_j(m)$ can be chosen as local coordinates in the fiber, related to the usual coordinates p_j through

$$P_j = A_j^{\ k} p_k,$$

where the $A_j^{\ k}$ are the same local functions as before (the components of the X_j).

Now consider the 2μ one-forms $\{\theta^{k'} \equiv (\tau_M^*)^*\theta^k, dP_k\}$, $k = 1, \dots, \mu$. The vector fields that make up the desired parallelization of T^*M are then defined to be the duals of these one-forms, denoted X_j^+ (as in TM) and $\partial/\partial P_k, j, k \in \{1, \dots, \mu\}$ (see Eq. (14) for similar notation on TM). These vector fields satisfy the same commutation relations as do those on TM. We shall not write the analogs of Eqs. (11) and (12).

It is extremely easy to show that $\theta_0 \equiv p_j dx^j = P_j\theta^{j'}$, and then $\omega = -d\theta_0$ can be written in the form

$$\omega = -dP_j \wedge \theta^{j'} + \tfrac{1}{2}P_j g_{hr}^j \theta^{h'} \wedge \theta^{r'}. \tag{16}$$

From equations such as this, one can write out Hamilton's canonical equations, Poisson brackets, etc. in terms of the P_j and $\theta^{j'}$. For example, $\{P_j, P_k\} = -g_{jk}^h P_h$.

4. Next we turn to the parallelization of TG, where G is a Lie group. Let $\{X_j\}$, $j = 1, \dots, \mu$ (we shall write $\dim G = \mu$ rather than γ) be a set of left-invariant vector fields on G which, in the identification of T_eG with the Lie algebra $\mathfrak{G}$ of G, correspond to a basis of $\mathfrak{G}$. Then the X_j are independent and thus yield a parallelization of G. (As we have occasionally written, right-invariant vector fields would do just as well.) Then the vector fields $\{X_k^T, \dot{X}_k \equiv \partial/\partial\xi^k\}$ are a parallelization $\{Y_j\}$ of TG, as they are obviously independent.

The commutation relations of the vector fields on TG are

$$[X_j^T, X_k^T] = c_{jk}^h X_h^T, \quad [X_j^T, \dot{X}_k] = c_{jk}^h \dot{X}_h, \quad [\dot{X}_j, \dot{X}_k] = 0, \tag{17}$$

where the c^h_{jk} are the structure constants of $\mathfrak{G}$ (in the particular basis being used). The first set of these relations followed from (15), and the third from (12). To obtain the second, one may proceed as follows. First one finds the one-forms dual to the $\{Y_j\}$. These are not, as might be guessed, $\theta^{k'} \equiv (\tau_G)^*\theta^k$ and $d\xi^k$, for

$$i_{X_j^T}\theta^{k'} = \delta^k_j, \quad i_{\dot{X}_j}\theta^{k'} = 0, \quad i_{\dot{X}_j}d\xi^k = \delta^k_j; \qquad (18)$$
$$i_{X_j^T}d\xi^k \equiv L_{X_j^T}\xi^k \neq 0.$$

The last equation forces a search for one-forms to replace the $d\xi^k$. Since the $\theta^{k'}$ and $d\xi^k$ are nevertheless independent, one can write the general one-form as a linear combination of them. We shall therefore put

$$\gamma^k = d\xi^k + K^k_j\theta^{j'}$$

and endeavor to find K^k_j such that $\{\theta^{k'}, \gamma^k\}$ form a suitable set dual to the $\{Y_k\}$. Note that this set will do for the first three equations of (18) independent of the choice of the K^k_j. The K^k_j need be chosen only to satisfy the fourth. By contracting γ^k with X^T_j and setting the result equal to zero, one arrives at

$$K^k_j = -i_{X_j^T}d\xi^k = -L_{X_j^T}\xi^k.$$

To calculate the Lie derivative here we note that the definition of the ξ^k implies that for any second-order vector field Δ,

$$i_\Delta\theta^{k'} = \xi^k.$$

Therefore one obtains

$$L_{X_j^T}\xi^k = L_{X_j^T}i_\Delta\theta^{k'} = i_\Delta L_{X_j^T}\theta^{k'} + i_{[X_j^T,\Delta]}\theta^{k'}.$$

The second term on the right-hand side vanishes. This is because, as was pointed out in the discussion of the Noether theorem (above Eq. (7) of Chapter 15), the commutator of a lifted vector field with a second-order vector field is a vertical vector field, and since $\theta^{k'}$ is by definition the pullback of a one-form on the base manifold, its contraction with a vertical vector field vanishes. Thus (here we use (8))

$$L_{X_j^T}\xi^k = i_\Delta i_{X_j^T}d\theta^{k'} = -\tfrac{1}{2}i_\Delta[c^k_{rh}i_{X_j^T}(\theta^{k'} \wedge \theta^{h'})]$$
$$\equiv i_{X_j^T}d\xi^k = -c^k_{jh}\xi^h.$$

In this way one arrives at

$$\gamma^k = d\xi^k + c^k_{jh}\xi^h\theta^j. \qquad (19)$$

Now Eq. (18) can be rewritten using the set $\{\theta^{k'}, \gamma^k\}$ instead of $\{\theta^{k'}, d\xi^k\}$. The third and fourth equations become

$$i_{\dot{X}_j}\gamma^k = \delta^k_j, \quad i_{X_j^T}\gamma^k = 0. \qquad (20)$$

We now return to the second of Eqs. (17) and calculate $[X^T_j, \dot{X}_k]$. Contract first with the $\theta^{k'}$.

$$i_{[X_j^T,\dot{X}_k]}\theta^{h'} = -L_{\dot{X}_k}(i_{X_j^T}\theta^{h'}) + i_{X_j^T}L_{\dot{X}_k}\theta^{h'}$$
$$= i_{X_j^T}i_{\dot{X}_k}d\theta^{h'} = 0.$$

Now contract with γ^h:

$$\begin{aligned} i_{[X_j^T,\dot{X}_k]}\gamma^h &= i_{[X_j^T,\dot{X}_k]}d\xi^h + c^h_{rn}\gamma^n i_{[X_j^T,\dot{X}_k]}\theta^{r'} \\ &= L_{X_j^T} i_{\dot{X}_k} d\xi^h - i_{\dot{X}_k} L_{X_j^T} d\xi^h = -i_{\dot{X}_k} dL_{X_j^T}\xi^h \\ &= i_{\dot{X}_k}(c^h_{jr}\xi^r) = c^h_{jk}. \end{aligned}$$

The second of Eqs. (17) follows immediately.

5. Finally, we construct a parallelization of T^*G, where G is a Lie group. Again, one starts from an independent set of left-invariant (or right-invariant) vector fields $X_j \in G$. The vector fields $\{X_k^*, \partial/\partial P_k\}$ form a parallelization $\{Y_k\}$ of T^*G, where the $X_k^* \in T^*G$ are the lifted vector fields defined at Eq. (20) of Chapter 11 and the $\partial/\partial P_k$ are defined in the discussion of parallelizing T^*M in §(3). As was the case for TG, the one-forms $\theta^{h'} \equiv (\tau_G^*)^*\theta^h$ and dP_k, which one might have guessed do so, actually do not form a set dual to the Y_j, and for a similar reason: $i_{X_j^*} dP_k \neq 0$. Roughly in the same way as was shown for the set $\{\theta^{k'}, \gamma^k\}$ on TG, it can be shown that the set $\{\theta^{k'}, \pi_k\}$ on T^*G is dual to the Y_j, where

$$\pi_k = dP_k - c^h_{jk} P_h \theta^{j'}. \tag{21}$$

The commutation rules of the vector fields can then be found, again roughly in the same way as they were found on TG to establish the validity of (17). They are

$$\begin{gathered} [X_j^*, X_k^*] = c^h_{jk} X_h^*, \quad [X_j^*, \partial/\partial P_k] = -c^k_{jh}\partial/\partial P_h, \\ [\partial/\partial P_j, \partial/\partial P_k] = 0. \end{gathered} \tag{22}$$

From the expressions at Eq. (16) for θ_0 and ω it can be shown that the X_k^* are Hamiltonian. In fact it is found that

$$i_{X_k^*}\omega = dP_k, \tag{23}$$

so that the P_k are the Hamiltonian functions of the X_k^*. The first of the commutation relations (22) shows, moreover, that the lifted action of $\mathfrak{G}$ on TG, that provided by the X_k^*, is strongly Hamiltonian. Finally, we write down the expression for ω in terms of the set $\{\theta^{k'}, \pi_k\}$:

$$\omega = -\pi_k \wedge \theta^{k'} - \tfrac{1}{2} P_k c^k_{hj} \theta^{h'} \wedge \theta^{j'}. \tag{24}$$

25.3 FORMALISMS OF DYNAMICS ON TG AND T^*G

The definitions and the relations of §25.2 will now be used to write down the Lagrangian formalism on TG and the Hamiltonian formalism on T^*G, where G is a Lie group. Eventually the results so obtained will be applied to the rigid rotator.

1. Let $\mathscr{L} \in \mathscr{F}(TG)$ be a Lagrangian function and Δ be its dynamical vector field. Then the Euler–Lagrange equations are given by

$$L_\Delta \theta_{\mathscr{L}} - d\mathscr{L} = 0, \tag{25}$$

where $\theta_{\mathscr{L}}$ is the Lagrangian one-form defined at Eq. (37) of Chapter 10 (and more formally in the Appendix to Chapter 10). Let the coordinates on G now be called q^k rather than x^k, for we intend to apply the results to dynamical systems in which G will play the role of the configuration manifold Q. We want to write out the Euler–Lagrange equations in terms of the vector fields X_k^T and $\dot{X}_k \equiv \partial/\partial\xi^k$, which parallelize TG, and the one-forms dual to them, as discussed in §25.2(4).

For $\theta_{\mathscr{L}}$ this is relatively easy to do from its expression in local coordinates:

$$\theta_{\mathscr{L}} = \frac{\partial\mathscr{L}}{\partial\dot{q}^k}dq^k = B_j^k\frac{\partial\mathscr{L}}{\partial\xi^k}dq^j = \frac{\partial\mathscr{L}}{\partial\xi^k}\theta^{k'}, \tag{26}$$

where we use the notation established above Eq. (11). Next, the one-form $d\mathscr{L}$ can be written (a similar equation holds for the exterior derivative of any function in $\mathscr{F}(TG)$)

$$d\mathscr{L} = (L_{X_k^T}\mathscr{L})\theta^{k'} + \frac{\partial\mathscr{L}}{\partial\xi^k}\gamma^k, \tag{27}$$

We shall use the fact that Δ is of second order; in terms of the variables now being employed, this means that

$$\Delta = \xi^k X_k^T + \varphi^k\partial/\partial\xi^k, \quad \varphi^k\in\mathscr{F}(TG), \tag{28}$$

which is easily found from the equation above (15) for the X_k^T and by remembering that in local coordinates a second-order vector field is of the form $\dot{q}^k\partial/\partial q^k + f^k(q,\dot{q})\partial/\partial\dot{q}^k$. To obtain $L_\Delta\theta_{\mathscr{L}}$ it will be helpful to calculate $L_\Delta\theta^{k'}$:

$$L_\Delta\theta^{k'} = i_\Delta d\theta^{k'} + di_\Delta\theta^{k'} = i_\Delta(-\tfrac{1}{2}c_{jh}^k\theta^{j'}\wedge\theta^{h'}) + d\xi^k,$$

where the last term comes from the expression for Δ as a vector field of second order. That expression is used again on the first term to yield

$$L_\Delta\theta^{k'} = c_{jh}^k\xi^h\theta^{j'} + d\xi^k = \gamma^k. \tag{29}$$

Thus

$$L_\Delta\theta_{\mathscr{L}} = \left(L_\Delta\frac{\partial\mathscr{L}}{\partial\xi^k}\right)\theta^{k'} + \frac{\partial\mathscr{L}}{\partial\xi^k}\gamma^k. \tag{30}$$

Finally, by combining (27) with (30), the Euler–Lagrange equations (25) can be put in the form

$$\left(L_\Delta\frac{\partial\mathscr{L}}{\partial\xi^k} - L_{X_k^T}\mathscr{L}\right)\theta^{k'} = 0,$$

or, for $k\in\{1,\ldots,\mu\}$,

$$\frac{d}{dt}\frac{\partial\mathscr{L}}{\partial\xi^k} - L_{X_k^T}\mathscr{L} = 0. \tag{31}$$

Recall that the ξ^k are variables on the fibers. Perhaps they could have been denoted $\dot{Q}^k$ rather than ξ^k, sort of generalized $\dot{q}^k$, but then there would exist no variables Q^k to which they correspond. Thus the statement that the dynamical system Δ is of second order, ordinarily written in the form $\dot{q}^k = dq^k/dt$, cannot be made in the notation obtained from the parallelization of TG. The only way to make this statement is

$$dq^k/dt = A_j{}^k \xi^j. \tag{32}$$

2. We now turn to the Hamiltonian formalism on T^*G. The equations of motion, Hamilton's canonical equations, are given on the cotangent bundle by

$$i_\Delta \omega = dH, \tag{33}$$

where ω is the natural symplectic form on T^*G, given in the terms obtained from the parallelization by Eq. (24), and $H \in \mathscr{F}(T^*G)$ is the Hamiltonian function of the dynamical system Δ. One may write

$$\Delta = f^k X_k^* + g_k \partial/\partial P_k, \quad f^k, g_k \in \mathscr{F}(T^*G), \tag{34}$$

and (as for any function in $\mathscr{F}(T^*G)$)

$$dH = (L_{X_k^*} H)\theta^{k'} + \frac{\partial H}{\partial P_k} \pi_k. \tag{35}$$

Eqs. (33)–(35) can be combined with (26) to yield expressions for the f^k and g_k; this leads to

$$\Delta = \frac{\partial H}{\partial P_k} X_k^* + \left(P_j c^j_{kh} \frac{\partial H}{\partial P_h} - L_{X_k^*} H \right) \frac{\partial}{\partial P_k}. \tag{36}$$

Eq. (36) represents Hamilton's canonical equations written in the variables obtained from the parallelization of T^*G. When $G = \mathbb{R}^\mu$ is the additive group, the c^j_{kh} vanish, X_k^* becomes $\partial/\partial q^k$, and P_k becomes p_k. Then (36) is transformed into the familiar

$$\Delta = \frac{\partial H}{\partial p_k} \frac{\partial}{\partial q^k} - \frac{\partial H}{\partial q^k} \frac{\partial}{\partial p_k}.$$

25.4 THE RIGID ROTATOR

We shall now apply some of the above results to the *rigid rotator*, a rigid body with one point fixed. The configuration manifold Q of the rotator is the Lie group $G = SO(3)$, for each of its positions, each *orientation*, can be obtained from a given orientation by a unique rotation. Thus the dynamical system of a rigid body is a dynamical system on a Lie group, and the above results are immediately applicable to it.

As is familiar, there are two systems of coordinates involved in the dynamics of a rotator, *body* and *space*. We intend to show that these two coordinate systems arise from parallelizations involving left-invariant and right-invariant vector fields, and for this reason we start with a discussion of the relation between the parallelizations obtained in these two ways.

1. Since we shall now have to distinguish between left-invariant and right-invariant vector fields, we shall denote the left-invariant ones as before by X_k and right-invariant ones by Z_k. As has been seen, a parallelization of a manifold, in particular of G, is a *trivialization* of the tangent bundle, exhibiting its diffeomorphism to a direct product. Let the trivialization of G associated with the left-invariant vector fields X_k be called λ. One may write

$$\lambda : TG \to G \times \mathbb{R}^{\mu} : v_g \mapsto (g; \xi^1, \ldots, \xi^{\mu}) \in G \times \mathbb{R}^{\mu}, \tag{37}$$

where $v_g \equiv \xi^k X_k(g) \in T_g G$ (see §25.2(2)). Similarly, let the trivialization associated with the right-invariant vector fields Z_k be called ρ. One then has

$$\rho : TG \to G \times \mathbb{R}^{\mu} : v_g \mapsto (g; \zeta^1, \ldots, \zeta^{\mu}) \in G \times \mathbb{R}^{\mu}, \tag{38}$$

where $v_g \equiv \zeta^k Z_k(g) \in T_g G$; the ζ^k play the same role with respect to the Z_k as the ξ^k do with respect to the X_k. As in §25.2(2), ξ^k is the component of v_g along X_k, and ζ^k is its component along Z_k. Note that the ξ^k and ζ^k vary from point to point $g \in G$, but we suppress this in our notation.

What is the relation between these two sets of components of v_g? In order to answer this question we return to the definition of left-translation (Eq. (1) of Chapter 23 and some of the discussion which follows it) and now add right-translation. The definitions are contained in

$$L_g : G \to G : h \mapsto gh, \quad R_g : G \to G : h \mapsto hg. \tag{39}$$

Given a basis $\{e_k\}$ of $\mathfrak{G}$, the vector fields X_k and Z_k may be defined by

$$\begin{aligned} X_k(g) &= (T_e L_g)(e_k), \\ Z_k(g) &= (T_e R_g)(e_k), \end{aligned} \tag{40}$$

where $T_e L_g$ and $T_e R_g$, of course, map $T_e G$ onto $T_g G$. Hence

$$v_g = \xi^k (T_e L_g)(e_k) = \zeta^k (T_e R_g)(e_k),$$

so that by applying $(T_e L_g)^{-1}$ one arrives at (we omit the parentheses around the e_k)

$$\xi^k e_k = \zeta^j (T_e L_g)^{-1} (T_e R_g) e_j = \zeta^j T_e (L_{g^{-1}} R_g) e_j.$$

It is easily shown from Eq. (39) above and Eq. (10) of Chapter 23 that $I_g = L_g R_{g^{-1}}$, so that

$$\xi^k e_k = \zeta^j [Ad_g]^{-1} e_j, \tag{41}$$

where $Ad_g = T_g I_g$ is the Lie group action defined at Eq. (11) of Chapter 23. Another way of writing (41) is

$$\xi^k(g) = [(Ad_g)^{-1}]^k_j \zeta^j(g); \quad \zeta^j(g) = [Ad_g]^j_k \xi^k(g). \tag{42}$$

Here we have indicated the g dependence of the ξ^k and ζ^k.

One can construct similarly two trivializations λ^* and ρ^* of T^*G with similar properties. We do not go into the details, but merely list some of the formulas, whose meaning is evident from the clear analogy with those for λ and ρ. Let α_g be a vector in T_g^*G, and write

$$\alpha_g = P_k \theta^{k'} = R_k \varphi^{k'},$$

where the $\varphi^{k'}$ are obtained from the Z_k in the same way as the $\theta^{k'}$ are obtained from the X_k. Then

$$\lambda^*: T^*G \to G \times \mathbb{R}^\mu : \alpha_g \mapsto (g; P_1, \ldots, P_\mu),$$
$$\rho^*: T^*G \to G \times \mathbb{R}^\mu : \alpha_g \mapsto (g; R_1, \ldots, R_\mu).$$

One arrives at

$$P_k(g) = [Ad_g^*]_k^j R_j(g); \quad R_j(g) = [(Ad_g^*)^{-1}]_j^k P_k(g). \tag{43}$$

2. Now consider a dynamical system whose configuration manifold is a group G. Let TG be the carrier manifold of the dynamics, which means that the point which we have called v_g moves in time; in fact the point can no longer be called v_g, for it does not lie always in the same fiber as g changes in time. Eq. (42) establishes the relations between the globally defined coordinates $\xi^k(g)$ and $\zeta^k(g)$. We now study the relation between their time derivatives.

According to Eq. (42)

$$\frac{d\zeta^j}{dt} = (Ad_g)_k^j \frac{d\xi^k}{dt} + \left[\frac{d}{dt}(Ad_g)_k^j\right]\xi^k. \tag{44}$$

As a first step in evaluating the derivative in the second term on the right-hand side, recall that the Lie group action Ad of G on $\mathfrak{G}$ is the integral of the Lie algebra action ad of $\mathfrak{G}$ on itself. That is,

$$\frac{d}{dt} Ad_{\exp(ta)}|_{t=0} a = [a, b], \quad a, b \in \mathfrak{G}, \tag{45}$$

where $\exp(ta)$ is the integral curve of X_a defined through Eq. (24) of Chapter 23 and the identification of T_eG with $\mathfrak{G}$. But (44) differs from (45) on the one hand because $g(t)$ represents an arbitrary curve in G, not necessarily the integral curve of a left-invariant vector field such as X_a, and on the other because the derivative is not necessarily taken at the identity element of G (at the origin). For our purpose we shall replace $g(t)$ by a convenient curve which has the same first derivative at the point of interest, in the process effectively moving the calculation to the origin. One first uses (40) to write

$$\frac{d}{d\tau} g(\tau)|_{\tau=t} \equiv \Omega^k Z_k(g(t)) = \Omega^k (T_e R_g) e_k,$$

defining the Ω^k. The new curve $\tilde{g}(\tau)$ is defined by

$$\tilde{g}(\tau) = [\exp(\Omega^k \tau e_k)] g(t).$$

Then

$$\frac{d}{d\tau}\tilde{g}(\tau)|_{\tau=0} = \Omega^k e_k g(t) = \Omega^k Z_k(g(t)) = \frac{d}{d\tau} g(\tau)|_{\tau=t},$$

so that g and $\tilde{g}$ do in fact have the same derivatives.

Now the required derivative can be calculated. One has

$$\frac{d}{d\tau} Ad_{g(\tau)} a|_{\tau=t} = \frac{d}{d\tau} Ad_{\exp(\Omega^k \tau e_k) g} a|_{\tau=0} = \frac{d}{d\tau}[Ad_{\exp(\Omega^k \tau e_k)}][Ad_{g(t)}] a|_{\tau=0}$$
$$= [\Omega^k e_k, Ad_{g(t)} a].$$

In Eq. (44) a as such does not appear, but if (44) is applied to the basis $\{e_k\}$ of the Lie algebra, a becomes $\xi^k e_k$, and $Ad_g a$ becomes $(Ad_g)^k_j \xi^j e_k = \zeta^k e_k$. Thus (44) becomes

$$\frac{d\zeta^j}{dt} = [Ad_{g(t)}]^j_k \frac{d\xi^k}{dt} + \Omega^h \zeta^k c^j_{hk}. \tag{46}$$

If G is $SO(3)$ and the dynamics on G is the motion of a rigid body, the Ω^k represent the components of the angular velocity vector; then the c^j_{hk}, the structure constants of $SO(3)$, make the last term look just like the vector cross-product. Eq. (46) will then be recognized as the formula connecting the time derivatives of the space and body coordinates in rigid-body motion, where the ξ^k are the body coordinates and the ζ^k are the space coordinates. This justifies our assertion that these two coordinate systems arise from parallelizations involving left-invariant and right-invariant vector fields.

3. Similar definitions and calculations can be made in T^*G. The result analogous to (46) is

$$\frac{dR_j}{dt} = [Ad^*_{g^{-1}}]^k_j \frac{dP_k}{dt} - \omega^k P_h [Ad_{g^{-1}}]^r_j c^h_{kr}, \tag{47}$$

where $\omega^k = Ad_{g^{-1}} \Omega^k$ is the kth component of the (angular) velocity in the body frame.

We shall not go through the derivation in any detail. The action of Ad^* is obtained from its definition at Eq. (12) of Chapter 23:

$$\langle Ad^*_{g^{-1}} \alpha, a\rangle = \langle \alpha, Ad_{g^{-1}} a\rangle, \quad \alpha \in \mathfrak{G}^*, \quad a \in \mathfrak{G}.$$

The derivative of $Ad^*_{g^{-1}}$ is thus obtained from that of $Ad_{g^{-1}}$, which in turn can be obtained from $Ad_g Ad_{g^{-1}} = \mathbf{1}$ and the Leibniz rule; this yields

$$\frac{d}{dt} Ad_{g^{-1}} = -Ad_{g^{-1}} \left[\frac{d}{dt} Ad_g\right] Ad_{g^{-1}}.$$

Some calculation then leads to

$$\frac{d}{dt} Ad_{g^{-1}} a = [\omega^k e_k, Ad_{g^{-1}} a].$$

In the present case α is $P_k\theta^{k\prime}$, and with a chosen to be e_j one obtains

$$\frac{d}{dt}\langle Ad^*_{g^{-1}}\alpha, e_j\rangle = \left[\frac{d}{dt}(Ad^*_{g^{-1}})^k_j\right]P_k$$

$$= -\langle\alpha, [\omega^k e_k, (Ad_{g^{-1}})^h_j e_h]\rangle = -\omega^k(Ad_{g^{-1}})^h_j c^r_{kh}P_r.$$

4. We now treat a Lagrangian dynamical system whose configuration manifold is a group G and whose Lagrangian is a function only of the coordinates on the fibers of TG. This clearly generalizes the dynamical system of a rigid body. Let the Lagrangian be given in the coordinates we have been discussing by

$$\mathscr{L} = \tfrac{1}{2}T_{jk}\xi^j\xi^k, \tag{48}$$

where the $T_{jk} = T_{kj}$ form a constant matrix. This form for $\mathscr{L}$ is suggestive in view of the remark made just after Eq. (31), in which the ξ^k were identified as sort of generalized $\dot{q}^k$, for if the ξ^k are taken to be velocities, $\mathscr{L}$ looks like a kinetic energy. In particular, it looks like the kinetic energy of a free rotator, and hence its Lagrangian.

The Euler–Lagrange equations are given by (31). The second term in that expression is easily calculated because the Lie derivatives of the ξ^k were found in obtaining (19). One arrives at

$$T_{jk}d\xi^k/dt = c^h_{rj}T_{hk}\xi^r\xi^k. \tag{49}$$

For the case of $SO(3)$, i.e. for the rigid rotator, these equations contain the same information as Euler's equations for the motion of a rigid body: they are a set of first-order differential equations for the motion of the angular velocity vector.

Assume now that the matrix of the T_{jk} is invertible, which makes it possible to go over to the Hamiltonian formalism on T^*G. The Hamiltonian function is given on T^*G by

$$H = \tfrac{1}{2}\tau^{jk}P_jP_k, \tag{50}$$

where the matrix of the $\tau^{jk} = \tau^{kj}$ is inverse to the T_{jk}. The equations of motion can be obtained from the Poisson brackets:

$$dP_j/dt = \{P_j, H\},$$

which can be found from the known $\{P_j, P_k\} = -c^h_{jk}P_h$. These are known either from a calculation such as the one suggested following Eq. (16) (the g^h_{jk} are now the structure constants c^h_{jk}) or from the first of (22) and the fact that the P_k are the Hamiltonian functions of the X^*_k. In this way one finds that

$$dP_j/dt = -c^h_{jk}\tau^{kr}P_hP_r. \tag{51}$$

The time derivatives of the R_k can be found by using Eqs. (43) and (47). We shall not go into the details, but merely state the result, namely $dR_k/dt = \{R_k, H\} = 0$. Essentially this is a reflection of the fact that the right-invariant vector fields commute with the left-invariant ones (which, of course, provides a trivial way to obtain this result).

26

Examples and applications

In this chapter we apply some aspects of the formalism, particularly of the second half of the book, to several examples of dynamical systems.

26.1 THE HARMONIC OSCILLATOR IN n DEGREES OF FREEDOM

1. We again study the isotropic harmonic oscillator. Let $Q = \mathbb{R}^n$, and let the Hamiltonian function be given on phase space $M \equiv T^*\mathbb{R}^n = \mathbb{R}^n \times \mathbb{R}^n$ by

$$H = \tfrac{1}{2}\sum_j [(p_j)^2 + \lambda^2(q^j)^2], \tag{1}$$

with the usual symplectic form $\omega = -dp_j \wedge dq^j$. (We again use the traditional n rather than ν for the dimension or the number of degrees of freedom. This is related to the involvement of the unitary group, which we continue to call $U(n)$.) It is convenient to introduce the complex variables (see §14.2(2))

$$z_j = \lambda x^j + ip_j \tag{2}$$

and to think of M as $\mathbb{C}^n$. In these terms Eq. (1) becomes

$$H = \tfrac{1}{2}z_j\bar{z}_j, \tag{3}$$

(the energy E is then half the *magnitude* $\sum|z_j|^2$ of $z\in\mathbb{C}^n$) and the symplectic form becomes

$$\omega = (2\lambda i)^{-1}d\bar{z}_j \wedge dz_j. \tag{4}$$

It is easily verified that if $F, G\in\mathscr{F}(M)$, their Poisson bracket is given by

$$\{F,G\} = -2\lambda i(\partial_j F\bar{\partial}_j G - \bar{\partial}_j F\,\partial_j G), \tag{5}$$

where $\partial_j \equiv \partial/\partial z_j$ and $\bar{\partial}_j \equiv \partial/\partial \bar{z}_j$. The Hamiltonian vector field $X_F\in\mathfrak{X}(M)$ whose Hamiltonian function is F is then found to be

$$X_F = 2\lambda i(\partial_j F)\bar{\partial}_j - 2\lambda i(\bar{\partial}_j F)\partial_j. \tag{6}$$

In particular, the dynamical vector field Δ whose Hamiltonian is given by (3) is

$$\Delta = \lambda i(\bar{z}_j\bar{\partial}_j - z_j\partial_j), \tag{7}$$

and its integral curves are given by

$$z_j(t) = e^{-i\lambda t} z_j(0). \tag{8}$$

By taking the Lie derivative with respect to Δ it is easily shown that each product of the form $z_j\bar{z}_k$ is a constant of the motion.

For this dynamical system it is as easy to start with an algebra of constants of the motion and integrate to an invariance group of the Hamiltonian as to start with the invariance group and obtain the algebra. We choose to start with the algebra. Since each product $z_j\bar{z}_k$ is a constant of the motion, so is every bilinear function of the form

$$F = (2\lambda i)^{-1} c_{jk}\bar{z}_j z_k, \tag{9}$$

where the c_{jk} form an arbitrary constant matrix C. The condition that F be real imposes the condition $\bar{c}_{jk} = -c_{kj}$, or the C matrices must be anti-Hermitian. Such constants of the motion are thus in one-to-one correspondence with the anti-Hermitian matrices. This correspondence is in fact a Lie-algebra isomorphism: let F and G be two bilinear constants of the motion of the form of (9) and let them correspond to the anti-Hermitian matrices C and D, respectively. Then a brief calculation will show that the Poisson bracket $\{F, G\}$ corresponds to the commutator bracket $[C, D]$. This is related to the following two facts. The Poisson bracket of two constants of the motion is itself a constant of the motion (in this case also bilinear), so that these functions close under the Poisson bracket to form a Lie algebra; the commutator of two anti-Hermitian matrices is itself anti-Hermitian, so that these matrices also close under the commutator to form a Lie algebra. This latter is $u(n)$, the Lie algebra of the unitary group $U(n)$ (see in this connection the example labeled (b) of §23.1(3)), for every unitary matrix can be obtained by exponentiating an anti-Hermitian one. Thus the Lie algebra of the constants of the motion is isomorphic to $u(n)$. The dimension of $u(n)$ as a Lie algebra over the reals is n^2; this comes from the $2n^2$ real numbers in the elements of a general complex matrix, which are then subjected to the n^2 conditions of antihermiticity (the diagonal elements are then all imaginary and the off-diagonal ones occur in negative complex conjugate pairs).

Remark. The condition of antihermiticity arose from the requirement that the constants of the motion be real. In fact the anti-Hermitian matrices are a vector space only over the reals, so the representation of $u(n)$ by these matrices restricts the Lie algebra to being over the reals in spite of the fact that the matrix elements are complex.

From Eq. (6) one finds that the vector fields generated by the constants of the form of (9) are linear and hence complete: with the F of (9), Eq. (6) yields

$$X_F = c_{kj} z_j \partial_k + \bar{c}_{kj}\bar{z}_j\bar{\partial}_k. \tag{10}$$

The integral curves are easily obtained. One arrives at

$$z(t)=e^{tC}z(0), \tag{11}$$

where $z=\{z_1,\ldots,z_n)\in\mathbb{C}^n$. Comparison of (11) and (8) shows that the dynamics corresponds to the matrix $C=\mathrm{i}\lambda\mathbf{1}$ (this can also be seen by comparing the Hamiltonian of (3) with the general bilinear constant of the motion of (9)), which defines a one-dimensional central subalgebra of $u(n)$. This subalgebra integrates to the one-parameter subgroup $U(1)\subset U(n)$.

Usually the symmetry group or algebra of the oscillator is defined with the dynamics, an obvious symmetry, removed. At the level of the Lie algebra this is accomplished by imposing the condition of zero trace on the anti-Hermitian matrices: this clearly excludes the dynamics and the single condition reduces the dimension by one to n^2-1. What remains is still a Lie algebra, for the trace of a commutator is necessarily zero. In fact the traceless anti-Hermitian matrices form the derived algebra of $u(n)$. This smaller Lie algebra is $su(n)$, for every unimodular unitary matrix can be obtained by exponentiating a traceless anti-Hermitian one.

2. The Hamiltonian vector fields generated by the constants of the motion corresponding to the matrices of $su(n)$ form a Lie algebra action of $su(n)$ on $M\equiv T^*\mathbb{R}^n=\mathbb{R}^{2n}=\mathbb{C}^n$. The n^2-1 bilinear traceless constants of the motion described above are the components of the momentum map

$$\mu: M\rightarrow su(n)^* \tag{12}$$

obtained from this action. That is, there is a way to order the n^2-1 constants of the motion $\{F_\kappa\}$ and there exists a basis $\{\varepsilon^\kappa\}$ in $su(n)^*$ such that $\mu(z)=F_\kappa(z)\varepsilon^\kappa$, $z\in M$ and $\kappa\in\{1,\ldots,n^2-1\}$.

It can now be shown that $\mu^{-1}(\alpha)$, $\alpha\in su(n)^*$, is a trajectory of the motion. We shall show this for $n=2$; the result extends to $n>2$ by pairing the z_k variables in as many ways as is necessary (including overlapping pairs) and using the results obtained in the way that follows for each subalgebra $su(2)\subset su(n)$. Thus for the present discussion the indices run only from 1 to 2 and $n^2-1=3$. Choose the three constants of Eq. (9), or rather the three traceless C matrices, to be

$$F_1\leftrightarrow C_1=\begin{vmatrix} i & 0\\ 0 & -i\end{vmatrix},\quad F_2\leftrightarrow C_2=\begin{vmatrix} 0 & 1\\ -1 & 0\end{vmatrix},\quad F_3\leftrightarrow C_3=\begin{vmatrix} 0 & i\\ i & 0\end{vmatrix}$$

and write

$$z_j=\rho_j\mathrm{e}^{i\theta},\quad j=1,2.$$

Then

$$\begin{aligned}\lambda F_1&=\tfrac{1}{2}[(\rho_1)^2-(\rho_2)^2],\\ \lambda F_2&=\rho_1\rho_2\sin(\theta_2-\theta_1),\\ \lambda F_3&=\rho_1\rho_2\cos(\theta_2-\theta_1).\end{aligned}$$

Thus F_1 determines $(\rho_1)^2-(\rho_2)^2$, and $(F_3)^3+(F_2)^2$ determines $(\rho_1\rho_2)^2$, which then determine ρ_1 and ρ_2. From these one finds the sine and the cosine of $\theta_2 - \theta_1$, so that the angles are determined by the F_j up to a common additive constant. But since the F_j are the components of the momentum map, their values are fixed by α. Therefore one can write

$$\mu^{-1}(\alpha) = e^{i\sigma}(z_1, z_2)$$

with only $\sigma \in \mathbb{R}$ arbitrary. One extends this procedure to all of M by choosing overlapping pairs of z_k, with the result that more generally

$$\mu^{-1}(\alpha) = e^{i\sigma} z, \tag{13}$$

where only σ is arbitrary and $z \in \mathbb{C}^n = M$ is fixed. Comparison with Eq. (8) shows that this is an orbit of the oscillator.

3. We now turn to the foliation obtained from the action of $SU(n)$ on $\mathbb{C}^n$ which is generated by the vector fields of Eq. (10). It will be shown first that $SU(n)$ acts transitively on the sphere

$$S^{2n-1} \equiv \{z \in \mathbb{C}^n \vdash \sum_k |z_k|^2 = \text{const.} \equiv 2E\}.$$

We will demonstrate this actually on a sphere of radius one; it then follows trivially for any other sphere. Let $\zeta^1 \in \mathbb{C}^n$ have magnitude one, $\sum |\zeta_j^1|^2 = 1$, but otherwise be arbitrary. Let it be the first element of an orthonormal set $\{\zeta^k\}$ in $\mathbb{C}^n$, so that

$$\zeta_j^h \bar{\zeta}_j^k = \delta^{hk}.$$

Then the matrix whose components are $c_{jk} = \zeta_j^k$ is unitary, but not necessarily of determinant one. If $\det C = e^{i\varphi}$, then the orthonormal set can be changed by multiplying one. If its elements, say ζ^2, by $e^{-i\varphi}$, and the resulting matrix C will be unimodular and hence in $SU(n)$. When this C is applied to the element $z_0 = (1, 0, \ldots, 0) \in \mathbb{C}^n$, also of magnitude one, it carries it into ζ^1. Since ζ^1 is arbitrary, this shows that there is an element of $SU(n)$ that carries z_0 on the unit sphere into any other point on the unit sphere.

To sum up, we refer to Fig. 22.1. In $M = \mathbb{R}^{2n} = \mathbb{C}^n$ the leaves of the $SU(n)$ action, that is the integral manifolds W_a of the distributions spanned by the Hamiltonian vector fields X_F obtained from the bilinear constants of the motion F, are spheres S^{2n-1}. One may take a to be the radius of the sphere, $a = 2E$, where E is the energy. As mentioned above, $N_\alpha = \mu^{-1}(\alpha)$ is an orbit of the oscillator determined by the point $\alpha \in su(n)^*$, or by a set of values for the $n^2 - 1$ bilinear constants of the motion. An orbit of the oscillator is a circle lying, of course, on one of the spheres S^{2n-1} of constant energy. What is missing so far is the lower part of Fig. 22.1, in particular $\mu(W_a)$. Recall (see, for example, the discussion of Fig. 22.1 in §22.3(3)) that each W_a is mapped into a coorbit in $\mathfrak{G}^*$, or for this case in $su(n)^*$.

Each $W_a = S^{2n-1}$ is carried by the momentum map to an orbit of dimension $2n-2$ in $su(n)^*$, for this orbit is obtained by quotienting S^{2n-1} with respect to circles S^1. We shall return to this point immediately. But first we point out that it is not difficult to understand which of the X_F vector fields are annihilated under this mapping. Consider a point $z \in S^{2n-1}$. Those of the X_F which generate one-parameter groups that leave z fixed, i.e. the generators of the isotropy group of z, all vanish at z, and hence so do the vector fields $\tilde{X}_F \in \mathfrak{X}(su(n)^*)$ in the coadjoint representation which correspond to them by equivariance of μ. In addition the one generator of $SU(n)$ which is tangent to the oscillator trajectory at z is also projected under μ to the null vector field on $su(n)^*$. Let us check the dimensions. The isotropy group of z is $SU(n-1)$, whose dimension is

$$(n-1)^2 - 1 = n^2 - 2n,$$

and this is thus the number of vector fields annihilated in this way by μ. In addition there is the one vector field along the oscillator trajectory. Hence the orbit of $\mu(z)$ under the coadjoint action, what has been called the coorbit, has dimension

$$n^2 - 1 - (n^2 - 2n) - 1 = 2n - 2,$$

which agrees with the above result.

Thus the coorbits $\mu(W_a)$ have the topology of S^{2n-1}/S^1: two points $\alpha, \beta \in su(n)^*$ are on the same orbit if their inverse images lie on the same oscillator trajectory, or if $\mu^{-1}(\alpha) = e^{i\varphi}\mu^{-1}(\beta)$. But S^{2n-1}/S^1 is by definition the complex projective space $\mathbb{C}P^n$, which thus describes the coorbits in $su(n)^*$. Each W_a sphere of radius a is mapped into a $\mathbb{C}P^n \subset su(n)^*$, and the set of these projective spaces is parametrized by the radii a of the spheres or equivalently by the energy $E = \frac{1}{2}a$.

Remark. For completeness we find the subgroup $H \in SU(n)$ which leaves $\mu(z)$ fixed, the isotropy group $G_{\mu(z)}$. An element $C \in SU(n)$ is in H iff $Cz = e^{i\varphi}z$, so that C is of the form

$$\left\|\begin{array}{c|c} e^{i\varphi} & 0 \\ \hline 0 & e^{-i\gamma}\mathbf{1} \end{array}\right\| \left\|\begin{array}{c|c} 1 & 0 \\ \hline 0 & \sigma \end{array}\right\|,$$

where the upper left-hand blocks are one-dimensional and the lower right-hand ones are $(n-1)$-dimensional. Here $\sigma \in SU(n-1)$ and $\gamma = \varphi/(n-1)$ is included to keep the determinant equal to one. Hence $H = U(1) \times SU(n-1)$.

26.2 THE NONRELATIVISTIC SPINNING PARTICLE

1. The classical nonrelativistic spinning particle is a dynamical system which illustrates the utility of the symplectic manifold formalism. We shall use the symplectic formalism together with an inversion of the reduction procedure which, by passing upward to a manifold of higher dimension, will allow a Lagrangian description of a dynamical system which is initially not Lagrangian. The Hamiltonian description of this system in local coordinates is known (Rijgrok and Van der Vlist, 1980; Balachandran *et al.*, 1983); we shall demonstrate some global features.

The carrier manifold for this system consists of two parts, $T^*\mathbb{R}^3$ for the coordinates and momenta of the particle, plus a spin space which will be constructed out of another $\mathbb{R}^3$. Let the spin be identified by a vector $\mathbf{s} = (s_1, s_2, s_3) \in \mathbb{R}^3$, but with the additional condition that $\mathbf{s} \cdot \mathbf{s} = l^2$, where $l \in \mathbb{R}$ is a constant, forcing $\mathbf{s}$ to lie on the sphere S^2 of radius l. Then the carrier manifold will be taken to be

$$M = T^*\mathbb{R}^3 \times S^2, \tag{14}$$

with symplectic form

$$\begin{aligned} \omega &= \omega_0 + \omega_s, \\ \omega_s &= l^{-1} \varepsilon_{hjk} s_h ds_j \wedge ds_k. \end{aligned} \tag{15}$$

What must be shown first is that ω_s is indeed a symplectic form on S^2, or that it is closed and nondegenerate. That it is closed follows immediately from the fact that it is a two-form on a two-dimensional manifold. To see that it is nondegenerate, consider the volume element $\Omega = ds_1 \wedge ds_2 \wedge ds_3$ on $\mathbb{R}^3$. Because Ω is a volume element, it will not vanish when contracted (or saturated) with three linearly independent vector fields. Choose one of these vector fields to be the normal $N = s_k \partial/\partial s_k$ to the sphere and the other two to be tangent to the sphere. Then $i_N \Omega = l\omega_s$, and it follows that ω_s fails to vanish when saturated with vector fields tangent to S^2, or with vector fields in $\mathfrak{X}(S^2)$. Thus ω_s is nondegenerate. For convenience we shall henceforth take $l = 1$. The Poisson brackets associated with the symplectic structure ω of (15) are then given by

$$\begin{aligned} &\{q^j, q^k\} = \{p_j, p_k\} = \{q^j, s_k\} = \{p_j, s_k\} = 0, \\ &\{q^k, p_j\} = \delta^k_j, \quad \{s_j, s_k\} = \varepsilon_{jkh} s_h, \end{aligned}$$

where the q^j and p_j are coordinates on $T^*\mathbb{R}^3$.

The Hamiltonian function $H \in \mathscr{F}(M)$ which is usually taken to describe the motion of a particle of mass m and magnetic moment γ moving in an external magnetic field $\mathbf{B} = (B_1, B_2, B_3)$ is

$$H = \mathbf{p} \cdot \mathbf{p}/(2m) + \gamma \mathbf{s} \cdot \mathbf{B}. \tag{16}$$

The dynamical vector field obtained from (15) and (16) is then

$$\Delta = (p_k/m)\partial/\partial q^k - \gamma s_j (\partial B_j/\partial q^k) \partial/\partial p_k + \gamma \varepsilon_{kjh} B_j s_h \partial/\partial s_k. \tag{17}$$

2. The dynamical system of (17) is obviously Hamiltonian by construction. Nevertheless M is not a cotangent bundle, for S^2 is not the cotangent bundle T^*Q of any manifold Q. Moreover, the symplectic form ω of (15), although closed, is not exact; indeed, ω_s is essentially the element of solid angle on S^2, so its integral over the sphere fails to vanish. Thus the dynamical system of (17), although Hamiltonian on M, has no Lagrangian counterpart, for ω cannot be written in the form $d\theta_{\mathscr{L}}$. Nevertheless, a Lagrangian description is often useful, and one would like to find a way to lend this system such a description. A way to do this is to move to a manifold of higher dimension with a dynamical system on it which projects down to the Δ of (17). What one wants, then, is a new manifold $\bar{M} \equiv T^*\mathbb{R}^3 \times T^*Q$ with a symplectic structure

$\omega = \omega_0 + \bar{\omega}_s$ and a Hamiltonian vector field $\bar{\Delta}$ with Hamiltonian function $\bar{H}$, such that there exists a projection $\pi: \bar{M} \to M$ which carries T^*Q into $S^2, \bar{\omega}$ into $\omega, \bar{H}$ into H, and $\bar{\Delta}$ into Δ. To be found are the manifold Q, the symplectic form $\bar{\omega}_s$ on T^*Q, and the Hamiltonian function $\bar{H} \in \mathscr{F}(\bar{M})$ from which one also obtains the vector field $\bar{\Delta}$. Finally, $\bar{\Delta}$ should also have a Lagrangian description, which imposes regularity conditions on $\bar{H}$.

This can all be accomplished with the aid of the treatment of the harmonic oscillator in §26.1 for the case $n = 2$. Let $Q = \mathbb{R}^2$, so that $T^*Q = T^*\mathbb{R}^2 = \mathbb{C}^2$, and let

$$\bar{\omega}_s = (2i)^{-1} d\bar{z}_k \wedge dz_k, \quad k = 1, 2. \tag{18}$$

The leaves W_a of the Hamiltonian action of $SU(2)$ on $\mathbb{C}^2$ are spheres S^3 of radii a, each of which is foliated by circles which are the trajectories of the oscillator. The Lie algebra $su(2)$ is diffeomorphic to $\mathbb{R}^3$, as is its dual $su(2)^*$, and the momentum map $\mu: \mathbb{C}^2 \to su(2)^*$ carries each S^3 sphere into its quotient with respect to the circles (this is the Hopf fibration; see §14.2(2)), a sphere $S^2 \subset \mathbb{R}^3$. The vector fields X_k which generate the action of $SU(2)$ on S^3 project onto vector fields $\tilde{X}_k$ which generate the coadjoint action of $SU(2)$ on S^2. Any vector field in $\mathfrak{X}(\mathbb{C}^2)$ which is tangent to an S^3 sphere and projectable with respect to the Hopf fibration yields a vector field on $S^2 \subset \mathbb{R}^3$. Now, the original dynamics Δ possesses a part which does in fact lie on an S^2 sphere (the last term in (17)) and which can therefore be interpreted as obtained by projection from $\mathbb{C}^2$. Thus the problem is now to lift that part of Δ (recall §19.3) to an appropriate vector field $\Delta^+ \in \mathfrak{X}(\mathbb{C}^2)$.

For Δ^+ to be tangent to the S^3 spheres it must be of the form $\Delta^+ = f^k X_k$, where $f^k \in \mathscr{F}(\mathbb{C}^2)$ and the X_k are the generators of the $SU(2)$ action. For it to be projectable with respect to the foliation whose leaves are the oscillator orbits, it must have the property that

$$[\Delta^+, \Delta_0] = g\Delta_0, \quad g \in \mathscr{F}(\mathbb{C}^2),$$

where Δ_0 is the oscillator vector field. This condition is certainly satisfied if the $f^k = \gamma B_k$ are three constants. One can write $B_k \in \mathscr{F}(\mathbb{R}^3)$ (the $\mathbb{R}^3$ here consists of the q^k, p_k variables), for the B_k are then still independent of the $\mathbb{C}^2$ variables. Then $\Delta^+ = \gamma B_k X_k$ is a Hamiltonian vector field whose Hamiltonian function is $H^+ = \gamma B_k F_k$, where F_k is the Hamiltonian function of the generator X_k.

So far we have not specified the X_k; that is, we have not named the particular vector fields to be chosen as a basis for the $su(2)$ action on $\mathbb{C}^2$. All that remains is to choose them so that they project down to the $\tilde{X}_k = \varepsilon_{kjh} s_j \partial/\partial s_h$, which then guarantees that Δ^+ will project down to the last term in (17). We leave this to the reader, as well as the task of writing out the full Hamiltonian H and obtaining the Lagrangian formulation.

26.3 THE ELECTRON–MONOPOLE SYSTEM

We now discuss the classical system consisting of a charged particle (say, an electron) moving in the magnetic field of an infinitely massive magnetic monopole. It will be seen that in this example the difference between global and local considerations is particularly relevant.

1. The configuration space must be taken as $Q = \mathbb{R}^3 - \{0\}$, with the magnetic pole located at the origin. The magnetic field will be described by the antisymmetric tensor

$$F_{jk} = \gamma \varepsilon_{jkh} x_h / r^3, \tag{19}$$

where r, as usual, is the radius vector from the origin and $\gamma = g/4\pi$ specifies the magnetic charge. The usual expression for the magnetic fieldstrength as a vector with components B_j is obtained by setting $B_j = \varepsilon_{jkh} F_{kh}$. Note that we continue to disregard conventions concerning upper and lower indices, as we have been in much of Chapter 26.

The dynamical vector field on TQ is

$$\Delta = \dot{x}_k \partial/\partial x_k + (e\gamma/m) r^{-3} \varepsilon_{kjh} x_j \dot{x}_h \partial/\partial \dot{x}_k, \tag{20}$$

for the coefficient of $\partial/\partial \dot{x}_k$ is just the kth component of the Lorentz force. It is easily verified that the generators of the action of $SO(3)$

$$R_k = \varepsilon_{kjh}(x_j \partial/\partial x_h + \dot{x}_j \partial/\partial \dot{x}_h) \tag{21}$$

(see §23.6(2)) are infinitesimal symmetries for Δ; i.e. $[R_j, \Delta] = 0$. Nevertheless, the components $J_k = m\varepsilon_{kjh} x_j \dot{x}_h$ of the angular momentum are not conserved: the angular momentum is not a constant of the motion.

Constants of the motion could be found through the Noether theorem if there were a suitable Lagrangian description of the motion. Because the forces are velocity-dependent in this case, however, it turns out that it is easier to proceed within the Hamiltonian formulation by finding a symplectic structure on TQ. We will return to the question of a Lagrangian formulation later.

Consider the symplectic structure on TQ defined by

$$\omega = m\,dx_j \wedge d\dot{x}_j + \tfrac{1}{2} e\gamma r^{-3} \varepsilon_{jkh} x_j\, dx_k \wedge dx_h. \tag{22}$$

Note that the last term of this expression is similar to the symplectic structure ω_s on S^2 in §26.2, but that now this is a two-form not on S^2, but on $\mathbb{R}^3$, and as such it requires its own verification of nondegeneracy and closure. We leave that to the reader. Then it is easily found that

$$i_\Delta \omega = d[\tfrac{1}{2} m \sum_j (\dot{x}_j)^2]. \tag{23}$$

The action of $SO(3)$ is Hamiltonian with the symplectic form of (22); indeed,

$$i_{R_k} \omega = d\left(m\varepsilon_{kjh} x_j \dot{x}_h + e\gamma \frac{x_k}{r} \right), \tag{24}$$

so that

$$L_{R_k} \omega = 0.$$

The last factor in (24) is the unit vector $\hat{x}_k = x_k/r$ along the radius vector from the monopole to the electron. The full expression in parentheses is the Hamiltonian function of the generator R_k of the rotations and is a conserved quantity because the R_k are symmetries for Δ. Therefore what is conserved is not simply the angular momentum as defined above, but that angular momentum with an additional term, usually interpreted as a *helicity* of magnitude $e\gamma$ directed along the radius vector.

Remark. It may be interesting to compute the Poisson brackets of the x_k and $\dot{x}_k$ with the ω of (22). They are

$$\{x_j, x_k\} = 0, \quad \{x_j, \dot{x}_k\} = \delta_{jk}/m,$$

$$\{\dot{x}_j, \dot{x}_k\} = -e\gamma\varepsilon_{jkh}x_h/(m^2r^3).$$

2. We now return to the question of a Lagrangian formulation. The problem is that the two-form in the second term of Eq. (22), although closed, is not exact (the first term is), so that neither is ω. This means that there is no $\theta_{\mathscr{L}}$ such that $\omega = d\theta_{\mathscr{L}}$ and hence that ω is not an $\omega_{\mathscr{L}}$. To see that the second term is not exact, it need only be understood (see the beginning of §26.2(2)) that it is the element of solid angle, whose integral over any closed surface about the origin fails to vanish. That it is not exact then follows from Stokes's theorem (the same reasoning was used implicitly in §26.2). The conclusion is that there is no Lagrangian function $\mathscr{L}$ for this dynamical system.

On the other hand, because the second term is closed, in every neighborhood $U \in \mathbb{R}^3$ which can be contracted to a point (essentially in every neighborhood which does not contain the origin) there is a one-form $A \in \mathfrak{X}^*(U)$ such that

$$dA = \tfrac{1}{2}\gamma r^{-3}\varepsilon_{jkh}x_j dx_k \wedge dx_h|U. \tag{25}$$

Let A be written $A = A_k dx_k$ in U (the dx_k form a local basis for one-forms). Then a local Lagrangian description is obtained with the Lagrangian function

$$\mathscr{L} = \tfrac{1}{2}m\sum_j (\dot{x}_j)^2 + eA_j\dot{x}_j, \tag{26}$$

for this gives

$$\theta_{\mathscr{L}} = m\dot{x}_j dx_j + eA_j dx_j \equiv m\dot{x}_j dx_j + A,$$

which in turn yields

$$d\theta_{\mathscr{L}} = m dx_j \wedge d\dot{x}_j - e dA \equiv \omega. \tag{27}$$

The local Lagrangian function that has been obtained may be recognized as the usual one for a charged particle in a magnetic field, with the three components of A understood as the components of the vector potential. A brief calculation will show also that the magnetic field is correctly obtained in terms of A. In the present context A appears more naturally as a one-form instead of a vector.

26.4 RELATIVISTIC INTERACTING PARTICLES

The so-called No-Interaction Theorem (Currie, Jordan and Sudarshan, 1963) demonstrates that under certain rather standard requirements concerning invariance, canonicity, and the Hamiltonian formalism it is impossible to obtain a relativistic dynamical system of nontrivially interacting particles. In this subsection, following a paper by Balachandran *et al.* (1982), we demonstrate the same result within the Lagrangian formalism by studying the relativistic invariance of a Lagrangian symplectic form. Then an alternative method is shown for constructing a relativistically invariant formalism.

1. Consider a system of N particles, whose configuration manifold is $Q = \mathbb{R}^{3N}$. The carrier manifold will be taken to be TQ with coordinates $(q^{\alpha k}, \dot{q}^{\alpha k})$, where the space indices, in Greek characters, range from 1 to 3, and the particle indices, in Latin letters, range from 1 to N. What is to be discussed is invariance under the action of the Poincaré algebra, or group, defined by (*all* sums over particle indices are indicated by summation signs; the summation convention is used for space indices)

$$
\begin{aligned}
P_\alpha &= -\sum \partial/\partial q^{\alpha k}, \\
J_\alpha &= \sum \varepsilon_{\alpha\beta\gamma}(q^{\beta k}\partial/\partial q^{\gamma k} + \dot{q}^{\beta k}\partial/\partial \dot{q}^{\gamma k}), \\
K^\alpha &= \sum (q^{\alpha k}\Delta^k - Y^{\alpha k}), \\
\Delta &= \sum \Delta^k;
\end{aligned}
\tag{28}
$$

here

$$
\begin{aligned}
Y^{\alpha k} &= (\delta^{\alpha\beta} - \dot{q}^{\alpha k}\dot{q}^{\beta k})\partial/\partial \dot{q}^{\beta k}, \\
\Delta^k &= \dot{q}^{\alpha k}\partial/\partial q^{\alpha k} + f^{\alpha k}(q, \dot{q})\partial/\partial \dot{q}^{\alpha k}
\end{aligned}
\tag{29}
$$

(each $f^{\alpha k}$ depends on all the variables; these functions are as yet not determined). The P_α generate the space translations, the J_α the rotations, the K^α the boosts, and Δ generates the time translations. For the specific choice of this action of the Poincaré group the reader is referred to Sudarshan and Mukunda (1974).

Now the following requirements will be imposed. There exists a Lagrangian function $\mathscr{L}$ whose symplectic form $\omega_{\mathscr{L}}$ is invariant under the action of the Poincaré algebra, and whose Euler–Lagrange equations describe the motion. Moreover, the generators of (28) must close under the Lie bracket to form the Poincaré Lie algebra. (This is not yet guaranteed, for the $f^{\alpha k}$ are still unspecified.) The procedure will be to compute the Lie derivatives of the various components of $\omega_{\mathscr{L}}$ with respect to the boosts and time translation. When these Lie derivatives are set equal to zero, certain restrictions will be placed on $\omega_{\mathscr{L}}$, and through it on $\mathscr{L}$. The first result will be that $\mathscr{L}$ necessarily splits up into separate Lagrangians $\mathscr{L} = \sum \mathscr{L}_k$, where each $\mathscr{L}_k$ depends only on the coordinates of the kth particle.

The components of the symplectic form whose derivatives will be taken are

$$\omega_{\mathscr{L}}(\partial/\partial\dot{q}^{\alpha k}, \partial/\partial\dot{q}^{\beta j}), \tag{30a}$$

$$\omega_{\mathscr{L}}(\partial/\partial\dot{q}^{\alpha k}, \partial/\partial q^{\beta j}), \tag{30b}$$

$$\omega_{\mathscr{L}}(\partial/\partial q^{\alpha k}, \partial/\partial q^{\beta j}). \tag{30c}$$

Remark. The expression in (30a) is identically zero. Nevertheless it will be useful to take its Lie derivatives and set them equal to zero. The conclusions arrived at can be understood by reading backward from the last equation to the first, i.e. to the Lie derivative of (30a), which must then vanish.

First consider the Lie derivative of (30a) with respect to K^{γ}. Setting it equal to zero yields

$$\omega_{\mathscr{L}}([K^{\gamma}, \partial/\partial\dot{q}^{\alpha k}],\ \partial/\partial\dot{q}^{\beta j}) + \omega_{\mathscr{L}}(\partial/\partial\dot{q}^{\alpha k}, [K^{\gamma}, \partial/\partial\dot{q}^{\beta j}]) = 0. \tag{31}$$

Now, from Eqs. (28) one obtains

$$[K^{\gamma}, \partial/\partial\dot{q}^{\alpha k}] = -q^{\gamma k}\partial/\partial q^{\beta k} + Z,$$

where Z involves only vertical terms. The vertical terms do not contribute in (31), for they give rise only to multiples of (30a), which vanish identically, so that (31) becomes

$$-q^{\gamma k}\omega_{\mathscr{L}}(\partial/\partial q^{\alpha k}, \partial/\partial\dot{q}^{\beta j}) + q^{\gamma j}\omega_{\mathscr{L}}(\partial/\partial q^{\beta j},\ \partial/\partial\dot{q}^{\alpha k}) = 0.$$

This is what one may conclude from the invariance of (30a) under the boosts (the chain of arguments, as mentioned above, can now be read backwards). If one now looks at the explicit local expression for $\omega_{\mathscr{L}}$ in terms of derivatives of $\mathscr{L}$ (see Eq. (38) of Chapter 10), one finds that the last equation can be written in the form

$$(q^{\gamma k} - q^{\gamma j})\partial^2\mathscr{L}/\partial\dot{q}^{\alpha k}\partial\dot{q}^{\beta j} = 0,$$

or that $\partial^2\mathscr{L}/\partial\dot{q}^{\alpha k}\partial\dot{q}^{\beta j}$ is antisymmetric for $k \neq j$. Since this second derivative is also symmetric, it must vanish, which means that $\mathscr{L}$ can be written in the form

$$\mathscr{L} = \sum_k \mathscr{L}_k(q, \dot{q}^k). \tag{32}$$

That is, $\mathscr{L}$ separates into a sum of functions each of which depends on the velocity of only one of the particles.

A similar strategy is now pursued with the other components of $\omega_{\mathscr{L}}$. We shall not go through all of the manipulations in detail. From

$$(q^{\gamma k} - q^{\gamma j})\omega_{\mathscr{L}}(\partial/\partial q^{\alpha k}, \partial/\partial q^{\beta j}) = 0$$

and Eq. (32) one obtains

$$\partial^2\mathscr{L}_k/\partial q^{\alpha j}\partial\dot{q}^{\beta k} = \partial^2\mathscr{L}_j/\partial q^{\beta k}\partial\dot{q}^{\alpha j}, \quad k \neq j. \tag{33}$$

The right-hand side does not depend on $q^{\beta k}$, and therefore neither does the left-hand side. Hence one can write

$$\partial\mathscr{L}_k(q, \dot{q}^k)/\partial q^{\alpha j} = \dot{q}^{\beta k}\partial G_{\beta k}(q)/\partial q^{\alpha j} + V_{k\alpha j}(q), \quad j \neq k.$$

If one writes a similar equation for $\partial \mathscr{L}_j/\partial q^{\beta k}$ and then returns to (32), one finds that the $G_{\beta k}$ functions satisfy the integrability conditions for the existence of a function $G(q)$ such that $G_{\beta k} = \partial G/\partial q^{\beta k}$. Then by adding up the expressions so obtained for the $\mathscr{L}_k$, one arrives at

$$\mathscr{L}(q, \dot{q}) = \sum_k [\hat{\mathscr{L}}_k(q^k, \dot{q}^k) + \dot{q}^{\beta k}\partial G(q)/\partial q^{\beta k}] + V(q).$$

The second term in the square brackets contributes a total time derivative, and so may be eliminated. The result is a further separation of $\mathscr{L}$:

$$\mathscr{L} = \sum \hat{\mathscr{L}}_k(q^k, \dot{q}^k) + V(q). \tag{34}$$

Each $\hat{\mathscr{L}}_k$ depends only on the position and velocity of the kth particle.

The rest of the calculation will show that $V(q)$ can also be separated into a sum of terms each of which involves only the variables of the kth particle. Take the Lie derivative with respect to Δ of (30c), using the expression for Δ obtained from (28) and (29). Setting the result equal to zero yields

$$\begin{aligned} &\omega_{\mathscr{L}}(\partial/\partial \dot{q}^{\gamma j}, \partial/\partial q^{\alpha h})\partial f^{\gamma j}/\partial q^{\beta k} + \omega_{\mathscr{L}}(\partial/\partial q^{\beta k}, \partial/\partial \dot{q}^{\gamma j})\partial f^{\gamma j}/\partial q^{\alpha h} = 0, \\ &\omega_{\mathscr{L}}(\partial/\partial \dot{q}^{\gamma h}, \partial/\partial q^{\alpha h})\partial f^{\gamma h}/\partial q^{\beta k} + \omega_{\mathscr{L}}(\partial/\partial q^{\beta k}, \partial/\partial \dot{q}^{\gamma k})\partial f^{\gamma k}/\partial q^{\alpha h} = 0 \end{aligned} \tag{35}$$

for $k \neq h$. The Lie derivative with respect to K^δ and use of (35) then leads to

$$(q^{\alpha k} - q^{\alpha j})\omega_{\mathscr{L}_k}(\partial/\partial q^{\beta k}, \partial/\dot{q}^{\gamma k})\frac{\partial f^{\gamma k}}{\partial q^{\delta j}} = 0.$$

The derivative with respect to $q^{\delta j}$ can be moved through the $\omega_{\mathscr{L}_k}$ term, which shows then that

$$\omega_{\mathscr{L}_k}(\partial/\partial q^{\beta k}, \partial/\partial \dot{q}^{\gamma k}) f^{\gamma k},$$

and hence also $f^{\gamma k}$, is independent of q^j for $j \neq k$.

Finally, from the Euler–Lagrange equation $L_\Delta \theta_{\mathscr{L}} = d\mathscr{L}$. One obtains

$$\frac{\partial V(q)}{\partial q^{\alpha k}} = \frac{\partial \hat{\mathscr{L}}_k}{\partial q^{\alpha k}}(q^k, \dot{q}^k) - \frac{\partial^2 \hat{\mathscr{L}}_k(q^k, \dot{q}^k)}{\partial \dot{q}^{\alpha k}\partial \dot{q}^{\beta k}} f^{\beta k} - \frac{\partial^2 \hat{\mathscr{L}}_k(q^k, \dot{q}^k)}{\partial \dot{q}^{\alpha k}\partial q^{\beta k}} \dot{q}^{\beta k}.$$

The right-hand side of this equation is independent of q^j for $j \neq k$, and hence so is the left-hand side. Thus $V(q)$ separates as asserted, and one finally obtains

$$\mathscr{L} = \sum [\hat{\mathscr{L}}_k(q^k, \dot{q}^k) - V_k(q^k)] \equiv \sum \mathscr{L}_k(q^k, \dot{q}^k). \tag{36}$$

What remains is to show that each $\mathscr{L}_k$ is the usual relativistic free-particle Lagrangian. Without going into any detail, we sketch a proof. Since $\mathscr{L}$ is the sum of single-particle Lagrangians $\mathscr{L}_k$, its symplectic form $\omega_{\mathscr{L}}$ is a similar sum of single-particle symplectic forms ω_k. The full symplectic form ω is invariant under the action of the Poincaré group, in particular under the diagonal action (in which each ω_k is acted upon individually). One can then discuss each ω_k or each $\mathscr{L}_k$ separately, dropping the subscript k, and then the derivation becomes the usual one. Among other things, it can be shown that the single-particle $\mathscr{L}$ can be selected entirely invariant, rather than invariant up to an additive time derivative (Levy-Leblond, 1969). We shall not go through the standard derivation which shows that $\mathscr{L}$ must be of the form

$$\mathscr{L} = c(1 - \dot{\mathbf{q}}\cdot\dot{\mathbf{q}})^{\frac{1}{2}}. \tag{37}$$

Thus we have arrived at the No-Interaction Theorem: Poincaré invariance and the Lagrangian formalism force the Lagrangian of a many-particle dynamical system to be the free-particle Lagrangian.

2. We now describe a method for constructing a relativistically invariant dynamics for N particles in interaction (Balachandran *et al.*, 1984). In order to bypass the No-Interaction Theorem one must abandon at least some of the assumptions that go into it. In particular, one may return to the concept of time as a parameter. Traditionally clock-time is considered the evolution parameter: a configuration is defined by the simultaneous specification of coordinates and velocities. This, of course, poses a problem for relativistic dynamics because distant simultaneity cannot be defined in a relativistically meaningful way. Moreover, the No-Interaction Theorem leads to undesirable results under this traditional view. One must therefore be prepared to consider other alternatives.

We shall start with the one-particle case, which itself presents no problems, in order to prepare the ground for generalization and to establish the notation. The configuration manifold Q is $\mathbb{R}^4$ with coordinates x^μ, $\mu \in \{0, \ldots, 3\}$, and diagonal metric tensor $g^{\mu\nu}$ for which $g^{kk} = -g^{00} = 1$, $k \in \{1, 2, 3\}$. The phase manifold is then $T^*\mathbb{R}^4 \equiv \mathbb{R}^8$, on which one defines the symplectic form

$$x^{\mu\prime} = \Lambda^\mu{}_\nu x^\nu + a^\mu, \tag{39}$$

The Poincaré group acts on Q (see §23.5(5)) according to

$$x^{\mu\prime} = \Lambda^\mu{}_\nu x^\nu + a^\mu, \tag{39}$$

and the action is lifted to T^*Q by adding to (39) the equations

$$p_\mu{}' = \Lambda_\mu{}^\nu p_\nu. \tag{40}$$

Note the difference between $\Lambda^\mu{}_\nu$ and $\Lambda_\mu{}^\nu$; the indices are raised and lowered by means of the metric tensor and its inverse. Only a seven-dimensional submanifold V_m of T^*Q is available to a particle of (rest) mass m, namely its *mass shell*, defined by

$$g^{\mu\nu} p_\mu p_\nu = m^2. \tag{41}$$

The mass shell is itself Poincaré-invariant.

The mass shell $V_m \subset T^*Q$, being the submanifold available to the particle, is therefore the actual carrier manifold of the dynamical system. In order to study the motion, one pulls back the symplectic form ω to V_m, writing $\omega_m \equiv i_m^*\omega \in \mathfrak{X}^*(V_m)$ where $i_m: V_m \to T^*Q$ is the natural injection. The two-form ω_m is degenerate with a one-dimensional kernel. It is easily verified that one of the vector fields in this kernel, which therefore specifies it completely, is

$$X = p_0 \partial/\partial x^0 + p_k \partial/\partial x^k. \tag{42}$$

Since V_m and ω are both Poincaré-invariant, the one-dimensional distribution defined by $\ker(\omega_m)$ is also Poincaré-invariant. This distribution determines a set of world lines which turn out to be the world lines of a free particle, as will be made explicit below. The set of world lines obtained in this way on the seven-dimensional manifold V_m is itself six-dimensional, in accord with the requirement that the motion be determined by something like the position and velocity.

To calculate the world lines, we shall use the vector field X of (42), although any vector field fX could also be used, $f \neq 0 \in \mathscr{F}(T^*Q)$. The advantages of X are that it is Poincaré-invariant and Hamiltonian, with Hamiltonian function $p^\mu p_\mu$. But different observers can use different functions f, obtaining different evolution parameters τ which need not have any invariant physical significance. Thus each observer can choose his own terms in which to describe the time development. In any case, by choosing X and calling τ the evolution parameter, one obtains

$$dx^\mu/d\tau = p_\mu, \quad dp_\mu/d\tau = 0,$$

which yields

$$x^\mu = p_\mu \tau + x^\mu(0), \quad p_\mu = p_\mu(0) = \text{const.}$$

These can be solved, if one wishes, for the x^k as functions of x^0, which one may call the time t. The result is the familiar

$$x^k(t; x(0), p) = p_k(t - t_0)/p_0 + x^k(0),$$

where $t_0 \equiv x^0(0)$. This is clearly the free particle. Note that this result has been achieved without any explicit reference to dynamical vector fields.

The procedure just described for the single free particle can be generalized in many ways to N particles in interaction. We describe briefly a method based on a scheme for two-particle interactions due to Komar (1978). The carrier manifold is now $M = T^*\mathbb{R}^{4N}$, of dimension $8N$, with the symplectic form (as before, all sums on particle indices are indicated by summation signs)

$$\omega = \sum_k dx^{\mu k} \wedge dp_{\mu k}, \tag{43}$$

where k is the particle index. The action of the Poincaré group on M is essentially the same as that defined by (39) and (40):

$$x^{\mu k\prime} = \Lambda^\mu{}_\nu x^{\nu k} + a^\mu, \quad p_{\mu k} = \Lambda_\mu{}^\nu p_{\nu k}. \tag{44}$$

Now a set of $2N - 1$ Poincaré-invariant constraints are imposed, analogous to (41), reducing the carrier manifold to one of dimension $6N + 1$ on which the time-evolution of the dynamical system takes place. The constraint functions chosen are

$$\begin{aligned} K_k &= g^{\mu\nu} p_{\mu k} p_{\nu k} + (m_k)^2 - U_k, \\ \chi_k &= p_{\mu k}(x^{\mu, k+1} - x^{\mu 1}), \quad k = 1, \ldots, N-1. \end{aligned} \tag{45}$$

The first N constraint functions generalize the single-particle constraint of (41); the interaction is contained in the potentials U_k, which must also be Poincaré-invariant. The second group of $N - 1$ constraint function χ_k is independent of the

U_k. They fix a section in M on which it is reasonable to consider the motion.

One now pulls back ω to the $(6N+1)$-dimensional submanifold $V_0 = (K,\chi)^{-1}(0)$, that is, to the submanifold on which all of the K_k and χ_k of (45) vanish. Let the two-form so obtained be called $\omega_0 \equiv i^*_{V_0}\omega \in \Omega(V_0)$. The dimension of $\ker \omega_0$ can be calculated from the Poisson brackets of the K_k and χ_k, as was done in Chapter 20, and it can be shown that the U_k can be chosen so that $\dim(\ker \omega_0) = 1$. In this way it becomes possible to construct a $6N$-parameter set of world lines for the N particles in a Poincaré-invariant way. The Poincaré group merely permutes the world lines among themselves.

Remark. Sudarshan and Mukunda (1981) impose the additional condition $\{K_j, K_k\} = 0 \ \forall\ k,j$. It can be shown that the U_k can be chosen so that $\dim(\ker \omega_0) = 1$ even under this condition.

These world lines, as in the single-particle case, may be taken as the trajectories of the dynamics. Each observer can parametrize them in his own way, using his own laboratory clock-time. Again the result has been obtained without any explicit reference to a dynamical vector field.

26.5 ACTION-ANGLE VARIABLES AND HAMILTON-JACOBI: A SECOND LOOK WITH LAGRANGIAN FOLIATIONS

In this subsection we return to the material of Chapter 14 in order to show how it fits naturally into the framework provided by foliations, more specifically by Lagrangian foliations.

1. Consider a symplectic manifold (M, ω) of dimension $\mu = 2\nu$ and a set of functions $f_1, \ldots, f_\nu \in \mathscr{F}(M)$ such that $df_1 \wedge \ldots \wedge df_\nu \neq 0$. This is a sufficient condition for

$$F \equiv \{f_k\} : M \to \mathbb{R}^\nu \tag{46}$$

to be a submersion (§8.2) and hence to generate a foliation (§18.1(2)). Assume further that the f_j are in involution, i.e. that $\{f_j, f_k\} = 0 \forall j, k$. Then the foliation is Lagrangian (§18.3). To prove this, one must show that its leaves are of dimension ν and that the pullback by the natural injection of ω onto the leaves vanishes. The dimensionality of the leaves is obvious. For the rest, let X_j be the Hamiltonian vector field obtained from f_j; i.e.

$$i_{X_j}\omega = df_j.$$

Then each $X_j(m)$ lies in $T_m F^{-1}(a) \equiv T_m M_a$, $a \in \mathbb{R}^\nu$, as follows from

$$i_{X_j} df_k = -\{f_j, f_k\} = 0.$$

Moreover, linear independence of the df_j and regularity of ω imply that the X_j are also linearly independent at each $m \in M$, and then dimensionality considerations show that they span $T_m M_a$. Since involutivity of the f_j implies that

$$i_{X_j} i_{X_k} \omega = 0,$$

the assertion follows.

The above argument can be used also for more general submersions of the type

$$\varphi : M \to N, \tag{47}$$

even if $N \neq \mathbb{R}^\nu$, by applying it to neighborhoods of N, so long as $\dim N = \nu$. Thus (47) will be called a *Lagrangian submersion* iff

$$\{\varphi^* f, \varphi^* g\} = 0 \tag{48}$$

for all $f, g \in \mathscr{F}(N)$. This will be useful in particular if N is a torus T^ν.

Remark. One could allow φ to be somewhat less than a submersion (i.e. any C^∞ mapping) and yield what we have called a singular foliation, still calling φ Lagrangian if (48) is satisfied. This is the sort of thing that is done in the description of *caustics* in wave optics.

2. Returning to a submersion such as (46) in terms of functions, assume further that at least one of the f_k, say f_r, is a *proper map*, i.e. that inverse images of compact sets are compact. This means in particular that the M_a are compact; assume further that they are connected (or one can restrict the discussion to connected components of the M_a). Then the vector fields X_k, tangent to f_r, exist on a compact manifold and are therefore complete. As they commute among themselves, they generate an action of the additive group $\mathbb{R}^\nu$, and compactness implies that the leaves of this action are ν-dimensional tori contained in M_a. In fact even more can be said: consider again $F : M \to \mathbb{R}^\nu$, and let U be some contractible neighborhood of $\mathbb{R}^\nu$. Then $F^{-1}(U)$, if connected, is diffeomorphic to $U \times T^\nu$. That is, locally M is a Kronecker product in which one of the factors is a ν-torus. This result relies on the compactness of the leaves, which makes F a proper map.

In accordance with §14.1, then, M admits action-angle variables locally. There exist closed one-forms dJ_k and dw^k (in spite of the notation, these need not yet be exact) such that

$$\omega = -dJ_k \wedge dw^k, \tag{49}$$

where the dJ_k are the pullbacks of one-forms on U and the dw^k on T^ν. Since U is by definition a contractible neighborhood, the J_k can be integrated (they are exact) to provide local action variables J_k. Since T^ν is multiply connected, the dw^k cannot be integrated, but locally (that is locally on T^ν) they provide what are called the angle variables. The complete set $\{J_k, w^k\}$ are the action-angle variables.

Remark. The Lie–Carathéodory version of the Darboux theorem states that given a set of ν functions in involution, one can always find a set of ν others such that the symplectic form can be written in its canonical Darboux form with the first set playing the role of the p_k and the second set the q^k (or vice versa). One might suppose that (49) is a consequence of this version of the theorem. But this version is valid only for a neighborhood of a point $m \in M$, whereas (49) holds in a neighborhood of a torus. The sharper version we are describing is due to Nehorošev (1972).

3. In preparation for a discussion of the Hamilton–Jacobi equation, we now restrict the discussion to a cotangent bundle $M = T^*Q$, and again consider a submersion defined by (46). This time a further assumption is added, namely that the leaves of the foliation generated by the submersion are transversal to the fibers of the natural projection $\tau_Q^*: T^*Q \to Q$. Transversality here means (see the paragraph following Eq.(26) of Chapter 24) that $T_m M_a$ and $\ker(T\tau_Q^*)$ together span the entire tangent space $T_m M \equiv T_m(T^*Q)$ at each $m \in T^*Q$, where $M_a = F^{-1}(a)$ as before.

We proceed for a while in canonical local coordinates $\{q^k, p_k\}$ for T^*Q; then transversality is equivalent to

$$\det\left|\frac{\partial f_j}{\partial p_k}\right| \neq 0 \tag{50}$$

at each $m \in M$. Eq. (50) implies that the equations

$$f_k(q, p) = a_k, \quad k \in \{1, \ldots, \nu\},$$

which define the Lagrangian foliation, can be solved for the p_k, yielding equations of the form

$$p_k - \alpha_k(q, a) = 0. \tag{51}$$

Clearly Eqs. (50) and (51) generate the same Lagrangian foliation, or the functions $p_k - \alpha_k$ (with the a_k treated as parameters) are as much in involution as are the f_k. In local coordinates this leads to

$$\begin{aligned} 0 &= \{p_j - \alpha_j, p_k - \alpha_k\} \\ &= \frac{\partial \alpha_k}{\partial q^j} - \frac{\partial \alpha_j}{\partial q^k}, \end{aligned}$$

which implies that

$$\alpha = \alpha_j(q, a) dq^j \tag{52}$$

is a closed one-form on Q depending on the parameter $a \in \mathbb{R}^\nu$.

Another, more direct way of obtaining this result is to recall that the pullback of ω (we continue to leave off the subscript 0 on the symplectic form) to the leaves of a Lagrangian foliation vanishes. That is, $i_a^* \omega = 0$, where $i_a: M_a \to T^*Q$ is the natural injection. If one writes $\omega = dq^k \wedge dp_k$, the pullback is given by $dq^k \wedge d\alpha_k$, and its vanishing implies that $\alpha_k dq^k$ is closed.

In conclusion, what has been demonstrated is that with each submersion $F: T^*Q \to \mathbb{R}^\nu$ which is transversal to the fibers of the natural projection τ_Q^* there is associated a closed a-dependent one-form on Q. We now sketch a proof of the converse. Consider any one-form $\alpha \in \mathfrak{X}^*(Q)$. It defines the C^∞ map

$$\alpha: Q \to T^*Q: q \mapsto (q, \alpha(q)).$$

Let M_α be the graph of α, a submanifold of T^*Q (the graph is obviously transversal to the fibers of τ_Q^*), and consider $\theta_0 \equiv p_k dq^k \in \mathfrak{X}^*(T^*Q)$. It is easily verified that $\alpha^*(\theta_0) = \alpha$. In fact if one writes $\alpha = \alpha_k dq^k$ in local coordinates, M_a is characterized by $p_k = \alpha_k$. The similarity to (51) is obvious. It follows that $\alpha^*(d\theta_0) = d\alpha$, and then M_α is a Lagrangian submanifold for $\omega = -d\theta_0$ iff α is closed.

4. For the next step to Hamilton–Jacobi theory, consider a product manifold of the form $Q \times N$, where $\dim N = \dim Q = \nu$, and a function $S \in \mathscr{F}(Q \times N)$. Let $dS \in \mathfrak{X}^*(Q)$ be the closed n-dependent one-form obtained from S by taking its exterior derivative with the N dependence frozen (i.e. at fixed points $n \in N$). Then from §(3) it follows that

$$dS: Q \times N \to T^*Q$$

provides a Lagrangian foliation of T^*Q which is transversal to the fibers of the natural projection $\tau_Q^*: T^*Q \to Q$. Under the regularity conditions described below, the quotient manifold of T^*Q with respect to this foliation is diffeomorphic to N.

Clearly $dS = (\partial S/\partial q^j)dq^j$ is now playing the role of $\alpha = \alpha_j dq^j$. In local coordinates one therefore defines the functions

$$p_j - \partial S(q,n)/\partial q^j \equiv g_j(p,q,n)$$

and solves them in local coordinates on N for the n_j. This is possible only if S depends on the N variables in an *essential way*, i.e. only if

$$\det\left|\frac{\partial^2 S}{\partial q^j \partial n_k}\right| = \det\left|\frac{\partial g_j}{\partial n_k}\right| \neq 0; \qquad (53)$$

it will henceforth be assumed that (53) holds. Let the solutions for the n_j be written

$$n_j = f_j(q,p). \qquad (54)$$

Then $\varphi = \{f_1, \ldots, f_\nu\}$ defines a Lagrangian submersion onto a manifold which is naturally identified with N.

5. So far in this subsection action-angle variables and Hamilton–Jacobi theory have been discussed from the point of view of Lagrangian foliations without bringing in the dynamics. Now we introduce the Hamiltonian dynamical vector field $\Delta \in \mathfrak{X}(M)$. For simplicity we shall limit these remarks to $M = T^*Q$.

Let $H \in \mathscr{F}(T^*Q)$ be the Hamiltonian function of Δ, and consider a Lagrangian submersion $\varphi: T^*Q \to N$, $\dim N = \dim Q = \nu$, which is a constant of the motion, i.e. is such that

$$L_\Delta \varphi^* f = 0 \ \forall f \in \mathscr{F}(N).$$

Equivalently, φ is a constant of the motion iff

$$\{H, \varphi^* f\} = 0 \ \forall f \in \mathscr{F}(N).$$

Now, since the submersion is Lagrangian, all functions of the form $\varphi^* f$ are in involution, and then dimensional considerations (there can be at most ν independent functions in involution on T^*Q) imply that there exists an $h \in \mathscr{F}(N)$ such that $H = \varphi^* h$.

An obvious consequence of this is that Δ is tangent to the leaves of the Lagrangian foliation and that H is constant on the leaves. If the foliation gives rise to action-angle variables, this means that the equations for the integral curves of Δ (these are Hamilton's canonical equations written in terms of the action-angle variables) become

$$\dot{J}_k = 0,$$
$$\dot{w}^k = \partial H / \partial J_k. \tag{55}$$

As is well known, the $\partial H / \partial J_k$ are constants of the motion, and these equations are trivially integrated. Such a dynamical system is called *completely integrable.* A Hamiltonian dynamical system is completely integrable if it admits a proper Lagrangian submersion which is a constant of the motion (or, less precisely, if there exist ν constants of the motion in involution).

Finally, we turn to Hamilton–Jacobi theory and demonstrate that any transversal Lagrangian foliation which is invariant under the Hamiltonian vector field Δ provides a complete integral for the Hamilton–Jacobi equation associated with the Hamiltonian function H of Δ.

The Hamilton–Jacobi equation, it may be recalled, is written in local coordinates in the form

$$H(q, \partial S / \partial q) - E = 0, \tag{56}$$

where $H(q, p)$ is the Hamiltonian function in local coordinates, $S \in \mathscr{F}(Q)$ is the function to be found, and $\partial S / \partial q$ stands for the set $\{\partial S / \partial q^j\}$. Eq. (56) can be written intrinsically in the form

$$(dS)^*(H - E) = 0. \tag{57}$$

A solution of the Hamilton–Jacobi equation (56) or (57) for S is desired in the form of a *complete integral*, that is in the form of a function depending on ν parameters $c_1, \ldots, c_\nu \in \mathbb{R}^\nu$ in addition to the q^j in such a way that

$$\det \left| \frac{\partial^2 S}{\partial q^j \partial c_k} \right| \neq 0.$$

The relation to §(4), especially to Eq. (53), is almost obvious. According to §(4) there is a transversal Lagrangian submersion associated with the one-form dS.

We assert that if dS can be found such that this submersion is a constant of the motion, then S is a solution of the Hamilton–Jacobi equation. Indeed, if the submersion is a constant of the motion, as was shown above, H can be written in the form $H = F^*(h)$, where $F = \{f_1, \ldots, f_\nu\}$ and the f_j are obtained in the same way as those of (54) (here the c_j play the role of the n_j). This means that for each set of values $f_j = c_j$, i.e. on each leaf $F^{-1}(c) \equiv M_c$, H takes on a fixed value, say $H = E$, and thus satisfies (57).

26.6 UNIQUENESS OF THE LAGRANGIAN

We now return to the discussion of §13.1(2), in which some dynamical systems were exhibited which admit more than one Lagrangian. There are other systems, as will be shown, whose Lagrangian functions are unique (of course up to the trivial ambiguity of an additive time derivative). Recall, for instance, the electron–monopole system of §26.3, for which it was claimed that there is no global Lagrangian function. It will be shown in §(3) in particular that the Lagrangian for a charged particle in a magnetic field is unique and therefore that having found one (nonglobal) Lagrangian description for the electron–monopole system, one can find no other.

1. Consider a dynamical system $\Delta \in \mathfrak{X}(M)$, $M = TQ$, admitting more than one Lagrangian description, so that there exist functions $\mathscr{L}_1, \mathscr{L}_2 \in \mathscr{F}(M)$ such that

$$\begin{aligned} i_\Delta \omega_1 &= dE_1, \\ i_\Delta \omega_2 &= dE_2, \end{aligned} \tag{58}$$

where ω_k and E_k are the symplectic form and energy function obtained from $\mathscr{L}_k$, $k = 1, 2$. In the usual way, each symplectic form, for instance ω_1, defines the map

$$\omega_1 : \mathfrak{X}(M) \to \mathfrak{X}^*(M) : X \to \omega_1(X, \quad).$$

Since the ω_1 and ω_2 are by assumption regular, they define an $\mathscr{F}(M)$-linear map

$$\Lambda \equiv (\omega_1)^{-1} \circ \omega_2 : \mathfrak{X}(M) \to \mathfrak{X}(M). \tag{59}$$

The map Λ can also be characterized by

$$\omega_1(\Lambda X, Y) = \omega_2(X, Y). \tag{60}$$

It is clear that Λ is well defined if ω_1 is nondegenerate and is an isomorphism if ω_2 is nondegenerate.

We list, essentially without proof, some properties of the Λ map (see Hojman and Urrutia, 1981).

(a) Consider the eigenvalue problem for Λ: to find those $X \in \mathfrak{X}(M)$ which satisfy

$$\Lambda X = \lambda X, \quad \lambda \in \mathscr{F}(M). \tag{61}$$

It turns out that the eigenvalues λ of Λ are at least doubly degenerate.

(b) From the definition (59) it is obvious that

$$L_\Delta \Lambda = 0, \tag{62}$$

which implies that the trace of Λ and all of its powers are constants of the motion, as are therefore the eigenvalues λ. It follows that $\Lambda L_\Delta X = \lambda L_\Delta X$ for X a solution of (61).

(c) The definition (60) implies immediately that

$$\omega_1(\Lambda X, Y) = \omega_1(X, \Lambda Y) \tag{63}$$

and hence that

$$\omega_2(\Delta, X) = \omega_1(\Lambda\Delta, X) = \omega_1(\Delta, \Lambda X).$$

and this means that

$$dE_1\Lambda = dE_2. \tag{64}$$

(d) The defining relation can be applied to vector-field couples of the form (i) $X = \partial/\partial q^j$, $Y = \partial/\partial q^k$, (ii) $X = \partial/\partial \dot{q}^j$, $Y = \partial/\partial q^k$, (iii) $X = \partial/\partial \dot{q}^j$, $Y = \partial/\partial \dot{q}^k$ to obtain an expanded expression for Λ in a natural chart on $M \equiv TQ$. The result is

$$\Lambda = m_j^{\ k}(dq^j \otimes \partial/\partial q^k + d\dot{q}^j \otimes \partial/\partial \dot{q}^k) + n_j^{\ k} dq^j \otimes \partial/\partial \dot{q}^k, \tag{65}$$

where $m_j^{\ k}, n_j^{\ k} \in \mathscr{F}(M)$. This expression will be useful in application to specific cases.

2. For example, consider the case in which one of the Lagrangian functions is of the common form

$$\mathscr{L} = \tfrac{1}{2}\delta_{jk}\dot{q}^j\dot{q}^k - V(q), \tag{66}$$

and write $\Delta = \dot{q}^j\partial/\partial q^j + f^j\partial/\partial \dot{q}^j$ (the f^j are essentially the components of the force and are obtained in the usual way from the potential V). It is found then that $m_j^{\ k} = m_k^{\ j}$ is symmetric and that $n_j^{\ k} = 0 \ \forall\, j,k$. Let m be the matrix of the $m_j^{\ k}$. Observe that if m is the unit matrix, Λ also becomes the unit matrix, and this means that $\omega_1 = \omega_2$. This may be interpreted to mean that the two Lagrangians are trivially equivalent (see §13.1(1) and recall that (58) is assumed).

In the local chart Eq. (62) leads to $L_\Delta m = 0$ (or $dm/dt = 0$) and $[B, m] = 0$, where B is the matrix with elements $B_j^{\ k} = \partial f^k/\partial q^j$. From these one then obtains the infinite set of algebraic conditions

$$[(d/dt)^k B, m] = 0, \quad k = 0, 1, 2, \ldots, \tag{67}$$

not all of which are independent. If $\{(d/dt)^k B\}$ is an irreducible set of matrices, Eq. (67) implies that m must be a multiple of the unit matrix. This happens in many situations, i.e. for many dynamical systems, and allows one then to conclude that the Lagrangian function is essentially unique.

The B matrix can be used also to show that if a dynamical system in two degrees of freedom admits two nontrivially equivalent Lagrangians at least one of which is of the type of (66), then there exists a linear change of coordinates which separates the potential; that is, the system can then be written as the sum of two noninteracting systems, each in one degree of freedom. Let $\Delta \in \mathfrak{X}(T\mathbb{R}^2)$ be the dynamical vector field obtained from the Lagrangian function

$$\mathscr{L}_1 = \tfrac{1}{2}(\dot{x}^2 + \dot{y}^2) - V(x, y), \tag{68}$$

and assume that there exists another nontrivially equivalent Lagrangian function $\mathscr{L}_2$ for Δ. Then write the B matrix in the form

$$B = \begin{vmatrix} \varphi_1 & \varphi_3 \\ \varphi_3 & \varphi_2 \end{vmatrix},$$

where $\varphi_1 = \partial^2 V/\partial x^2$, $\varphi_2 = \partial^2 V/\partial y^2$, and $\varphi_3 = \partial^2 V/\partial x \partial y$, and the m matrix in the form

$$m = \begin{vmatrix} \alpha & \beta \\ \beta & \gamma \end{vmatrix}.$$

Note that if $\varphi_3 = 0$, the potential is already separated; thus we assume that $\varphi_3 \neq 0$. The commutation relation $[B, m] = 0$ now becomes

$$(\alpha - \gamma)\varphi_3 + \beta(\varphi_2 - \varphi_1) = 0.$$

If $\beta = 0$, then $\alpha = \gamma$, and m is the unit matrix multiplied by a function, and $dm/dt = 0$ implies that the function is a constant of the motion; this means that ω_2 is ω_1 multiplied by a constant of the motion. Then by using (68) to obtain an explicit expression for ω_1 and imposing closure on ω_2 one can show that the constant of the motion is in fact a numerical constant. That, in turn, means that the two Lagrangians are trivially equivalent.

Assume, therefore, that $\beta \neq 0$. One now has

$$(\varphi_2 - \varphi_1)/\varphi_3 = (\gamma - \alpha)/\beta.$$

The right-hand side is a constant of the motion, and therefore so is the left-hand side. Since the φ_k are functions only of x and y, the left-hand side must be a numerical constant; call this constant ρ. Then by putting in the expressions for the φ_k in terms of the partial derivatives of V, one arrives at

$$\frac{\partial^2 V}{\partial x^2} - \frac{\partial^2 V}{\partial y^2} = \rho \frac{\partial^2 V}{\partial x \partial y}.$$

This is a partial differential equation of hyperbolic type, and a linear change of coordinates reduces it to the usual wave equation. Through new coordinates obtained in this way, then, V is separated into the sum of two functions, each depending on one variable, as asserted.

3. It will now be shown that if a dynamical system Δ on $T\mathbb{R}^3$ has the Lagrangian function

$$\mathscr{L} = \delta_{jk}\dot{q}^j\dot{q}^k + A_j(q)\dot{q}^j, \tag{69}$$

then there is no other (nontrivially) equivalent Lagrangian. The Lagrangian of (69) is that for a charged particle in a magnetic field.

Since now the Lagrangian is not of the form of (66), it is no longer true that the n matrix must vanish. If the full expression (65) for Λ is now inserted into $L_\Delta \Lambda = 0$, one obtains

$$L_\Delta m = n, \tag{70a}$$

$$L_\Delta n = nC + [m, B], \tag{70b}$$

$$L_\Delta m = \tfrac{1}{2}[C, m], \tag{70c}$$

where B has the same meaning as above and C is the matrix whose elements are $c_j^{\ k} = \partial f^k / \partial \dot{q}^j$. Recall that the f^k are the components of the force. The nature of the magnetic force, as is easily verified, forces C to be antisymmetric. In addition, (63) continues to imply that m is symmetric, and then (70a) implies that n is also symmetric.

Clearly Eqs. (70a) and (70c) imply that $n = \tfrac{1}{2}[C,m]$. Lengthy algebraic calculations involving Eqs. (70) lead to a generalization of (67). One then finds that n must vanish, just as in the case of a potential function, and this means that $[C,m] = 0$. When the symmetry properties of C and m are taken into account, it is found that this commutation condition leads to a set of equations for the six independent components of m whose solution depends on two arbitrary constants. Without going into the details of the calculation, we present the solution:

$$m_j^{\ k} = \nu h_j h_r \delta^{rk} + \mu \delta_j^k,$$

where the h_k are the components of the magnetic field. Note that nondegeneracy implies that $\mu \neq 0$. One then uses the further commutation relations of (67) and the similar ones for C, that is $[(d/dt)^k C, m] = 0$, to show that in general ν must vanish. Thus m is a multiple of the unit matrix, and hence Λ is a multiple of the identity mapping, so that every equivalent Lagrangian is trivially equivalent to the Lagrangian of (69).

26.7 DYNAMICAL SYSTEMS WITH CONSTRAINTS

The canonical formalism for constrained dynamical systems was devised initially by Dirac as a method for dealing with systems whose Lagrangians may be singular (Dirac, 1950). The motivation was to develop methods that would allow one to put generally covariant and gauge invariant field theories into canonical form. Here we will indicate how naturally Dirac's formalism fits into the framework of foliations and reduction of dynamical systems. We shall not go into the details, avoiding technicalities and making whatever assumptions are needed to simplify the treatment. For a much more thorough discussion see Marmo, Mukunda and Samuel (1983).

1. Recall that a regular Lagrangian system on TQ is one whose Lagrangian function $\mathscr{L} \in \mathscr{F}(TQ)$ yields an $\omega_\mathscr{L}$ which is a symplectic structure on TQ, which means that $\omega_\mathscr{L}$ must be regular. Then the algebraic equation

$$i_\Delta \omega_\mathscr{L} = dE_\mathscr{L} \tag{71}$$

defines a unique vector field $\Delta \in \mathfrak{X}(TQ)$, and Δ is always of second order.

We now turn to the situation in which the two-form $\omega_{\mathscr{L}}$ obtained from $\mathscr{L}$ is degenerate. This situation can be characterized in local coordinates by the statement that the Hessian $\partial^2 \mathscr{L}/\partial \dot{q}^j \partial \dot{q}^k$ of $\mathscr{L}$ has less than the maximum rank $\nu = \dim Q$. If the rank of the Hessian is less than maximum, so is that of $\omega_{\mathscr{L}}$. It has been seen on occasion that changes in the rank of $\omega_{\mathscr{L}}$ or of two-forms derived from it can lead to complications. We shall therefore make the simplifying assumption that $\rho \equiv \text{rank}(\omega_{\mathscr{L}})$ is constant on TQ.

One may now follow Dirac to investigate the transition from TQ to T^*Q in these circumstances. Recall that the transition is accomplished with the aid of the fiber derivative (§11.1) $F\mathscr{L}: TQ \to T^*Q$. From the assumption that $\rho = \text{const.}$ it follows that the rank of $TF\mathscr{L}: T(TQ) \to T(T^*Q)$ is also constant. One may then assume further that $F\mathscr{L}(TQ)$ is a submanifold Σ of T^*Q, whose dimension is $\kappa = \nu + \rho$. Let this manifold be determined by a set of functions $\Phi = (\varphi_1, \ldots, \varphi_{\nu-\rho}): T^*Q \to \mathbb{R}^{\nu-\rho}$ such that

$$\Sigma \equiv F\mathscr{L}(TQ) = \Phi^{-1}(0). \tag{72}$$

Dirac calls the φ_k the *primary constraints.*

It is easily shown from its definition that $E_{\mathscr{L}}$ is $F\mathscr{L}$*-projectable*, i.e. that it is the pullback by the fiber derivative of a function $\tilde{E}_{\mathscr{L}} \in \mathscr{F}(\Sigma)$. Moreover, the fiber derivative is defined so that $\omega_{\mathscr{L}} = (F\mathscr{L})^* \omega_0$, where ω_0 is the natural symplectic form on T^*Q. One may now ask for the two-form induced on Σ by ω_0 through the natural injection $i_\Sigma: \Sigma \to T^*Q$, defined by

$$\omega = (i_\Sigma)^* \omega_0. \tag{73}$$

In general ω will be degenerate, so that the equation

$$i_X \omega = d\tilde{E}_{\mathscr{L}} \tag{74}$$

may have no solutions $X \in \mathfrak{X}(\Sigma)$. No solution will exist if $E_{\mathscr{L}}$ is not in the image $\omega(\mathfrak{X}\Sigma)$ when ω is treated as the map $\omega: \mathfrak{X}(\Sigma) \to \mathfrak{X}^*(\Sigma)$. A necessary and sufficient condition for solutions to exist is that $i_Y d\tilde{E}_{\mathscr{L}} = 0$ for all $Y \in \ker(\omega)$. Generally, however, we shall be obliged to solve (74) on a subset $\Sigma' \subset \Sigma$ where $i_Y d\tilde{E}_{\mathscr{L}}$ does in fact vanish for all $Y \in \ker(\omega)$. The relations $i_Y d\tilde{E}_{\mathscr{L}} = 0$ are essentially what Dirac calls the *secondary constrants.*

To proceed one makes the further simplifying assumption that Σ' is a submanifold of Σ and again considers a natural injection, this time $i_{\Sigma'}: \Sigma' \to \Sigma$, and the pullbacks

$$\begin{aligned} dE' &= (i_{\Sigma'})^* d\tilde{E}_{\mathscr{L}}, \\ \omega' &= (i_{\Sigma'})^* \omega. \end{aligned} \tag{75}$$

Again one attempts to solve an equation such as (74), now on Σ', namely

$$i_X \omega' = dE', \quad X \in \mathfrak{X}(\Sigma'), \tag{76}$$

and again this equation may have solutions only on a restricted domain of Σ'. One may therefore be led in this way to a set of *tertiary constraints*, and this process could continue to yield constraints of even higher order. In this way one arrives either at inconsistency or at a submanifold on which equations such as

(74) and (76) can be solved. In the case of inconsistency the Lagrangian $\mathscr{L}$ is said simply to fail in defining a consistent set of equations of motion and hence not to define a dynamical system.

To proceed, we shall assume that no tertiary constraints arise. Then on $\Sigma' \subset \Sigma$ there are now given a two-form ω' and a function E' such that Eq. (76) admits solutions X. If $\operatorname{rank}(\omega')$ is constant on Σ', the distribution obtained from $\ker(\omega')$ is, as has been seen in similar situations, involutive. Assuming that the foliation it defines is regular and has a smooth quotient manifold N, one obtains a symplectic formulation of the dynamical system on N in the way that has been described in discussing the reduction of Hamiltonian dynamical systems.

Because in general the original q and p variables will not be projectable onto N, the Hamiltonian vector field $\Delta_N \in \mathfrak{X}(N)$ obtained in this way will not have a clear physical interpretation. This is a manifestation of the problem of lifting reduced vector fields, discussed in §19.3.

For further details the reader is referred again to Marmo, Mukunda and Samuel (1983).

Conclusion

We have now arrived at the end of this book. Let us begin our closing statement by saying what the book is not, where its limits lie: we name three such limits.

First, the physics we deal with is classical in the sense that it is nonrelativistic. Time plays a special role: it is simply an evolutionary parameter which does not interfere with the topological or differential properties of Q. Even the Appendix to Part I does not alter this, for in it time is always introduced by taking a direct product of Q with $\mathbb{R}$. Section 26.4, in which relativistic applications are briefly discussed, barely begins to remedy this situation.

Second, the physics we deal with is classical in the sense that it is nonquantum. The relevance of our treatment to quantum mechanics, even within the context of what is called geometric quantization, has yet to be established.

Third, the manifolds we deal with are all of finite dimensions. Systems of infinite dimensions are surely of great interest and importance, but it is not clear that they can be dealt with in ways that are usefully similar to ours.

There are, of course, many other things the book is not. What it is, however, or rather what we have attempted to make it, is a mathematical treatment of classical physical dynamical systems, a subject which lies somewhere near the foundation of physics.

A mathematical, as opposed to a physical, definition of a *dynamical system* says essentially (stripped, that is, of many cautionary and technical details) that a dynamical system is a vector field on a differentiable manifold; the manifold is treated as given *a priori*, and the system is a field on that manifold. A *physical* dynamical system, on the other hand, is composed of both the vector field and the manifold. The manifold does not exist *a priori*, waiting for a dynamical system to be laid on it by observation, but observation discovers both the manifold and the vector field, and these together constitute the dynamical system. In spite of the fact that we approach the subject from physics rather than mathematics, our treatment nevertheless focuses on the mathematics and we work generally within the mathematical definition, writing often that it is the vector field Δ which is the dynamical system. In other words, like true theorists, we assume that the heavy work of observation has already been performed and that the manifold has been established. But establishing the manifold and modelling observed physical phenomena by a mathematical dynamical system is not a procedure that has well defined rules. The theorist does not simply discover the manifold and the vector field. He is guided, as was discussed in the early chapters of this book, by concepts already constructed (e.g. continuity and differentiability, even such considerations as a preference for second-order vector fields). It is through the complex interplay between such concepts and observations (or experimental data) that the dynamical systems on which this book concentrates are achieved. Even as we lose ourselves in the delights and elegance, such as it is, of the mathematics, we should remember how we got here and be aware that although these complicated matters were touched on in the early sections of this book, there is much that yet remains to be understood.

It is at least in part because it arises naturally in the interplay just mentioned between observation and theoretical considerations, that TQ, rather than T^*Q, is so important in our development of the formalism. The simplest way to separate trajectories, the most direct way to obtain a lifting, is by going from Q to the $(q, \dot{q})$ manifold, the tangent bundle. In this sense T^*Q plays a subsidiary role, which is perhaps seen particularly clearly in the discussion of ambiguity (Section 13.2). In this we differ from some other authors (e.g. Zeeman), who see T^*Q as the natural carrier space for the dynamics.

In more ways also, our book does not always agree with others and sometimes takes routes which are not traditional. One departure from tradition is, of course, our treatment of the relation between Lie groups and algebras, more or less the reverse of the usual one, which places the algebras before the groups. Another type of departure is the introduction of certain operations independently, before that of the Lie groups and algebras which usually come first. *Reduction* of a dynamical system is an example. Although reduction exists on its own and has been applied independently of groups and algebras by generations of physicists, its formal treatment in the past has usually involved groups and algebras. We believe that our treatment exhibits more clearly the nature of reduction and shows more explicitly how Lie algebraic and group properties enrich the reduction process by increasing what can be said and demonstrated about it.

Reduction depends on *foliation*, and a foliation need not be associated with the action of a Lie group or algebra. This leads to a different view of *symmetry*, for symmetry is traditionally associated with such actions. In our treatment, a foliation can in a sense be a symmetry for a dynamics (that is, when it is left invariant by a dynamical system) without the existence of a group action which is a symmetry for the dynamics (i.e. which leaves the dynamics invariant). This is not to say that group actions are not important, for they are indeed often available and they do enrich reduction, but one may now wonder how fundamental the traditional formulations are of the so-called symmetry laws of nature.

We have attempted in this book to develop some aspects of the application of differential geometry to the theory of dynamical systems. It is our hope that we have succeeded in demonstrating some of the beauty inherent both in the field of dynamical systems and in the intrinsic, geometrical approach. The field is not closed; it has interfaces, for example, with such studies as chaos and quantization, it is not yet applied unambiguously to special relativity. There is a lot of other unfinished business, undoubtedly some not even started. This is a wide open field, and we hope that readers will be interested enough, will have been sufficiently attracted, to pursue the subject further, continuing to enjoy its delights.

References

Abraham, R., and Marsden, J. E. (1978). *Foundations of Mechanics.* Benjamin, Reading, Mass.

Abraham, R., and Robbin, J. (1967). *Transversal Mappings and Flows.* Benjamin-Cummings, Reading, Mass.

Ado, I. D. (1947). The representation of Lie algebras by matrices. *Usp. Mat. Nauk. (n.s.)* **2**, 159.

Aristotle. *The Physics* 184 A. Wicksteed's translation, Loeb Classical Library, London (1929).

Arnol'd, V. I. (1978). *Mathematical Methods of Classical Mechanics.* Springer, New York.

Arnol'd, V. I., and Avez, A. (1967). Théorie Ergodique. Gauthier Villars, Paris.

Balachandran, A. P., Marmo, G., Stern, A. (1982). A Lagrangian approach to the No-Interaction Theorem. *Nuovo Cimento* **69** A, 175.

Balachandran, A. P., Marmo, G., Skagerstam, B. S., Stern, A. (1983). *Gauge Symmetries and Fibre Bundles. Applications to Particle Dynamics.* Springer Verlag, Berlin.

Balachandran, A. P., Marmo, G., Mukunda, N., Nilsson, J. S., Simoni, A., Sudarshan, E. C. G., and Zaccaria, F. (1984). Unified geometrical approach to relativistic particle dynamics. *J.M.P.* **25**, 167.

Behrends, R. E., Dreitlein, J., Fronsdal, C., and Lee, W. (1962). Simple groups and strong interaction symmetries. *Rev. Mod. Phys.* **34**, 1.

Bishop, R. L., and Crittenden, R. J. (1964). *Geometry of Manifolds. Academic Press, New York.*

Boothby, W. M. (1975). *An Introduction to Differentiable Manifolds and Riemannian Geometry.* Academic Press, New York.

Bott, R. (1972). Lectures in characteristic classes and foliations. In Bott, R., Gitler, S., and James, I. M.: *Lectures on Algebraic and Differential Geometry.* Springer Verlag, Berlin.

Bourbaki, N. (1967). Variétés Différentielles et Analytiques. Fascicule de Résultats, 1 à 7 (*Éléments de Mathématique*, 33). Hermann, Paris.

Bourbaki, N. (1971). Variétés Différentielles et Analytiques. Fascicule de résultats, 8 à 15 (*Élements de Mathématique*, 36). Hermann, Paris.

Brickell, F., and Clark, R. S. (1970). *Differentiable manifold. An Introduction.* Van Nostrand. London.

Candotti, E., Palmieri, C., and Vitale, B. (1972). Universal Noether's nature of infinitesimal transformations in Lorenz-covariant field theories. *Nuovo Cimento* **70** A, 233.

Čapek, M., ed., (1976). *Godel, K.: Static Interpretation of Space.* Reidel, Dordrecht.

Caratú, G., Marmo, G., Simoni, A., Vitale, B., and Zaccaria, F. (1976). Homogeneous canonical transformations in phase space: on necessary conditions for the existence of symplectic actions of simple Lie groups on differentiable manifolds. *Nuovo Cimento* **19** B, 228.

Caratú, G., Marmo, G., Simoni, A., Vitale, B., and Zaccaria, F. (1976). Lagrangian and Hamiltonian formalism: an analysis of classical mechanics on tangent and cotangent bundles. *Nuovo Cimento* **31** B, 152.

Cartan, H. (1957). Variétés analytiques réelles et variétés complexes. *Bull. Soc. Math. France* **85**, 77.

Chillingworth, D. R. J. (1976). *Differential Topology with a View to Applications.* Pitman, London.

Chu Bon-Yao (1974). Symplectic homogeneous spaces. *Trans. Am. Math. Soc.* **197**, 145.

Currie, D. G., Jordan, T. F., and Sudarshan, E. C. G. (1963). Relativistic invariancce and Hamiltonian theories of interacting particles. *Rev. Mod. Phys.* **35**, 350.

Cushman, R. (1974). The momentum mapping of the harmonic oscillator. *Symposia Mathematica*, Vol. XIV. Academic Press, New York.

Darboux, G. (1982). Sur le problèm de Pfaff. *Bull. Sci. Math.* **6**, 14, 49.

Dieudonné, J. (1971). *Elements d'Analyse*, Vol. IV. Gauthier-Villars, Paris.

Dirac, P. A. M. (1950). Generalized Hamiltonian dynamics. *Canad. J. Math.* **2**, 129.

Dodson, C. T. J., and Poston T. (1977). *Tensor Geometry.* Pitman, London.

Dogen. In Nakamura, H. (1964). *Ways of Thinking of Eastern People*, East-West Center, Honolulu.

Donaldson, S. K. (1983a). Proceedings of the 1982 International Congress of Mathematicians, Warsaw.

Donaldson, S. K. (1983b). An Application of Gauge Theory to Four-Dimensional Topology, J. Diff. Geometry **18**, 279–315.

Eells, J. (1974). Fibre bundles. In *International Course on Global Analysis and its Applications* (Trieste 1972), Vol. 1. AIEA, Trieste.

Flanders, H. (1969). *Differential Forms with Applications to the Physical Sciences*, Academic Press, New York.

Forsyth, A. R. (1959). *Theory of Differential Equations.* Dover, New York.

Fort, M. K., ed. (1962). *Topology of 3-manifolds and Related Topics.* In particular: Fox, R. H. A quick trip through knot theory. Prentice-Hall, Englewood Cliffs, N.J.

Fradkin, D. M. (1967). Existence of the dynamical symmetries $O(4)$ and $SU(3)$ for all classical central potential problems. *Progr. Theor. Phys.* **37**, 789.

Godbillon, C. (1969). *Géométrie Différentielle et Mécanique Analytique.* Hermann, Paris.

Grünbaum, A. (1973). *Philosophical Problems of Space and Time.* Reidel, Dordrecht.

Guillemin, V., and Pollack, A. (1974). *Differential Topology.* Prentice-Hall, Englewood Cliffs, N.J.

Hartshorne, R. (1977). *Algebraic Geometry.* Springer Verlag, Berlin.

Hausner, M., and Schwartz, J. T. (1968). *Lie Groups, Lie Algebras.* Gordon & Breach, New York.

Helgason, S. (1962). *Differential Geometry and Symmetric Spaces.* Academic Press, New York.

Henneaux, M. (1982). Equations of motion, commutation relations and ambiguities in the Lagrangian formalism. *Annals of Physics* **140**, 45.

Hirsch, M. W., and Smale, S. (1974). *Differential Equations, Dynamical Systems and Linear Algebra.* Academic Press, New York.

Hojman, S., and Urrutia, L. F. (1981). On the inverse problem of the calculus of variations. *J. Math. Phys.* **22**, 1896.

Hopf, H. (1931). Über die Abbildungen der 3-Sphäre auf die Kugelflache. *Math. Annalen* **104**, 637.

Houtappel, R. M. F., Van Dam, H., and Wigner, E. P. (1965). The conceptual basis and use of the geometric invariance principles, *Rev. Mod. Phys.* **37**, 547.

Husemoller, D. (1975). *Fibre Bundles.* Springer, New York.

Jacobson, N. (1962). *Lie Algebras.* Interscience, New York.

Kamber, F. W., and Tondeur, P. (1975). *Foliated Bundles and Characteristic Classes.* Springer Verlag, Berlin.

Kasner, E. K. (1913). *Differential Geometric Aspects of Dynamics.* American Mathematical Society, New York.
Kervaire, M. (1960). Manifold which does not admit any differentiable structure, *Comm. Math. Helv.* **34**, 257.
Kirillov, A. A. (1962). Unitary representations of nilpotent Lie groups. Russian Mathematical Surveys **17**, 53.
Komar, A. (1978). Constraint formalism of classical mechanics. *Phys. Rev. D* **18**, 1881.
Komar, A. (1978a). Interacting relativistic particles. *Phys. Rev. D* **18**, 1887.
Kostant, B. (1965). Orbits, Symplectic Structures, and Representation Theory. In *Proceedings of the United States Japan Seminar on Differential Geometry, Kyoto, 1965.* Nippon Hyoronisha, Tokyo, p. 71.
Kostant, B. (1970). Quantization and unitary representations. *Lecture Notes in Math.* **170**, 83.
Kreck, M. (1984). Manifolds with Unique Differentiable Structure. Topology **23**, 219–232. (Received October, 1981.)
Iwasawa, K. (1948). On the representations of Lie algebras. *Japan J. Math.* **19**, 405.
Lawson, H. B. (1974). Foliations. *Bull. Am. Math. Society*, **80**, 369.
Lawson, H. B. (1977). The quantitative theory of foliations. *Am. Math. Soc., CBMS Series*, Vol. 27. American Mathematical Society, Providence, R.I.
Levy-Leblond, J. M., (1969). Group Theoretical Foundations of Classical Mechanics; the Lagrangian Gauge Problem. *Comm. Math. Phys.* **12**, 64.
Lutzky, M. (1981). Non-canonical symmetries and isospectral representations of Hamiltonian systems. *Phys. Letters A* **87**, 274.
MacLane, S. (1968). Geometrical mechanics (2 parts). Department of Mathematics, University of Chicago, Chicago.
Mandal, K. (1968). *A Comparative Study of the Concepts of Space and Time in Indian Thought.* Varanasi, Bombay.
Marle, C. M. (1975). Symplectic manifolds, dynamical groups, and Hamiltonian mechanics. Cahen-Flato, ed., *Differential Geometry and Relativity.* Reidel, Dordrecht, pp. 249–269.
Marmo, G., Mukunda, N., and Sudarshan, E. C. G. (1984). Relativistic particle dynamics. Lagrangian proof of the No-Interaction Theorem. *Phys. Rev. D* **30**, 2110.
Marmo, G., and Saletan, E. J. (1977). Ambiguities in the Lagrangian and Hamiltonian formalism: transformation properties. *Nuovo Cimento* **40** B, 67.
Marmo, G., and Simoni, A. (1976). *Q*-dynamical systems and constants of motion. *Lett. al Nuovo Cimento Serie 2* **15**, 179.
Marmo, G., Saletan, E. J., and Simoni, A. (1979). Reduction of symplectic manifolds through constant of the motion. *Nuovo Cimento* **50** B, 21.
Marmo, G., Saletan, E. J., Simoni, A., and Zaccaria, F. (1981). Liouville dynamics and Poisson brackets. *J.M.P.* **22**, 835.
Marmo, G., Mukunda, N., and Samuel, J. (1983). Dynamics and symmetry for constrained systems: a geometrical analysis. *La Rivista del Nuovo Cimento* **6**, 2.
Milnor, J. (1956). On manifolds homeomorphic to the 7-sphere. *Ann. Math.* **64**, 399.
Milnor, J. (1959). Differentiable structures on spheres. *Am. J. Math.* **81**, 962.
Milnor, J. (1963). Morse theory. *Annals of Math. Studies*, Princeton University Press, Princeton.
Mukunda, N., and Sudarshan, E. C. G. (1981). Form of relativistic dynamics with world lines. *Phys. Rev. D* **23**, 2210.
Needham, J. (1962). *Science and Civilization in China.* Cambridge University Press, London.
Nehorošev, N. N. (1972). Action-angle variables and their generalization. *Trans. Moscow Math. Soc.* **26**, 180.
Palais, R. (1957). A global formulation of the Lie theory of transformation groups. *Memoirs of the American Math. Soc.*, no. 22. American Mathematical Society, Providence, R.I.
Palis, J., and do Carmo, M., eds., (1977). *Geometry and Topology—Lecture Notes in Mathematics*, Springer Verlag, Berlin.

Piaget, J., ed., (1973). *L'épistemologie du Temps*. Presses Universitaires de France, Paris.

Plato. Timaeus 34 A. Translated by R. G. Bury, Loeb Classical Library, London (1929).

Rijgrok, T. W., and Van der Vlist, H. (1980). On the Hamiltonian and Lagrangian formulation of classical dynamics for particles with spin. *Physic* **101** A, 571–580.

Robbin, J. W. (1973). Relative equilibria in mechanical systems. In (M. Peixoto, ed.), 425. *Dynamical Systems*. Academic Press, New York.

Robbin, J. W. (1974). *Symplectic Mechanics, in Global Analysis and its Applications*, Vol. III. I.A.E.A., Vienna.

Rolsen, D. (1976). *Knots and Links*. Publish or Perish. Boston.

Saletan, E. J., and Cromer, A. H. (1971). *Theoretical Mechanics*. John Wiley, New York.

Sarlet, W., and Cartrijn, F. (1981). Generalizations of Noether's theorem in classical mechanics. *SIAM Rev.* **23**, 467.

Sarlet, W. and Cartrijn, F. (1983). Symmetries, first integrals, and the inverse problem of Lagrangian mechanics: II. *J. Phys. A: Math. and Gen.* **16**, 1383.

Schreiber, M. (1977). *Differential Forms, a Heuristic Introduction*. Springer, New York.

Simoni, A., Vitale, B., and Zaccaria, F. (1967). Conserved quantities and symmetry groups for the Kepler problem. *Nuovo Cimento Serie X* **50**, 95.

Sklar, L. (1974). *Space, Time and Spacetime*. California University Press, Berkeley.

Smirnov, V. I. (1961). *Linear Algebra and Group Theory*. McGraw-Hill, New York (reprinted by Dover, New York, 1970).

Souriau, J. M. (1966). Qantification géométrique. *Comm. Math. Phys.* **1**, 374.

Souriau, J. M. (1970). *Structure des Systèmes Dynamiques*. Dunod, Paris.

Steenrod, N. (1951). *The Topology of Fibre Bundles*. Princeton University Press, Princeton.

Stehle, P., Han, M. Y. (1967). Symmetry and degeneracy in classical mechanics. *Phys. Rev.* **159**, 1075.

Sternberg, S. (1963). *Lectures on Differential Geometry*. Prentice-Hall, Englewood Cliffs, N.J.

Sudarshan, E. C. G., and Mukunda, N. (1974). *Classical Dynamics: A Modern Perspective*. John Wiley, New York.

Takens, F. (1977). Symmetries, conservation laws and variational principles. In Palis, J. and do Carmo (eds.) *Lecture Notes in Math. 597, Geometry and Topology*. Springer Verlag, Berlin.

Thirring, W. (1978). *A Course in Mathematical Physics. I Classical Dynamical Systems*. Springer, New York.

Vitale, B. (1966). Invariance and noninvariance dynamical groups. Madras Lecture Notes, and Lecture Notes at the Latin American School of Physics, Caracas.

Warner, F. (1971). *Foundations of Differentiable Manifolds and Lie Groups*. Scott and Foreman, Glenview.

Weinstein, A. (1977). Lectures on symplectic manifolds. In *C.B.M.S. Conf. series, Am. Math. Soc.*, no. 29. American Mathematical Society, Providence, R.I.

Whitney, H. (1936). Differentiable manifolds. *Ann. Math.* **37**, 645.

Whittaker, E. T. (1904) *A treatise on the analytical dynamics of particles and rigid bodies*. Cambridge University Press., London (4th edn, 1959).

Willard, S. (1968). *General Topology*. Addison-Wesley, Reading, Mass.

Winfree, A. T. (1980). *The Geometry of Biological Time*. Springer, New York.

Wintner, A. (1941). The analytical foundations of celestial mechanics. Princeton University Press, Princeton.

Wittgenstein, L. Translated by Pears, D. F., and McGuinness, B.F. Routledge & Kegan Paul, London (1961).

Yano, K., and Ishihara, S. (1973). *Tangent and Cotangent Bundles*. Dekker, New York.

Further reading

CHAPTER 1

Arnol'd, V. I., (1983). *Geometrical Methods in the Theory of Ordinary Differential Equations.* Springer, New York.

Born, M. (original edition 1926). *Problems of Atomic Dynamics.* The M.I.T. Press, Cambridge, Mass. (Paperback edn, 1970).

Doubrovine, B., Novikov, S., and Fomenko, A. (1982). *Géométrie Contemporaine, Méthodes et Applications.* Mir, Moscow.

Guillemin, V., and Sternberg, S. (1977). *Geometric Asymptotics. Mathematical Surveys, 14.* American Mathematical Society, Providence, R. I.

Hawking, S. W. and Ellis, G. F. R. (1973). *The Large Scale Structure of Space Time.* Cambridge University Press, Cambridge.

Helleman, R. H. G., (ed.), (1980). *Non-linear Dynamics. Annals of the New York Academy of Sciences, 357.*

Hurt, N. E. (1983). *Geometric Quantization in Action. Applications of Harmonic Analysis in Quantum Field Theory,* D. Reidel, Boston.

Marle, C. M. (1983). Géométrie symplectique, première partie. *Publications Mathématiques de l'Université Pierre et Marie Curie 32,* Paris.

Trautman, A. (1984). *Differential Geometry for Physicists. Informal Lectures.* Bibliopolis, Napoli.

CHAPTER 2

De Rham, G. (1960). *Variétés Différentiables.* Hermann, Paris.

Malliavin, P. (1972). *Géométrie Différentielle Intrinsique.* Hermann, Paris.

Satake, I. (1956). On a generalization of the notion of manifold. *Proc. N.A.S.* 359.

Spivak, M. (1979). *A Comprehensive Introduction to Differential Geometry.* Publish or Perish, Berkeley.

CHAPTER 4

Smale, S. (1980). *The Mathematics of Time. Essay on Dynamical systems, economic processes, and related topics.* Springer, New York.

CHAPTER 5

Cartan, H. (1967). *Formes différentielles. Collection Méthodes,* Hermann, Paris.

Cartan, H. (1967a). *Calcul différentiel. Collection Méthodes,* Hermann, Paris.

Choquet-Bruhat, Y. (1968). Géométrie differentelle et systèmes exterieurs. *Monographies Universitaires de Math.* No. 28. Dunod, Paris.

Choquet-Bruhat, Y., and De Witt-Morette, C., with Dillard-Bleick, M. (1982). *Analysis, Manifolds and Physics*, revised edn, North-Holland, Amsterdam.

De Rham, G. (1961). La théorie des formes différentielles extérieures et l'homologie des variétés différentiables. *Rendiconti di Matematica* **20**, 105.

Eiseman, P. R. and Stone, A.P. (1975). A note on a differentiable concomitant. *Proceedings of the American Math. Soc.* **53**, 179.

Frolicher, A., and Nijenhuis, A. (1956). Theory of vector valued differential forms. *Nederl. Akad. Wetensch. Proc. Ser. A59 (Indag. Math.)* **18**, 338.

Kahler, E. (1962). Der innere differentialkalkul. *Rendiconti di Matematica* **21**, 425.

Koszul, J. L. (1960). *Lectures on Fibre Bundles and Differential Geometry.* Tata Institute of fundamental Research, Bombay.

Nelson, E.(1967). *Tensor Analysis. Mathematical Notes*, Princeton University Press, Princeton.

CHAPTER 6

Breitenecker, M., and Thirring, W. (1979). Scattering theory in classical dynamics. *Rivista del Nuovo Cimento* **2**, 1.

Godbillon, C. (1983). *Dynamical Systems on Surfaces.* Springer Verlag, Berlin.

Koshorke, U. (1981). *Vector Fields and other Vector Bundle Morphisms. A Singularity Approach.* Springer Verlag, Berlin.

Vitale, B. (1980). From orbits to vector fields and symmetries. In *Group Theoretical Methods in Physics*, Nauka, Moscow.

CHAPTER 8

Haefliger, A. (1961). Plongements différentiables de variétés dans variétés. *Comm. Math. Helv.* **36**, 46.

Thom, R. (1954). Quelques propriétés globales des variétés différentiables. *Comm. Math. Helv.* **28**, 17.

CHAPTER 9

Crampin, M. (1983). Tangent bundle geometry for Lagrangian dynamics. *J. Phys. A: Math. Gen.* **16**, 3755.

Dombrowski, P. (1962). On the geometry of the tangent bundle. *J. Reine Angew. Math.* **210**, 73.

Grifone, J. (1972). Structure presque-tangente et connexions, I. *Ann. Inst. Fourier, Grenoble* **22**, 287.

Klein, J. (1962). Éspaces variationelles et méchanique. *Thèse Ann. Inst. Fourier, Grenoble* **12**, 1–124.

CHAPTER 10

Bluman, G. W., and Cole, J. D. (1974). *Similarity Methods for Differential Equations.* Springer, New York.

Klein, J. (1963). Les Systèmes dynamiques abstraits. *Ann. Inst. Fourier, Grenoble* **13**, 191–202.

Klein, J., and Voutier, A. (1968). Formes extérieures génératrices de sprays. *Ann. Inst. Fourier, Grenoble* **18**, 241.

Losco, L. (1972). Solutions particulières et invariants integraux en mécanique céleste. Thèse, Université de Besancon.

Markus, L., and Meyar, K. R. (1974). Generic Hamiltonian dynamical systems are neither integrable nor ergodic. *Memoirs of the American Math. Soc. Number 144*. American Mathematical Society Providence, R.I.

Sarlet, W., Cartrijn, F., and Crampin, M. (1984). A new look at second order equations and Lagrangian mechanics. *J. Phys. A: Math. Gen.* **17**, 1999.

CHAPTER 12

Arnol'd, V. I., and Avez, A. (1968). *Ergodic Problems of Classical Mechanics*. Benjamin, New York.

Balachandran, A.P., Mukunda, N., Nilson, Y. S., Sudarshan, E. C. G., and Zaccaria, F. (1981). Evolution, symmetry and canonical structure in dynamics. *Phys. Rev.* **D 23**, 2189.

Cantrijn, F. (1974). Symplectic approach to non-conservative mechanics. *Journal Math. Phys.* **25**, 271.

Lichnerowicz, A. (1975). Variétés canoniques et transformations canoniques. *C.R. Acad. Sci. Paris*, **280** A, 37–40.

Magri, F. (1976a). An operator approach to Poisson brackets. *Annals of Physics* **99**, 196.

Magri, F. (1976b). An operator approach to symmetries. *Nuovo Cimento* **3** B, 334.

Sudarshan, E. C. G., and Mukunda, N. (1976). The relation between Nambu and Hamiltonian mechanics. *Phys. Rev. D* **13**, 2846.

Zaccaria, F., Sudarshan, E. C. G., Nilson, J.S., Mukunda, N., Marmo, G., and Balachandran, A.P. (1983). Universal unfolding of Hamiltonian systems: from sympletic structure to fiber bundles. *Phys. Rev. D* **27**, 2327.

CHAPTER 13

Fuchssteiner, B. (1982). The Lie algebra structure of degenerate Hamiltonian and bihamiltonian systems. *Progress of Theoretical Physics* **68**, 1082.

Giandolfi, F., Marmo, G., and Rubano, C. (1981). Equivalent descriptions of classical dynamical systems: some differential geometric remarks. *Nuovo Cimento* **66** B, 34.

Marmo, G., and Rubano, C. (1983). Equivalent lagrangians and Lax representation. *Nuovo Cimento* **78** B, 70.

CHAPTER 14

Babu, J. K., and Mukunda, N. (1975). The Hamilton–Jacobi equation revisited. *Pramana* **4**, 1.

De Filippo, S., Marmo, G., Salerno, M., and Vilasi, G. (1984). A new characterization of completely integrable systems. *Nuovo Cimento* **83** B, 97.

Guillemin, V., and Sternberg, S. (1983). On the method of Symes for integrating systems of the Toda type. *Letters in Math. Phys.* **7**, 113.

Lychagin, V. V. (1975). Local classification of non-linear first order partial differential equations. *Russian Math. Surveys* **30**, 105.

Magri, F. (1978). A simple model of the integrable Hamiltonian equation. *J. Math. Phys.* **19**, 1156.

Marle, C. M. (1980). Action de groupes et separation de variables pour les systèmes Hamiltoniens. *J. Math. Pures et Appl.* **59**, 133–144.

Marle, C. M. (1982). A propos des systèmes completement integrables et des variables actions-angles. *Journées Relativistes*, Lyon, 23–25 April 1982.

Marle, C. M. (1983). Normal forms generalizing action-angle coordinates for Hamiltonian actions of Lie groups. *Lett. Math. Phys.* **7**, 55.

Minura, F., and Takayuki, Nono (1983). Symmetries of generalized Hamilton–Jacobi equation under generalized contact transformation. *Tensor* **40**, 163.

Mukunda, N. (1978). Phase space methods and the Hamilton–Jacobi form of dynamics. *Proc. Indian Academy of Sciences* **87** A, 85.

Reyman, A. G., and Semenov Tian Shansky, M. A. (1979). Reduction of Hamiltonian systems, affine Lie algebras and Lax equations. *Inventiones Math.* **54**, 81.

Weinstein, A. (1972). The invariance of Poincaré's generating function for canonical transformations. *Inventiones Math.* **16**, 202.

CHAPTER 15

Komorowsky, J. (1970). A geometrical formulation of the general free boundary problems in the calculus of variations and the theorems of E. Noether connected with them. *Rep. Math. Phys.* **1**, 105.

Trautman, A. (1956). On the conservation theorems and equations of motion in covariant field theories. *Bull. Acad. Pol. Sci. Cl. III* **4**, 675.

Trautman, A. (1967). Noether equations and conservation laws. *Comm. Math. Phys.* **6**, 248.

Trautman, A., Pirani, F. A. E., and Bondi, H. (1964). *Lectures on General Relativity*, Vol. I. Prentice-Hall, Englewood Cliffs, N.J.

Trautman, A. (1962). Conservation laws. In Witten, L., ed., *Gravitation. An Introduction to current research.* John Wiley, New York.

CHAPTER 16

Marsden, J. E. (1981). Lectures on geometric methods in mathematical physics. CBMS-NSF Regional conference series in Applied Mathematics SIAM, Philadelphia.

Modugno, M., ed., (1983). *Proceedings of the International Meeting on Geometry and Physics.* Pitagora Editrice, Bologna.

CHAPTER 18

Molino, P. (1977). Théorie des G-Structures: le problème d'équivalence. *Lecture Notes in Mathematics 588.* Springer Verlag, Berlin.

Pittie, H. V. (1976). Characteristic classes of foliations. *Research Notes in Mathematics 10.* Pitman, London.

Reeb, G. H. (1974). *Feuilletage. Résultats anneaux et niveaux.* Les Presses de l'Université de Montreal.

Reinhart, B. L. (1983). *Differential Geometry of Foliations. The fundamental integrability problem.* Springer Verlag, Berlin.

Schweitzer, P. A., ed., (1978). Differential topology, foliations and Gelfand–Fuks cohomology. Proceedings, Rio de Janeiro, 1976. *Lecture Notes in Mathematics 552.* Springer Verlag, Berlin.

Wu, Wen-Tsun, and Reeb, G. (1952). Sur les éspaces fibres et les variétés feuilletées. Hermann, Paris.

CHAPTER 20

Lichnerowicz, A. (1975). Variété symplectique et dynamique associée a une sous-variété. *C.R. Acad. Sci. Paris Ser. A.* **280**, 523.

Marle, C. M. (1982a). Sous-variétés de range constant d'une variété symplectique. Paris, preprint.

Marle, C. M. (1982b). Poisson manifolds in mechanics. Paris, preprint.

Marle, C. M. (1982c). Sous-variétés de rang constant et sous-variétés symplectiquement regulières d'une variété symplectique. *C.R. Acad. Sci. Paris Ser. I Math.* **295**, 119.

Marmo, G. (1982). Function groups and reduction of Hamiltonian systems. Proceed-

ings of the IUTAM-ISIMM Symposium on Modern Developments in Analytical Mechanics. Turin.
Marmo, G. Saletan, E.J., and Simoni, A. (1979a). Reduction of symplectic manifolds through constant of the motion. *Nuovo Cimento.* **50** B, 21.
Sniatycki, J., and Weinstein, A. (1983). Reduction and quantization for singular momentum mappings. *Letters in Mathematical Physics* **7**, 155.
Weinstein, A. (1973). Lagrangian submanifolds and Hamiltonian systems. *Annals of Math.* **98**, 3, 377.
Weinstein, A. (1973a). Normal modes for nonlinear Hamiltonian systems. *Inventiones Math.* **20**, 47.

CHAPTER 21

Kurosh, A. G. (1965). *Lectures on General Algebras.* Pergamon, Oxford.
Serre, J. P. (1965). *Lie algebras and Lie groups.* Benjamin, New York.
Sussmann, H. J. (1973). Orbits of families of vector fields and integrability of distributions. *Trans. Am. Mat. Soc.* **180**, 171.
Varadarajan, V. S. (1974). Lie groups, Lie algebras and their representations. Prentice-Hall, Englewood Cliffs, N.J.
Yutze Chow (1978). *General Theory of Lie Algebras.* Gordon & Breach, New York.

CHAPTER 22

Aguirre, E., Doebner, H. D., and Hennig, J. D. (1983). All local classical symmetries in Hamiltonian mechanics. *Lett. Math. Phys.* **7**, 85.
Bunchaft, F. (1975). Mukunda's conjecture and regularity on realizations of semi-simple Lie algebras. *Nuovo Cimento* **25** A, 110.
Bunchaft, F. (1976). On the basis of a complex-symplectic formalism on the (real) finite-dimensional phase space of classical mechanics. *Nuovo Cimento Lettere* **16**, 399.
Chand, P., Metha, C. L., Sudarshan, E. C. G., and Mukunda, N. (1967). Realizations of Lie algebras by analytic functions of generators of a given Lie algebra, *J.M.P.* **8**, 2048.
Dothan, Y. (1970). Finite dimensional spectrum generating algebras. *Phys. Rev.* **20**, 2944.
Fradkin, D. M. (1965). 3-dimensional isotropic harmonic oscillator and *SU* (3). *A.J. Phys.* **33**, 207.
Guillemin, V., and Sternberg, S. (1980). The moment map and collective motion. *Ann. of Physics* **127**, 220.
Hermann, R. (1972). Spectrum generating algebras and symmetries in mechanics. I, II. *J. Math. Phys.* **13**, 833.
Meyer, K. R. (1973). Symmetries and integrals in mechanics. In M. Peixoto, ed., *Dynamical systems.* Academic Press, New York, pp. 259–273.
Mukunda, N. (1967). Realizations of Lie Algebras in classical mechanics. *J.M.P.* **8**, 1069.
Mukunda, N. (1967a). Dynamical symmetries and classical mechanics. *Phys. Rev.* **155**, 1383.
Onofri, E., and Pauri, M. (1973). Constants of motion and degeneration in Hamiltonian systems. *J. Math. Phys.* **14**, 1106.
Pauri, M., and Prosperi, G. M. (1966). Canonical realizations of Lie symmetry groups. *J. Math. Phys.* **7**, 366.
Rawnsley, J. H. (1975). Representations of semi-direct product by quantization. *Math Proc. Camb. Phil Soc.* **78**, 345.
Simms, D. J. (1973). Bohr–Sommerfeld orbits and quantizable symplectic manifolds. *Proc. Camb. Phil. Soc.* **73**, 489.
Souriau, J. M. (1966). Modèles classiques quantifiables pour les particules élémentaires. *C.R. Acad. Sc. Paris* **263**, 1191.

Souriau, J. M. (1967). Réalisations d'algèbres de Lie au moyen de variables dynamiques. *Nuovo Cimento S* **49**, 197.

Souriau, J. M. (1967a). Quantification geometrique. Applications *Ann. Inst. H. Poincaré*, **VI**, 4,311.

Weinstein, A. (1983). The Local structure of Poisson manifolds. *J. Diff. Geometry* **18**, 523.

CHAPTER 23

Freudenthal, H. (1964). Lie groups in the foundations of geometry. *Advances in Math.* **1**, 145.

Helgason, S. (1978). *Differential Geometry, Lie Groups and Symmetric Spaces.* Academic Press, New York.

Kirillov, A. A. (1976). *Elements of the Theory of Representations.* Springer, New York.

Lichnerowicz, A. (1958). *Géométrie des Groupes de Transformations.* Dunod, Paris.

Tondeur, P. (1965). Introduction to Lie groups and transformation groups. *Lectures Notes in Math. no. 7*, Springer Verlag, Berlin.

CHAPTER 24

Andrie, M., and Simms, D. J. (1972). Constants of motion and Lie group actions. *J. Math. Phys.* **13**, 331–336.

Bacry, H., Rueg, H., and Souriau, J. M. (1966). Dynamical groups and spherical potentials in classical mechanics. *Comm. Math. Phys.* **3**, 323.

Crampin, M., McCarthy, P. J. (1983). A lifting theorem for compact symplectic manifolds. *J. Phys. A: Math. Gen.* **16**, 3949.

Iacob, A. (1973). *Topological Methods in Classical Mechanics* (in Rumanian). Editura Academiei Romania, Bucharest.

Kazhdan, D., Kostant, B., and Sternberg, S. (1978). Hamiltonian group actions and dynamical systems of Calogero type. *Comm. Pure Appl. Math.* **31**, 481.

Ligon, T., and Schaaf, M. (1976). On the global symmetry of the classical Kepler problem. *Reports on Math. Phys.* **9**, 281–300.

Marle, C. M. (1976). Symplectic manifolds, dynamical groups and Hamiltonian mechanics. In Cahen, M., and Flato, M., eds., *Differential Geometry and Relativity*, D. Reidel, Dordrecht, p. 244.

Moser, J. (1970). Regularization of Kepler's problem and the averaging method on a manifold. *Comm. on Pure and Applied Math.* **23**, 609.

Onofri, E., and Pauri, M. (1969). Separation coordinates and dynamical symmetries. *Lettere Nuovo Cimento serie I* **2**, 607.

Onofri, E., and Pauri, M. (1972). Dynamical quantization, *J. Math. Phys.* **13**, 533.

O'Raifertaigh, L., Sudarshan, E.C.G., and Mukunda, M. (1965). Group theory of the Kepler problem, *Phys. Lett.* **19**, 322,

O'Raifertaigh, L., Sudarshan, E.C.G., and Mukunda, N. (1965a). Characteristic noninvariance groups of dynamical systems. *Phys. Rev. Lett.* **15**, 1041.

Symes, W. W. (1980). Hamiltonian group actions and integrable systems. *Physica* **1** D, 339.

Simms, D. J. (1973a). Invariance groups in classical and quantum mechanics. *Zeitschrift Naturf.* **28**a.

Souriau, J. M. (1974). Sur la variété de Kepler. In Convegno di Geometria Simplettica e Fisica Matemetica. Ist Naz. di Alta Matematica, Roma, 1973. *Symposia Mathematica, XIV*, Academic Press, London.

Sternberg, S. (1975). Symplectic homogeneous spaces. *Trans. Am. Math. Soc.* **212**, 113.

Sudarshan, E. C. G., and Mukunda, N. (1966). The harmonies in the motion of celestial bodies under Newtonian attraction. *Lectures in Theoretical Physics*, Vol. VIII-B. University of Colorado Press, Boulder.

CHAPTER 25

Anderson, J. L., and Bergman, P. G. (1951). Constraints in covariant field theories. *Phys. Rev.* **83**, 1018.

Arens, R. (1971). Classical Lorentz invariant particles. *J. Math. Phys.* **12**, 2415.

Arens, R. (1973). Hamiltonian formalism for non-invariant dynamics. *Comm. Math. Phys.* **34**, 91.

Balachandran, A. P., Marmo, G., Skagerstam, B. S., and Stern, A. (1980). Magnetic monopoles with no strings. *Nucl. Phys. B.* **162**, 385.

Balachandran, A.P., Marmo, G., Skagerstam, B. S., and Stern, A. (1980a). Supersymmetric point particles and monopoles with no strings. *Nucl. Phys. B.* **164**, 427.

Barducci, A., Casalbuoni, R., and Lusanna, L. (1977). Classical spinning particles interacting with external gravitational fields. *Nuclear Physics B.* **124**, 521.

Benenti, S., Tulczyjew, W. M. (1980). The geometrical meaning and globalization of the Hamilton–Jacobi methods. Proceedings 'Differential geometrical methods in mathematical physics' P. L. Garcia, A. Perez-Rendon, J. M. Souriau, eds., *Lecture Notes in Math. 836, 484.* Springer Verlag, Berlin.

Benenti, S., Tulczyjew, W. M. (1982). Sur le Théoreme de Jacobi en mécanique analytique. *C.R. Acad. Sci. Paris*, **294**, 677.

Chistodoulou, D., and Francaviglia, M. (1975). A geometric approach to Dirac's constraints of quantum gravitation. *Atti Accademia delle Scienze di Torino* **110**, 379.

Cognola, G., Vanzo, L., Zerbini, S., and Soldati, R. (1981). On the Lagrangian formulation of a charged spinning particle in an external electromagnetic field. *Phys. Lett.* **104** B, 67.

Darboux, G. (1878). Problème de Mécanique. *Bull. Sc. Math.* p. 443.

Dirac, P. A. M. (1931). Quantized singularities in the electromagnetic field. *Proc. R. Soc. London A* **133**, 60.

Dirac, P. A. M. (1964). Lectures on quantum mechanics. Belfer Graduate School of Science, Yeshiva University, New York.

Francaviglia, M. (1978). Application of infinite dimensional differential geometry to general relativity. *Rivista del Nuovo Cimento* **1**, 1.

Friedman, J. L., and Sorkin, R. (1979). Dyon spin and statistics: a fiber-bundle theory of interacting magnetic and electric charges. *Phys. Rev. D* **20**, 2511.

Friedman, J. L., and Sorkin, R. (1980). A spin-statistic theorem for composites containing both electric and magnetic charges. *Comm. Math. Phys.* **73**, 161.

Goldberg, J. N., Sudarshan, E. C. G., and Mukunda, N. (1981). Relativistically interacting particles and world lines. *Phys. Rev. D* **23**, 2231.

Hanson, A., Regge, T., and Teitelbom, C. (1976). Constrained Hamiltonian systems. Accademia Nazionale dei Lincei, Roma.

Kihlberg, A., Marnelius, R. and Mukunda, N. (1981). Relativistic potential models as systems with constraints and their interpretation. *Phys. Rev. D* **23**, 2201.

Kunzle, H. P. (1969). Degenerate Lagrangian systems. *Ann. Inst. H. Poincaré Ser. A* **11**, 393.

Kunzle, H. P. (1972). Canonical dynamics of spinning particles in gravitational and electromagnetic fields. *J. Math. Phys.* **13**, 739.

Leutwyler, H. (1965). A no-interaction theorem in classical relativistic Hamiltonian particle mechanics. *Nuovo Cimento* **37**, 556.

Maison, D., and Orfanidis, S. J. (1976). Some geometrical aspects of monopole theories. *Phys. Rev. D* **15**, 3608–3614.

Menzio, M. R., and Tulczyjew, W. M. (1978). Infinitesimal symplectic relations and generalized Hamiltonian dynamics. *Ann. Inst. H. Poincaré Ser. A* **23**, 349.

Miller, J. G. (1976). Charge quantization and canonical quantization. *J.M.P.* **17**, 643–646.

Mukunda, N. (1976). Symmetries and constraints in generalized Hamiltonian mechanics. *Ann. Phys. (N.Y.)* **99**, 408.

Mukunda, N. (1980). Time dependent constraints in classical dynamics. *Physica Scripta* **21**, 801.

Mukunda, N. (1980a). Generators of symmetry transformations for constrained Hamiltonian systems. *Physica Scripta* **21**, 783.

Mukunda, N., Sudarshan, E. C. G. (1968). Structure of the Dirac bracket in classical mechanics. *J. Mat. Phys.* **9**, 411.

Poincaré H. (1896). Remarques sur une expérience de M. Birkeland. *C.R. Acad. Sci.* **123**, 530.

Samuel, J., and Mukunda, N. (1984). Relativistic particle interactions. A comparison of independent and collective variable models. *Pramana* **22**, 131.

Sniatycki, J. (1974). Dirac brackets in geometric dynamics. *Ann. Inst. H. Poincaré Ser. 20*, 365.

Sudarshan, E. C. G., Mukunda, N., and Goldberg, J. N. (1981). Constraint dynamics of particle world line. *Phys. Rev. D* **23**, 2218.

Sudarshan, E. C. G., and Mukunda, N. (1983). Forms of relativistic dynamics with world line conditions and separability. *Foundations of Physics* **13**, 385.

Sundermeyer, K. (1982). Constrained dynamics. *Lecture Notes in Physics*, vol. 169. Springer Verlag, Berlin, New York.

Index